AF774574

THE SHAPE OF SOCIAL INEQUALITY: STRATIFICATION AND ETHNICITY IN COMPARATIVE PERSPECTIVE

RESEARCH IN SOCIAL STRATIFICATION AND MOBILITY

Series Editor: Kevin T. Leicht

Volumes 1–3:	Edited by Donald J. Treiman and Robert V. Robinson
Volumes 4–6:	Edited by Robert V. Robinson
Volumes 7–9:	Edited by Arne L. Kalleberg
Volumes 10–13:	Edited by Robert Althauser and Michael Wallace
Volumes 14 & 15:	Edited by Michael Wallace
Volumes 16–20:	Edited by Kevin T. Leicht
Volume 21:	Edited by Arne L. Kalleberg, Stephen L. Morgan, John Myles and Rachel A. Rosenfeld

RESEARCH IN SOCIAL STRATIFICATION AND MOBILITY VOLUME 22

THE SHAPE OF SOCIAL INEQUALITY: STRATIFICATION AND ETHNICITY IN COMPARATIVE PERSPECTIVE

EDITED BY

DAVID B. BILLS

The University of Iowa, Iowa, USA

2005

ELSEVIER
JAI

Amsterdam – Boston – Heidelberg – London – New York – Oxford
Paris – San Diego – San Francisco – Singapore – Sydney – Tokyo

ELSEVIER B.V.
Radarweg 29
P.O. Box 211
1000 AE Amsterdam
The Netherlands

ELSEVIER Inc.
525 B Street, Suite 1900
San Diego
CA 92101-4495
USA

ELSEVIER Ltd
The Boulevard, Langford Lane, Kidlington
Oxford OX5 1GB
UK

ELSEVIER Ltd
84 Theobalds Road
London
WC1X 8RR
UK

© 2005 Elsevier Ltd. All rights reserved.

This work is protected under copyright by Elsevier Ltd, and the following terms and conditions apply to its use:

Photocopying
Single photocopies of single chapters may be made for personal use as allowed by national copyright laws. Permission of the Publisher and payment of a fee is required for all other photocopying, including multiple or systematic copying, copying for advertising or promotional purposes, resale, and all forms of document delivery. Special rates are available for educational institutions that wish to make photocopies for non-profit educational classroom use.

Permissions may be sought directly from Elsevier's Rights Department in Oxford, UK; phone: (+44) 1865 843830, fax: (+44) 1865 853333, e-mail: permissions@elsevier.com. Requests may also be completed on-line via the Elsevier homepage (http://www.elsevier.com/locate/permissions).

In the USA, users may clear permissions and make payments through the Copyright Clearance Center, Inc., 222 Rosewood Drive, Danvers, MA 01923, USA; phone: (+1) (978) 7508400, fax: (+1) (978) 7504744, and in the UK through the Copyright Licensing Agency Rapid Clearance Service (CLARCS), 90 Tottenham Court Road, London W1P 0LP, UK; phone: (+44) 20 7631 5555; fax: (+44) 20 7631 5500. Other countries may have a local reprographic rights agency for payments.

Derivative Works
Tables of contents may be reproduced for internal circulation, but permission of the Publisher is required for external resale or distribution of such material. Permission of the Publisher is required for all other derivative works, including compilations and translations.

Electronic Storage or Usage
Permission of the Publisher is required to store or use electronically any material contained in this work, including any chapter or part of a chapter.

Except as outlined above, no part of this work may be reproduced, stored in a retrieval system or transmitted in any form or by any means, electronic, mechanical, photocopying, recording or otherwise, without prior written permission of the Publisher. Address permissions requests to: Elsevier's Rights Department, at the fax and e-mail addresses noted above.

Notice
No responsibility is assumed by the Publisher for any injury and/or damage to persons or property as a matter of products liability, negligence or otherwise, or from any use or operation of any methods, products, instructions or ideas contained in the material herein. Because of rapid advances in the medical sciences, in particular, independent verification of diagnoses and drug dosages should be made.

First edition 2005

Library of Congress Cataloging in Publication Data
A catalog record is available from the Library of Congress.

British Library Cataloguing in Publication Data
A catalogue record is available from the British Library.

ISBN: 0-7623-1178-9
ISSN: 0276-5624 (Series)

♾ The paper used in this publication meets the requirements of ANSI/NISO Z39.48-1992 (Permanence of Paper). Printed in The Netherlands.

Working together to grow
libraries in developing countries

www.elsevier.com | www.bookaid.org | www.sabre.org

ELSEVIER BOOK AID International Sabre Foundation

CONTENTS

PART III: COMPARATIVE APPLICATIONS

LIST OF CONTRIBUTORS

David B. Bills	The University of Iowa, Iowa, USA
Claudia Buchmann	The Ohio State University, Columbus, USA
M. D. R. Evans	University of Melbourne, Victoria, Australia
Danielle Cireno Fernandes	Federal University of Minas Gerais, Belo Horizonte, Brazil
Patricia Fernández-Kelly	Princeton University, Princeton, USA
Glenn V. Fuguitt	University of Wisconsin-Madison, USA
David B. Grusky	Stanford University, Stanford, USA
William J. Haller	Clemson University, Clemson, USA
Charles Hirschman	University of Washington, Seattle, USA
Jonathan Kelley	The Australian National University, Canberra, Australia
Meredith Kleykamp	Princeton University, Princeton, USA
Lawrence R. Landerman	Duke University, Durham, USA
Monica Murero	International Institute of Infonomics, DC Heerlen, The Netherlands
Jorge Alexandre Neves	Federal University of Minas Gerais, Belo Horizonte, Brazil
Alejandro Portes	Princeton University, Princeton, USA
Bam Dev Sharda	University of Utah, Utah, USA
C. Matthew Snipp	Stanford University, Stanford, USA
Kenneth I. Spenner	Duke Unversity, Durham, USA

Marta Tienda	Princeton University, Princeton, USA
Kim A. Weeden	Cornell University, New York, USA
Joseph Woelfel	State University of Buffalo, New York, USA
Krzysztof Zagorski	Centrum Badania Opinii Spolcznej, Warszawa, Poland

EDITORIAL BOARD

Robert Althauser
Department of Sociology, Indiana University, USA

Thomas DiPrete
Department of Sociology, Columbia University, USA

Scott Eliason
Department of Sociology, University of Minnesota, USA

Jennifer Glass
Department of Sociology, The University of Iowa, USA

David Grusky
Department of Sociology, Stanford University, USA

David S. Hachen
Department of Sociology, University of Notre Dame, USA

Joan Huber
Department of Sociology, The Ohio State University, USA

Arne L. Kalleberg
Department of Sociology, University of North Carolina, USA

Robert L. Kaufman
Department of Sociology, The Ohio State University, USA

Charles W. Mueller
Department of Sociology, The University of Iowa, USA

Barbara Reskin
Department of Sociology, University of Washington, USA

Rachel Rosenfeld†
Department of Sociology, University of North Carolina, USA

Arthur Sakamoto
Department of Sociology, University of Texas-Austin, USA

Kenneth I. Spenner
Department of Sociology, Duke University, USA

Michael Wallace
Department of Sociology, University of Connecticut, USA

PREFACE: ARCHIBALD ORBEN HALLER, AN INTELLECTUAL PORTRAIT

Sören Kierkegaard once reflected that "the door to happiness" opened by pushing outwards, never inwards. By this, the Danish philosopher acknowledged a fundamental truth about the human condition: its social character and the limits of solipsism and the selfish pursuit of gain. Science, like literature and other enterprises of the mind, is a field populated by individualists. Despite its seemingly cooperative character, the ultimate mark of a successful career is, for many, the public recognition accorded to one's personal work in an intellectual marketplace that confines others to oblivion.

Kuhn once likened this marketplace to a field in which each scientist stands on his or her own patch and calls attention to its merits, while disparaging others'. Social science is an integral part of that metaphor and its practitioners strive, accordingly, for that sort of personalized recognition to the detriment of everything else. "Happiness" in this pursuit comes from the printed pages and the discoveries ascribed to an individual and their citation and recognition by others. Kierkegaard would object.

A career spanning half a century of uninterrupted intellectual production, pursuing consistent thematic lines and widely recognized in the discipline provides ample basis for distinction, according to these criteria. However, a career that joins this achievement with solicitous care for others, as shown in the mentoring of dozens of younger scholars and the sponsoring of their growth beyond their graduate years is truly extraordinary. In such a model, the door always opens to the outside and always lets others in. The scholar whose career this book celebrates fits that model to perfection.

To the more than one hundred scholarly publications and reports to his credit, Archibald Haller adds the production of seventy doctoral dissertations and masters' theses written under his supervision. Most of these students became his friends and then his collaborators. Not surprisingly, close to half of Haller's research publications feature one or more former students as co-authors. Several of the most prominent are represented in this volume.

The numbers say something, but they do not convey the quality of the experience. Working under Arch was not a distant affair since, whether you wanted it or not, he made you an integral part of his own career and of his life. He brought you into his office and his house, took you to lunch and to research conferences, alerted you to opportunities and deadlines, and went to bat for you when you were finally ready to enter the academic marketplace. With his own mentor, William Sewell, and his collaborators Arch perfected a training machine that yielded an extraordinary output of both research and researchers. For a good while, much of what was American sociology pivoted around that Wisconsin powerhouse.

The industrial analogy is faulty, however, because it suggests rigid authority lines and standardized production. For those who, as students, partook of the experience, it was nothing of the sort. It was more like joining a family. Not a family that wallowed in the contemplation of its own brilliance, but rather one that permanently strained toward the next level of innovation and achievement. You felt protected, but also constantly challenged. The empirical rigor of a Wisconsin training, coupled with the warmth of our mentor marked a distinct experience. Other students at Wisconsin underwent the same stress of what was, undoubtedly, one of the most demanding Sociology graduate programs of the day. But we had Arch on our side.

Let me illustrate with a personal note. As I prepared to launch my dissertation fieldwork on political orientations in low-income settlements of Santiago, Chile, the project was denounced by an overzealous fellow student as a possible "covert" operation by the CIA and the Pentagon against the left in that country. News traveled fast to Chilean militants, and I was soon confronted with articles in the local media denouncing my modest study as a new plot of the Pentagon against Chilean national sovereignty. Most of the higher-ups at Wisconsin were of the opinion that, while the accusations were unfounded, I should leave the country to avoid further trouble. Years of preparation for the project would have gone to waste and, as a mere graduate student, I would have had no defense against such a decision. But Arch stood firm. Thanks to him, the project was allowed to continue. Eventually it was completed. The original data were left in Chile along with a preliminary report of findings and the dissertation was completed ten months later. To date, results of that study, conducted over 30 years ago, are still used in Chile as a point of reference for contemporary analyses of urban social movements.

Haller, himself a graduate of Wisconsin, began his career at Michigan State as a rural sociologist and a student of American patterns of social stratification and mobility. Never a macro-sociologist, he focused on empirical analyses of the social psychology of educational and occupational achievement, first among farm youth and then in the general population. The role of families, of peer influences, and of career aspirations loomed large in these studies. The Occupational Aspiration Scale, developed by Haller with his student Irwin W. Miller dates from this period. Throughout these years, Haller worked closely with William Sewell, producing an uninterrupted flow of articles on status aspirations and achievements of rural youth published in the best sociology journals.

Haller's return to Wisconsin in 1965 coincided with the advent of path analysis, a statistical method for complex theory-testing developed by geneticist Sewall Wright since the 1920s and introduced into Sociology by Otis D. Duncan. Haller and I collaborated with Duncan in an early application of this method, developing a non-recursive model of peer influences on aspirations. The advent of path analysis was important because it allowed scholars to integrate, into a single model, the complex causal sequence shaping the educational and early occupational career of young people. This integrative effort led to the "status attainment" model, first introduced into the sociological literature in 1969.

The Wisconsin status attainment model served as a complement and a counterweight to the inter-generational mobility model developed by Peter Blau and Otis Duncan in their classic study *The American Occupational Structure*. Unlike the latter, which focused on objective measures of occupational status among parents and sons, the Wisconsin model sought to flesh out the process by considering the intervening effects of significant others – such as parents, peers and teachers – school grades, and the character and level of individual aspirations. This model became a platform for a generation of studies in social stratification that involved not only Wisconsin scholars, but researchers worldwide. Under the leadership of Robert Hauser, David Featherman, Karl Alexander, and others, this literature reached extraordinary levels of methodological sophistication. Nevertheless, the basic ideas advanced by the original model remained unaltered.

While subsequently criticized for its emphasis on individual rather than contextual factors, the fundamental predictions of the status attainment model have stood the test of time. This is due to the fact that it was not grounded on speculation but on the solid body of research built by Sewell, Haller, and their students over a decade. Appropriately, this literature culminated in another *festschrift*, *Social Structure and Behavior: Essays in Honor of William Hamilton Sewell*, co-edited by Haller and published in 1982.

While all of this was going on, Arch was developing a parallel set of interests abroad. Since the 1960s, he had been traveling to Brazil and collecting data on that country. He did so less as a full-time "Brazilianists" than as a student of stratification with a strong comparative interest. But as the trips and the data collection projects multiplied, Arch increasingly became familiar with the culture and language of the country and started to make his influence felt. There were a growing number of Brazilian students coming to Wisconsin under the sponsorship of the Land Tenure Center. Many started gravitating toward Agriculture Hall where the Rural Sociology Department and Arch's office were. He took them in and embarked with them in a new line of comparative stratification and development research that lasts to our day. The first results of this collaboration appeared in articles co-authored by Haller and Helcio Ulhoa Saraiva and published in the *American Journal of Sociology* and in *Rural Sociology* in 1972. They were followed, in quick succession, by a series of articles and reports on Brazilian stratification and mobility, co-authored by Saraiva, Jose Pastore, Tarcizio Quirino, and Hernando Gomez Buendia.

During those years, Haller traveled to Brazil with increasing frequency becoming acquainted with the diverse regions of that vast nation. In the course of the next three decades, he was going to lecture all around the country, hold numerous visiting professorships, and implement a large research program, always in collaboration with his former and present students. During this period, he published no less than forty articles and research reports on Brazilian occupational prestige hierarchies and status attainment.

The result of these efforts was the emergence of a school of sociology in Brazil at variance with the then dominant disciplinary currents. Mainstream Brazilian sociology was at the time heavily historical, deeply influenced by the dependency paradigm, and quite averse at complex statistical analyses of quantitative data. Sociological social psychology was in its infancy. What Haller and his students did was not so much to challenge the dominant paradigm as to show that there was another way of doing sociology where data could be brought to test, in a rigorous way, hypotheses about Brazilian society. While opposed by those more adept at historical speculation, the new brand of modern scientific sociology carved a space for itself by repeatedly showing its accuracy and usefulness. The leaders of the country did not ignore this achievement and, in 1981, Haller was decorated with the Brazilian Order of Merit of Work for his contributions to sociological research and teaching in the nation.

In closing, I wish to return to the Kierkegaard theme. Many scholars make significant contributions to their discipline only to end their careers in isolation. They did not realize that while intellectual work is often a lonely pursuit, science is still a human enterprise and, as such, inevitably social. It consists of ideas *and* the people who have them, discuss them, and advance them. In contrast to careers that end in solitude, the ceremonies at the University of Wisconsin on occasion of Haller's retirement in 1995 were a veritable apotheosis of warmth and recognition. Former students and colleagues came from everywhere to recognize not only the scholar, but also the friend and the man. The many who had benefited from his guidance and support during the early years of their own careers now had the occasion to signal their appreciation for their mentor.

This book is, in a sense, a continuation of that event embodying, in lasting form, the same sentiments of enduring affection and intellectual respect. While

I leave to the editor the task of explaining and summarizing the contents, I believe that I can interpret the common motivation underlying these essays: we are proud to be Arch's former students and friends and glad to honor him in this manner.

Alejandro Portes
Princeton University, August 2003

INTRODUCTION

The concepts that sociologists use to examine social stratification are at any given point an uneasy mix of the old and new. Many durable categories reach back to the founders of sociology. The most notable are the class analysis of Marx and Weber's analyses of social honor. Other founders too have left enduring marks – Pareto on the circulation of elites, Durkheim on associative groups, and Sorokin on social mobility. Other concepts have to be invented or reinvented as times, circumstances, and understandings change. While concepts rarely arise ex nihilo, cultural capital, social capital, social closure, status attainment, occupational hoarding, and maximally maintained inequality come to mind as conceptual breakthroughs that redirected and refocused theoretical and empirical work in important ways.

Likewise, empirical methods develop over time. While researchers typically offer these methods to permit the more effective testing of extant hypotheses, in many cases these new methodologies lead to new hypotheses that could not have been assessed, or even thought of, in their absence. Such applications as path analysis, log linear models, and two-sided logit models both helped resolve old questions and pointed the way to new ones.

Archibald O. Haller has been engaged in these sorts of concerns throughout a long, productive, and memorable career. This volume commemorates his many contributions to theory and research in social stratification. Without knowing he was doing so, Arch himself set the agenda for this volume over a decade ago:

> A full program aimed at understanding stratification requires: first, that we know what stratification structures consist of and how they may vary; second, that we identify the individual and collective consequences of the different states and rates of change of such structures; and third, seeing that *some* degree of stratification seems to be present everywhere, that we identify the factors that make stratification structures change (1992/2000, p. 2864).

This festschrift for Arch Haller, *The Shape of Social Inequality: Stratification and Ethnicity in Comparative Perspective*, is organized into three sections. The contributors to Part I, "Concepts for Social Stratification," wish to develop and elaborate a range of concepts that they believe have the potential to significantly advance our understanding of social stratification. Kim Weeden and David Grusky's "Are There Any Big Classes At All?" returns to themes that they have been elaborating in a series of papers. Weeden and Grusky's project is to develop a conceptual case for a realist formulation of class. The key to their proposal is, put somewhat simply, dissagregation. In the present paper, the authors construct a "hybrid" class model that draws on both the influential "big class schemes" of such researchers as Erikson and Goldthorpe (1992) and a far more disaggregated occupationalized conception of class. They hypothesize that the degree of occupationalization will vary throughout the class structure.

Weeden and Grusky proceed to examine the external distinctiveness and internal homogeneity of the class structure of the United States. Based on their innovative analyses of the Current Population Survey and General Social Survey, Weeden and Grusky conclude that a fully-saturated occupational model best captures the class structure of the contemporary United States. They stop short of jettisoning the value of big class models, but conclude that "the vast majority of the present-day labor force can instead be found in classes where occupational distinctions are alive and well."

Joe Woelfel and Monica Murero's "Spaces and Networks: Concepts for Social Stratification" takes on quite a different set of questions. The authors begin with a forceful claim: "Perhaps the great achievement of the social sciences in the 20th century was the discovery that reference frames and conceptual systems influence our perceptions of reality, and that these reference frames and conceptual systems are themselves socially constructed. Perhaps the great failure of the social sciences in the 20th century was the failure to develop formal technical criteria for evaluating reference frames and constructing 'better' ones." This is the challenge that they set for themselves.

Woelfel and Murero focus their attention on the communication aspects of social structure and social processes. They see the applicability of communication processes to social stratification in the expectations that define statuses and roles, how these expectations are encoded, where they are stored, and how they are communicated. Most importantly, perhaps, they wish to extricate sociology from its preoccupation with "goals" as the causal mechanism underlying social processes. They seek instead a system in which goals can be understood as themselves derivative of broader causal forces.

The authors conceptualize the status attainment process – the Wisconsin Model developed over thirty years ago by Haller, Sewell, and colleagues (Sewell, Haller & Portes, 1969) – as a communication process. They use the "Wisconsin model" as a vehicle to present their own "Galileo Model." The conceptual apparatus offered by Woelfel and Murero is a challenging one for stratification researchers. It does, however, hold the potential for a serious reappraisal of many of the assumptions of mainstream research on stratification in ways that treat both theory and measurement seriously.

Glenn Fuguitt's chapter, "Some Demographic Aspects of Rurality," asks researchers to revisit some recurrent but too often overlooked questions about social stratification. For Fuigitt, sociologists are too often unreflective about what is at stake when employing notions of rurality and urbanity. He redirects our attention to the demographic aspects of rurality. The concepts here, he notes, are deceptively simple and extremely powerful ones – population, age cohorts, the life course, fertility, mortality, migration, and shifts in status. Fuigitt sees the demographic perspective as a means by which the macro and micro levels of society can be bridged.

Fuguitt is critical of both those who claim too much for the concept of rurality and those who claim too little. "Rural" is not "a lifestyle, a state of mind, a communication network, or a self-identification." Neither, though, can it be reduced to something merely to be "controlled for" in empirical analyses of other social processes. Fuguitt puts forward an alternative position that takes seriously the tenets of demography and the now often neglected field of human ecology. This position focuses on areas with small population and low density (but not, significantly, isolation). Fuguitt bring a skeptical eye to much of what has passed as "demographic analysis" over the past couple of decades. He emphasizes the persistent heterogeneity among rural areas, and demonstrates the applicability of this insight to a number of substantive issues. He is persuasive that stratification researchers have yet to exploit the demographic perspective.

Theories and concepts do not stand apart from empirical research. Part II of the volume turns to "Applications in U.S. Society." The section begins with C. Mathew Snipp and Charles Hirschman's "Assimilation in American Society: Occupational Achievement and Earnings for Ethnic Minorities in the United States, 1970 to 1990." The authors set themselves the important task of examining "the social and economic inequality that exists among the major racial and ethnic groups in American society." They are particularly interested in how these patterns changed in the rapidly changing American society of 1970 to 1990.

Snipp and Hirschman's organizing concept is "assimilation." They acknowledge that the history of assimilation research is a contentious one and that the concept applies with great or lesser weight to different racial and ethnic groups. They use these insights to good advantage. Building on the earlier work of Hirschman and Wong (1984), Snipp and Hirschman provide a careful analysis of both the current status and recent experience of several racial and ethnic groups – Whites, African-Americans, Latinos, Japanese, Chinese, Filipinos, and American Indians.

The authors present a careful and comprehensive analysis of trends in racial and ethnic differences in occupational status and earnings among working age males. (Their decision to restrict the sample to men is driven by both practical and theoretical reasons.) Their findings are too extensive to summarize here, but on balance they show strong and enduring barriers to socioeconomic assimilation among the most historically marginalized groups in American society. They conclude with a discussion of the relevance of their findings for a variety of policy initiatives.

William Haller's chapter, "Changes in the Structure of Status Systems: Employment Shifts in the Wake of Deindustrialization," turns our attention to the consequences for the status system of the extensive industrial restructuring of the Pittsburgh region during the 1970s and 1980s. The outlines of the economic transformations that so disrupted the labor markets of Pittsburgh and other industrial cities are well known – job displacement, high unemployment, the growth of an urban underclass, seemingly endless plant closings. Haller's innovation is to analyze various features of Pittsburgh's emerging status system using the "Changes in the Structure of Status Systems" scheme offered in Haller (1970).

Haller painstakingly constructs a data set from tract-level data from the Decennial Censuses of 1970, 1980, and 1990 for the Pittsburgh Standard Metropolitan Statistical Area. His goal is to investigate tract-level changes in such indicators of status systems as employment, poverty, and other unequally distributed resources. He reports that deindustrialization in the Pittsburgh region during the 1980s had deep effects on joblessness and poverty, and that these effects were experienced with particular force for minority populations. At the same time, Haller finds far more modest effects on such "underclass" behaviors as dropping out of school, welfare receipt, or unwed parenting. Haller deftly ties these empirical findings back in to his conceptual discussion of status systems.

The next contribution is Meredith Kleykamp and Marta Tienda's "Physical and Mental Health Status of Adolescent Girls: A Comparative Ethnic Perspective." The authors of this chapter understand clearly that young people are particularly vulnerable to unequal systems of power and material well-being. This vulnerability is compounded by gender, race/ethnicity, and social class. Kleykamp and Tienda focus on the physical and mental well-being of Hispanic adolescent girls. In particular, they consider the mental health indicators of self-esteem, depression, stress, and suicidal ideation and the physical health indicators of exposure to physical and sexual abuse, experiences with violence, and perceived safety.

The authors report that family structure is of immense consequence to the well-being of Hispanic adolescent girls. Their susceptibility to a range of adverse outcomes is heightened by living in parent-absent homes, and these effects are greater for girls than they are for boys. Further, the girls in their sample are placed at higher risk of poor physical and mental well-being when they come from homes of low socioeconomic status. The difficulties faced by these girls extend to their behavioral responses as well. The authors note "minority youth's unequal access to the medical system because they are significantly more likely to be uninsured, to lack a regular provider, and to receive their medical care outside of the private health care delivery system."

Kenneth Spenner, Claudia Buchmann, and Lawrence Landerman's contribution is entitled "The Black-White Achievement Gap in the First College Year: Evidence from a New Longitudinal Case Study." Their concern is one of great moment for both sociological theory and public policy – the remarkably persistent achievement gap between whites and blacks that takes place from the earliest to the most advanced levels of schooling. They observe that despite the attention this gap has attracted, both the causal mechanisms underlying it and the policy remedies to alleviate it are poorly understood.

Spenner and his colleagues bring to this debate a rich new longitudinal study of racio-ethnic differentials in educational performance in higher education. This is the "Campus Life and Learning Project," which they describe as "a prospective panel study of two cohorts, the graduating classes of 2005 and 2006, at Duke University." The authors report that the achievement gap emerges virtually as soon as students enter postsecondary education. Interpreting their findings through an impressive synthesis of sociological and psychological theory, Spenner et al. present an empirically complex case, but one which they believe is amenable to policy intervention to support those "students who are struggling academically [and who] are more likely to drop out or experience other academic episodes detrimental to their college performance and completion."

Part III, "Comparative Applications," demonstrates the applicability of some of these emerging concepts to the study of comparative stratification. Haller's comments indicate that any viable program of stratification research must attend to structural variations in stratification systems across time and space. The trend toward comparative research would seem to be inexorable, given the growth of large-scale and nationally representative data sets, the technological and intellectual infrastructures for their analyses, and the construction and maintenance of international research collaborations. Abbott has recently argued that "internationalization would seem to be the main social structural event in sociology's future" (Abbott, 2000, p. 298). Indeed, many contemporary large-scale analyses of social stratification have constructed strongly comparative and cross-national designs (Kerckhoff, 1995; Shavit & Muller, 1995).

Bam Dev Sharda's "Status Allocation in Village India" challenges what were for years two "givens" in the sociological canon. The first of these is the existence of a monolithic and completely closed Indian caste system (Dumont, 1970), and the second is the belief that all societies will eventually converge on an essentially meritocratic and universalistic system of mobility, or the "thesis of industrialization" (Kerr et al., 1964). Sharda demonstrates that neither model accurately depicts status allocation in Village India. He proposes instead that explaining social stratification in India demands attention to features of both agrarian society (specifically family land ownership) and modern society (schooling).

Sharda's careful empirical analysis suggests to him that "stratification in village India is more closely associated with agrarian relations of production rather than industrial relations of production." That is, social stratification in village India is driven less by the idiosyncrasies of Indian society (which Sharda believes were largely a Western construct to begin with) and more by the characteristics of agrarianism. As such, status allocation in this setting has many commonalities with rural settings elsewhere. Sharda's proposed program of research thus directs attention to the need to creatively combine the best of discounted theories in search of one able to account for emergent stratification regimes.

Patricia Fernandez-Kelly's contribution, "The Future of Gender in Mexico and the United States: Economic Transformation and Changing Definitions," is a fascinating inquiry into how large-scale economic change can transform the ways that social actors think about gender. She bases her analysis on a comparison between a core (United States) and peripheral (Mexico) nation. Fernandez-Kelly begins with a socio-historical account of how transformations of economic systems have co-evolved with changing definitions of masculinity and femininity. She shows, for instance, how the roles of "breadwinner" and "housewife" emerged out of a variety of conflicts and strains that accompanied the transformation of the American economy in the early years of the twentieth century, and how a very different set of economic changes led to a distinctive Mexican system of patriarchy.

Fernandez-Kelly goes on to show how the more recent era of economic internationalization has redefined gender yet again. She builds on this to present several specific instances of how the relationship between economic change and gender identities has developed in the United States and Mexico over the last century. Her analysis demonstrates convincingly that "gender is not a secondary process but a central aspect in the articulation of class hierarchies."

In "Do Ethnic Enclaves Benefit or Harm Linguistically Isolated Employees?", Mariah Evans tries to resolve a question that has seriously challenged our thinking about ethnicity and social stratification – why "partially separated ethnic sub-economies, or 'ethnic enclaves,' might thrive and benefit their workers and employers." She takes as her point of departure what had become nearly a sociological truism, namely that social disadvantage inevitably follows from segregation and social closure. Evans insists instead that we examine the conditions under which expected disadvantage might be transformed into actual advantage.

For Evans, the key to understanding why some evidently socially isolated employees prosper economically has to do with linguistic fluency. She develops a "communications costs" conceptual framework that focuses on the costs borne by the dominant-language employers of linguistically weak employees. The outcome of this is that entrepreneurs in ethnic enclaves can benefit by enticing "employees out of the broader labor market by making job offers at least slightly better than the job offers available in the broader market." In doing so, subordinate – language employers and employees both stand to benefit. Evans's analysis of Australian data largely confirms a number of hypotheses suggested by this line of reasoning.

In "Economic Change and the Legitimation of Inequality: The Transition from Socialism to the Free Market in Central-East Europe," Jonathan Kelley and Krzyszto Zagorski use an exceptional resource, the World Inequality Study, to examine the relationship between national economic structure and its citizens' normative judgments about income inequality. The authors take advantage of what they portray as a great natural experiment, the collapse of Communism in Central-East Europe. The issue is of practical as well as sociological import. The ability of newly marketized and democratized societies to accommodate shifting regimes of distribution will have much to say about their prosperity and security over the coming decades.

Kelley and Zagorski's findings show how rapidly normative systems can respond to material change. As they state, "the transition from a Communist command economy led the public abruptly to change its view about inequality, at least in the larger Central-East European nations and most, but not all, of the smaller nations." They observe that the norms of Central-East European nations will in due time converge with those of most Western nations, but from the direction of less rather than more tolerance of inequality.

It is fitting to close with two chapters on Brazil. Brazil is a particularly interesting case for several reasons. Not only is the gap between rich and poor individuals and families high by any standard; but regional inequalities, especially between the impoverished Northeast and the more developed South, are legendary. Understanding these patterns is made even more challenging by Brazil's complex patterns of race and ethnic relations, educational access, and class structure.

Danielle Fernandes' focus in "Race, Socioeconomic Development, and the Educational Stratification Process in Brazil" is on two questions. She states these as "What impact does economic development have on educational stratification; and what is the role of race in this process?" She recognizes that even by the standards of Latin American nations, Brazil was a latecomer to educational expansion. Fernandes notes, however, that Brazil's remarkable economic growth of the past few decades has forced many structural changes in its educational system, but not necessarily in ways that have expanded access for marginalized racial groups.

Fernandes raises serious doubts that traditional modernization theory can explain the changing relationship between socioeconomic origins and educational attainment. She turns instead to the important re-conceptualization of this issue developed by Mare (1980, 1981). After an exhaustive analysis of the "persistent barriers to educational equality in Brazil," Fernandes' conclusions mirror those found in much of the world. As she states, "socioeconomic transformations brought about by the process of industrialization *have lessened neither the effects of socioeconomic origins nor of race. Indeed there is compelling evidence that the negative effects of being Black or Mulatto have increased*."

Jorge Alexandre Neves offers a comprehensive analysis of earnings determination in the huge agricultural sector of Brazil in his "Labor Force Classes and the Earnings Determination of the Farm Population in Brazil: 1973, 1982, and 1988." He situates his analysis in the "conservative modernization" that characterized Brazilian agriculture in the Post World War II era. This sort of economic transformation, which Neves portrays as combining "technological improvements with the absence of social reforms," led to fundamental changes in the class structure of the Brazilian agricultural sector.

Neves brings three waves of the Brazilian National Household Sample Survey (PNAD) to bear on a number of hypotheses that he derives from a broad literature on social stratification and economic development. Applying a carefully developed scheme of the agricultural class structure in Brazil, Neves finds substantial but often surprising effects of modernization on income determination. Of particular importance is his finding that agricultural workers in Brazil, contrary to some analysts, do in fact benefit from enhancements of their human capital.

Over half a century ago, Robert Merton published his classic dyad, "The Bearing of Sociological Theory on Empirical Research" and "The Bearing of Empirical Research on Sociological Theory" (1949/1968). Merton insisted on the close interplay of sociological theory and sociological research. He maintained that theory could not be mere conceptual elaboration, nor could research be limited to the gathering and reporting of facts (even in the form of hypothesis testing). The papers in this volume are all firmly in the tradition of the mutual dependence of theory and research. All are concerned with understanding the shape of inequality in real times and real places, and doing so by discovering and clarifying the tools to better understand some of the perennial issues in social stratification. Collectively, these chapters provide a powerful statement on the accomplishments and possibilities of stratification research.

It has been my privilege to work on this volume. I offer my thanks to Alex Portes for initiating the idea and to Series Editor Kevin Leicht for his willingness to turn over an issue of *Research in Social Stratification and Mobility* to this project. I am also grateful for the excellent work of colleagues who reviewed manuscripts.

They are Eric Hanley, Erin Kaufman, David Plank, Eric Reed, Jill Stevens, and Tor Wynn. As always, Karen Bixby was enormously helpful in beating a monstrous manuscript into submission.

David B. Bills
Editor

REFERENCES

Abbott, A. (2000). Reflections on the future of sociology. *Contemporary Sociology*, *29*, 296–300.

Dumont, L. (1970). *Homo hierarchicus: The caste system and its implications*. Chicago: University of Chicago Press.

Erikson, R., & Goldthorpe, J. H. (1992). *The constant flux: A study of class mobility in industrial societies*. Oxford: Clarendon Press.

Haller, A. O. (1970). Changes in the structure of status systems. *Rural Sociology*, *35*, 469–487.

Haller, A. O. (2000). Societal stratification. In: E. F. Borgatta & R. J. V. Montgomery (Eds), *Encyclopedia of Sociology* (pp. 2864–2874). New York: MacMillan Library Reference.

Kerckhoff, A. C. (1995). Institutional arrangements and stratification processes of industrial societies. *Annual Review of Sociology*, *21*, 323–347.

Kerr, C., Dunlop, J. T., Harbison, F., & Myers. C. A. (1964). *Industrialism and industrial man: The problems of labor and management in economic growth*. New York: Oxford University Press.

Mare, R. (1980). Social background and the school continuation decision. *Journal of the American Statistical Association*, *75*, 295–305.

Mare, R. (1981). Changes and stability in educational stratification. *American Sociological Review*, *46*, 72–87.

Merton, R. K. (1968). The bearing of sociological theory on empirical research and The bearing of empirical research on sociological theory. In: R. K. Merton (Ed.), *Social Theory and Social Structure* (Chaps 4 and 5). New York: Free Press.

Sewell, W. H., Haller, A. O., & Portes, A. (1969). The educational and early occupational status attainment process. *American Sociology Review*, *34*, 82–92.

Shavit, Y., & Muller, W. (Eds) (1995). *From school to work: A comparative study of educational qualifications and occupational destinations*. Oxford: Clarendon Press.

PART I: CONCEPTS FOR SOCIAL STRATIFICATION

ARE THERE ANY BIG CLASSES AT ALL?

Kim A. Weeden and David B. Grusky

ABSTRACT

The postmodernist critics of class analysis continue to claim that the life chances, attitudes, and behaviors of individuals are no longer strongly determined by their social class. We argued in a prior paper that class analysis is vulnerable to this claim because conventional big-class and gradational models are statistical constructions that ignore the deeply institutionalized occupational boundaries at the site of production. We then demonstrated that much variability in life chances, attitudes, and behaviors could be captured by disaggregating big classes into detailed occupations (i.e. "micro-classes") that better correspond to institutionalized boundaries. The present paper addresses the possibility that the resulting micro-class scheme is too extreme in presuming that all *conventional big classes are occupationally fractured and therefore untenable. We test the hypothesis that the craft, professional, and service classes are more poorly formed than other putative big classes because the forces of occupationalization are especially prominent in these three regions of the class structure. This hypothesis is examined for each of 55 outcomes culled from the Current Population Survey and the General Social Survey. We find that a partial big-class scheme, plausible though it may be, is empirically unattractive because occupationalizing forces are at work throughout the class structure.*

The Shape of Social Inequality: Stratification and Ethnicity in Comparative Perspective
Research in Social Stratification and Mobility, Volume 22, 3–56
© 2005 Published by Elsevier Ltd.
ISSN: 0276-5624/doi:10.1016/S0276-5624(04)22001-7

Many critics of class analysis maintain that class-based social divisions cannot adequately account for large-scale social and political change, contemporary forms of collective action, or individual identities and lifestyles (e.g. Beck, 2000; Clark & Lipset, 2001, 1991; Hechter, 2004; Inglehart, 1997; Kingston, 2000; Pahl, 1989; Pakulski & Waters, 1996). In the wake of such criticism, the defenders of class analysis have typically conceded that macro-level historical change cannot be understood with simple big-class models, but they insist that such models do serve well for the more modest task of explaining voting behavior, social attitudes, consumption practices, and other individual-level outcomes (e.g. Holton, 1996; Holton & Turner, 1989; cf. Sørensen, 2000). These defenders of class analysis continue to justify their preferred big-class model by arguing that a particular type of differentiation at the site of production (e.g. type of employment contract, level of authority) is especially "fundamental" or "theoretically crucial" and then carving up the site of production into the categories that this form of differentiation implies. The resulting "analytical" categories are subsequently used for the purpose of predicting individual-level outcomes of all kinds.

As we have argued in prior work, this preference for analytically derived groupings has, ironically, increased the vulnerability of class analysis to the post-modernist critique (see Grusky & Sørensen, 1998, 2001; Grusky & Weeden, 2001, 2002; Weeden & Grusky, 2004, forthcoming). The proliferation of class maps, each corresponding to a different set of analytic principles, lends credence to the claim that classes are mere statistical constructions that reflect mainly the theoretical predilections of the class analyst. Notably, the classes emerging out of these analytic exercises tend to be weakly institutionalized, resting as they do on abstract arguments about which classificatory principles are fundamental rather than on formal (e.g. licenses, apprenticeship and training programs) and informal (e.g. interactional) mechanisms of closure that actually generate *gemeinschaftlich* communities at the site of production. As a result, such classes capture but a relatively modest proportion of the structure at the site of production, offering further ammunition for those who argue that class no longer much matters.

We have instead offered a "third road" for class analysis in which the structure at the site of production is represented by detailed occupations rather than by big classes. The main rationale for this third road is that occupations are more likely than big classes to show evidence of social closure, collective action, class awareness, and other group-level properties that class analysts typically regard as defining features of classes (see Grusky & Sørensen, 1998, 2001; Grusky & Weeden, 2001, 2002; Grusky, Weeden & Sørensen, 2000; Weeden, 2002; Weeden & Grusky, 2004, forthcoming). In making the empirical case for our micro-class approach, we have demonstrated that: (a) two of the more prominent big-class schemes (i.e. Erikson & Goldthorpe, 1992; Featherman & Hauser, 1978) account for only a modest proportion of the total bivariate association between detailed occupations and a host of individual-level outcomes (e.g. life chances, political behaviors, and social attitudes); (b) there is nearly as much heterogeneity within big classes as between them; and (c) the alternative micro-class approach remains explanatorily powerful even in the context of multivariate models that control for a great many correlates of class (Weeden & Grusky, forthcoming). We accordingly concluded that sociologists seeking to understand individual-level behavior and attitudes would be well served by abandoning big-class formulations in favor of a micro-class approach that takes seriously the institutionalized occupational barriers that form within the division of labor.

We recognize, however, that our proposed alternative diverges so far from the canon that it may be as difficult for defenders of the class-analytic faith to embrace as the postmodernist critique. Indeed, Goldthorpe (2002, p. 214) characterizes our approach as a "remedy . . . worse than the disorder diagnosed," while Portes (2000, p. 250) notes that "supporters of Marxist theories may justifiably respond that, with friends like these, who needs enemies?" Are these commentators correct in implying that our proposed alternative is too radical? In this paper, we consider a middle-ground solution that represents the class structure in hybrid form, with conventional big classes appearing in some sectors of the division of labor and more differentiated "micro-classes" appearing elsewhere. This solution is credible because the division of labor appears to have become "occupationalized" to varying degrees, with jurisdictional settlements more firmly ensconced in some sectors than in others. The professional and skilled craft classes are likely to be quite heterogeneous in lifestyles, consumption practices, social attitudes, and political behaviors because the component occupations and their closure-inducing mechanisms (e.g. associations, licenses, credentials) are well-formed. It follows that incumbents are exposed to especially variable cultures and socializing forces. By contrast, the disruptive forces of occupationalization may have been successfully held at bay in some sectors of the division of labor, implying that at least some of the big classes identified in conventional class schemes are well formed. As plausible as it is, this account has not been pitted against any number of alternatives, including the null hypothesis that academics are simply more sensitive to occupational distinctions in the big classes (e.g. professionals) with which they are most familiar (see Bourdieu, 1987, p. 10).

The purpose of our analyses, therefore, will be to consider whether at least some of the conventionally posited big classes have become "social realities . . . manifest in the formation of common patterns of behavior and attitude" (Giddens, 1973, p. 111). There are, we argue, two ways that such class "structuration" is revealed. The first, external distinctiveness, refers to the degree to which the posited class differs from other classes with respect to working conditions, life chances, political behaviors, social attitudes, and consumption practices. If a class is characterized by especially distinctive practices, the boundaries separating it from other classes should become more salient to members and non-members alike. The second form of structuration, internal homogeneity, refers to the extent to which the postulated class encompasses occupations that cohere with respect to attitudes, behaviors, or working conditions. If such coherence is minimal, there is little that unites class members and transforms a class defined at the site of production into a status group in the Weberian sense. These two forms of structuration can, in principle, vary independently. It is altogether possible, for example, that the mean income of a posited big class (e.g. professionals) will be distinctively high even as the occupations that comprise that class have quite disparate incomes. Conversely, the working conditions within a big class may be relatively homogeneous (i.e. high internal homogeneity), but also differ only trivially from the conditions prevailing in other classes (i.e. low external distinctiveness).

In light of this analytic distinction, it is troubling that the few available tests of big-class schemes have focused exclusively on the external distinctiveness criterion, thereby ignoring the question of whether the postulated classes are even minimally homogeneous. Worse yet, these conventional tests have typically been carried out for a limited set of criterion variables, usually voting patterns and working or employment conditions (e.g. Evans, 1999; Evans & Mills, 1998, 2001; Hout, Brooks & Manza, 1993; Manza & Brooks, 1999; Marshall et al., 1997; but see Halaby & Weakliem, 1993). These analyses, although clearly useful, ignore the possibility that within-class heterogeneity is substantial, perhaps so substantial as to raise questions about the appropriateness of big-class formulations. Although the implicit (or sometimes explicit) assumption is that residual heterogeneity in life chances, working conditions, and other criterion variables can be safely ignored when attempting to explain outcomes (e.g. political behavior), the empirical evidence suggests, to the contrary, that such intra-class distinctions can in fact be explanatorily powerful (Weeden, 2004; Weeden & Grusky, forthcoming). It remains to be seen, though, whether all big classes are equally heterogeneous.

Our goal, then, is to extend prior work by assessing both the internal homogeneity and external distinctiveness of conventional big classes. Moreover, whereas past analysts have tended to fixate on their preferred criterion variables (e.g. life chances, employment relations), we will consider how conventional big classes fare when evaluated comprehensively across the many possible criterion variables that have been used to define classes. Under the typical class-analytic strategy outlined above, an analyst identifies a single variable as "fundamental" (e.g. type of employment contract), and the posited class map is then ' 'validated" by showing that it captures variability in that fundamental variable (or set of variables). When Evans and Mills, for example, attempt to assess the criterion validity of the Erikson-Goldthorpe class scheme, they examine the association between class location and "theoretically relevant job attributes" pertaining to the type of employment contract (2000, p. 643; see also Evans, 1992; Evans & Mills, 1998). This approach leaves open the possibility that the Erikson-Goldthorpe scheme conceals substantial heterogeneity in criterion variables that other class theorists have regarded of equal or greater theoretical centrality.

We will assess the external distinctiveness and internal homogeneity of conventional big classes in terms of three types of criterion variables that have historically been of interest to class analysts: (a) life chances (e.g. income, education, working conditions); (b) lifestyles (e.g. consumption practices, institutional participation); and (c) sentiments and dispositions (e.g. political preferences and behaviors, social attitudes). We will also ask whether conventional big classes are demographically well formed (e.g. racially and ethnically homogeneous). In devising their schemes, class analysts have not typically sought to maximize the demographic homogeneity of the posited classes (cf. Bourdieu, 1984), but intra-class fragmentation by race and ethnicity has nonetheless long been of special concern to scholars interested in class formation (e.g. Bonacich, 1972; Bradley, 1996). Although our analysis allows for an unusually comprehensive assessment of class structuration, we obviously cannot claim that it exhausts the long list of variables that class analysts have regarded as important.

WHERE IS STRUCTURATION FOUND?

As indicated above, our main objective is to assess whether there is more structuration in some big classes than in others, where structuration is understood to depend on both internal homogeneity and external distinctiveness. We will first consider the sources of internal homogeneity by examining the mechanisms through which the occupations in a putative big class come to differ in their life chances, lifestyles, and sentiments and dispositions. In our prior paper (Weeden & Grusky, forthcoming), we outlined three main mechanisms of interest (i.e. allocation, social conditioning, and institutionalization of conditions), each of which can sharpen occupational distinctions and thereby undermine intra-class homogeneity. We will examine these mechanisms in the context of the Featherman-Hauser class scheme (Featherman & Hauser, 1978) because it is more detailed, and captures more of the available structure at the site of production, than competing schemes (see Weeden & Grusky, forthcoming). When we turn to empirical analyses, however, we will consider both the Featherman-Hauser and Erikson-Goldthorpe (Erikson & Goldthorpe, 1992, pp. 35–47) schemes.

Table 1. Mechanisms Generating Structuration in FH Classes.

	Profs.	Managers	Sales Workers	Clerical Workers	Craft Workers	Operatives	Service Workers	Laborers	Farmers	Farm Laborers
A. Allocation										
1. Supply	High	Low	Low	Low	High	Low	High	Low	Low	Low
2. Demand	High	Low	Low	Med.	High	Low	High	Low	Low	Low
B. Social conditioning										
1. Training	High	Low	Low	Low	High	Low	High	Low	Low	Low
2. Interactional closure	High	Low	Low	Low	High	Med.	High	Med.	Low	Low
3. Interest formation	Med.	Med.	Med.	Med.	Med.	Med.	Med.	Med.	Med.	Med.
4. Learning generalization	Med.	Med.	Med.	Med.	Med.	Med.	Med.	Med.	Med.	Med.
C. Institutionalization of conditions	Med.	Med.	Med.	Med.	Med.	Med.	Med.	Med.	Med.	Med.

The first mechanism of interest, allocation, refers to processes that affect the types of individuals who are selected into particular locations in the site of production (see Table 1). On the supply side, workers tend to opt for occupations that are consistent with their self-conceptions, presumably preferring positions that best express their pre-existing tastes for certain types of work and job conditions. For example, individuals with liberal political values are especially likely to be attracted to the profession of sociology, given its reputation as a haven for left-leaning politics and lifestyles. This reputation thus operates as a self-fulfilling prophecy that draws in workers attracted to that reputation (Caplow, 1954). We might similarly expect journalists to be self-selected for inquisitiveness, lawyers for argumentativeness, social workers for empathy, religious workers for spirituality, printers for radicalism, nurses for nurturance, and bartenders for volubility. In addition to such dispositional reputations, occupations may also have demographic or lifestyle "reputations" (e.g. the female-typing of nursing, the staidness of accountants) that similarly serve as self-fulfilling prophecies by selecting for workers who find those reputations attractive given their own ascriptive traits, lifestyle predilections, and human capital.

On the demand side, employers and other gatekeepers filter applicants on the basis of individual-level attributes, typically by matching the traits of new recruits with those of current employees. That is, employers and other gatekeepers are well aware of the dispositional, demographic, and related reputations of occupations, and they are often motivated to recruit in accord with those reputations because of discriminatory practices (pure or statistical) or because workplace harmony and productivity is assumed to be best secured by maintaining homogeneity. In some cases, such demand-side filtering is formalized via explicit selection devices (e.g. licensing boards, unions, certifying organizations) that establish whether the attributes, training, and experience of potential employees are consistent with expectations for the position being filled. This demand-side filtering also occurs indirectly and informally whenever employers recruit new workers through the homophilous social networks of current employees.

Are such occupation-specific selective processes likely to be especially pronounced in some big classes? In addressing this question, it bears noting that our illustrative examples were all drawn from the professional (e.g. lawyer), craft (e.g. printer), or service classes (e.g. bartender), thus suggesting that occupation-based allocative processes may be more prominent in these sectors of the class structure. Indeed, because occupationalization has a long history in professional, craft, and service classes and detailed occupational reputations are accordingly well established, there is good reason to expect selective processes to be especially pronounced in these classes. By contrast, the managerial, operative, and laboring classes should be more nearly homogeneous, as class-wide dispositional reputations (e.g. the aggressive manager) tend to dominate such minor occupation-specific ones as might be identified (e.g. the punctilious government official). To be sure, recruitment in the operative and laboring occupations is often network-based (and hence homophilous), but allocative forces on the supply side are nonetheless weak because well-developed dispositional reputations are lacking. In the clerical class, most of the constituent occupations (e.g. secretary, stock clerk) also lack strong dispositional reputations, yet they do typically have well-known demographic reputations (i.e. "sex-typing") that generate quite distinctive gender profiles across the constituent occupations. We have therefore hypothesized that demand-side structuration is of "medium" strength in this class.

If the allocative mechanism evokes the imagery of social classes as vessels for like-minded workers, the "social conditioning" mechanism (Bourdieu, 1984, p. 101) refers to the causal and transformative effects of the conditions of work. These conditions shape the development of classwide and local political interests (Dahrendorf, 1959; Krause, 1996; Marx, [1869] 1963), alter the attributes that workers value both on and off the job (Kohn, [1980] 2001, pp. 539, 540; Kohn & Schooler, 1983), affect lifestyles and patterns of family interactions (Zablocki & Kanter, 1976, p. 276; see also Menaghan, 1991), and motivate workers to learn particular skills (Becker, 1993). The effects of social conditioning are generated through four sub-mechanisms: training, interactional closure, interest formation, and learning generalization. We review each of these in turn.

The first two sub-mechanisms, training and interactional closure, draw on the classical sociological forces of socialization and normative control. The training sub-mechanism becomes relevant, for example, whenever employees complete lengthy occupation-specific training (e.g. apprenticeships, police and military academies, graduate and professional schools) that solidifies preexisting attitudes, instills explicit codes of behavior, and otherwise generates homogeneity among new recruits (e.g. Caplow, 1954). Although the task of socializing incumbents can be directly undertaken through formal training, it also occurs more informally as incumbents interact with like-minded colleagues and are exposed to specialized political beliefs, social attitudes, perceived interests, and consumption practices. In some cases, these processes generate true occupational communities at the site of production, not merely because shared attitudes crystallize out of frequent social interactions, but also because groups impose sanctions against members who deviate from normative beliefs or behaviors (e.g. Berelson, Lazarsfeld & McPhee, 1954; Park, 1952, p. 196; Wilensky & Ladinsky, 1967).

There is presumably much variability across big classes in the extent to which formalized occupation-specific training and interactional closure is found. The professional class is, arguably, the home ground of such occupationalization, given that it: (a) frequently relies on arduous training regimens (e.g. graduate schools, professional schools) that instill occupationally distinctive beliefs and behaviors; and (b) entails substantial intra-occupational interaction on the job and within occupational associations and communities. Similarly, the craft class is characterized by well-developed vocational schools, apprenticeships, and occupational unions, all of which socialize workers and inculcate occupation-specific codes of behavior. Although scholars have long appreciated that professional and craft classes are fractured by occupationalizing forces (e.g. Wilensky, 1965), we argue that the service class is also likely to be deeply differentiated along occupational lines (see Table 1). This differentiation is generated because some of the constituent occupations have well-developed training regimens (e.g. law enforcement officers, firefighters) while others are carried out in "closed" workplaces that inhibit contact with workers in other occupations (e.g. wait staff). Similarly, some operative and laboring occupations (e.g. textile operatives, gardeners) are relatively closed by virtue of the social organization of the workplace, even though the occupations themselves lack the formal closure devices prevalent in the professional, craft, and service classes.

We appreciate that these various forces for differentiation are sometimes countered by classwide homogenizing effects. For example, post-secondary schools provide generalized training for members of a broadly defined professional class, and some attitudinal homogeneity may therefore be generated at the classwide level (by virtue of shared socialization and interactional closure). This homogenizing force should generate a common appreciation within the professional class for tolerance, critical discourse, and associated liberal values and behaviors that colleges inculcate. Moreover, residential segregation generates further classwide interactional closure, given that neighborhoods typically are segregated at the level of big classes rather than occupations. Although such classwide processes may be especially prominent in the professional sector, we are nonetheless hypothesizing in Table 1 that ongoing professional occupational closure will overcome the homogenizing effects of the college experience and produce differentiation in excess of what prevails in other classes.

The remaining two sub-mechanisms, interest formation and learning generalization, become relevant insofar as occupational categories are homogeneous with respect to working conditions, opportunities, and the resulting "logic" of the occupational situation. As Bourdieu puts it, "homogenous conditions of existence impose homogenous conditionings and produce homogenous systems of dispositions capable of generating similar practices" (1984, p. 104; see also Becker & Carper, 1956). This homogeneity of practice is established either because the underlying opportunities, constraints, and logic of the situation generate a characteristic set of interests (Goldthorpe, 2002) or because "there is a direct translation of the lessons of the job to outside-the-job realities" (Kohn, [1980] 2001,

p. 539). These two sub-mechanisms differ in the emphasis placed on instrumental calculation. In the interest-based account, incumbents adopt beliefs or pursue courses of action that, given the logic of the situation, allow them to best realize their objectives. In the learning generalization account, incumbents unconsciously come to appreciate and value salient features of their job or workplace, thereby motivating them to express those preferences more generally in their attitudes and behaviors. For example, workers involved in substantively complex tasks should learn to value complexity and intellectual prowess both on and off the job, thus creating a preference for intellectually demanding leisure activities. In either case, it is homogeneous conditions on the job that leads to correspondingly homogeneous interests or tastes, with these in turn generating homogeneous behaviors and attitudes.

The critical question, then, for our present purposes is whether some big classes have especially heterogeneous working conditions and therefore generate diverse interests or tastes. This question is difficult to answer given the many different dimensions of working conditions that are potentially relevant either to interests or tastes. In fact, we refrain from advancing any hypotheses here, not only because we lack direct measurements of the working conditions in big classes (and the amount of occupational heterogeneity therein), but also because we know so little about how these working conditions affect tastes and interests. To be sure, some of our readers might argue that, because occupationalization is so well-developed in the professional, craft, and service classes, we should expect much occupational variability in working conditions within these classes. This argument is unconvincing, though, because the occupations that comprise other classes, while perhaps not as deeply institutionalized, are constructed (by statisticians and sociologists) in order to capture heterogeneity in working conditions. It is unclear whether "natural" occupationalization will account for more variability in working conditions than "constructed" occupationalization reflecting these classificatory efforts. We therefore assigned the same "medium" score on these mechanisms for all big classes in Table 1.

The final mechanism listed in Table 1 ("institutionalization of conditions") refers explicitly to the organizational processes by which work is typically structured and rewarded. We have included this mechanism because some of the outcomes in our life chances domain (see Appendix A, Table A.1) refer to on-the-job conditions, such as working hours and income, that are established through union bargaining, intra-organizational institutions (e.g. internal labor markets), and federal legislation (e.g. minimum wage laws). For these particular conditions, we again wish to ask whether some big classes encompass especially heterogeneous occupations, thus rendering the classes less cohesive or unitary. As with the prior mechanisms of interest formation and learning generalization, there is no a priori reason to believe that either natural or constructed occupations will best capture heterogeneity in working conditions. We therefore assigned a score of "medium" to all classes on this mechanism (see Table 1).

The foregoing discussion can be readily summarized. As indicated in Table 1, the forces for occupationalization are relatively well-developed in the professional, craft, and service classes, thus making a big-class formulation especially implausible in those sectors of the division of labor, at least on an internal homogeneity criterion. Because the remaining classes are, by contrast, better candidates for big-class structuration, we might posit a partial big-class model that allows for intra-class differentiation only in the professional, craft, and service sectors. We will also consider in the following analyses whether internal homogeneity and external distinctiveness tend to occur together and generate a limited number of big classes (e.g. laborers, operatives) that are well structured on both dimensions. We will not, however, review here the large body of literature on issues of external distinctiveness, given that conventional "tests" of class mappings are routinely prosecuted with an external distinctiveness criterion (e.g. Evans & Mills, 2000; Goldthorpe & McKnight, forthcoming). Although we will evaluate external distinctiveness for a far larger set of outcomes than has previously been attempted, the logic of this part of our analysis should be familiar.

DATA

We assess class differences in structuration with respect to 55 individual-level indicators of life chances, consumption practices, institutional participation, political attitudes and behaviors, social attitudes and dispositions, and demographic structuration. We chose these 55 variables based on available sample sizes, coverage across survey years, and consistent item wording over time. In a few cases, a substantial number of potential items on a given topic (e.g. abortion) remained after we imposed the above restrictions, and we opted to choose a few representative items rather than overweight the analysis with items pertaining to that topic. We then combined multiple years of the GSS and CPS surveys to obtain adequate sample sizes for each of these items.[1] The indicators of life chances and demographic structuration are principally from the March Current Population Survey (CPS) of 1972–2002 (Bureau of Labor Statistics, varies), while the rest are from the General Social Survey (GSS) of 1972–2002 (Davis, Smith & Marsden, 2004). We have listed in Appendix A all variables and their source questions, response categories, and contributing surveys.

We define class and occupation schemes with 1970 Standard Occupation Classification (SOC) codes. Unfortunately, data from post-1991 GSS and post-1982 CPS files are only published in 1980 or 1990 SOC schemes, forcing us to reconcile classifications. This inconsistency is resolved through back-coding the more recent data into the 1970 scheme by: (1) translating the 1990-basis data into the 1980 scheme; (2) multiplying each 1980-basis record by the number of 1970-SOC codes that contribute to the 1980 code (U.S. Bureau of the Census, 1989); and (3) assigning sex-specific weights to each record in the resulting expanded data set.[2] This weight equals the proportion of the 1980 code that is drawn from the constituent 1970 code, multiplied by the survey weight and, for CPS data, a deflation factor that retains the original sample size. The analyses are restricted to adult respondents of age 25–64 in the civilian labor force.[3]

We proceeded by translating the 1970-basis SOC codes into various class maps, with our featured map being a highly disaggregate scheme of 126 occupations (see Appendix B, Table B.1). In constructing this scheme, we capitalized on Weeden's (2002) archive of occupation-level data on occupational associations, unions, certifications, and licensing arrangements to capture the institutionalized boundaries in the division of labor (see Weeden & Grusky, forthcoming for details). We evaluate this scheme against two well-known representations of class structure at the site of production.[4] The first, the "Featherman-Hauser" (FH) scheme, is constructed by cross-classifying Census major occupations and employment status, yielding the following 12 categories: self-employed professionals, employed professionals, employed managers, self-employed managers, sales workers, clerical workers, craft workers, operatives, service workers, laborers, farmers, and farm laborers (see, e.g. Featherman & Hauser, 1978).[5] The second, the Erikson-Goldthorpe (EG) scheme, contains seven categories: service workers, routine nonmanuals, petty bourgeoisie, skilled craft workers, unskilled manual workers, farmers, and agricultural workers.[6] In translating the 1970 SOC codes into this scheme, we relied extensively on the Erikson-Goldthorpe protocol for recoding 1960 SOC codes, and we further checked our results against ISCO-based protocols developed by Ganzeboom and Treiman.[7] The relationships between the FH, EG, and disaggregate class maps are specified in Appendix B.[8]

MODELS AND METHODS

We begin our analyses with 54 four-way tables formed by cross-classifying detailed occupation, employment status, sex, and outcome. In addition, because one of our variables (i.e. veteran status) is not consistently available for women, we also analyze a single three-way table formed by cross-classifying detailed occupation, employment status, and veteran status.[9] We first evaluate a hybrid class model that only allows for detailed occupational effects within the professional, craft, and service classes (in the FH scheme). Although we do not fit the corresponding model for the EG scheme, our prior work (Weeden & Grusky, forthcoming) revealed that the FH scheme performs better than the EG, at least for U.S. data. We suspect, then, that our FH results provide a lower bound estimate of the costs of aggregating occupations. We initially assess our hypotheses by decomposing log-likelihood test statistics, but then examine the parameters of association because, unlike the log-likelihood decompositions, these are unaffected by degrees of freedom and the power of those degrees of freedom. These parameters allow us to quantify the extent of internal homogeneity and external distinctiveness in big classes. For this parametric analysis, we consider both FH and EG big classes.

Before carrying out these analyses, we smooth the data in two ways, which allows us to increase the power of our tests while remaining true to the logic of the various class schemes. The first type of smoothing eliminates any possible three-way interactions between sex, occupation, and outcome. Although we would prefer to analyze tables for men and women separately, the three-way cross-classification of sex, occupation, and outcome is quite sparse for the GSS outcomes, thereby amplifying the risk that the analysis will unfairly capitalize on noise. By smoothing, we can retain the pooled sample size, halve the number of models to be presented, and nonetheless allow for sex-specific distributions of responses and occupations (i.e. occupational segregation).

The second type of data smoothing is necessary because the EG and FH schemes, unlike our detailed occupation scheme, take employment status into account in some classes. The detailed and big-class schemes are therefore nested only if detailed occupations are further disaggregated by employment status. However, such extreme disaggregation is not only inconsistent with our conceptual approach, but makes the GSS tables unacceptably sparse. The solution to this problem differs by class scheme. In analyses using the FH scheme, we disaggregate professional and managerial occupations by employment status, but constrain the three-way interaction between occupation, employment status, and outcome to be the same for all occupations within the professional class and for all occupations within the managerial class. This approach maintains consistency with the FH approach by allowing for interactions with employment status at the big-class level (but not the detailed occupational level).

The FH tables are smoothed, then, by fitting to each disaggregate table a model that: (a) constrains the association between occupation and outcome to be identical for men and women while allowing two-way interactions between sex and occupation and between sex and outcome; and (b) constrains the association between employment status and outcome to be same for all occupations in the professional class and for all occupations in the managerial class. The following model implements these restrictions:

$$m_{ijgp} = \alpha_i \varphi_j \nu_g \mu_p \beta_{ij} \varepsilon_{ig} \delta_{jg} \gamma_{ip} \eta_{jp} \lambda_{gp} \theta_{igp}, \quad (1)$$

where i indexes occupation, j indexes outcome, g indexes sex, and p indexes employment status.[10] The employment status variable has three levels: self-employed professionals, self-employed managers, and all other occupations. The η_{jp} term therefore allows: (a) self-employed professionals to have different responses on the outcome than employed professionals; and (b) self-employed managers to have different responses on the outcome than employed managers. These differences take the form, however, of classwide "shift effects" that pertain equally to all detailed occupations within the professional or managerial category. The expected values from this model become the data to which we fit all subsequent models (for a related approach, see Featherman & Hauser, 1978, pp. 86, 131, 167, 173).[11]

The logic of the EG scheme does not allow a similar treatment of employment status. Because Erikson and Goldthorpe (1992) assume that occupation is irrelevant within the petty bourgeoisie, occupations as diverse as child-care attendant and ship's officer are, for self-employed respondents, aggregated together. Our approach, by contrast, privileges occupation over employment status, with the result being that the two classification schemes are not nested. We solve this problem by carrying out two separate analyses of the EG tables. The first set of results will pertain to arrays in which the petty bourgeoisie, as defined by Erikson and Goldthorpe, have simply been excluded, thereby rendering the EG scheme nested within our own. The second set of results pertains to the petty bourgeoisie alone and assesses the extent to which it is indeed a distinctive and homogeneous class.

The smoothing model for the EG tables is therefore applied to the three-way array of sex, occupation, and outcome from which the petty bourgeoisie has been excised. As with the FH analysis, we purge the three-way association between these variables, thereby removing any sex differences in the occupation-outcome association from the pool of total association to be explained. This yields the following model:

$$m_{ijg} = \alpha_i \varphi_j \nu_g \beta_{ij} \varepsilon_{ig} \delta_{jg}, \quad (2)$$

where all symbols are defined as above. The fitted values from this model are used for all subsequent EG analyses. It bears emphasizing that these data smoothing procedures merely exclude residual forms of association that neither the big-class nor micro-class approaches predict.

With the smoothed data in hand, we fit models that decompose the total occupation-by-outcome association in the FH tables into: (a) the component generated between big classes; (b) the component generated within the professional, craft, and service classes; and (c) the component generated within all remaining FH classes. The first task is, of course, to specify the total occupation-outcome association at the site of production. This is represented by the log-likelihood statistic of a model that excludes the terms β_{ij} and η_{jp} from Eq. (1):

$$m_{ijgp} = \alpha_i \varphi_j \nu_g \mu_p \varepsilon_{ig} \delta_{jg} \gamma_{ip} \lambda_{gp} \theta_{igp}. \tag{3}$$

The portion of the total association generated by big-class effects is estimated with a model that allows an interaction between the FH classes and the response categories of the outcome variable:

$$m_{ijgp} = \alpha_i \varphi_j \nu_g \mu_p \varepsilon_{ig} \delta_{jg} \gamma_{ip} \lambda_{gp} \theta_{igp} \zeta_{jc}, \tag{4}$$

where c indexes aggregate class, and the remaining symbols are defined as before. The mapping of detailed occupations (and employment status) into the class variable, c, is described in Appendix B. In the FH tables, p is nested within c because p identifies self-employed professionals and managers, two of the twelve FH classes.

We fit our hybrid model by allowing an interaction between occupations and the response categories in the two professional classes, the craft class, and the service class:

$$m_{ijgp} = \alpha_i \varphi_j \nu_g \mu_p \varepsilon_{ig} \delta_{jg} \gamma_{ip} \lambda_{gp} \theta_{igp} \zeta_{jw}, \tag{5}$$

where w indexes disaggregate occupations in the professional, craft, and service classes and big classes elsewhere. As discussed above, we impose equality constraints on the ζ_{jw} in the two professional classes, but allow a single shift effect for employment status. The log-likelihood test statistic from this model quantifies the association between outcomes and occupations within the remaining big classes (i.e. those that we hypothesize to be "true" big classes).

The parametric analyses rely on a log-multiplicative association model (e.g. Goodman, 1979) that distinguishes between: (a) the strength of the class-outcome association *between* big classes; and (b) the strength of the occupation-outcome association *within* big classes. This produces the following model:

$$m_{ijg} = \alpha_i \varphi_j \nu_g \varepsilon_{ig} \delta_{jg} e^{(\rho_i \chi_j + \kappa_c \chi_j)}, \tag{6}$$

where κ_c are scale values for classes (constrained to sum to zero), ρ_i are scale values for detailed occupations (constrained to sum to zero within each class), χ_j are scale values for response categories, and all other symbols are defined as before.

The class scale parameters, κ_c, indicate the direction and distance of each class from the "average class" and hence can be interpreted as a measure of external distinctiveness.

The occupational scale values estimated under Eq. (6) can be used to construct an index, A_C, that characterizes the amount of internal heterogeneity within each of the C classes in the FH scheme:

$$A_C = \exp\left\{\left[\frac{1}{(O_C-1)}\right] \times \sum_{i=S_c}^{E_c} \rho_i^2\right\}^{1/2}, \tag{7}$$

where C refers to the total number of classes, S_c refers to the first occupation in the cth class, E_c refers to the last occupation in the cth class, and O_c refers to the total number of occupations in the cth class (see Grusky & Charles, 1998, Eq. (4)).[12] Throughout this chapter, we will present the inverse of A_C, meaning that high values correspond to high internal homogeneity.

The calculation of A_C is further complicated because our association model (Eq. (6)) estimates two sets of occupational scale values in the professional and managerial classes, one for self-employed respondents, and another for employed respondents. We calculate A_C in this case by taking the mean of: (a) the within-class index excluding the two self-employed classes (A_{CEM}); and (b) the within-class index excluding the two employed classes (A_{CSE}). Because the three-way interaction between occupation, employment status, and outcome has been purged from the data in the smoothing procedure, A_{CSE} and A_{CEM} are identical for binary outcomes, given that the model is saturated. However, for outcomes with more than two response categories, the values of A_{CSE} and A_{CEM} are not equivalent. We have found, fortunately, that the disparities are both trivial and nonsystematic, and our conclusions are therefore unaffected by resorting to the mean.

When the model of Eq. (6) is applied to the EG tables, we secure directly analogous measures of internal homogeneity for all the EG classes save the petty bourgeoisie (which was excluded from these tables). The extent of homogeneity within the petty bourgeoisie can, in turn, be assessed with a simplified parameterization of Eq. (6) fit to sex-smoothed tables containing data for the petty bourgeoisie only:

$$m_{ijk} = \alpha_i \varphi_j \nu_g \varepsilon_{ig} \delta_{jg} e^{(\rho_i \chi_j)}, \tag{8}$$

where i now indexes the 91 occupations in the petty bourgeoisie, and all other symbols are defined as above. The occupational scale values from this model can then be used to define A_C for the petty bourgeoisie (see Eq. (7)). To quantify the external distinctiveness of the petty bourgeoisie relative to other classes, we combine the EG and petty bourgeoisie tables and apply the following model:

$$m_{ijk} = \alpha_i \varphi_j \nu_k \beta_{ik} \delta_{jk} e^{(\kappa_c^* \chi_j)}. \tag{9}$$

Unlike the κ_c estimated under Eq. (6), these class scale values, κ_c^*, are not purged of occupational composition effects. However, we found no systematic differences in κ_c and κ_c^* for the six classes other than the petty bourgeoisie, so we have little reason to believe that lower-order compositional effects are influencing our estimates of the petty bourgeoisie's external distinctiveness.

We next illustrate our parametric approach by choosing a single outcome and graphing the FH class and occupation scale values estimated under Eq. (6) for that outcome (see Fig. 1). For the purposes of this exercise, we arbitrarily choose RICHWORK, a GSS item that asks respondents if they would continue working if they were rich. This item is useful for illustrative purposes because its A_C values are close to the mean values of A_C (across all items). The class scale values, indicated in Fig. 1 with a horizontal line (and attached value), show that self-employed

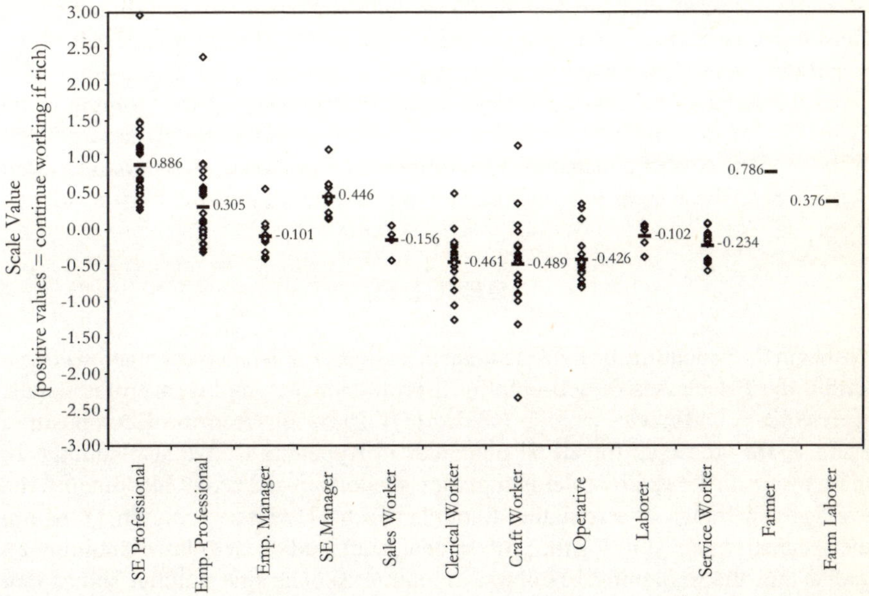

Fig. 1. Scale Values Estimated for the Association of Occupation, FH Class, and Work Orientation. *Note:* Scale values are estimated from a RC model (see Eq. (6)) applied to 1972–2002 GSS data ($n = 14{,}599$). Diamonds represent occupations, horizontal bars indicate class-specific means.

professionals are quite distinctive ($\kappa_c = 0.89$), as are farmers ($\kappa_c = 0.79$). We find that members of these two classes are rather more likely to claim that they would continue working after becoming rich than are members of other classes.

The occupation scale values, ρ_i, are plotted as points dispersed around each class mean. As Fig. 1 shows, there is substantial dispersion of these scale values around the class means, particularly in the professional and craft classes. For example, we find that religious workers are very devoted to their chosen occupation (self-employed $\rho_i = 2.96$, employed $\rho_i = 2.38$), whereas engineers are not (self-employed $\rho_i = 0.26$, employed $\rho_i = -0.32$). The extent of this within-class variance defines the values of A_C. These variances are constrained to be equal in the two professional classes (i.e. employed and self-employed) and, similarly, in the two managerial classes, but the underlying class means can vary (see Fig. 1).

Informative though it would be, we obviously do not have the space to present graphs analogous to Fig. 1 for each of the other 54 outcomes. We instead proceed by summarizing, for each big class, the extent of internal homogeneity, the extent and direction of external distinctiveness, and the joint distribution of these two measures across all domains and within each domain or sub-domain (life chances, consumption practices, institutional participation, political attitudes and behaviors, social attitudes and dispositions, and demographic structuration). These summaries avoid the pitfalls of assessing a class model based solely on the strength of its association with a single outcome (e.g. vote choice) or a small set of outcomes, but at the obvious cost of concealing variability across outcomes. Readers interested in examining these measures for a particular outcome are directed to Appendix C.

RESULTS

We begin by evaluating the hybrid model that allows for detailed occupation effects within the FH classes of self-employed professionals, employed professionals, craft workers, and service workers (see Eq. (5)). The results from the decomposition exercise are presented for all 55 outcomes in Appendix C. We also summarize these results in Table 2 by calculating, for each domain and across all domains, the average percentage of association found between FH classes (column 1), within the occupations constituting the professional, craft, and service classes (column 2), and within the remaining FH classes (column 3). The first column shows that the FH class model captures, on average, half (49.9%) of the total association at the site of production, although there is some variation in this statistic across domains (see Weeden & Grusky, forthcoming). Of more interest for our present purposes is column 2, which shows that 30% of the total association in the table

Table 2. Decomposition of Association in FH Tables, by Domain.

Domain	Between Big Classes	Within Professional, Craft, Service Classes	Within Remaining Classes
Life chances	63.69%	22.13%	14.18%
Lifestyles			
Consumption practices	45.41	31.20	23.39
Institutional participation	38.18	40.22	21.61
Class-based sentiments			
Political attitudes and behaviors	45.57	31.31	23.12
Social attitudes and dispositions	52.63	29.76	17.61
Demographic structuration	59.17	24.71	16.11
All domains	49.87	30.49	19.64

Note: Fit statistics and decompositions of association for individual items are provided in Appendix C, Table C.1.

can be captured by disaggregating in the professional, craft, and service classes, leaving under 20% of the total association within the managerial, clerical, sales, operative, and labor classes. This *prima facie* evidence in support of the hybrid model is tempered by the larger expenditure of degrees of freedom for association (df = 63) in the professional, craft, and service sector than in the remaining classes (df = 53). That is, if the two sets of classes captured the same association per degree of freedom, we would expect the professional, craft, and service classes to explain 27% ([63/116] × 0.50 = 0.27) of the total association in the table, only slightly less than the 30% that we observe.

This result suggests that there is somewhat more occupationalization in the professional, craft, and service classes than in other classes (just as we anticipated), but also that the cross-class differences in the extent of occupationalization are by no means overwhelming. Indeed, conventional significance tests show that the hybrid model is preferred for only 10 of the 55 outcomes, the big-class model is preferred for two outcomes, and the saturated model is preferred for the remaining 43 outcomes (see Appendix C, Table C.1). Ironically, the BIC criterion universally prefers the saturated model for the CPS outcomes, where sample sizes are larger, and the big-class model for all GSS outcomes, where sample sizes are smaller. As we have discussed elsewhere (Weeden & Grusky, forthcoming), the BIC results differ across surveys not because CPS outcomes are structured in micro-class terms and GSS outcomes in big-class terms, but because BIC in the small-sample context (i.e. GSS) tends to prefer a parsimonious model relative to an overparameterized one, even where the former is "sociologically unacceptable" (Raftery, 1995, pp. 152, 153).

These results suggest, then, that the hybrid model is a viable solution for only a limited number of outcomes. The results are nonetheless ambiguous enough that final judgment should be withheld until we examine the parameter estimates themselves. After all, the case for the saturated model is both assisted by the large number of degrees of freedom that it expends and harmed by the relatively low power of those degrees of freedom, given that a small number of workers appear in any given occupation. We cannot know which of these two offsetting effects is more important without turning to the parameter estimates.

We begin by assessing the extent of internal homogeneity in the FH classes by presenting the geometric mean of $1/A_C$ within each of the domains and across all domains for the eight FH classes for which A_C is defined (see Table 3).[13] The "all domain" means reveal that internal homogeneity ranges from 0.56 to 0.73.[14] As shown here, internal homogeneity is lowest in the professional and service classes, just as Table 1 suggested. Conversely, internal homogeneity is greatest in the laborer class, consistent with the Marxian view that deskilling leads to an equalization of "the various interests and conditions of life within the ranks of the proletariat" (Marx & Engels, [1848] 1978, p. 480). We did not anticipate, however, the relatively high internal homogeneity in the craft sector, which shows no greater occupation-based differentiation than the clerical or operative classes. More generally, the cross-class differences in internal homogeneity appear modest, although we have to concede that such a characterization is difficult to justify without some agreed-upon yardstick specifying what constitutes modest or substantial variability in A_c.

Can the hybrid model at least be rescued with respect to some types of outcomes? If there is any irony in the domain-specific results in Table 3, it is that internal homogeneity is lowest with respect to life chances, the very outcomes for which big class approaches are defended most vigorously (e.g. Goldthorpe & McKnight, forthcoming). At the same time, internal homogeneity is slightly greater for political and social attitudes than for other outcomes, implying that a big-class formulation is somewhat more defensible in these domains. In particular, political attitudes in the professional and managerial classes are only weakly differentiated by occupation, perhaps reflecting the effects of the shared college experience in liberalizing all professionals and the effects of shared economic interests in "conservatizing" all managers. By contrast, political attitudes in the operative and laboring classes are more highly differentiated by occupation, a result that contradicts the conventional view that these big classes are relatively homogeneous. Similarly, although some big classes are quite homogenous in race and ethnic composition (e.g. managerial, laborer), others are less so. The service class, in particular, registers a relatively

Table 3. Internal Homogeneity in FH Classes, by Domain.

Domain	Professionals (SE and Emp.)	Managers (SE and Emp.)	Sales Workers	Clerical Workers	Craft Workers	Operatives	Service Workers	Laborers
Life chances	0.32	0.49	0.52	0.42	0.45	0.47	0.26	0.67
Lifestyles								
Consumption practices	0.60	0.69	0.71	0.60	0.59	0.58	0.58	0.66
Institutional participation	0.50	0.74	0.71	0.58	0.63	0.61	0.51	0.75
Class-based sentiments								
Political attitudes and behaviors	0.65	0.76	0.73	0.62	0.62	0.58	0.62	0.60
Social attitudes and dispositions	0.61	0.76	0.68	0.70	0.69	0.68	0.68	0.81
Demographic structuration	0.73	0.80	0.77	0.74	0.73	0.71	0.62	0.78
All domains	0.56	0.71	0.68	0.61	0.62	0.61	0.56	0.73

Note: Values in the table are 1/Ac (see Eq. (7)). Farmers and farm laborers are single-occupation classes and hence are excluded.

low index value, presumably because network-based allocation generates much occupational recruitment that is segregated along racial and ethnic lines.

It might at this point be argued that the FH class scheme, based as it is on purely "administrative" categories, does not well represent the big-class tradition. Does the same pattern of results obtain when a "theoretically informed" scheme, such as EG, is considered? To address this question, Table 4 presents an analogous set of measures as Table 3, but now for the five EG classes that contain more than one component occupation (i.e. farmers and farm laborers, both single-occupation classes, are excluded). The mean across all domains reveals that occupational differentiation is, if anything, yet greater in the EG classes and that again there are only modest inter-class differences in internal homogeneity. Although we had expected that the nonskilled class would be relatively homogeneous, the values in Table 4 indicate that occupationalizing forces are nearly as well developed in this class as in the service class. The only notable outlier is the petty bourgeoisie. This class, which is one of the defining features of the EG classification, is exceedingly poorly developed by our measure of internal homogeneity. Evidently, the EG assumption that self-employment status trumps occupational location is problematic, thus belying long-standing efforts to represent self-employment as a defining work condition. Finally, the domain-specific statistics again reveal the greatest occupation-based differentiation in the life chances domain and the least in the demographic domain, just as was the case with the FH scheme.

Table 4. Internal Homogeneity in EG Classes, by Domain.

Domain	Service	Routine Nonmanual	Petty Bourg.	Skilled Workers	Nonskilled Workers
Life chances	0.37	0.34	0.36	0.44	0.34
Lifestyles					
Consumption practices	0.60	0.60	0.33	0.55	0.56
Institutional participation	0.51	0.57	0.39	0.62	0.56
Class-based sentiments					
Political attitudes and behaviors	0.63	0.61	0.37	0.58	0.58
Social attitudes and dispositions	0.60	0.71	0.32	0.67	0.69
Demographic structuration	0.72	0.68	0.40	0.71	0.70
All domains	0.56	0.60	0.34	0.60	0.58

Note: Values in the table are 1/Ac. For all classes except the petty bourgeoisie, these are calculated from Eq. (7); for the petty bourgeoisie, they are calculated from scale values from Eq. (8). EG classes IV (farmers) and VII (farm laborers) are single-occupation classes and hence are excluded from this table.

Table 5. External Distinctiveness of FH Classes, by Domain.

Domain	SE Professionals	Emp. Professionals	Emp. Mgrs	SE Mgrs	Sales Workers	Clerical Workers	Craft Workers	Operatives	Service Workers	Laborers	Farmers	Farm Laborers
Life chances	1.28	1.77	1.53	0.43	1.06	0.14	0.00	−0.80	−1.16	−1.26	−0.78	−2.21
Lifestyles												
Consumption practices	0.66	0.42	0.34	0.44	0.37	−0.17	−0.24	−0.59	−0.27	−0.56	0.20	−0.60
Institutional participation	0.41	0.35	0.11	0.27	0.05	0.19	0.13	−0.05	−0.13	−0.10	−0.40	−0.70
Class-based sentiments												
Political attitudes and behaviors	0.58	0.24	0.36	0.40	0.38	−0.14	−0.40	−0.69	−0.37	−0.43	0.36	−0.28
Social attitudes and dispositions	0.73	0.61	0.35	0.17	0.38	−0.01	−0.25	−0.40	−0.10	−0.34	−0.58	−0.56
Demographic structuration	0.60	0.27	0.27	0.29	0.55	−0.23	−0.12	−0.52	−0.44	−0.70	0.93	−0.90
All domains	0.72	0.61	0.44	0.31	0.42	−0.03	−0.18	−0.48	−0.30	−0.49	−0.22	−0.76

Note: Values in the table are the arithmetic means (across outcomes) of the class scale values estimated under Eq. (6).

We turn next to considerations of external distinctiveness. Whereas the internal homogeneity criterion speaks to the fracturing effect of occupationalizing forces, the external distinctiveness criterion considers whether a given class assumes, on average, a value on the outcome item that differs much from the values assumed by other classes. We again present both overall and domain-specific averages for the two big-class schemes (Tables 5 and 6). As shown in Table 5, the overall averages of the FH class scale values (see Eq. (6)) predictably follow a socioeconomic gradient, with professionals (average $\kappa_c = 0.72$ or 0.61) and farm laborers (average $\kappa_c = -0.76$) falling at the two extremes and clerical workers (average $\kappa_c = -0.03$) falling in roughly the center of the gradient. The farm class is of course unusually complicated, because unlike other classes it cannot be understood in socioeconomic terms. Although the life chances and social attitudes of the farm class are consistent with socioeconomic disadvantage, its consumption practices, political attitudes, and demographic composition are consistent with socioeconomic advantage. A similar pattern of results is evident for the EG classes (see Table 6).

These all-domain averages mask variation in the extent to which big classes are distinctive in different domains. The contrast between the life chances domain and all others is especially dramatic: For both the FH and EG schemes, the life chances domain reveals very dispersed scale values, whereas all other domains have comparatively less dispersed values. This result is consistent with the claim that big-class schemes succeed in capturing differences in life chances (e.g. Goldthorpe & McKnight, forthcoming). However, our results for other domains imply that this success is the exception, not the rule. Moreover, we have also found that these substantial between-class differences in life chances are joined with equally substantial within-class differences (see Tables 3 and 4), meaning that even class analysts with an exclusive interest in life chances cannot safely rely on big-class models.

We conclude this section by examining the joint distribution of external distinctiveness and internal homogeneity for the FH classes (see Fig. 2). Because external distinctiveness is signaled by either high or low κ_c values, the strongest big-class candidates will show up in either the top right or bottom right quadrants of Fig. 2. It is notable that the laboring and managerial classes often appear in these quadrants. If one wished, then, to press the case for a hybrid model, Fig. 2 suggests that aggregation is most defensible in these two classes. Although laborers have quite differentiated political attitudes, this result can be traced to a single outcome, membership in political clubs (MEMPOL), and might therefore be overlooked (see Appendix C, Tables C.2–C.5). Otherwise, these two classes combine considerable internal homogeneity with at least moderate levels of external distinctiveness, thus rendering them the best approximations to true "big classes" available in the contemporary U.S. labor market. Granted, it is always possible that these

Table 6. External Distinctiveness of EG Classes, by Domain.

Domain	Service	Routine non-manual	Petty Bourg.	Farmers	Skilled Workers	Non-Skilled Workers	Farm Laborers
Life chances	2.28	0.06	0.07	−0.34	0.65	−0.51	−2.13
Lifestyles							
Consumption practices	0.59	−0.01	0.25	0.32	−0.12	−0.31	−0.41
Institutional participation	0.51	0.07	0.33	0.07	−0.17	−0.28	−0.20
Class-based sentiments							
Political attitudes and behaviors	0.47	−0.06	0.30	0.52	−0.31	−0.44	−0.18
Social attitudes and dispositions	0.78	0.18	0.15	−0.37	−0.05	−0.13	−0.41
Demographic structuration	0.46	−0.16	0.24	1.08	−0.03	−0.49	−0.86
All domains	0.79	0.06	0.21	0.05	−0.04	−0.29	−0.56

Note: Values in table are arithmetic means (across outcomes) of the class scale values estimated under Eq. (6) and, for the petty bourgeoisie, Eq. (9).

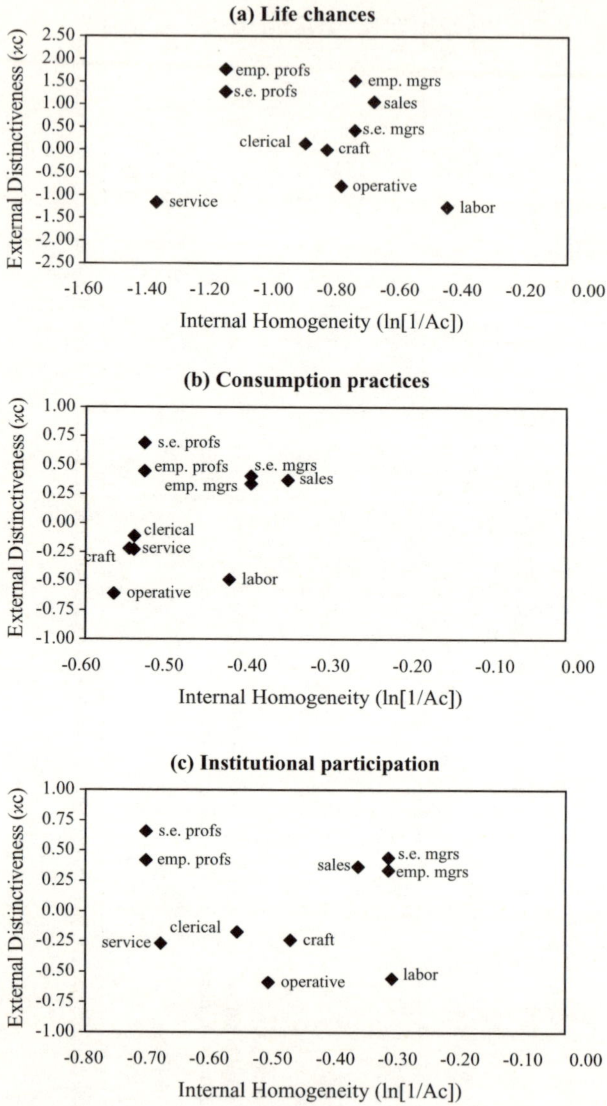

Fig. 2. External Distinctiveness and Internal Homogeneity of FH Classes.

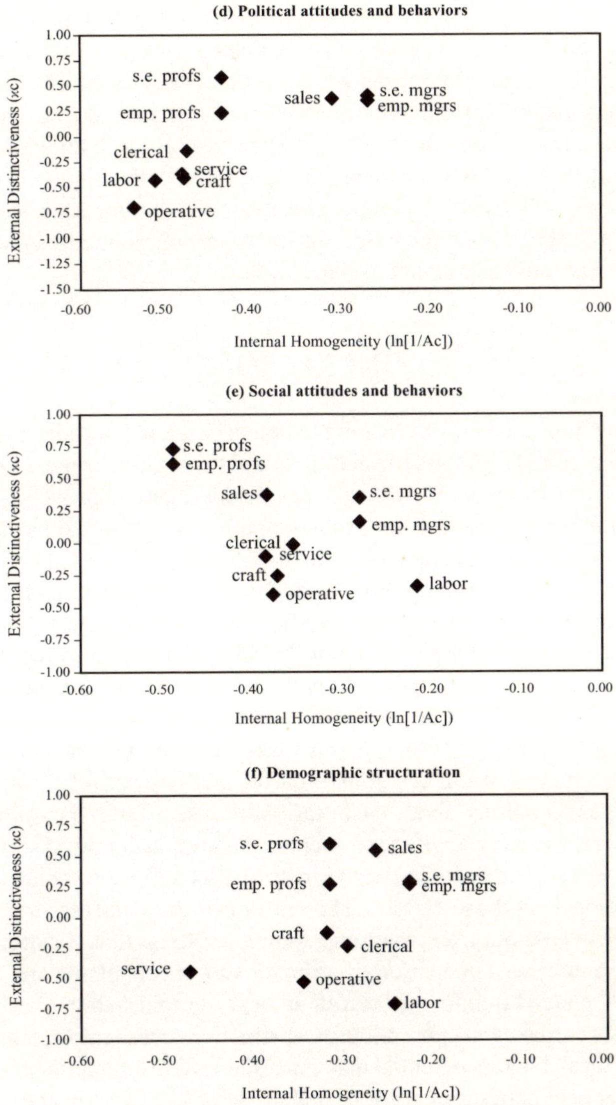

Fig. 2. (*Continued*)

classes exhibit homogeneity not because occupational differentiation is truly lacking, but because the Census occupational categories are notoriously poor and fail to pick up the occupation-based structure that better categories would reveal. The more conservative interpretation, however, and the one to which we default, is that for these two classes there is an "obliteration of all distinctions" (Marx & Engels, [1848] 1978, p. 480) that makes the aggregation of component occupations relatively costless. In all remaining classes, such aggregation comes at the substantial price of masking much structure at the site of production, and a big-class formulation is accordingly called into question.

CONCLUSIONS

We motivated this analysis by asking whether a viable middle ground exists between conventional big-class formulations and the highly disaggregate approach that we have to date favored. After all, our disaggregate approach is admittedly profligate with degrees of freedom, thus making it worth considering a middle-range solution in which occupational distinctions are glossed over in those big classes where occupationalizing forces are (putatively) undeveloped. This middle-range solution proves to be unattractive. By conventional significance tests, the hybrid model is preferred for only 10 of the 55 outcomes analyzed here and the big-class model for only two, leaving 43 outcomes for which our micro-class model is preferred.

Nonetheless, there is *some* cross-class variability in the extent of occupationalization, and our results are partly, but not entirely, consistent with conventional expectations about such variability (see Table 1). As anticipated, occupationalization is especially prominent in the professional and service classes, presumably because the differentiating forces of allocation and social conditioning are well developed in those classes. The craft class, by contrast, is relatively undifferentiated, even though conventional arguments stress that vocational training, craft unions, and other occupation-specific closure mechanisms are prevalent in that class. In understanding this result, it bears emphasizing that occupation-specific closure makes it possible for distinctive cultures to emerge at the disaggregate level, but this potential may go unrealized insofar as there is nothing in the underlying occupational conditions (e.g. life chances, working conditions) that support such differentiation. There is nothing to preclude socially closed groupings from independently settling on similar cultural solutions. That is, the potential for craft differentiation may not have been fully realized because of the shared middle-class "logic" of the craft situation (e.g. middle-class income, relatively high security), much as rational action theorists suggest (e.g. Goldthorpe, 2000).

Although occupationalizing forces are suppressed in the craft sector, at least relative to what we anticipated, the craft class is by no means the best candidate for true big-class status. As Fig. 2 revealed, the laboring and managerial classes are especially well structured, showing both greater external distinctiveness and more substantial internal homogeneity than other classes. These two classes are in this sense tailor-made for a big-class formulation. Detailed occupation-specific reputations are poorly developed in both cases; training regimens are either classwide (i.e. the MBA degree) or virtually nonexistent (i.e. minimal occupation-specific training among laborers); and the underlying logic of the class situation is one of distinctive privilege in the case of managers and distinctive disadvantage in the case of laborers. It is fitting that these two classes, which may be viewed as the contemporary remnants of the Marxian two-class model (i.e. "business" class, unskilled laboring class), are the home ground of a big-class formulation.

This result should, however, provide little solace for the big-class theorist, given that a mere 13% of the contemporary U.S. labor force can be found in these two classes (U.S. Equal Employment Opportunity Commission, 2004). The vast majority of present-day workers can instead be found in classes where occupational distinctions are alive and well. The main problem, then, with our proposed hybrid solution was that it gave over far more terrain to the big-class formulation than was merited. The professional, service, and craft sectors should not be viewed as idiosyncratic outposts of occupationalization; rather, the forces for occupationalization have diffused throughout the division of labor, and only the managerial and laboring classes have to this point resisted them.

The question that then emerges is whether these two classes will continue to resist occupationalization. Are they merely "late adopters" and destined to occupationalize like other big classes? Or will they remain undifferentiated over the long term? Could it be argued, pace Marx, that the laboring and managerial classes reveal our future and that other big classes will ultimately become as undifferentiated as they are? In addressing these questions, the trajectories of the laboring and managerial classes are best considered separately, because they would appear to be idiosyncratic for very different reasons. The distinctive feature of the laboring class is that it is undifferentiated almost by definition; after all, if some segment of this class accumulates specialized (i.e. occupation-specific) human capital, the usual methods of occupational closure become more viable and the occupation is likely to be reallocated (by statisticians) into one of the high-skill big classes. This is precisely the mechanism by which the laboring class has declined in size over the course of the last century. The ongoing effects of differentiation are revealed, therefore, in the progressive decline in the number of laborers, not via the emergence of occupational distinctions among the ranks of laborers.

The managerial class, by contrast, is formally at risk of occupationalization, but inertial forces within it nonetheless make occupationalization a slower and more sporadic process. In particular, the managerial class is the one and only big class that is characterized by true class-wide training and credentialing (in the form of the "business school" and the MBA). This organizational form, which has become deeply institutionalized over the last century, instills classwide attitudes, values, and orientations and legitimates all forms of intra-class mobility in the labor market. Undoubtedly, there will be continuing efforts to develop more specialized managerial training and degrees (e.g. degrees in "public administration"), but these efforts will run up against a generalist organizational form that does not exist in any other putative big class.

We suspect, then, that both the managerial and laboring classes will remain less differentiated than other classes over the long term, albeit for different reasons. The laboring class is, by definition, a mere residual of undifferentiated unskilled labor, a residual that has shrunk over the last century and will likely continue to shrink. The managerial class is, by contrast, relatively well-formed, but the mechanisms of class formation and social closure have in this case developed at the big class level, quite idiosyncratically. Although the managerial class may therefore be understood as a true big class, there is little reason to believe that other putative big classes will ever develop similarly class-wide forms of social closure. To the contrary, we suspect the march toward increasingly differentiated occupation-based training, licenses, certification programs, and apprenticeships will continue unabated, and with it the solidification of distinctive, occupation-based beliefs and practices.

NOTES

1. We have disaggregated the data by decade in another paper (Weeden & Grusky, 2004) and found considerable stability in the strength of the association between occupations and outcomes. This analysis ignores the hybrid model and hence is not directly analogous to the analyses we present here, but it nonetheless suggests that temporal shifts in the occupation-outcome association are not substantial enough to affect the conclusions we draw from pooled data.

2. For a given 1980 occupation, suppose that 90% of incumbents would have been coded into occupation X in the 1970 scheme, while 10% would have been coded into occupation Y. Each person with this 1980 occupation contributes two records to the expanded data set: one record receives code X and a weight of 0.9, and the other receives code Y and a weight of 0.10.

3. The CPS samples are further restricted to households in months 1–4 of the sampling rotation, thereby preventing a household from contributing observations in successive years.

4. Although we would have liked to assess Wright's influential neo-Marxian class map, the variables needed to implement this map faithfully are not consistently available in

the GSS or CPS, and Wright's own data (e.g. 1997) contain too few cases to sustain an occupation-level analysis.

5. We opt against a 17-category version of the FH scheme because the industrial distinctions that it adds to the 12-class version (e.g. retail sales workers, wholesale sales workers) typically fall outside the purview of class analysis.

6. Although this scheme has a detailed variant, we use the more commonly applied seven-class version because it can be implemented faithfully with CPS and GSS data (see esp. Erikson & Goldthorpe, 1992, pp. 35–47).

7. We thank Walter Mueller, John Goldthorpe, Harry Ganzeboom, and Donald Treiman for sharing the conversion protocols.

8. To maintain consistency across class schemes, we assigned 14 (of 428) sparsely populated SOC occupations to FH categories that diverge from the major occupation group. These inconsistencies occur when technically similar occupations are coded into different SOC major groups. For example, milliners are coded as operatives in the SOC, while tailors are coded as craft workers. There are not enough milliners to justify a separate occupational category, so we combine milliners with tailors and, to be consistent with the EG scheme, assign the resulting occupation to the FH class of "craft worker." We privileged EG over FH because the former is gaining ascendancy in the day-to-day practice of sociology, even in North America (see, e.g. Manza & Brooks, 1999).

9. The 1988 CPS was the first to ask the veteran status question of women. Because our data series begins in 1972, we were forced to exclude women from all analyses of this outcome.

10. We identify the parameters of this and all subsequent models by imposing standard constraints. For brevity's sake, we will note the identifying constraints only when they are unclear or affect the interpretation of the model.

11. We have added a constant (0.1) to empty cells (e.g. Agresti, 1990).

12. As we have indicated, Eq. 6 is saturated for binary outcomes, but not for outcomes with more than two response categories. Consequently, we also fit a multidimensional association model (with J–1 sets of occupation scale values) to each array with more than two response categories, and we then calculated the corresponding values of A_C. The results from this exercise indicate that, if anything, the single-dimensional model leads to a more conservative estimate of A_C. Because the multidimensional models yield substantially more parameters and are more sensitive to sparse cell counts, we present the decompositions from the unidimensional model in this paper.

13. The index A_C is not defined for the two farm classes (which contain one detailed occupation each) and is constrained to be equal in the two professional and two managerial classes.

14. The "all-domain" means in Tables 3-6 weight each outcome equally, thereby giving more weight to the domains with more constituent outcomes. If we instead weight each domain equally, our conclusions are unaffected.

ACKNOWLEDGMENTS

The research reported here was supported by a National Science Foundation research grant (SBS-9906419) and with discretionary funds from Cornell

University, the University of Chicago, and Stanford University. We wish to thank David Bills and an anonymous reviewer for their comments.

REFERENCES

Agresti, A. (1990). *Categorical data analysis*. New York: Wiley.

Beck, U. (2000). *Brave new world of work* (trans. P. Camiller). Cambridge, UK: Polity Press, and Malden, MA: Blackwell.

Becker, G. (1993). *Human capital: A theoretical and empirical analysis, with special reference to education*. Chicago: University of Chicago Press.

Becker, H. S., & Carper, J. (1956). The elements of identification with an occupation. *American Sociological Review*, *21*(3), 341–348.

Berelson, B. R., Lazarsfeld, P. F., & McPhee, W. N. (1954). *Voting: A study of opinion formation in a presidential campaign*. Chicago: University of Chicago Press.

Bonacich, E. (1972). A theory of ethnic antagonism: The split labor market. *American Sociological Review*, *37*(5), 547–559.

Bourdieu, P. (1984). *Distinction: A social critique of the judgement of taste* (trans. by R. Nice). New York: Cambridge University Press.

Bourdieu, P. (1987). What makes a social class? On the theoretical and practical existence of groups. *Berkeley Journal of Sociology*, *32*, 1–17.

Bradley, H. (1996). *Fractured identities: Changing patterns of inequality*. Cambridge, UK: Polity Press.

Bureau of Labor Statistics (Years vary). *March demographic files of the current population surveys*. Machine-readable data file. Bureau of the Census, Washington, DC.

Caplow, T. (1954). *The sociology of work*. Minneapolis: University of Minnesota Press.

Clark, T. N., & Lipset, S. M. (1991). Are social classes dying? *International Sociology*, *6*, 397–410.

Clark, T. N., & Lipset, S. M. (Eds) (2001). *The breakdown of class politics*. Washington, DC: Woodrow Wilson Center Press, and Baltimore: Johns Hopkins University Press.

Dahrendorf, R. (1959). *Class and class consciousness in industrial society*. Stanford, CA: Stanford University Press.

Davis, J. A., Smith, T. W., & Marsden, P. (2004). *GSS (General Social Survey) 1972–2002 cumulative data file*. Chicago: NORC.

Erikson, R., & Goldthorpe, J. H. (1992). *The constant flux: A study of class mobility in industrial societies*. Oxford: Clarendon Press.

Evans, G. (1992). Testing the validity of the Goldthorpe class schema. *European Sociological Review*, *8*(3), 211–232.

Evans, G. (1999). *The end of class politics*. Oxford: Oxford University Press.

Evans, G., & Mills, C. (1998). Identifying class structure: A latent class analysis of the criterion-related and construct validity of the Goldthorpe class schema. *European Sociological Review*, *14*(1), 87–106.

Evans, G., & Mills, C. (2000). In search of the wage-labour/service contract: New evidence on the validity of the Goldthorpe class schema. *British Journal of Sociology*, *51*(4), 641–666.

Featherman, D. L., & Hauser, R. M. (1978). *Opportunity and change*. New York: Academic.

Giddens, A. (1973). *The class structure of the advanced societies*. New York: Harper.

Goldthorpe, J. H. (2000). *On sociology: Numbers, narrative, and the integration of research and theory*. Oxford: Oxford University Press.

Goldthorpe, J. H. (2002). Occupational sociology, yes, class analysis, no: A comment on Grusky and Weeden's research agenda. *Acta Sociologica*, *45*(3), 211–217.

Goldthorpe, J. H., & McKnight, A. (forthcoming). *The economic basis of social class*. In: S. L. Morgan, O. B. Grosky & G. S. Fields (Eds), *Mobility and Inequality: Frontiers of Research from Sociology and Economics*. Stanford: Stanford University Press.

Goodman, L. (1979). Simple models for the analysis of association in cross-classifications having ordered categories. *Journal of the American Statistical Association*, *74*, 537–552.

Grusky, D. B., & Charles, M. (1998). The past, present, and future of sex segregation methodology. *Demography*, *35*(4), 497–504.

Grusky, D. B., & Sørensen, J. B. (1998). Can class analysis be salvaged? *American Journal of Sociology*, *103*, 1187–1234.

Grusky, D. B., & Sørensen, J. B. (2001). Are there big social classes? In: D. Grusky (Ed.), *Social Stratification: Class, Race, and Gender in Sociological Perspective* (2nd ed., pp. 183–194). Boulder: Westview.

Grusky, D. B., & Weeden, K. A. (2001). Decomposition without death: A research agenda for the new class analysis. *Acta Sociologica*, *44*(3), 203–218.

Grusky, D. B., & Weeden, K. A. (2002). Class analysis and the heavy weight of convention. *Acta Sociologica*, *45*(3), 229–236.

Grusky, D. B., Weeden, K. A., & Sørensen, J. B. (2000). The case for realism in class analysis. *Political Power and Social Theory*, *14*, 291–305.

Halaby, C., & Weakliem, D. (1993). Class and authority in the earnings function. *American Sociological Review*, *58*, 16–30.

Hechter, M. (2004). From class to culture. *American Journal of Sociology*, *110*(2), 400–446.

Holton, R. J. (1996). Has class analysis a future? In: D. J. Lee & B. S. Turner (Eds), *Conflicts About Class: Debating Inequality in Late Industrialism* (pp. 26–41). London: Longman.

Holton, R. J., & Turner, B. S. (1989). *Max Weber on economy and society*. London: Routledge & Kegan Paul.

Hout, M., Brooks, C., & Manza, J. (1993). The persistence of classes in post-industrial societies. *International Sociology*, *3*, 259–277.

Inglehart, R. (1997). *Modernization and postmodernization: Cultural, economic, and political change in 43 societies*. Princeton: Princeton University Press.

Kingston, P. W. (2000). *The classless society*. Stanford: Stanford University Press.

Kohn, M. L. ([1980] 2001). Job complexity and adult personality. In: D. B. Grusky (Ed.), *Social Stratification: Class, Race, and Gender in Sociological Perspective* (2nd ed., pp. 532–541). Boulder: Westview Press.

Kohn, M. L., & Schooler, C. with the collaboration of J. Miller (1983). *Work and personality: An inquiry into the impact of social stratification*. Norwood, NJ: Ablex Publishing Company.

Krause, E. A. (1996). *Death of the guilds: Professions, states, and the advance of capitalism, 1930 to the present*. New Haven: Yale University Press.

Manza, J., & Brooks, C. (1999). *Social cleavages and political change*. Oxford: Oxford University Press.

Marshall, G., Swift, A., & Roberts, S. (1997). *Against the odds? Social class and social justice in industrial societies*. New York: Oxford University Press.

Marx, K. ([1869] 1963). Eighteenth brumaire of Louis Bonaparte. In: R. C. Tucker (Ed.), *The Marx-Engels Reader* (2nd ed.). New York: Norton.

Marx, K., & Engels, F. ([1848] 1978). Manifesto of the Communist Party. In: R. C. Tucker (Ed.), *The Marx-Engels Reader* (2nd ed.). New York: Norton.

Menaghan, E. G. (1991). Work experiences and family interaction processes: The long reach of the job? *Annual Review of Sociology*, *17*, 419–444.

Pahl, R. E. (1989). Is the emperor naked? Some questions on the adequacy of sociological theory in urban and regional research. *International Journal of Urban and Regional Research*, *13*, 709–720.

Pakulski, J., & Waters, M. (1996). *Death of class*. Thousand Oaks, CA: Sage.

Park, R. E. (1952). *Human communities: The city and human ecology*. Glencoe, IL: Free Press.

Portes, A. (2000). The resilient importance of class: A nominalist interpretation. *Political Power and Social Theory*, *14*, 249–284.

Raftery, A. E. (1995). Bayesian model selection in social research. *Sociological Methodology*, *25*, 111–163.

Sørensen, A. B. (2000). Toward a sounder basis for class analysis. *American Journal of Sociology*, *105*(May), 1523–1558.

U.S. Bureau of the Census (1989). The relationship between the 1970 and 1980 industry and occupation classification systems (technical paper 59). Washington, DC: U.S. Government Printing Office.

U.S. Bureau of the Census (2004). U.S. equal employment opportunity commission. 2000 Special EEO File, Table 3. Available http://www.eeoc.gov/stats/census.

Weeden, K. A. (2002). Why do some occupations pay more than others? Social closure and earnings inequality in the United States. *American Journal of Sociology*, *108*(1), 55–101.

Weeden, K. A. (2004). *Big-class politics or occupation politics*? Unpublished manuscript. Department of Sociology, Cornell University.

Weeden, K. A., & Grusky, D. B. (2004). Are social classes decomposing? Unpublished manuscript. Department of Sociology, Cornell University.

Weeden, K. A., & Grusky, D. B. (forthcoming). The case for a new class map. *American Journal of Sociology*.

Wilensky, H. (1965). The professionalization of everyone? *American Journal of Sociology*, *70*(2), 137–158.

Wilensky, H. S., & Ladinsky, J. (1967). From religious community to occupational group: Structural assimilation among professors, lawyers, and engineers. *American Sociological Review*, *32*(4), 541–561.

Wright, E. O. (1997). *Class counts: Comparative studies in class analysis*. Cambridge: Cambridge University Press.

Zablocki, B. J., & Kanter, R. M. (1976). The differentiation of life-styles. *Annual Review of Sociology*, *2*, 269–298.

APPENDIX A

Table A.1. Variable Mnemonics, Sources, Description, and Response Categories, by Domain.

Mnemonic	Source	Description and Response Categories (in Parentheses)
Life chances		
educ	CPS	Highest grade completed. (Less than high school, high school, some college, college, some graduate school.)
ftpt	CPS	Full-time/part-time status. (Usually works full-time, usually works part time.)
ftincome	CPS	Income 1: Wage, self-employment, and farm income of currently working respondents who usually work 35+ hours per week. (Year-specific quintiles.)
income	CPS	Income 2: Wage, self-employment, and farm income of currently working respondents. (Year-specific quintiles.)
finrela	GSS	Subjective position: Income compared with American families in general. (Below average, average, above average.)
tenure	CPS	Wealth: Ownership of living quarters. (Rent, own.)
Lifestyles		
Consumption practices		
news	GSS	Intellectual pursuits 1: "How often do you read the newspaper?" (Less than once a week or never, once a week, a few times per week, daily.)
tvhours	GSS	Intellectual pursuits 2: "On the average day, about how many hours do you personally watch television?"
memlit	GSS	Intellectual pursuits 3: "Are you a member of literary, art, discussion, or study groups?" (No, yes.)
satfam	GSS	Family orientation 1: "How much satisfaction do you get from your family life?" (Little or none, quite a bit, a great deal, or a very great deal.)
socrel	GSS	Family orientation 2: "How often do you spend a social evening with relatives?" (At least one a week, one to "several" times a month, less.)
socommun	GSS	Friendship orientation 1: "How often do you spend a social evening with someone who lives in your neighborhood?" (At least once a week, one to "several" times a month, once or twice a year, never.)
socfrend	GSS	Friendship orientation 2: "How often do you spend a social evening with friends who live outside the neighborhood?" (At least one a week, one to "several" times a month, less.)
memserv	GSS	Service 1: "Are you a member of service clubs?" (No, yes.)
memfrat	GSS	Service 2: "Are you a member of [non-scholastic] fraternal groups?" (No, yes.)
memsport	GSS	Sports and hobbies 1: "Are you a member of sports groups?" (No, yes.)
memhobby	GSS	Sports and hobbies 2: "Are you a member of hobby or garden clubs?" (No, yes.)

Table A.1. (*Continued*)

memnum	GSS	Communitarianism: Number of group or club memberships. (0, 1, 2, 3 or more.)
satjob	GSS	Work orientation 1: "On the whole, how satisfied are you with the work you do?" (Very, moderately, dissatisfied.)
richwork	GSS	Work orientation 2: "If you were to get enough money to live as comfortably as you would like for the rest of your life, would you continue working or stop working?" (Stop, continue.) Employed respondents only.
hours	CPS	Work orientation 3: Hours worked last week. (1–34, 35–40, 41–60, 61 or more.) Working respondents only.
Institutional participation		
marstat	CPS	Marriage 1: Current marital status (Never married, separated, divorced, widowed, married.)
divorce	GSS	Marriage 2: "Have you ever been divorced or legally separated?" (Yes, no.) Ever-married respondents only.
child	GSS	Children: "How many children have you ever had?" (4+, 3, 2, 1, 0.)
relig	GSS	Religion 1: "What is your religious preference?" (Protestant, Catholic, Jewish, none, other.)
attend	GSS	Religion 2: "How often do you attend religious services?" (Never, 1–3 per year, 1–3 per month, 1+ per week.)
union	CPS	Union membership: "On this job, are you [is household member] a member of a labor union or of an employee association similar to a union?" (Yes, no.) Wage and salary workers in 1983–2002.
vet	CPS	Veteran status. (Yes, no.) Men only.
Class-based sentiments		
Political attitudes and behaviors		
partyid	GSS	Party identification: "Do you usually think of yourself as a . . ." (Strong Democrat, Democrat, Independent, Republican, strong Republican.)
polviews	GSS	Political ideology 1. "Where would you place yourself on a scale . . .?" (Extremely/conservative, slightly conservative, moderate, slightly liberal, extremely/liberal.)
helpnot	GSS	Political ideology 2: Assess the federal government's intervention into our country's problems. (Government doing too much, government should do more, both.)
mempolit	GSS	Collective action: "Are you a member of political clubs?" (No, yes.)
Social attitudes and dispositions		
spkath	GSS	Tolerance 1: "Should [an atheist] be allowed to make a speech in your community?" (No, yes.)
spkcom	GSS	Tolerance 2: "Should [an admitted Communist] be allowed to make a speech in your community?" (No, yes.)
homosex	GSS	Tolerance 3: "Are sexual relations between two adults of the same sex wrong?" (Always, sometimes, not at all.)

Table A.1. (*Continued*)

pornlaw	GSS	Tolerance 4: "Which of these statements comes closest to your feelings about pornography laws?" (Should be laws against the distribution whatever the age; should be laws against the distribution to persons under 18; should be no laws forbidding distribution.)
cappun	GSS	Crime 1: "Do you favor or oppose the death penalty for persons convicted of murder?" (Favor, oppose.)
courts	GSS	Crime 2: "Do you think the courts in this area deal too harshly or not harshly enough with criminals?" (Too harshly, not harshly enough, about right.)
prayer	GSS	Church and state separation: The court has ruled that governments may not require the reading of the Bible in public schools. (Disapprove, approve.)
racmar	GSS	Racial attitudes 1: "Do you think there should be laws against marriages between blacks and whites?" (Yes, no.) Non-blacks only.
racopen	GSS	Racial attitudes 2: "Which law would you vote for?": (A) a homeowner can decide to whom to sell his house, (b) a homeowner cannot refuse to sell his house to a black. (Law a, law b.) Non-blacks only.
helpblk	GSS	Racial attitudes 3: Do you think the government is responsible for redressing past discrimination? (Government shouldn't give special treatment to blacks, government is obligated to help blacks, both.)
fework	GSS	Gender attitudes 1: "Do you approve or disapprove of a married woman earning money in business or industry if she has a husband capable of supporting her?" (Disapprove, approve.)
fepol	GSS	Gender attitudes 2: "Most men are better suited emotionally for politics than are most women." (Agree, disagree.)
abnomore	GSS	Abortion attitudes 1: "[Should abortion be legal if a woman] is married and does not want any more children?" (No, yes.)
abrape	GSS	Abortion attitudes 2: "[Should abortion be legal if a woman] became pregnant as a result of rape?" (No, yes.)
anomia	GSS	Anomia: "It's hardly fair to bring a child into this world with the way things look for the future." (Agree, disagree.)
chldidel	GSS	Family attitudes: "What do you think is the ideal number of children for a family to have?" (As many as want, 4+, 3, 2 or fewer.)
obey	GSS	Values for children 1: "If you had to choose, which thing . . . would you pick as the most important for a child to learn to prepare him or her for life? To obey." (Most important, 2nd or 3rd, 4th or 5th.)
thnkself	GSS	Values for children 2: "To think for himself or herself." (Most important, 2nd or 3rd, 4th or 5th.)
helpothr	GSS	Values for children 3: "To help others." (Most important, 2nd or 3rd, 4th or 5th.)
class	GSS	Subjective identification: What is your social class? (Lower or working, middle, upper.)
Demographic composition		
race	CPS	Race. (Black, white, other.)

Table A.1. *(Continued)*

ethnic	GSS	Ethnicity: Country or part of the world from which respondent's ancestors came. (Eastern Europe, Southern Europe, Northern Europe, other.)
spneth	CPS	Spanish ethnicity. (Mexican/Chicano(a), Puerto Rican, Cuban, other Spanish, not Spanish.)

Note: Data are from 1972 to 2002 surveys, although not all questions were asked in all years (see Davis, Smith, and Marsden, 2004; BLS, years vary).

APPENDIX B

Table B.1. EG and FH Class Codes by Occupation and Employment Status.

Occupation	EG Class		FH Class		Occupation	EG Class		FH Class	
	Emp.	Self-Emp.	Emp.	Self-Emp.		Emp.	Self-Emp.	Emp.	Self-Emp.
101 architects	1	1	2	1	201 government officials	1	1	3	4
102 engineers	1	1	2	1	202 financial managers	1	1	3	4
103 natural scientists	1	1	2	1	203 buyers	1	3	3	4
104 engineering and science techs	1	1	2	1	204 sales managers	1	3	3	4
105 physicians and dentists	1	1	2	1	205 office managers, n.e.c.	1	3	3	4
106 other health professionals	1	1	2	1	206 building managers	1	3	3	4
107 nurses and dental hygienists	1	1	2	1	207 restaurant managers	1	3	3	4
108 therapists	1	1	2	1	208 health administrators	1	3	3	4
109 health technicians	1	1	2	1	209 school administrators	1	1	3	4
110 social scientists	1	1	2	1	210 managers, n.e.c.	1	3	3	4
111 religious workers	1	1	2	1	301 insurance agents	1	3	5	5
112 social workers	1	1	2	1	302 real estate agents	1	3	5	5
113 professors and instructors	1	1	2	1	303 agents, n.e.c.	1	3	5	5
114 primary, secondary teachers	1	1	2	1	304 salespersons	2	3	5	5
115 jurists	1	1	2	1	401 clerical supervisors	1	3	6	6
116 librarians and curators	1	1	2	1	402 estimators and investigators	2	3	6	6
117 creative artists	1	1	2	1	403 insurance adjusters	1	1	6	6
118 authors and journalists	1	1	2	1	404 cashiers	2	3	6	6
119 designers and decorators	1	1	2	1	405 bank tellers	2	3	6	6
120 accountants	1	1	2	1	406 counter clerks, except food	2	3	6	6
121 computer specialists	1	1	2	1	407 secretaries	2	3	6	6
122 personnel workers	1	1	2	1	408 accounting clerks	2	3	6	6
123 public relations professionals	1	1	2	1	409 office machine operators	2	3	6	6
124 applied research workers	1	1	2	1	410 tabulation clerks	2	3	6	6

Table B.1. (*Continued*)

Occupation	EG Class		FH Class		Occupation	EG Class		FH Class	
	Emp.	Self-Emp.	Emp.	Self-Emp.		Emp.	Self-Emp.	Emp.	Self-Emp.
125 professionals, n.e.c.	1	1	2	1	411 postal clerks	2	2	6	6
412 mail carriers	6	6	6	6	514 aircraft mechanics	5	3	7	7
413 mail distribution clerks	6	3	6	6	515 automobile mechanics	5	3	7	7
414 telephone operators	2	3	6	6	516 small electronics mechanics	5	3	7	7
415 expediters	2	3	6	6	517 heavy equipment mechanics	5	3	7	7
416 stock clerks & storekeepers	6	3	6	6	518 mechanics, n.e.c.	5	3	7	7
417 warehouse clerks	2	3	6	6	519 electricians	5	3	7	7
418 teacher aides	2	3	6	6	520 brickmasons	5	3	7	7
419 clerks, n.e.c.	2	3	6	6	521 carpenters	5	3	7	7
501 supervisors of manual labor	5	3	7	7	522 painters	5	3	7	7
502 inspectors	5	3	7	7	523 plumbers	5	3	7	7
503 metal processors	5	3	7	7	524 construction crafts, n.e.c.	5	3	7	7
504 machinists	5	3	7	7	525 craft workers n.e.c.	5	3	7	7
505 structural metal workers	5	3	7	7	601 graders and sorters	6	3	8	8
506 stationary engine operators	5	3	7	7	602 launderers	6	3	8	8
507 heavy machinery operators	5	3	7	7	603 sewers	6	3	8	8
508 power & phone line workers	5	3	7	7	604 textile operatives	6	3	8	8
509 railroad conductors & engineers	5	5	7	7	605 precision machine operatives	6	3	8	8
510 printers	5	3	7	7	606 finishing machine operatives	5	3	8	8
511 tailors	5	3	7	7	607 assemblers	6	3	8	8
512 bakers	5	3	7	7	608 welders	5	3	8	8
513 heating and cooling mechanics	5	3	7	7	609 meat cutters	6	3	8	8
610 packagers	6	3	8	8	802 bartenders	2	3	9	9
611 machine operatives, n.e.c.	6	3	8	8	803 wait staff	2	3	9	9

612 miners	6	3	8	8	804 cooks	2	3	9	9
613 lumbermen and sawyers	6	3	8	8	805 kitchen helpers	6	3	9	9
614 fork lift operatives	6	3	8	8	806 practical nurses	2	3	9	9
615 home delivery workers	6	3	8	8	807 health aides	2	3	9	9
616 mass transit drivers	6	3	8	8	808 child care workers	2	3	9	9
617 taxicab drivers and chauffeurs	6	3	8	8	809 hair stylists	2	3	9	9
618 truck drivers	6	3	8	8	810 attendents, n.e.c.	6	3	9	9
619 garage workers	6	3	8	8	811 law enforcement officers	1	3	9	9
620 operatives, n.e.c.	6	3	8	8	812 guards	2	3	9	9
701 freight handlers	6	3	10	10	813 firefighters	5	5	9	9
702 retail stock handlers	6	3	10	10	814 housekeepers, ex. Private	2	3	9	9
703 construction laborers	6	3	10	10	815 food counter workers	2	3	9	9
704 gardeners	6	3	10	10	816 private household workers	6	3	9	9
705 laborers, n.e.c.	6	3	10	10	901 farmers	4	4	11	11
801 cleaners	6	3	9	9	902 farm laborers	7	4	12	12

Note: The EG classes are as follows: 1 = service class; 2 = routine non-manual workers; 3 = petty bourgeoisie; 4 = farmers; 5 = skilled workers and foremen; 6 = non-skilled workers; 7 = employed farm laborers. The FH classes are as follows: 1 = self-employed professionals; 2 = employed professionals; 3 = employed managers; 4 = self-employed managers; 5 = sales workers; 6 = clerical workers; 7 = craft workers; 8 = operatives; 9 = service workers; 10 = laborers; 11 = farmers; 12 = farm laborers.

APPENDIX C

Table C.1. Fit Statistics and Decompositions of Association from FH Models, Applied to 55 Outcome Tables.

Domain & Variable	Cond. Ind. (Eq. (3))			FH (Eq. (4))				Hybrid (Eq. (5))				% w/in Non-PCS Classes	Preferred Model (Sig. Tests)
	N	L^2	df	L^2	df	% Between FH Classes	BIC	L^2	df	BIC	% w/in PCS		
Life chances													
educ	866,958	506,828.1	500	118,844.1	456	76.55	112,609	27,996.3	208	25,152	17.92	5.52	saturated
ftpt	789,767	47,310.9	125	25,314.3	114	46.49	23,766	14,694.1	52	13,988	22.45	31.06	saturated
ftincome	695,253	218,303.0	500	71,507.3	456	67.24	65,373	25,234.7	208	22,437	21.20	11.56	saturated
income	866,956	275,833.7	500	94,330.5	456	65.80	88,096	35,105.7	208	32,262	21.47	12.73	saturated
finrela	22,608	3,375.4	250	983.1	228	70.87	−1,303	315.0	104	−728	19.79	9.33	saturated
tenure	783,721	25,293.6	125	11,338.0	114	55.17	9,791	3,769.4	52	3,064	29.92	14.90	saturated
Lifestyles													
Consumption practices													
news	15,333	899.8	375	517.2	342	42.52	−2,779	204.4	156	−1,299	34.76	22.72	saturated
tvhours	14,782	1,249.7	375	531.7	342	57.46	−2,752	239.6	156	−1,258	23.37	19.17	saturated
memlit	9,927	657.2	125	218.3	114	66.78	−831	58.8	52	−420	24.27	8.95	hybrid
satfam	12,305	473.6	375	404.6	342	14.56	−2,816	195.4	156	−1,274	44.18	41.25	saturated[a]
socrel	14,191	652.2	375	416.9	342	36.08	−2,853	160.9	156	−1,331	39.25	24.67	hybrid
socommun	14,168	720.6	375	453.2	342	37.11	−2,816	206.0	156	−1,285	34.31	28.58	saturated
socfrend	14,189	578.3	375	388.1	342	32.89	−2,882	158.9	156	−1,333	39.64	27.47	hybrid
memserv	9,931	519.2	125	193.0	114	62.83	−856	88.6	52	−390	20.10	17.07	saturated
memfrat	9,934	334.3	125	234.7	114	29.79	−814	84.2	52	−394	45.03	25.18	saturated
memsport	9,944	321.8	125	152.8	114	52.52	−897	58.4	52	−420	29.35	18.13	hybrid
memhobby	9,926	199.4	125	129.8	114	34.92	−919	68.7	52	−410	30.65	34.43	big-class
memnum	10,032	1,540.3	375	740.4	342	51.93	−2,411	287.8	156	−1,150	29.38	18.69	saturated
satjob	22,494	1,106.6	250	524.4	228	52.61	−1,760	295.5	104	−747	20.68	26.71	saturated
richwork	14,599	443.5	125	209.4	114	52.79	−884	83.9	52	−415	28.29	18.92	saturated
hours	789,768	96,378.9	375	42,095.0	342	56.32	37,451	18,260.4	156	16,142	24.73	18.95	saturated

Institutional participation													
marstat	866,958	27,295.5	500	12,408.0	456	54.54	6,173	4,773.4	208	1,929	27.97	17.49	saturated
divorce	15,038	323.5	125	182.7	114	43.54	−914	76.5	52	−424	32.80	23.66	saturated
childs	23,503	1,484.5	500	786.1	456	47.04	−3,803	347.5	208	−1,746	29.55	23.41	saturated
relig	23,512	1,356.7	500	758.6	456	44.08	−3,831	258.8	208	−1,835	36.84	19.08	saturated
attend	23,319	1,125.5	375	905.3	342	19.57	−2,534	276.9	156	−1,292	55.83	24.60	saturated
union	294,895	35,433.7	125	24,082.7	114	32.03	22,647	6,878.3	52	6,223	48.55	19.41	saturated
vet	489,698	11,318.5	125	8,327.2	114	26.43	6,834	2,670.8	52	1,990	49.97	23.60	saturated
Class-based sentiments													
Political attitudes and behaviors													
partyid	23,106	1,329.3	500	729.1	456	45.15	−3,853	271.0	208	−1,819	34.47	20.38	saturated
polviews	20,243	1,118.7	500	682.9	456	38.96	−3,839	272.5	208	−1,790	36.68	24.36	saturated
helpnot	11,466	616.9	250	300.9	228	51.22	−1,830	144.9	104	−827	25.29	23.49	saturated
mempolit	9,938	254.5	125	135.0	114	46.95	−914	61.7	52	−417	28.79	24.26	big-class

Table C.1. *(Continued)*

Domain & Variable	Cond. Ind. (Eq. (3))			FH (Eq. (4))				Hybrid (Eq. (5))				% w/in Non-PCS Classes	Preferred Model (Sig. Tests)
	N	L^2	df	L^2	df	% Between FH Classes	BIC	L^2	df	BIC	% w/in PCS		
Social attitudes and dispositions													
spkath	15,569	948.1	125	204.2	114	78.46	−896	75.9	52	−426	13.53	8.01	saturated
spkcom	15,414	1,237.6	125	223.8	114	81.92	−876	79.5	52	−422	11.66	6.42	saturated
homosex	14,094	1,228.7	250	436.0	228	64.52	−1,742	145.2	104	−848	23.57	11.82	saturated
pornlaw	14,652	452.2	250	378.1	228	16.37	−1,809	132.0	104	−866	54.44	29.18	saturated
cappun	20,034	489.8	125	321.6	114	34.34	−808	68.8	52	−446	51.61	14.05	hybrid
courts	20,738	530.9	250	354.8	228	33.16	−1,911	112.9	104	−921	45.58	21.26	hybrid
prayer	12,297	581.0	125	219.0	114	62.32	−855	70.2	52	−420	25.61	12.07	saturated
racmar	14,098	1,120.8	125	230.4	114	79.45	−859	126.9	52	−370	9.23	11.32	saturated
racopen	11,070	333.5	125	189.7	114	43.12	−872	82.6	52	−402	32.10	24.78	saturated
helpblk	11,932	443.4	250	312.5	228	29.52	−1,828	120.2	104	−856	43.37	27.11	hybrid
fework	13,151	426.9	125	151.4	114	64.54	−930	72.5	52	−421	18.48	16.99	saturated[a]
fepol	12,808	358.6	125	207.4	114	42.17	−871	96.0	52	−396	31.06	26.77	saturated
abnomore	17,319	693.8	125	285.9	114	58.80	−827	99.9	52	−408	26.80	14.40	saturated
abrape	17,307	395.7	125	260.3	114	34.23	−852	88.2	52	−419	43.48	22.29	saturated
anomia	10,331	916.9	125	176.1	114	80.79	−878	88.6	52	−392	9.55	9.66	saturated
chldidel	15,252	610.5	375	422.9	342	30.72	−2,871	153.1	156	−1,350	44.20	25.08	hybrid
obey	9,204	733.0	250	323.3	228	55.89	−1,758	121.6	104	−828	27.53	16.59	hybrid
thnkself	9,204	615.5	250	341.3	228	44.55	−1,740	139.6	104	−810	32.77	22.69	saturated
helpoth	9,211	401.6	250	249.9	228	37.77	−1,831	97.6	104	−852	37.92	24.31	hybrid
class	22,751	4,292.1	250	864.1	228	79.87	−1,423	319.5	104	−724	12.69	7.44	saturated
Demographic structuration													
race	866,957	43,420.7	250	16,906.3	228	61.06	13,789	6,161.5	104	4,740	24.75	14.19	saturated
ethnic	18,164	1,171.6	375	615.0	342	47.51	−2,739	237.9	156	−1,292	32.19	20.31	saturated
spneth	846,717	27,360.6	500	8,494.7	456	68.95	2,271	3,785.4	208	946	17.21	13.84	saturated

Note: PCS = Professional, craft, and service classes (see Table 1). Sample sizes reflect the addition of 0.1 to zero cells. See Appendix A for variable definitions.
[a] For these outcomes, the big-class model is preferred to the hybrid model, but the saturated model is preferred to both.

Table C.2. Internal Homogeneity in FH Classes, All Outcomes.

Domain & Variable	*N*	Professionals (SE and Emp.)	Managers (SE and Emp.)	Sales Workers	Clerical Workers	Craft Workers	Operatives	Service Workers	Laborers
Life chances									
educ	866,958	0.05	0.23	0.35	0.40	0.39	0.41	0.24	0.54
ftpt	789,767	0.53	0.61	0.53	0.45	0.39	0.36	0.32	0.69
ftincome	695,253	0.33	0.47	0.54	0.26	0.35	0.37	0.10	0.57
income	866,956	0.33	0.48	0.47	0.26	0.35	0.43	0.10	0.56
finrela	22,608	0.54	0.63	0.64	0.59	0.61	0.64	0.58	0.89
tenure	783,721	0.71	0.72	0.69	0.75	0.72	0.73	0.61	0.85
Lifestyles									
Consumption practices									
news	15,333	0.62	0.79	0.69	0.80	0.71	0.63	0.75	0.90
tvhours	14,782	0.55	0.75	0.63	0.60	0.69	0.71	0.67	0.78
memlit	9,927	0.49	0.64	0.69	0.32	0.39	0.43	0.45	0.40
satfam	12,305	0.73	0.71	0.72	0.78	0.68	0.75	0.78	0.68
socrel	14,191	0.68	0.78	0.85	0.83	0.76	0.77	0.76	0.70
socommun	14,168	0.71	0.76	0.77	0.74	0.71	0.51	0.77	0.91
socfrend	14,189	0.73	0.86	0.64	0.82	0.69	0.79	0.77	0.78
memserv	9,931	0.61	0.37	0.65	0.42	0.46	0.46	0.34	0.33
memfrat	9,934	0.49	0.65	0.63	0.42	0.43	0.32	0.37	0.53
memsport	9,944	0.61	0.76	0.82	0.70	0.65	0.58	0.62	0.70
memhobby	9,926	0.64	0.75	0.73	0.33	0.49	0.34	0.54	0.62
memnum	10,032	0.55	0.54	0.59	0.56	0.64	0.63	0.54	0.74
satjob	22,494	0.68	0.78	0.75	0.76	0.74	0.79	0.64	0.69
richwork	14,599	0.56	0.75	0.82	0.68	0.54	0.72	0.82	0.83
hours	789,768	0.50	0.64	0.71	0.61	0.54	0.65	0.34	0.66

Table C.2. *(Continued)*

Domain & Variable	*N*	Professionals (SE and Emp.)	Managers (SE and Emp.)	Sales Workers	Clerical Workers	Craft Workers	Operatives	Service Workers	Laborers
Institutional participation									
marstat	866,958	0.79	0.89	0.78	0.74	0.79	0.77	0.65	0.93
divorce	15,038	0.54	0.78	0.80	0.63	0.70	0.69	0.61	0.76
childs	23,503	0.71	0.78	0.62	0.62	0.75	0.68	0.66	0.75
relig	23,512	0.26	0.84	0.60	0.63	0.51	0.39	0.50	0.64
attend	23,319	0.30	0.74	0.65	0.69	0.68	0.62	0.57	0.75
union	294,895	0.42	0.47	0.74	0.27	0.42	0.52	0.24	0.63
vet	489,698	0.75	0.81	0.80	0.64	0.66	0.70	0.49	0.81
Class-based sentiments									
Political attitudes and behaviors									
partyid	23,106	0.63	0.74	0.73	0.69	0.62	0.71	0.66	0.77
polviews	20,243	0.73	0.82	0.82	0.82	0.77	0.79	0.74	0.82
helpnot	11,466	0.74	0.81	0.85	0.74	0.67	0.68	0.78	0.89
mempolit	9,938	0.51	0.69	0.57	0.35	0.46	0.30	0.38	0.23

Social attitudes and dispositions									
spkath	15,569	0.57	0.79	0.74	0.69	0.71	0.74	0.67	0.79
spkcom	15,414	0.55	0.77	0.80	0.68	0.75	0.75	0.73	0.87
homosex	14,094	0.61	0.87	0.65	0.77	0.67	0.70	0.67	0.83
pornlaw	14,652	0.68	0.79	0.78	0.79	0.71	0.76	0.68	0.85
cappun	20,034	0.63	0.79	0.86	0.76	0.66	0.68	0.64	0.81
courts	20,738	0.70	0.76	0.88	0.84	0.80	0.79	0.68	0.81
prayer	12,297	0.57	0.78	0.68	0.70	0.71	0.71	0.76	0.75
racmar	14,098	0.39	0.67	0.73	0.56	0.72	0.56	0.63	0.70
racopen	11,070	0.73	0.78	0.64	0.66	0.70	0.68	0.61	0.82
helpblk	11,932	0.74	0.79	0.83	0.73	0.73	0.69	0.73	0.87
fework	13,151	0.49	0.78	0.38	0.71	0.73	0.69	0.76	0.67
fepol	12,808	0.66	0.81	0.73	0.69	0.66	0.67	0.73	0.74
abnomore	17,319	0.63	0.82	0.75	0.75	0.75	0.71	0.66	0.89
abrape	17,307	0.55	0.72	0.49	0.75	0.61	0.44	0.60	0.81
anomia	10,331	0.72	0.70	0.66	0.71	0.70	0.75	0.66	0.79
chldidel	15,252	0.73	0.75	0.50	0.43	0.51	0.74	0.64	0.86
obey	9,204	0.54	0.69	0.65	0.71	0.69	0.71	0.65	0.90
thnkself	9,204	0.70	0.69	0.64	0.71	0.65	0.64	0.70	0.86
helpoth	9,211	0.72	0.78	0.80	0.83	0.72	0.71	0.75	0.82
class	22,751	0.48	0.65	0.66	0.67	0.64	0.68	0.66	0.77
Demographic structuration									
race	866,957	0.73	0.78	0.76	0.73	0.77	0.72	0.60	0.86
ethnic	18,164	0.68	0.77	0.76	0.73	0.73	0.70	0.61	0.82
spneth	846,717	0.78	0.85	0.78	0.77	0.68	0.69	0.65	0.69

Note: Values in the table are 1/Ac (see Eq. (7)). Farmers and farm laborers are single-occupation classes and hence are excluded from the table. Sample sizes reflect the addition of 0.1 to zero cells.

Table C.3. Internal Homogeneity in EG Classes, All Outcomes.

Domain & Variable	Service	Routine Non-Manual	Petty Bourg.	Skilled Workers	Non-Skilled Workers	*N*, No Petty Bourg.	*N*, Petty Bourg.
Life chances							
educ	0.07	0.32	0.28	0.33	0.31	813,911	53,054
ftpt	0.47	0.46	0.49	0.41	0.34	741,400	48,368
ftincome	0.43	0.18	0.25	0.37	0.19	695,265	39,442
income	0.44	0.16	0.29	0.37	0.18	813,910	53,055
finrela	0.58	0.55	0.39	0.56	0.60	20,432	2,194
tenure	0.70	0.69	0.54	0.68	0.69	735,685	48,038
Lifestyles							
Consumption practices							
news	0.65	0.79	0.36	0.69	0.63	13,827	1,536
tvhours	0.57	0.67	0.34	0.69	0.68	13,317	1,495
memlit	0.49	0.33	0.28	0.39	0.42	8,961	983
satfam	0.71	0.74	0.32	0.65	0.68	11,110	1,229
socrel	0.70	0.80	0.41	0.74	0.74	12,796	1,424
socommun	0.71	0.74	0.36	0.70	0.57	12,775	1,422
socfrend	0.69	0.77	0.37	0.68	0.74	12,794	1,424
memserv	0.49	0.45	0.26	0.34	0.40	8,963	986
memfrat	0.44	0.48	0.28	0.38	0.38	8,968	981
memsport	0.64	0.71	0.26	0.66	0.45	8,976	983
memhobby	0.64	0.36	0.28	0.37	0.37	8,960	983
memnum	0.54	0.62	0.35	0.60	0.59	9,049	1,016
satjob	0.70	0.73	0.41	0.68	0.78	20,324	2,186
richwork	0.54	0.71	0.29	0.55	0.74	13,157	1,453
hours	0.57	0.50	0.45	0.46	0.51	741,401	48,373
Institutional participation							
marstat	0.79	0.70	0.65	0.77	0.70	813,911	53,062
divorce	0.53	0.59	0.33	0.71	0.69	13,418	1,631
childs	0.69	0.62	0.38	0.72	0.65	21,236	2,301
relig	0.31	0.62	0.29	0.52	0.38	21,243	2,309
attend	0.35	0.60	0.37	0.68	0.67	21,077	2,266
union	0.37	0.36	n/a	0.42	0.34	294,895	n/a
vet	0.74	0.56	0.40	0.64	0.61	454,541	35,157
Class-based sentiments							
Political attitudes and behaviors							
partyid	0.63	0.66	0.40	0.59	0.71	20,879	2,261
polviews	0.71	0.77	0.38	0.75	0.80	18,268	2,011
helpnot	0.75	0.74	0.43	0.60	0.71	10,363	1,123
mempolit	0.47	0.36	0.28	0.44	0.28	8,969	986

Table C.3. (*Continued*)

Domain & Variable	Service	Routine Non-Manual	Petty Bourg.	Skilled Workers	Non-Skilled Workers	*N*, No Petty Bourg.	*N*, Petty Bourg.
Social attitudes and dispositions							
spkath	0.59	0.70	0.29	0.70	0.64	14,062	1,518
spkcom	0.55	0.78	0.31	0.74	0.72	13,913	1,512
homosex	0.64	0.73	0.44	0.67	0.71	12,705	1,412
pornlaw	0.71	0.70	0.42	0.68	0.76	13,245	1,428
cappun	0.60	0.73	0.35	0.61	0.72	18,071	1,974
courts	0.71	0.77	0.24	0.75	0.76	18,734	2,028
prayer	0.58	0.75	0.35	0.71	0.71	11,108	1,202
racmar	0.36	0.67	0.30	0.71	0.59	12,644	1,466
racopen	0.74	0.64	0.28	0.67	0.69	9,985	1,098
helpblk	0.71	0.71	0.40	0.66	0.70	10,790	1,164
fework	0.42	0.74	0.34	0.66	0.67	11,861	1,302
fepol	0.66	0.72	0.22	0.67	0.70	11,551	1,269
abnomore	0.67	0.72	0.32	0.73	0.71	15,647	1,683
abrape	0.53	0.68	0.29	0.61	0.50	15,643	1,674
anomia	0.68	0.67	0.27	0.68	0.69	9,280	1,065
chldidel	0.61	0.72	0.42	0.56	0.69	13,800	1,483
obey	0.55	0.69	0.38	0.64	0.74	8,335	890
thnkself	0.69	0.70	0.39	0.63	0.66	8,336	890
helpoth	0.71	0.77	0.25	0.70	0.74	8,341	890
class	0.51	0.63	0.29	0.64	0.68	20,559	2,212
Demographic structuration							
race	0.71	0.64	0.36	0.77	0.73	813,911	53,054
ethnic	0.68	0.66	0.31	0.72	0.72	16,400	1,793
spneth	0.78	0.73	0.57	0.64	0.66	795,104	51,631

Note: Values are 1/Ac. EG classes IV (farmers) and VII (farm laborers), both single-occupation classes, are excluded from the table. Sample sizes reflect the addition of 0.1 to zero cells.

Table C.4. External Distinctiveness of FH Classes, All Outcomes.

Domain & Variable	SE Profs.	Emp. Profs.	Emp. Managers	SE Managers	Sales Workers	Clerical Workers	Craft Workers	Operatives	Service Workers	Laborers	Farmers	Farm Laborers
Life chances												
educ	5.09	4.93	2.51	1.01	2.07	0.13	−1.90	−3.09	−1.47	−2.85	−1.63	−4.80
ftpt	−0.99	0.41	1.02	−0.29	0.10	0.35	0.78	0.58	−0.51	−0.56	−0.43	−0.46
ftincome	1.25	2.27	2.29	0.12	1.51	0.22	0.59	−0.62	−1.88	−1.22	−1.45	−3.08
income	0.94	2.15	2.30	0.18	1.41	0.33	0.58	−0.53	−1.88	−1.30	−1.24	−2.93
finrela	1.05	0.85	0.84	0.92	0.96	−0.14	−0.16	−0.85	−0.83	−1.09	−0.65	−0.89
tenure	0.34	0.02	0.22	0.64	0.33	−0.03	0.09	−0.29	−0.40	−0.56	0.70	−1.07
Lifestyles												
Consumption practices												
news	0.41	0.44	0.36	0.42	0.53	0.24	−0.24	−0.42	0.02	−0.73	0.07	−1.10
tvhours	0.88	0.81	0.46	0.71	0.46	−0.26	−0.33	−0.77	−0.65	−0.88	0.05	−0.49
memlit	1.53	1.16	0.76	0.99	0.46	−0.32	−0.71	−1.34	−0.26	−1.62	0.19	−0.83
satfam	0.14	0.02	0.03	−0.01	0.33	−0.02	0.05	−0.05	−0.19	−0.26	0.14	−0.18
socrel	0.24	0.39	0.22	−0.04	0.15	−0.15	−0.18	−0.32	−0.06	−0.16	−0.28	0.19
socommun	0.45	0.15	0.19	0.29	0.23	−0.11	−0.18	−0.59	−0.23	−0.46	0.14	0.13
socfrend	0.44	0.36	0.12	0.26	0.43	0.18	−0.01	−0.22	−0.09	−0.01	−0.58	−0.86
memserv	1.44	0.74	0.57	0.96	0.98	−0.06	−0.17	−0.94	−0.56	−0.97	0.83	−2.82
memfrat	0.81	0.20	0.51	0.48	0.45	−0.35	−0.20	−0.88	−0.05	−0.69	0.09	−0.36
memsport	0.29	0.23	0.46	0.15	0.51	0.00	−0.15	−0.53	−0.25	−0.22	−0.32	−0.16
memhobby	0.67	0.37	0.32	0.33	0.24	−0.19	−0.09	−0.68	0.02	0.17	−0.02	−1.15
memnum	1.04	0.78	0.42	0.33	0.50	−0.10	−0.24	−0.70	−0.36	−0.67	0.47	−1.45
satjob	0.82	0.17	0.13	0.67	0.08	−0.38	−0.11	−0.53	−0.25	−0.53	0.37	−0.41
richwork	0.82	0.24	−0.17	0.38	−0.22	−0.53	−0.56	−0.49	−0.30	−0.17	0.72	0.31
hours	−0.10	0.21	0.67	0.68	0.39	−0.51	−0.44	−0.30	−0.78	−1.14	1.08	0.24

Institutional participation												
marstat	0.16	−0.13	0.13	0.45	0.11	−0.12	0.16	0.12	−0.33	−0.61	0.62	−0.33
divorce	0.25	0.45	−0.17	−0.33	−0.13	0.05	−0.19	−0.11	−0.25	−0.36	0.68	0.11
childs	0.27	0.88	0.36	−0.21	0.25	0.22	−0.13	−0.36	−0.08	0.11	−0.44	−0.86
relig	1.35	0.91	0.70	1.14	1.11	0.16	−0.34	−0.84	−0.04	−0.11	−2.70	−1.35
attend	0.23	0.24	0.06	0.19	0.01	0.08	−0.24	−0.23	−0.19	−0.38	0.33	−0.11
union	n/a	0.15	−0.52	n/a	−1.20	0.67	1.43	1.07	0.16	0.89	−1.20	−1.44
vet	0.21	−0.07	0.22	0.35	0.23	0.29	0.26	−0.02	−0.17	−0.25	−0.10	−0.95
Class-based sentiments												
Political attitudes and behaviors												
partyid	0.24	−0.02	0.44	0.67	0.82	−0.18	−0.41	−0.83	−0.46	−0.63	0.51	−0.14
polviews	0.53	0.49	0.16	−0.05	0.09	−0.18	−0.31	−0.26	0.00	−0.14	−0.45	0.11
helpnot	0.44	0.05	0.32	0.59	0.35	−0.04	−0.26	−0.50	−0.33	−0.55	0.50	−0.57
mempolit	1.11	0.43	0.49	0.40	0.24	−0.15	−0.63	−1.19	−0.68	−0.39	0.88	−0.52

Table C.4. *(Continued)*

Domain & Variable	SE Profs.	Emp. Profs.	Emp. Managers	SE Managers	Sales Workers	Clerical Workers	Craft Workers	Operatives	Service Workers	Laborers	Farmers	Farm Laborers
Social attitudes and dispositions												
spkath	1.18	1.03	0.62	0.35	0.49	0.14	−0.31	−0.66	−0.18	−0.52	−0.87	−1.27
spkcom	1.63	1.09	0.58	0.26	0.46	0.13	−0.42	−0.68	−0.10	−0.63	−1.11	−1.21
homosex	0.96	0.73	0.43	0.10	0.40	0.04	−0.49	−0.45	0.06	−0.26	−0.94	−0.59
pornlaw	0.09	0.27	0.16	0.02	0.06	0.01	0.05	0.00	0.09	−0.05	−0.32	−0.37
cappun	0.43	0.42	−0.18	−0.14	−0.19	−0.15	−0.25	0.22	0.02	0.16	−0.43	0.07
courts	0.46	0.27	−0.06	−0.19	−0.10	−0.08	−0.11	−0.03	0.12	0.12	−0.25	−0.17
prayer	0.71	0.62	0.23	0.03	0.38	−0.16	−0.32	−0.46	−0.20	−0.28	−0.37	−0.17
racmar	1.70	1.43	0.80	0.31	0.42	−0.01	−0.51	−0.86	−0.16	−0.82	−1.26	−1.04
racopen	0.34	0.40	0.28	−0.11	0.17	0.18	−0.24	−0.08	0.23	0.11	−1.21	−0.07
helpblk	0.22	0.38	−0.04	−0.33	−0.08	0.03	−0.13	0.21	0.30	0.24	−0.65	−0.14
fework	0.94	0.55	0.43	0.20	0.74	0.08	−0.26	−0.48	−0.24	−0.55	−0.73	−0.69
fepol	0.34	0.53	0.31	0.04	0.38	0.04	−0.04	−0.29	0.01	−0.17	−0.54	−0.62
abnomore	0.83	0.40	0.28	0.18	0.44	−0.03	−0.21	−0.58	−0.10	−0.25	−0.53	−0.43
abrape	0.24	0.26	0.39	0.30	0.89	−0.01	−0.02	−0.25	−0.11	−0.33	−0.05	−1.31
anomia	0.99	0.81	0.62	0.38	0.78	−0.20	−0.47	−0.81	−0.41	−0.80	−0.14	−0.76
chldidel	0.14	0.45	0.36	−0.11	0.54	0.34	0.23	−0.43	−0.22	−0.23	−0.38	−0.68
obey	0.73	0.75	0.45	0.39	0.35	−0.11	−0.32	−0.59	−0.14	−0.54	−0.63	−0.33
thnkself	0.48	0.46	0.30	0.44	0.40	−0.15	−0.28	−0.48	−0.21	−0.38	−0.31	−0.27
helpoth	0.34	0.39	0.26	0.22	−0.01	−0.05	−0.13	−0.10	−0.08	−0.36	−0.39	−0.10
class	1.82	1.05	0.82	0.98	0.97	−0.28	−0.83	−1.20	−0.71	−1.22	−0.43	−0.96
Demographic structuration												
race	0.64	0.12	0.21	0.37	0.50	−0.40	−0.12	−0.62	−0.58	−0.83	1.27	−0.55
ethnic	0.43	0.14	0.17	0.26	0.52	−0.31	−0.16	−0.51	−0.43	−0.52	0.87	−0.44
spneth	0.74	0.56	0.44	0.25	0.62	0.01	−0.08	−0.44	−0.30	−0.76	0.65	−1.71

Note: Values in the table are the class scale values estimated under Eq. (6).

Table C.5. External Distinctiveness of EG Classes, All Outcomes.

Domain & Variable	Service	Routine Non-Manual	Petty Bourg.	Farmers	Skilled Workers	Non-Skilled Workers	Farm Laborers
Life chances							
educ	5.33	0.98	0.80	−0.33	−0.69	−1.52	−3.77
ftpt	0.43	−0.35	−0.62	−0.55	1.00	0.00	−0.54
ftincome	2.83	−0.15	−0.15	−0.77	1.37	−0.43	−2.86
income	2.70	−0.18	−0.14	−0.63	1.32	−0.47	−2.74
finrela	1.25	−0.06	0.54	−0.26	0.15	−0.54	−0.53
tenure	0.28	−0.11	0.44	0.85	0.22	−0.22	−1.02
Lifestyles							
Consumption practices							
news	0.62	0.29	0.15	0.21	−0.06	−0.16	−0.91
tvhours	0.92	−0.24	0.20	0.27	−0.19	−0.56	−0.20
memlit	1.27	−0.02	0.54	0.44	−0.46	−0.75	−0.48
satfam	0.08	−0.08	−0.06	0.24	0.05	−0.08	−0.21
socrel	0.34	−0.02	0.10	−0.21	−0.16	−0.22	0.27
socommun	0.25	−0.05	0.07	0.22	−0.12	−0.49	0.19
socfrend	0.54	0.28	0.27	−0.35	0.20	0.08	−0.75
memserv	1.20	0.31	1.00	1.26	−0.10	−0.39	−2.29
memfrat	0.51	−0.04	0.25	0.23	0.00	0.61	−0.10
memsport	0.46	0.14	0.17	−0.14	−0.01	−0.41	−0.04
memhobby	0.64	0.18	0.44	0.24	0.01	−0.28	0.79
memnum	0.93	0.03	0.14	0.67	−0.01	−0.38	−1.24
satjob	0.36	−0.23	0.40	0.48	−0.02	−0.39	−0.20
richwork	0.18	−0.33	0.13	0.79	−0.57	−0.37	0.30
hours	0.37	−0.74	0.43	1.09	−0.34	−0.64	0.26
Institutional participation							
marstat	0.03	−0.21	0.39	0.69	0.16	−0.31	−0.36
divorce	0.15	−0.17	−0.29	0.59	−0.26	−0.29	−0.04
childs	0.81	0.26	−0.05	−0.30	0.02	−0.06	−0.73
relig	1.48	0.76	1.47	−2.00	0.23	−0.04	−0.43
attend	0.20	0.01	−0.08	0.37	−0.26	−0.13	−0.19
union	0.06	−0.25	n/a	1.18	−1.52	−0.89	1.42
vet	0.23	0.17	0.24	0.05	0.39	0.06	−0.90
Class-based sentiments							
Political attitudes and behaviors							
partyid	0.34	−0.10	0.47	0.65	−0.36	−0.60	0.08
polviews	0.45	0.02	0.04	−0.30	−0.20	−0.11	0.13
helpnot	0.37	0.02	−0.64	0.68	−0.21	−0.32	−0.54
mempolit	0.60	−0.20	0.29	1.01	−0.46	−0.67	−0.28

Table C.5. *(Continued)*

Domain & Variable	Service	Routine Non-Manual	Petty Bourg.	Farmers	Skilled Workers	Non-Skilled Workers	Farm Laborers
Social attitudes and dispositions							
spkath	1.25	0.38	0.24	−0.50	0.07	−0.19	−1.02
spkcom	1.34	0.43	0.29	−0.71	−0.02	−0.21	−0.83
homosex	0.94	0.34	0.27	−0.62	−0.18	−0.08	−0.40
pornlaw	0.27	0.09	−0.03	−0.32	0.14	0.01	−0.18
cappun	0.21	0.01	−0.04	−0.36	−0.19	0.27	0.06
courts	0.22	0.06	0.01	−0.18	−0.11	0.12	−0.10
prayer	0.64	0.05	0.11	−0.18	−0.15	−0.28	−0.09
racmar	1.69	0.42	0.26	−0.83	−0.15	−0.42	−0.71
racopen	0.47	0.33	0.00	−1.09	−0.10	0.22	0.17
helpblk	0.21	0.24	−0.11	−0.63	−0.10	0.31	−0.02
fework	0.88	0.29	0.28	−0.45	−0.05	−0.20	−0.46
fepol	0.59	0.27	0.15	−0.36	0.12	−0.08	−0.54
abnomore	0.64	0.16	0.19	−0.30	−0.01	−0.26	−0.23
abrape	0.64	0.19	0.21	0.21	0.22	−0.07	−1.18
anomia	1.04	−0.01	0.22	0.15	−0.19	−0.46	−0.52
chldidel	0.66	0.08	0.10	−0.19	0.32	−0.21	−0.65
obey	0.85	0.13	0.12	−0.37	−0.08	−0.30	−0.24
thnkself	0.58	0.07	0.13	−0.11	−0.10	−0.27	−0.17
helpoth	0.38	0.09	0.06	−0.22	0.01	−0.05	−0.22
class	1.49	0.00	0.54	0.07	−0.39	−0.61	−0.56
Demographic structuration							
race	0.25	−0.32	0.31	1.32	−0.09	−0.65	−0.52
ethnic	0.31	−0.25	0.13	0.99	−0.11	−0.49	−0.44
spneth	0.76	0.11	0.32	0.87	0.09	−0.27	−1.55

Note: Values in the table are the class scale values estimated under Eq. (6) (all classes except the petty bourgeoisie) or Eq. (8) (petty bourgeoisie).

SPACES AND NETWORKS: CONCEPTS FOR SOCIAL STRATIFICATION

Joseph Woelfel and Monica Murero

ABSTRACT

While early stratification students were deeply categorical in their theory and methods, concerning themselves with discrete categories or "bins" such as status, strata, classes, layers and such, Haller already conceived of stratification as a continuous, comparative process. While older theory held that each individual occupied a discrete status to which were attached role expectations which guided the individual's behavior, Haller conceived of status as a continuous hierarchy, through which individuals moved in a career trajectory over time.

The Galileo System of measuring social objects as points in a multidimensional Riemannian continuum is a deliberate effort to realize Haller's theoretical conception of a continuous array of statuses, occupations and positions through which individuals move in continuous arcs, impelled by the expectations of others and other forces.

In this paper we describe the fundamental premises of the Galileo model and present data showing the utility of the model.

The Shape of Social Inequality: Stratification and Ethnicity in Comparative Perspective
Research in Social Stratification and Mobility, Volume 22, 57–71
© 2005 Published by Elsevier Ltd.
ISSN: 0276-5624/doi:10.1016/S0276-5624(04)22002-9

CONCEPTS AND REFERENCE FRAMES

Perhaps the great achievement of the social sciences in the 20th century was the discovery that reference frames and conceptual systems influence our perceptions of reality, and that these reference frames and conceptual systems are themselves socially constructed. Perhaps the great failure of the social sciences in the 20th century was the failure to develop formal technical criteria for evaluating reference frames and constructing "better" ones. Of course, reference frames and conceptual systems are routinely evaluated on political, moral and esthetic grounds, and activists commonly develop and implement new frames which benefit their cause (e.g. by insisting on terms such as "right to life" and "pro choice" instead of "anti-abortion" and "pro abortion"). Few social scientists, however, attempt to evaluate reference frames and conceptual systems from an information theoretic perspective, by questioning how well they serve as precise and efficient encoding systems.

With few exceptions, and in spite of frequent repetitions of words like "continuing" and "ongoing" in the works of theorists such as G. H. Mead (1934), even the most contemporary social theory and social research utilize the same conceptual system used by Aristotle: that is, a reference frame in which experience is "chopped up" into discrete categories, and in which behavior is described as discrete jumps from one categorical state to another. Moreover, contemporary social science continues to work within Aristotle's notion of entelechy, where each and every of these discrete jumps from one state to another is motivated by a goal or drive or need.

Before Galileo, the dominant model for understanding motion and change was Aristotle's entelechy.[1] Within this philosophy, goal or intention is the primary mover of all things animal, vegetable and mineral. Fire rises because it seeks its proper place at the periphery of the world. Heavy objects fall because they are seeking their proper place at the center of the world. In the entelechy, water seeks its own level, nature abhors a vacuum, and everything seeks its proper place, each thing following goals in an unbroken chain established once and for all in the mind of the original "unmoved mover."

The physical sciences adhered to the Aristotelian categorical, teleological model until Galileo Galilei, who renounced the concept of goals for physical objects, and viewed physical motion and change instead as the response of matter to impartial "forces." Galileo also abandoned the categorical model of discrete jumps from one categorical state to another, and replaced it with a continuous model of motion based on ratios to an arbitrary standard unit of measure.

STATUS ATTAINMENT AS A COMMUNICATION PROCESS

The social sciences, however, never passed through the scientific revolution, and the notion of goal is still preeminent as an explanation for human and social behavior. Indeed, the most common model of status attainment within sociology is one in which individuals are born into status which provide them with resources and impediments which facilitate or impede their goal-driven quest for achievement. However predominant such a view might be, it is possible to implement a model in which goals play a subordinate role, and indeed are themselves produced by impartial forces in the social and physical environment.

One particularly useful approach emphasizes the communication aspects of social structure and social processes. Viewed in this light, the social structure of any society consists in sets of definitions and beliefs of individuals about the way their lives are organized. A status, for example, can be defined as a set of expectations about the roles appropriate to its incumbents, as a role can be considered expectations about the behaviors appropriate to its holders. These definitions must be communicated to those who will occupy the status and execute the role as well as to those in other statuses and roles who will interact with those occupants. Indeed, every hundred years or so the entire population of the world dies, and must communicate the entire information structure that defines the society to another worldful of people.

A communication model, whatever the substantive context, focuses on the systems that encode, store, transmit and manipulate information. For stratification, communication research focuses on what are the expectations that define statuses and roles, how are these expectations encoded, where are they stored, and how are they communicated.

The theoretical model underlying the Wisconsin Significant Other Project ("Other than what?" you might ask) is largely a communication model. Based on what has been called "The Wisconsin Model" (Sewell et al., 1969), it assumes that individuals are born into statuses or locations in the society. These statuses present their incumbents with resources and encumbrances, and expose them to certain other people who observe their circumstances and communicate to them expectations for their activities. Of these "other people," some play a particularly important role in defining an individual's expectations, and these are called "significant others" in the model.

Based on the expectations which others communicate to them, along with their own self-reflexive observations of their situations and past actions, individuals form expectations about their own behaviors. These expectations, constrained by physical circumstances, genetic factors, good or ill fortune and the like, determine the activities of the individual. The word "expectation" is chosen here specifically to indicate that individuals behave in ways consistent with their understanding of who and what they are rather than in response to goals, needs, drives or other affective motivators (Lemert, 1950).

While this model is entirely consistent with an interactionist perspective, at the inception of the project there was almost no empirical data which addressed the effects of significant others on the attitudes, beliefs and expectations of individuals, particularly in a stratification context. The most significant data available were presented by Sewell et al. (1969) who showed that the likelihood that a high school student would choose to attend college was correlated with an index of three self-reported dichotomous measures: whether the student believed most of his/her friends, teachers and parents expected them to attend college.

Although a groundbreaking study of seminal importance, the Sewell et al. (1969) study was hampered by its underlying categorical conceptualization. All major concepts in the study are categorical: significant others' expectations are dichotomous (expect college, do not expect college), as are the students' own aspirations (expect to attend, don't expect to attend) and attainments (attend, not attend).[2] Problems with this categorical model were understood particularly among the Wisconsin status attainment workers, and a major part of the motivation behind the Wisconsin Significant Other Project was to replace these categorical variables with numerical concepts.

Haller, who conceived and organized the Wisconsin Significant Other Project, understood that status could be viewed as a continuous variable, and worked to extend the range of status that could be precisely measured, particularly to very low levels in rural Brazil. He also developed the Occupational Aspiration Scale (OAS) based on occupational prestige scores from the NORC scale (Haller & Miller, 1971). Haller also understood that status attainment was not a dichotomous, before/after measure, but rather each individual's status attainments represented a trajectory over time. The OAS measured short range and long range aspirations to tap the concept of the trajectory to some extent. Overall, the OAS rated adolescent's aspirations on an 80 point scale, considerably advanced over the dichotomous categorical measure in the Sewell et al. study (Woelfel & Haller, 1971).

Based on the OAS, the Wisconsin Significant Other Battery (WISOB) measured both significant others' expectations and the adolescents' own aspirations on numerical scales. Moreover, the WISOB did not assume that certain categories of persons (e.g. parents, teachers, peer friends) were always significant, nor that they were the only significant persons. Rather, efforts were made to identify the exact significant others for each adolescent based on a simple theoretical model (Woelfel, 1967a, b). As a result, the Significant Other project replaced the three dichotomous significant others' expectations variables from the Sewell et al. (1969) study with a numerical level of each significant other's expectation for each significant other identified. These numerical expectation levels could be averaged across all significant others for each adolescent to yield a mean level of expectation.

These mean levels of significant others' expectations worked extremely well, explaining between half and two thirds of the variance in the adolescents' own numerical aspirations measured on identical scales, which was about double the variance explained in the Sewell et al. study (Woelfel & Haller, 1971).

While the average expectations were initially calculated as a heuristic device to deal with the variable number of significant others per case, subsequent analysis showed that the procedure made theoretical sense, since the mean represented that point at which all the "forces" expressed by the expectations of the significant others would balance so that the net force acting on the adolescent would be zero (Woelfel & Hernandez, 1973). The underlying simple theory suggests that each significant other's expectation can be represented as a force vector pulling the adolescent in a specific direction, and that, over time, the individual's own attitude will tend toward the point at which all such forces are balanced.

Several studies showed that this theory worked quite well in several other contexts in addition to status aspirations, including political radicalism (Gillham & McPhail, 1974), cigarette smoking (Mettlin, 1973), jury decisions (Mistretta et al., 1973), and others (e.g. Danes et al., 1984; Saltiel & Woelfel, 1975). Within the area of status attainment, empirical support for the model has been consistently strong (Lin et al., 1981; Picou & Campbell, 1975; Saltiel, 1975, 1978, 1983).

THE GALILEO MODEL

A fundamental aspect of the "linear force aggregation" theory implied by the averaging model is the inherent interdependence of the theoretical calculus and its measurement system. The idea that an individual's attitudes and beliefs tend toward the mean of the information received pertinent to those attitudes and beliefs assumes directly that attitudes, beliefs and the information environment can all be measured on continuous numerical scales. At the time, however, (and to a considerable extent even today) most social scientists believed that variables, by their inherent nature, were measurable only at certain levels, i.e. ratio, interval, ordinal or nominal. The averaging theory could be expected to work only for the first two of these types, leaving important stratification variables, such as occupational choice, outside the theory. If a child's mother, for example, wants her to finish college (16 years of schooling) and her father expects her finish high school (12 years), the average is clearly 14. But what if her mother expects her to become an accountant and her father expects her to become a stenographer? What is the average of "accountant" and "stenographer?" Within the "force aggregation theory" the mean of accountant and stenographer is undefined.

Meanwhile, Woelfel (1980a) and his students at the University of Illinois had developed procedures for representing discrete "objects" in a continuous spatial array. Usually referred to as the "Galileo"[3] model, it takes as its elementary unit the "object," which, following Blumer, is "anything that can be designated or referred to . . ." (Blumer, 1966) and assumes that the perceived dissimilarity between any two "objects" can be measured on a continuous scale.

The original scale used by Woelfel and his students was a ratio paired comparisons design, where respondents were given a "criterion pair": (e.g. "If 'Bank Teller' and 'Postal clerk' are 100 units apart . . ."), then asked to estimate the differences among all possible pairs of the other n objects scaled: (". . . how far apart are a and b, a and c, . . . a and n, . . . $n - 1$ and n).

Although an unusual format for scaling at the time, the complete paired comparison ratio estimation task is generally considered the single most precise form of measurement known to psychometricians, and substantial evidence in the case specifically of Galileo scales indicates that they can attain considerable precision even at small sample sizes (Barnett, 1972; Gillham & Woelfel, 1977; Gordon, 1976; Gordon & DeLeo, 1975; Woelfel et al., 1980).

This procedure was applied to the area of occupational attainment by Saltiel (1983). Using a device based on the Occupational Aspiration Scale (Haller & Miller, 1971), he identified the 34 most frequently chosen occupations for high school students in a consolidated rural school district in Montana. The perceived dissimilarities among these occupations were measured using a ratio-scaled pair comparison Galileo scale; 1/3 of all possible pairs chosen at random were estimated by each student and by each of the student's significant others, who were identified by the Wisconsin Significant Other Battery.

These measurements produced a multidimensional space, averaged across all students and their significant others, within which are arrayed the 34 occupations.

Figure 1 represents a subset of 13 of the 34 Saltiel occupations. Earlier one-dimensional arrays of occupations can be related to this figure easily. The Duncan Socioeconomic Index (SEI) correlates about 0.9 (corresponding to an angle of about 26 degrees) with the first principle (left-right) axis of this space, for example.

Position in this space has no absolute significance, but relative position is important. Occupations arrayed close to each other are perceived to be similar by respondents, while those far apart are perceived to be different in proportion to their distance apart. Most important, since each occupation's position in the space is given by a vector of numerical coordinates, it is possible to take the average of

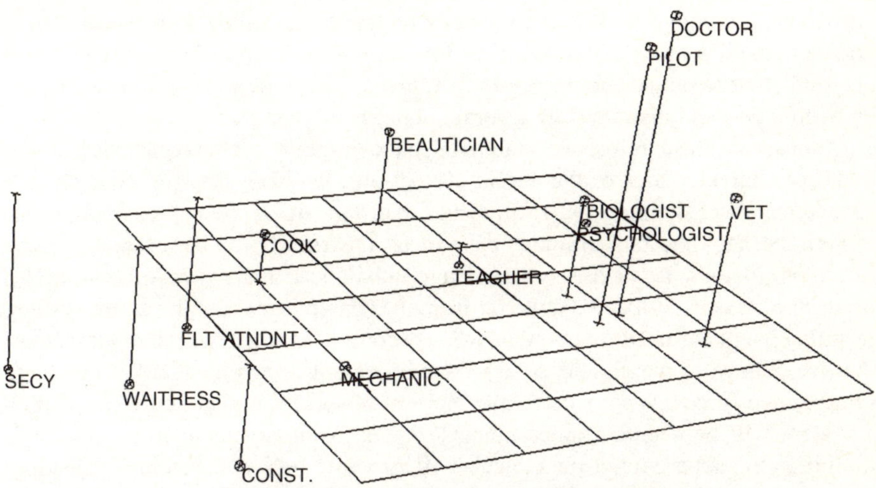

Fig. 1. View of a Subset of Saltiel's 34 Occupations.

2 or more occupations by taking the average of their coordinates. The results will be the coordinates of a point in the space.

This makes it possible to generalize the equations of the "linear force aggregation theory" to discrete choice situations in general and to occupational choice in particular. Simply put, the general theory suggests that the occupation chosen by any given student will lie close to the point defined by the average of the coordinates of the specific occupations expected for the child by its significant others.[4] Indeed, a variant of the Woelfel and Haller regression model, augmented by the mean coordinates of the significant others' discrete expectations, does even better in the choice situation than the original educational and occupational aspiration model, accounting for 84% of the variance in the average job choice on the first dimension, 93% of the variance on the second, and averages about 60% explained variance on the remaining (much smaller) dimensions.

The Galileo Model not only makes it possible to generalize the Wisconsin Status Attainment Model to the case of discrete occupational choice, but it provides a particularly favorable frame of reference for visualizing Haller's notion of a lifetime status trajectory. To do this, it is necessary to generate a sequence of such spaces over time, provide a common orientation, and then simply note the trajectory through the series of occupations occupied by any individual over time.

The development of a sequence of spaces is straightforward, and simply requires repeated measurements over time. Providing a common orientation, however, is not so straightforward, since occupational space, like physical space, has no "privileged" orientation. It is not given by nature, for example, that North should appear at the top of all maps, and East to the right, with altitude represented as a perpendicular to the north-south east-west plane. These are conventions arrived at by negotiation and dispute over several hundreds of years.

Choices of these reference standards have important consequences for the dynamic characteristics of the trajectories of objects. Consider, for example, the consequences of defining the earth rather than the sun as the center of the solar system, perhaps the most famous case of an unwise choice of reference frame in human history. In such a system, huge unbalanced masses revolve around a miniscule mass itself located quite far from the center of mass of the entire system, resulting in equations of motion that are very complicated and which require either that we believe in magic (the solar system does not obey physical law) or huge unexplained forces. What's more, the motions observers will actually see in such a system will be peculiar, since planets will occasionally seem to reverse their motion for no apparent reason. Objects will appear to behave capriciously, leading serious observers to conclude that phenomena are in fact not law governed, or, at the least, too complicated to be understood in a naturalistic scheme.

The mathematical solution for aligning reference systems (known as a Galilean transformation after Galileo) has been known since Galileo for three dimensional Euclidean spaces, but a general mathematical solution and computer algorithm for high dimensional non Euclidean spaces was developed by Woelfel and his coworkers (Woelfel, 1980b; Woelfel & Barnett, 1992; Woelfel & Fink, 1980; Woelfel et al., 1976, 1979, 1986, 1989). The only transformations allowed are "rigid body transformations," that is, spaces may be rotated and/or translated in any arbitrary way, so long as the interpoint distances in any space are not altered. If a two-dimensional space were drawn on a piece of paper, for example, the paper could be moved, turned upside down, reflected in a mirror, put top to bottom, or any such motion, but it could not be folded or bent.

Woelfel and Barnett (1992) illustrate the principles underlying the choice of a reference frame with a clock viewed over several intervals of time. Under normal circumstances, clocks are affixed to walls, set atop furniture or otherwise arrayed against objects we believe to be stable and unmoving. As time passes, the distances among the hands and the numbers on the clock face change, as do the distances among the clock hands and everything else in the reference frame. But if we consider the clock in otherwise empty space it will not be evident as time passes whether the hands are moving relative to the numbers, the numbers relative to the hands, or some combination of both. Indeed, any of these solutions is as good as any other, but only one of them is the simplest to perceive and remember: that is, the solution in which the numbers and pivot point remain fixed and only the hands move.[5]

The clock numbers and pivot provide fixed reference points against which the motion of the hands can be gauged. In the case of the clock example, these fixed reference points are well known, and can be specified as constraints on the rotation; in words, the constraints mean "rotate and translate each of the spaces in this time sequence until the motion of these fixed points is minimized."

In less familiar areas, such may not be the case. Woelfel et al. (1989), considered the case of people's conception of the days of the week, themselves and certain kinds of activity. 414 randomly selected telephone subscribers in the Albany, NY area were asked to respond to a Galileo-type paired comparison instrument assessing their perceptions of the differences among the seven days of the week (Sunday through Saturday), themselves, and activities such as work, recreation, and the like. Data were collected for each of 23 days.

The authors tried several rotations schemes, but found one in which the days and activities were held as close to motionless as possible and the respondent's self point was allowed to move freely. In this model, the days of the week form a roughly elliptical figure, the recreational activities are closest to the weekend, while "work" is located closest to Monday. As days pass, the self point orbits around the ellipse formed by the days, being closest to Sunday on Sunday, Monday on Monday, etc. As the self approaches the weekend, it grows closer to the recreational activities, and, as it moves closer to Monday, it moves correspondingly closer to work.

LAWS OF MOTION AND CHANGE

Woelfel and Barnett generalize beyond establishing fixed reference points to constrain reference frame selection, and suggest the possibility of constraining the transformation algorithm to produce a reference frame in which a given set of "laws of motion" will hold.[6] Kincaid et al. (1981) used the Galileo model to measure the beliefs of Korean immigrants to Hawaii, and that of the general Hawaiian public. They showed that the cultural space of the Koreans converged on the general Hawaiian cultural space as a function of the length of time they had lived in Hawaii, and further that the rate of convergence fit the generalized Newtonian equation for convergence of equilibrium thermodynamic systems to within measurement error. Becker (1993) showed that the convergence of Brazilian Japanese into Japan followed a similar model.

Woelfel et al. (1986) showed a randomly generated paragraph describing six imaginary people to 75 undergraduates at the University at Albany, then had each of them fill out a Galileo type instrument assessing the perceived differences among the six people after a randomly chosen waiting period of between one and 178 hours. Results of an earlier pretest had shown the space increased in size across the first hour; these results showed the size of the space exhibited a substantial rise at 9 hours, falling back to baseline at around 14 hours. An exact replication a year later, this time with 557 students participating, yielded very similar results.

Foldy and Woelfel (1985) showed in a similar experiment, with one hour time intervals, that the spaces of people who heard persuasive messages not only grew after reading the message, but exhibited oscillations consistent with a perturbed equilibrium dynamic system. This work is itself consistent with independent research by Fink and Kaplowitz (1993) and Kaplowitz and Fink (1996, 1997) showing that, even at very small and precisely measured time intervals, observed attitude changes are well modeled as damped harmonic oscillating systems.

NEURAL NETWORKS AS A BASIS FOR GALILEO SPACES

Recent work in neuroscience, computer science, psychology, communication and other areas has had an important impact on our understanding of cognitive processes. Somewhat simplified, neurons are cells which can be "activated" by stimulation. When photons fall on the retina of an eye, for example, they stimulate receptor cells which transmit electrochemical energy through the optic nerve to a series of neurons. Different visual patterns on the retina will lead to different patterns of activation among the neurons; these patterns of activation of neurons represent the brain's mechanism for representing external images internally.

When the photons stop impacting the retinal receptor cells, the neurons deactivate, and the internal representation of the pattern is lost. But if the same pattern is repeated with some frequency, the neurons which collectively represent the pattern tend to become connected to each other. When these connections are of sufficient strength, the pattern is remembered. Since the cells are interconnected, activation of a sufficient subset of them will result in activation of the rest via the interconnections.

This collection of perceptual elements into an interconnected pattern represents a solid physiological basis for Mead's concept of an "object." Moreover, each such cluster can be interconnected with still other clusters, and these, in turn can be connected with still others.[7] Each of these "objects" is defined entirely in terms of its relationship to the other concepts of which individuals are aware. Not only is this model consistent with Mead's understanding, but it is the foundation of the Galileo model as well (Woelfel, 1993a, 1997; Woelfel et al., 1993).

The most widely known technology based on the integration of neural models with the Galileo model is CATPAC©. Because of its underlying neural technology,

CATPAC (CATegory PACkage) is able to serve as a bridge between categorical conceptualizations and the continuous Galileo representation (Woelfel, 1993b).

The logic of CATPAC is identical to the logic of situated meaning in Mead. In CATPAC, a moving window (default size is 7 words) sweeps through a text. If we consider the window to be a "situation," then, initially, the first seven words in the text will be in the situation. Each of them is represented by an artificial neuron, which is activated when the word is in the situation (window). Connections among all active neurons are then incremented by a small amount. The window then slides one word to the right, and generates a second "situation." Once again, those neurons representing the words found in the new situation are activated, connections among all active neurons are incremented, and the window slides again.

At each "cycle" of this system, all connections among all neurons are slightly decremented to simulate forgetting. As a result, connections that happen infrequently will be lost, but those which co-occur with more frequency will be strengthened, so that words which often are found in the same situation will be tightly connected. The situated meaning of each word is given by its relationships to the other words in that situation. Thus CATPAC will define "Mustang" one way when situated in a context which includes Morgan, Palomino, etc., but another way in a context which includes Camero, Challenger, Firebird, and the like.

The Galileo model underlying CATPAC is completely general, and applies to any kind of objects found in any kind of situation. The program Oresme, for example, reads lists of objects found in any "bunch," such as items in a grocery cart, foods eaten at a given meal, movies favored by a person, and learns to relate them to each other following the procedures described above. The result is situated definitions of objects based on their interrelationship to other objects within situations. As with CATPAC, the interrelationships among the objects can be taken as the basis for plotting those objects in Galileo space, such that the more similar objects will be near each other, and different objects further apart.

CURRENT RESEARCH

The Galileo model and its newer associated neural model have developed widespread applications, and are used around the world in the public and private sector wherever procedures to influence attitudes, beliefs and behavior are used. Typical uses include worldwide studies of attitudes toward rain forest use in developing nations, land, park, logging, fishing, hunting and other uses among stakeholders in the Yukon, election campaigns, commercial advertising and market research, and academic research.

One of the more interesting areas of pure research focuses on the core notion in the Galileo model that meaning lies in the interrelationship among elements in a system, rather than in the elements themselves. This provides a renewed interest in the notion of the Collective Consciousness, particularly as the explosive development of the Internet expands the network of connections among people and groups worldwide. Langhorne (2000), for example, has made maps of individual's and groups perception of the Internet by reading the co-occurrences of site visits during Internet sessions into CATPAC; the resulting maps show websites whose distances apart are based on the actual usage by people. These "meanings" exist not only within the minds of individuals, but also in the patterns of interconnections among the individuals – the "collective" has knowledge and meanings which are unknown to the constituent individuals.

Woelfel (2000) has read descriptions of scenes to large assemblies of individuals, who are then asked to estimate the distances among the objects in the scenes. They are also asked to draw the scenes. Evidence so far from several hundred individuals indicates that virtually none of them can draw the scenes accurately after hearing the descriptions. The average of the distance estimates of all the individuals, however, reproduces the scenes with considerable precision; random split halves match each other; but systematic differences appear when, for example, the average of all males is compared to the average of all females. This provides suggestive evidence that the collection of people knows something that none of its constituents know, and that that knowledge can be measured by Galileo procedures.

There is as yet no specific theory available to indicate whether the collective consciousness has any capacity to develop self reflexivity, although the increasing connectedness of the Internet and the extensive focus of attention on its development resemble in general the kinds of communication mechanisms which develop self awareness in individuals. Moreover, an exponential increase in the technology of communication and display might at least in theory generate the kind of virtual experience that might parallel Durkheim's totemistic ritual on a much large scale. The consequences of this explosive development of the interconnectedness of people and organizations, along with an understanding that collective concepts consist of patterns in the matrix of interconnections brings on many fascinating opportunities for theory and research in the Galileo model.

NOTES

1. Aristotle was left to dominate human understanding of motion and change because his great rival Plato believed that motion and change were illusions not worthy of study.

2. To be sure, Sewell, a strong advocate of precise measurement of stratification variables, was well aware of these difficulties, but the data used for the analysis were collected by another investigator for another purpose.

3. These procedures have been described by a variety of names over the years, including multidimensional scaling, metric multidimensional scaling, and others, but these nomenclatures conflict with standard usage in psychometrics and mathematics. The only precise, unambiguous term in use is probably The Galileo Model, or, equivalently, The Galileo System.

4. Each significant other's expectation will be represented by a point in the occupation space. The set of all expectations for each student will form a geometric figure, e.g. any two will form a line segment, any three a triangle, any for a quadrilateral, and so on. The average of the coordinates will describe the exact geometric center of that figure. As in the Force Aggregation model, choosing an occupation near this point will minimize the total discrepancy between the student's choice and the set of all significant other expectations.

5. If this does not seem obvious, consider how much more difficult it would be to tell the time in a single glance if the clock face rotated as well as the hands.

6. It is important to realize that, while it is possible to generate an infinite number of reference frames by this method, it is not possible to generate reference frames which will produce any arbitrary outcome.

7. When represented mathematically, these interconnections can be seen to be a matrix of "connection strengths." This matrix can be seen as a kind of "similarities matrix" because neurons that are very frequently active in the same patterns will be tightly connected, and represented by a high numerical value. "Centering" this matrix, i.e. expressing it as deviation scores from the mean connection strength, will cause similar nodes to be represented by relatively large positive values, and dissimilar nodes to be represented by relatively large negative numbers. This similarities matrix (technically a scalar products matrix) is formally equivalent to the centroid scalar products matrix from which Galileo spaces are calculated, and serve as the basis for Galileo mappings.

REFERENCES

Barnett, G. A. (1972). *Reliability and metric multidimensional scaling*. East Lansing: Department of Communication, Michigan State University.

Becker C. B. (1993). *A communication based theory of cultural reproduction and transformation. Information and attitude change among Brazilian immigrants in Japan*. Ph.D. Dissertation. Presented to the State University of New York at Buffalo, Department of Communication.

Blumer, H. (1966). Commentary and debate. *American Journal of Sociology*, *71*, 535–547.

Danes, J., Hunter, J. E., & Woelfel, J. (1984). Belief change and accumulated information. In: J. E. Hunter, J. E. Danes & H. S. Cohen (Eds), *Mathematical Model of Attitude Change* (Human Communication Research Series, Vol. 1). Orlando, FL: Academic Press.

Fink, E. L., & Kaplowitz, S. A. (1993). Oscillation in beliefs and cognitive networks. In: G. A. Barnett & W. D. Richards (Eds), *Progress in Communication Science* (pp. 247–272). Norwood, NJ: Ablex Publishing Corporation.

Foldy, J., & Woelfel, J. (1985). Conceptual structures as damped harmonic oscillators. Paper presented to the International Communication Association, Honolulu.

Gillham, J. R., & McPhail, T. L. (1974). Political radicalization as a communication process. *Communication Research*, *1*, 243–263.

Gillham, J. R., & Woelfel, J. (1977). The Galileo system of measurement: Preliminary evidence for precision, stability, and equivalence to traditional measures. *Human Communication Research*, *3*(3), 243–263.

Gordon, T. F. (1976). Subject abilities to use MDS: Effects of varying the criterion pair. Paper presented to the Association for Education in Journalism, College Park, MD.

Gordon, T. F., & DeLeo, H. C. (1975). *Structural variations in 'Galileo' space: Effects of varying the criterion pair in metric multidimensional scaling*. Unpublished Manuscript. Philadelpha: Temple University.

Haller, A. O., & Miller, I. W. (1971). *The occupational aspiration scale: Theory, structure and correlates*. New York: Shenkman.

Kaplowitz, S. A., & Fink, E. L. (1996). Cybernetics of attitudes and decisions. In: J. Watt & C. A. Van Lear (Eds), *Dynamic Patterns in Communication Research* (pp. 277–300). Newbury Park, CA: Sage.

Kaplowitz, S. A., & Fink, E. L. (1997). Message discrepancy and persuasion. In: G. A. Barnett & W. D. Richards (Eds), *Progress in Communication Science* (Vol. 12). Norwood, NJ: Ablex Publishing Corporation.

Kincaid, D. L., Yum, J. O., Woelfel, J. & Barnett, G. A. (1981). The cultural convergence of Korean immigrants in Hawaii: An empirical test of a mathematical theory. Paper presented to the International Communication Association, Minneapolis.

Langhorne, A. (2000). Mapping the internet: Using neural networks to develop graphical internet navigation profiles in organizations. Doctoral dissertation. State University of New York at Buffalo.

Lemert, E. (1950). *Social pathology*. New York: McGraw-Hill.

Lin, N., Ensel, W. M., & Vaughn, J. C. (1981). Social resources and strength of ties: Structural factors in occupational status attainment. *American Sociological Review*, *46*(4), 393–405.

Mead, G. H. (1934). *Mind, self and society*. Chicago: University of Chicago Press.

Mettlin, C. (1973). Smokin as behavior: Applying social psychological theory. *Journal of Health and Social Behavior*, *14*, 145–152.

Mistretta, M., Miles, W., & Barnett, G. A. (1973). Jury and judicial decision-making: A research design and preliminary report. Paper presented to the American Sociological Association, New York.

Picou, S. J., & Campbell, R. E. (1975). *Career behavior of special groups*. Columbus: Merrill Publishing Company.

Saltiel, J. M. (1978). Predicting occupational choice: A multidimensional scaling approach. Paper presented at the First Annual "Metric Multidimensional Scaling Workshop". International Communication Association, Chicago.

Saltiel, J. M. (1983). *An application of the Wisconsin model of status attainment to the occupational choice process*. Unpublished manuscript. Department of Sociology, Montana State University, Bozeman, MT.

Saltiel, J. M., & Woelfel, J. D. (1975). Inertia in cognitive processes: The role of accumulated information in attitude change. *Human Communication Research*, *1*, 333–334.

Sewell, W. H., Haller, A. O., & Portes, A. (1969). The educational and early occupational status attainment process. *American Sociology Review*, *34*, 82–92.

Woelfel, J. D. (1967a). A paradigm for research on significant others. Unpublished Working Paper. Madison: University of Wisconsin.

Woelfel, J. D. (1967b). Others' expectations and their effects on the attitudes of an individual. Unpublished Working Paper. Madison: University of Wisconsin.

Woelfel, J. D. (1980a). Foundations of cognitive theory. In: D. P. Cushman & R. Mcphee (Eds), *Explorations in the Message Attitude Behavior Relationship*. New York: Academic Press.

Woelfel, J. D. (1980b). Variational principles of communication. Paper presented to the International Communication Association, Acapulco, Mexico.

Woelfel, J. D. (1993a). Cognitive processes and communication networks: A general theory. In: G. A. Barnett & W. D. Richards (Eds), *Progress in Communication Science* (Vol. 12). Norwood, NJ: Ablex Publishing Corporation.

Woelfel, J. D. (1993b). Artificial neural networks in policy research: A current assessment. *Journal of Communication*, *43*(1), 63–80.

Woelfel, J. D. (1997). Attitudes as nonhierarchical clusters in neural networks. In: G. A. Barnett & W. D. Richards (Eds), *Progress in Communication Science* (Vol. 12). Norwood, NJ: Ablex Publishing Corporation.

Woelfel, J. D. (2000). Collective representation and distributed encoding. In progress paper. State University of New York at Buffalo.

Woelfel, J. D., & Barnett, G. A. (1992). Procedures for the comparison of time dependent data to a theoretical criterion. *Quality and Quantity: A European Journal of Methodology*, *26*, 367–381.

Woelfel, J. D., Barnett, G. A., Pruzek, R., & Zimmelman, R. (1989). Rotation to simple processes: The effect of alternative rotation rules on observed patterns in time-ordered measurement. *Quality and Quantity*, *23*, 3–20.

Woelfel, J. D., & Fink, E. L. (1980). *The measurement of communication processes: Galileo theory and method*. New York: Academic Press.

Woelfel, J. D., & Haller, A. O. (1971). Significant others, the self-reflexive act, and the attitude formation process. *American Sociological Review*, *36*(1), 74–87.

Woelfel, J. D., & Hernandez, D. (1973). *A theory of linear force aggregation*. Unpublished monograph. Champaign: University of Illinois.

Woelfel, J. D., Holmes, R. A., Cody, M., & Fink, E. L. (1976). Message strategies in Riemman space. Paper presented before Joint Session of the Psychometric Society and the Mathematical Psychology Group, Durham, NC.

Woelfel, J., Holmes, R., Kincaid, D. L., & Barnett, G. A. (1980). *How to do a Galileo study*. Troy, NY: Good Books.

Woelfel, J. D., Newton, B., Holmes, R., Kincaid, D. L., & Lee, J. (1986). Effects of compound messages on global characteristics of Galileo spaces. *Quality and Quantity*, *20*, 133–145.

Woelfel, J. D., Richards, W. D., Jr., & Stoyanoff, N. J. (1993). Conversational networks. In: G. A. Barnett & W. D. Richards (Eds), *Progress in Communication Science* (Vol. 12). Norwood, NJ: Ablex Publishing Corporation.

SOME DEMOGRAPHIC ASPECTS OF RURALITY

Glenn V. Fuguitt

ABSTRACT

The distinction between rural and urban is one of the oldest concerns in sociology, and is an important basis of the subfield of rural sociology. In this paper I would like to consider this distinction from the demographic perspective. I will argue that the perspective should be central to the definition of rural and urban, and provides an essential basis for understanding rural areas.

Demographers are concerned about the size, geographic distribution and composition of the population. They are particularly interested in the changes in these over time as mediated through the processes of fertility, mortality, migration, and shifts in status (Hauser & Duncan, 1959). Central to this is the concept of a *population* – an aggregate of individuals that changes over time by the additions and subtractions of individuals through births, deaths, and movement in space or across status boundaries (Ryder, 1964).

These deceptively simple statements may make demographers appear to be little more than social accountants, but their consideration of populations and their changes can bear a close relation to macro-sociological concerns. The concept of a population, moreover, represents one way to move from the micro to the macro level, since population shifts come through individual acts of birth, death, and movement. The expansion of these simple elements by considering age groups

The Shape of Social Inequality: Stratification and Ethnicity in Comparative Perspective
Research in Social Stratification and Mobility, Volume 22, 73–90
© 2005 Published by Elsevier Ltd.
ISSN: 0276-5624/doi:10.1016/S0276-5624(04)22003-0

and their changes led to the powerful concept of the age cohort which has now come into common everyday use, and also to a major line of work by sociologists today on the life course.

In this brief paper, I can only touch on a few of the ways the demographer's perspective has been applied to concerns about rurality or rural-urban differences. Much remains to be done, moreover, to realize the potential of this approach. I will touch first on the definition of rural, and the complex issue of whether the rural-urban dichotomy continues to be useful today. Then I will consider the analysis of rural-urban differences, migration, and changes in composition through the demographic processes – particularly age structure. Although most of the rural population by far is found in developing countries, here I have concentrated on the United States. Many of the concerns and findings discussed here, however, may also apply to other parts of the world, particularly in the more developed countries.

Needless to say, there is an important need for comparative research across societies relating to the issues discussed here. And, a major recent development in work on the United States which I have not considered has been the analysis of international migration and the integration of newcomers from other lands.

THE RURAL DEFINITION

In reviewing the features of the social demographic perspective, Sweet and Bumpass (1992) listed first that the demographer is concerned with what is going on in a clearly defined population – usually a spatially defined population. The implication here is that for a demographer, whatever rural is, it is a spatially-defined population. It is not a lifestyle, a state of mind, a communication network, or a self-identification, though these might be associated with such spatially-defined populations. This leads us to a demographic/ecological definition of rural: *Rural consists of areas having small population size and low density.* These two conditions are not unambiguous. Density depends greatly on just where the boundaries are drawn, and there are no intrinsic cutting points between rural and urban for size and density. The importance of stopping here, however, with the definition and not including sociocultural or occupational elements as well was very eloquently stated by Hope Tisdale (1942) many years ago:

> ... Cities have been defined as ways of life, states of mind, collections of traits, types of occupations and the like. Such definitions are bound to get us into trouble sooner or later because none of the attributes named are constants of the city and all of them spill over into

> other areas. Traits change, political organization changes, the economic system changes. The only trait that is constant is that the city is different from what is not the city. The nature of this difference varies. If we say that the city is a collection of traits, we cut ourselves loose from the only solid base on which we can set up definite criteria and neck-deep in a sea of difficulties connected with the isolation of urban traits. It means that whatever we find more of in the city is an urban trait. But what is the city? Why, it is a collection of urban traits. How do we identify these traits? By their high intensity in the city. But what is the city? It is a collection of urban traits. And so forth. The only way to break this deadlock is to go back to population concentration. The criterion must be in terms of population. Then we can study traits, relationships, and characteristics to our hearts content.

Note that isolation is not included in the size/density definition. Although it is reasonable to conclude that areas of low size and density generally are relatively isolated, contact and integration with urban has always existed. We talk about "rural society," but by any all-inclusive definition there isn't any. An implication of this is that observed rural-urban differences may and probably do reflect interdependence rather than separateness.

Much early work in rural sociology was concerned with the rural-urban dimension. From an empirical perspective, Sorokin and Zimmerman's *Principles of Rural-Urban Sociology* (1929) was a monumental effort from the late 1920s to document empirically differences between rural and urban areas. By the 1950s, however, there were important criticisms of the rural-urban dimension, or continuum, particularly as a surrogate for the development process, and questions about the Wirth (1938) hypothesis that size and density lead to "urban-like" personality and social relationships (Gans, 1962; Lewis, 1952; Pahl, 1966). Somewhat later, such criticisms formed one basis for arguing that the important issues for rural sociologists were instead in the sociology of agriculture (Friedland, 1982; Newby, 1983) though more important than this empirical argument was the rise of the critical perspective in sociology. In a recent paper, Friedland (2002) continues to assert that the rural-urban continuum is a blind alley conceptually. But he accepts the argument of Bell (1992) that the rural-urban continuum is "real" as a strongly-held popular belief. And he illustrates that similar problems are emerging in the definition of agriculture.

Nevertheless rural areas are different, and how and why continues to intrigue us and forms a basis for much empirical research in rural sociology. (This may be confirmed by a quick review of the contents of recent issues of Rural Sociology. (See Garkovich & Bell, 1995, p. 573.) Fundamental here in any sociological perspective is concern about the consequences of small size and low density. Despite the common notion that the Wirth hypothesis (noted above) has been demolished, careful empirical work continues on this issue and with few exceptions shows residential differences in the nature of social ties (Beggs et al., 1996; Fischer, 1972, 1975; Hofferth & Iceland, 1998; Tittle & Stafford, 1992; White &

Guest, 2003; Wilson, 1993). In rural sociology "Density of Acquaintanceship" is a concept that has been utilized in examining social ties in recent community studies (Freudenburg, 1986; Freudenburg & Jones, 1991; see also Salamon, 2003). The effect of low size and density on institutional structures, for example school systems (Warner et al., 1992), medical services (Slesinger, 1991), and local government has also received attention, as has concern about overcoming the social and economic cost of distance through communication and other means (Dillman, 1985; Salant et al., 1997).

A major sociological perspective that relates to the rural-urban dimension is human ecology. This approach is closely identified with demography, particularly as it developed within sociology beginning in the 1950s. Indeed, one of the four major elements in human ecology theory is the population. Human ecology has always had its critics, and although there are recent adherents in rural sociology (Albrecht & Murdock, 1990), my impression is that it has fallen out of favor at least as a formal theory. This is not the place (nor am I the person) to critique formal theories. I believe, however, that human ecology has nevertheless made a lasting contribution to the empirical study of rurality. Though explanations of the phenomena observed may differ profoundly, the focus on rural-urban interdependence, metropolitan dominance and sustenance organization in human ecology is at least congruous with concern about rural-urban differences. Its taxonomy, including the urban hierarchy has, along with central place theory, been an important basis for understanding the structure of rural and urban communities and how rural areas fit in. (Conversely, our pioneer rural sociologist C. J. Galpin (1915) anticipated central place theory with his empirical studies, and influenced later human ecologists). From a change perspective, the concern of human ecologists with metropolitan expansion highlighted what Galpin called the rurban community. Although human ecologists have been criticized for ignoring the sociopolitical bases for improvements in transportation and communication, their work on the consequences of such technological changes on community structure have pinpointed a major aspect of the transformations of rural areas everywhere.

This approach also has contributed to an important extension of the rural-urban concept in the designation of metropolitan and nonmetropolitan areas. Though often mistakenly considered synonymous with rural and urban, it reflects the recognition that such boundaries are diffuse, in that rural and urban areas near big cities are different in important ways from those outside the range of everyday contact. (See Appendix.).

A third sociological perspective that relates to the rural-urban dimension concerns the socioeconomic consequences of peripheral status. Rural areas have always been economically interdependent with urban, and typically have been the sources of products from extractive industries. This means that the loci of power are elsewhere. Though the lack of local autonomy in rural areas has been exaggerated (Richards, 1978), the inequalities engendered through peripheral status, and economic restructuring, as we join the world economy, and as trade and public and private services become more large-scale and specialized, are important elements of the changes taking place in contemporary society. This perspective has contributed to a renewed interest in geographic location in the research done by rural sociologists and others on labor markets (e.g. Horan & Tolbert, 1984; Singleman & Deseran, 1993; Summers et al., 1990) and on spatial inequality (Lichter, 1992; Lobao & Saenz, 2002; Tickamyer, 2000). Demographers are contributing to this perspective, and research by others often rests on demographic data and methods.

To sum up, regardless of theoretical perspective, most demographers would think of rural areas as spatially defined populations of small size and density. The extent to which occupational and sociocultural differences have been and are associated with rural-urban by this definition then becomes an empirical question, the answer to which is important for understanding what is happening in rural areas. That rural America has become more heterogeneous and less differentiated from urban areas on some sociocultural dimensions is a truism, though research shows that this process has not proceeded as far or as consistently has some seem to believe, and in any event does not mean that there isn't any rural any more.

AGGREGATION AND DISAGGREGATION

A harder question to answer, however, is whether the rural-urban distinction is the most appropriate one to use in research. Some have argued no, because at least in western countries there is now more variation within rural and within urban areas than between (Zelinsky, 1991). That in itself is an empirical question, but we should always remain open to consider alternative residence categorizations. Yet to my knowledge none has been submitted as a serious alternative to some form of the rural-urban distinction, including those incorporating the metropolitan concept.

On the other hand, for many research problems, much of the variation within low density and high density areas may need to be captured by turning to a smaller geographic scale. No one would argue that it is best to restrict consideration to the simple rural-urban dichotomy at the national level, unless forced to by data limitations.

We have a long history of considering subareal variation in rural America. Efforts began in the 1920s to delineate type-of-farming areas (Baker, 1926, 1927, 1928, 1929, 1930; Beck & Forster, 1935; Elliott, 1933), and Bogue and Beale (1961) prepared a system of homogeneous multicounty regions that covered the entire nation. From this, Beale (Fuguitt & Beale, 1978) developed a set of 26 larger nonmetropolitan regions, and more recently the Economic Research Service of USDA (Bender et al., 1985) prepared the functional classification of nonmetro counties that is in wide use. Bogue's (1949) delineation of metropolitan communities, the EDA multicounty units from the U.S. Department of Commerce, and more recently the 1980 Public Use Sample D and 1990 PUMS-L for Labor Market Areas, used by rural sociologists among others, are examples of nodal regions oriented around cities.

All of these delineations may pose the danger of reification, in the sense of being considered the one best way of partitioning rural-urban space. Demographers, geographers and others have demonstrated that where basic interest is in the population and not the individual, then no geographic level is necessarily the "right" one (see Farmer et al., 1992). All conceal variation that would be revealed at a lower level, but that doesn't make the lowest the best. As Haggett (1965, p. 3) points out, all we can do with the unique case (the ultimate disaggregation) is to contemplate its uniqueness. Duncan and associates (1961) showed that from 1900 through 1950 U.S. population was deconcentrating across regions and across states, but it was concentrating across small multi-county areas (SEAs) and counties. Then Vining and Straus (1977) showed that in later decades there was deconcentration within all of these sets of units, and on this basis declared that the upturn in rural growth in the 1960s was a "clean break" with the past. It appears they were wrong in their inference about the long-term importance of this change, but this line of work demonstrates that the population redistribution process is understandable only in terms of several different areal delineations and not just one delineation at a particular geographic scale. In analyses with sets of areas, the rural-urban or metropolitan-nonmetropolitan distinction may be appropriate within each area (i.e. regions by metropolitan status) or between areas (i.e. metropolitan vs. nonmetropolitan counties). Or the units themselves may be designed to reflect differences in size-density or change in density as in the works by Duncan and associates and by Vining and Straus described above.

Applied or policy-oriented research needs to stress local variability. Nevertheless, the simple rural-urban or metro-nonmetro distinction, even at the national or state level, may be an appropriate starting point, for the following reasons: (1) There is always concern for making concise generalizations: "What do we know about Rural America?" We know at the extreme that every person and every community has unique qualities, but we seek to abstract from these what is common. (2) For policy and planning at the national and state level, questions are often posed about the total picture, and patience may be lacking to hear about all the qualifications and exceptions! (3) At the same time, an appropriate analysis strategy may be to start with rural and urban, since that is often where the problem is thought to reside, and then through areal and subpopulation disaggregation demonstrate that more precise targets need to be fixed.

An illustration of the latter circumstance is the widespread belief in the late 1960s and early 1970s that major social and economic problems in our large cities had as their basis the recent migration of blacks from the South to the North. In general this proposition was *not* supported, and on the contrary a fair amount of research suggested that race, jobs and discrimination was more important than rural-urban migrant status (Duncan, 1968; Long, 1988).

THE DEMOGRAPHIC APPROACH TO RURAL-URBAN DIFFERENCES

An important aspect of the approach of demographers to research is concern with demographic components of change (Sweet & Bumpass, 1992). As part of the definition of demography, population change may be separated into the components of fertility, mortality and migration. Similarly differences among subpopulations may be divided into that due to differences in population composition variables and to other variables. In one sense, the latter may be thought of as part of the quest for conclusions about what causes something, whereas the former raises issues about doing something about the outcome (i.e. the observed difference) regardless of the cause.

Fertility and Mortality

The difference between this component analysis and the usual practice of controlling for "demographic variables" is subtle but important. For example, fertility analyses in the U.S. today show that most of the rural-urban difference in fertility is due to age at marriage and socioeconomic status. After taking these variables into account, Slesinger (1974) found little rural-urban difference left. With a simpler analysis of census data, Beale (Fuguitt et al., 1989) finds some residual after controlling in turn for whether or not married, and educational status for blacks and nonblacks. As she recognized, Slesinger's finding does *not* mean there is (in her data) no rural-urban difference in fertility as was initially observed. It means rather that the difference found is due to differences in the composition of the population, which is significant in and of itself. If rural fertility is higher, it will have consequences for rural population growth and distribution, even if observed differences in levels are due to the fact that women in rural areas are more likely to marry earlier and to have lower educational status. This, indeed, leads to further questions such as why women in rural areas are more likely to be married, and to broader questions about rural-urban differences in family structure.

What does Beale's finding mean? Since there is a residual after controlling on some demographic variables are rural-urban differences "real?" No, these differences were real before all the controls were carried out. Rather, he showed the differences observed are not all due to age, race, marital status, education, etc., at least the way he measured them. What is that something else? Going back to the fundamental size-density definition of rural-urban, do we think low density causes people to have more babies? If not, what? Further analysis needs to introduce new variables to explicate how rural-urban residence may effect fertility levels.

A cautionary tale on the need to disaggregate, in this case by age: We showed (Fuguitt et al., 1991) that by the 1980s metropolitan-nonmetropolitan differences in total fertility rates had almost disappeared. The general trend of convergence in rural and urban fertility has long been noted, so this finding should not be too surprising. Yet, a further examination of birth rates by age showed this was due to a tradeoff of the continued decline in young nonmetro fertility, and an *increase* in metro fertility for older women. So within age groups metropolitan-nonmetropolitan fertility differences actually increased! With the components approach, we found another paradox with these data, in that the metro-nonmetro *convergence* in birth rates was an important component in the *divergence* in population growth rates during the 1980s, as nonmetro growth fell well below metro growth.

Recent mortality differences in the U.S. favor urban areas, but this difference almost disappears when age differences are controlled (Miller et al., 1987). Rather than simply concluding, however, that rural and urban areas are equally healthy today, the authors point to the need to consider special health and service needs of the aging population. As Morton (2004) notes, there are many aspects of the typical composition of rural populations that could lead to higher mortality, including a higher proportion of elderly persons, higher rates of poverty, more substandard housing, lower levels of education, lower likelihood of employment in white collar occupations, and lower access to quality medical care. How these and other compositional factors affect mortality rates is an important empirical question with policy implications. Moreover, Morton points out that compositional variations may lead to more variation within rural areas distinguished by location and/or level of urban influence, than that between rural and urban areas.

Migration

Migration has always been a major concern of rural sociologists. Historically rural-urban migration has been the key component in urbanization and population concentration. Today in developing countries the growth of large cities is due also to natural increase to a large degree, and in highly-urbanized nations like the United States differential migration among urban and metropolitan areas is more important for these areas than rural-urban migration. Migration is the demographic variable which requires the consideration of geographic subareas, since its very definition is movement from an area of origin to a destination. Here most clearly, we have the macro-micro interrelation of demography since we are concerned with the individual mover and with the areas of origin and destination.

Almost all studies have shown that migrants tend to better themselves in their moves. The general socioeconomic convergence noted between rural and urban areas has been cited as evidence that migration is an equilibrating process at the macro level as well, although sociologists and rural sociologists have always expressed concern about the consequences of selective outmigration for rural communities, at least since the time that E. A. Ross (1916) characterized many small towns in southern Michigan, Illinois, Wisconsin and Missouri as "fished out ponds populated chiefly by bullheads and suckers." The problems of this selective outmigration attracted little attention during the turnaround decade of the 1970s when nonmetropolitan areas grew more rapidly than metropolitan areas in the United States and many other western countries. Instead, focus tended to be on the problems of rapid growth, including the integration of newcomers and oldtimers (Frankena, 1980). This was despite the fact that there continued to be a net outmigration of young people and of those with higher socioeconomic status from most rural areas. But the subsequent trends of widespread absolute rural decline in the 1980s meant that concerns related to outmigration for rural areas resurfaced in the 1980s and continued in the 1990s, even with the milder nonmetropolitan "rebound" of the latter decade (Fuguitt & Beale, 1996). Further, there is the added assertion that with the depressed economic situation of many metropolitan areas, particularly those located in the North, migration may no longer be an avenue of social mobility for many persons of rural origin. More recent trends have shown that this most recent "rebound" has about run its course (Hamrick, 2003). As Lichter (1992) contends, there is a continual need for a substantial research effort on migration, with greater emphasis on small area analysis and the relation of population movements to locational inequalities as well as to individual achievement.

POPULATION AGING

A further explication of the importance of the demographic approach is in terms of the aging of the population. Rural and nonmetropolitan populations have generally had older populations than urban and metropolitan populations. Aging is a societal trend of extreme importance which challenges governmental programs at the national level such as medicare and social security, but which impacts on local rural communities with particular implications for providing services in low density areas.

At the local level, however, one cannot simply attribute the increased number and proportion of elders to past trends in fertility and improvement in mortality for elders. Internal migration is quite important, and it results in aging processes which may have quite different consequences for different areas. The absolute numbers reaching elderly status in rural areas is influenced by the pattern of migration for that aged cohort throughout the age span. In areas like the Great Plains, there is general population decline, but the number of elders may be declining more slowly than the total, or even increasing. Change in the proportion elderly is influenced by changes in these numbers but also by change in the remainder of the population. So the proportion of the population that is of advanced years may have increased markedly in some instances and declined in others. Nonmetropolitan recreation and retirement areas have gained older people though migration, so the numbers are increasing, but these areas have also gained younger people, so that the proportion elderly may not be increasing.

There are, of course, other important socioeconomic differences between types of nonmetropolitan areas. Nevertheless, the characteristics of elders and the capacity of communities to deal with problems of the older population may be quite distinct due to these different demographic processes.

THE NEW INTEREST IN SPACE

An important recent trend in social science has been the increased recognition of the importance of space in methods, theory and research. First there is the growth of work in Geographic Information Systems (GIS), which makes it possible to map data easily and look at spatial relationships. This has come though the greater availability of geocoded data for small areas, as in the 2000 Census of Population, and is augmented by developments in statistical procedures to examine the effects of "spatial dependence" among observations. Of importance here to all research using small areas has been the recognition that results of analysis using conventional linear models may be distorted by the clustering of observations (Anselin, 2000; Voss, 2003). But the greater ease in manipulating small area data has already enabled government agencies to improve the fit of data classifications to the concept of population concentration (see Appendix).

Concern about space has also found its way into more general or theoretical statements in sociology (Gieryn, 2000; Lobao, 1994; Silber, 1995) and related social science fields such as urban geography (Wilson & Moss, 1997). Space would appear to be intrinsic to the two subfields of urban and rural sociology and has always been an important basis for research in these fields. Yet in both concerns about residence and its consequences, and the approach of human ecology were rejected by many scholars in the 1960s and 1970s in favor of the critical approach, as exemplified by the "new urban sociology" and the "sociology of agriculture." LaGory (1993) argues that the most promising future for urban sociology is the development of a theoretically focused approach that builds on both of the perspectives of the new urban sociology and the old (human ecology). Lobao (1996) urges the modification of the critical approach by incorporating space so as to make residence more explicit and theoretically informed in research. Similar to the approach of this paper, she argues that residence classifications, such as metropolitan and nonmetropolitan should not be implicit proxies for assumed social relationships, which should instead be the basis of empirical investigation. We need to move away from static comparisons, moreover, and give more attention to the connections between rural and urban economic and social change.

Voss (2003) cautions that concern about space is a new emphasis but not a new departure. As already noted, most rural sociologists always have been concerned about residence and the study of rural-urban differences. The study of stratification through the lens of spatial inequality also has important antecedents in rural sociology. For example, the honoree of this volume, A. O. Haller, early on did research on occupational and educational aspirations and attainment in rural settings (for example, Haller, 1957, 1958; Sewell & Haller, 1965), and more recently related levels of development in Brazil to rural and urban geographic subregions (Bills et al., 1985; Haller, 1982). Nevertheless, the new emphasis in the social sciences on space should make the demographic approach to rurality a more central concern in our field.

CONCLUSIONS

The demographic perspective has made an important contribution to the study of rurality. Starting with a spatially-based population, a definition of rural based on size and density makes it possible to consider occupational and sociocultural differences within and among rural and urban areas, thereby monitoring the process of rural-urban convergence and increasing interdependence, even as multivariable analyses at smaller geographic scales show rural and urban heterogeneity. The focus on change, and components of change can help to move us away from static comparisons toward the study of territorial processes. The explication of macro and micro relations though the demographic variables can make an important contribution to better understanding rural people and places. As noted above, the new emphasis on space in sociology and the other social sciences should make the demographic approach to the study of rurality even more central to our field.

But we should do much more in exploiting the full potential of demographic analysis. For example, multiregional demographic models represent a way to consider migration and population change as a total system, and systematic projections can help us to better understand the implications of recent changes. Cohort analyses can contribute to the understanding of rural life and rural communities, and help us to develop future scenarios. Currently there is much concern about the family in the United States, and a great deal of work is being done by demographers. Summary variables show that rural areas continue to have a higher proportion of more traditional families. But many more detailed analyses are needed to elaborate and better understand this rudimentary difference and how it is changing, and what the implications are for rural America.

Finally, I should note that the long historical series of census data for the United States and many other countries, the increased capabilities of working with small area data, as well as with very complicated files showing interrelations among areas, particularly in migration and commuting, also contributes to the potential of the demographic approach to rurality. The number of able scholars now engaged in work in this area cannot help but make us optimistic about the future. The rest is up to them, and to those who come after.

ACKNOWLEDGMENTS

This work contributes to a cooperative project between the College of Agricultural and Life Sciences, University of Wisconsin-Madison, and the Economic Research Service, U.S. Department of Agriculture. The support of the Center for Demography and Ecology of the University of Wisconsin-Madison, through a grant from the National Institute of Child Health and Human Development, also is gratefully acknowledged. I am grateful for the assistance of Calvin Beale, Tim Heaton, Stanley Lieberson, Dan Lichter and others in preparing the manuscript, and to John Carlson, Senior Librarian of the Center for Demography and Ecology, for help with the bibliography. I am especially indebted to David Bills, editor of this volume, for his patience and encouragement.

REFERENCES

Albrecht, D. E., & Murdock, S. M. (1990). *The sociology of agriculture: A human ecological perspective*. Ames: Iowa State University Press.

Anselin, L. (2000). GIS, spatial econometrics and social science research. *Geographical Systems, 2*, 11–15.

Baker, O. E. (1926, 1927, 1928, 1929, 1930). Agricultural regions in the United States. *Economic Geography, 2*, 459–493; *3*, 50–86, 309–339; *4*, 44–73, 399–433; *5*, 36–69; *6*, 166–190.

Beck, P. G., & Forster, M. C. (1935). *Six rural problem areas*. Washington, DC: Federal Emergency Relief Administration, Research Monograph No. 1.

Beggs, J. J., Haines, V. A., & Hurlburt, J. S. (1996). Revisiting the rural-urban contrast: Personal networks in nonmetropolitan and metropolitan settings. *Rural Sociology, 61*, 306–325.

Bell, M. M. (1992). The fruit of difference: The rural-urban continuum as a system of identity. *Rural Sociology, 57*, 65–82.

Bender, L. D., Green, B. L., Hady, T. F., Kuehn, J. A., Nelson, M. K., Perkinson, L. B., & Ross, P. J. (1985). *The diverse social and economic structure of nonmetropolitan America*. Washington, DC: USDA Economic Research Service Rural Development Research Report No. 49.

Bills, D. B., Haller, A. O., Kelley, J., Olson, M. B., & Pastore, J. (1985). Class, class origins, regional socioeconomic development and the status attainment of Brazilian men. In: R. V. Robinson (Ed.), *Research in Social Stratification and Mobility* (Vol. 4, pp. 89–127). Greenwich, CT: JAI Press.

Bogue, D. J. (1949). *The structure of the metropolitan community: A study of dominance and subdominance*. Ann Arbor: University of Michigan Press.

Bogue, D. J., & Beale, C. L. (1961). *Economic areas of the United States*. New York: The Free Press of Glencoe, Inc.

Bureau of the Census (2001). Urban area criteria for census 2000-proposed criteria: Notice. *Federal Register, 66*(60/Wednesday, March 28/Notices).

Cromartie, J. B., & Brown, D. L. (2003). Nonmetropolitan population change, 1960–2000: A comparison of micropolitan and noncore areas. Presented at the Annual Meeting of the Association of American Geographers, March 5–8.

Dillman, D. A. (1985). The social impacts of information technologies in rural North America. *Rural Sociology, 50*, 1–26.

Duncan, O. D. (1968). Statement for the committee on government operations, United States senate. In: *The Rural-Urban Population Shift a National Problem* (pp. 99–101). 90th Congress 2nd Section, Committee Print. Washington, DC: U.S. Government Printing Office.

Duncan, O. D., Cuzzort, R. P., & Duncan, B. (1961). *Statistical geography*. Glencoe: Free Press.

Elliott, F. F. (1933). *Types of farming in the United States*. Washington, D.C.: United States Government Printing Office.

Farmer, F. L., Luloff, A. E., Ilvento, T. W., & Dixon, B. L. (1992). Rural community studies and secondary data: Aggregation revisited. *Journal of the Community Development Society, 23*, 57–70.

Fischer, C. S. (1972). 'Urbanism as a way of life' A review and an agenda. *Sociological Methods and Research, 1*, 187–215.

Fischer, C. S. (1975). Toward a subcultural theory of urbanism. *American Journal of Sociology, 80*, 1319–1341.

Frankena, F. (1980). *Community impacts of rapid growth in nonmetropolitan areas: a literature survey*. East Lansing: Michigan State University Agricultural Experiment Station Rural Sociology Series No. 9.

Freudenburg, W. R. (1986). The density of acquaintanceship: An overlooked variable in community research? *American Journal of Sociology*, *92*, 27–63.

Freudenburg, W., & Jones, R. E. (1991). Criminal behavior and rapid community growth: Examining the evidence. *Rural Sociology*, *56*, 619–645.

Friedland, W. H. (1982). The end of rural society and the future of rural sociology. *Rural Sociology*, *47*, 598–608.

Friedland, W. H. (2002). Agriculture and rurality: Beginning the 'final separation'? *Rural Sociology*, *67*, 350–371.

Fuguitt, G. V., & Beale, C. L. (1978). Population trends in nonmetropolitan cities and villages in subregions of the United States. *Demography*, *15*, 605–620.

Fuguitt, G. V., & Beale, C. L. (1996). Recent trends in nonmetropolitan migration: Toward a new turnaround? *Growth and Change*, *27*, 156–174.

Fuguitt, G. V., Beale, C. L., & Reibel, M. (1991). Recent trends in metropolitan-nonmetropolitan fertility. *Rural Sociology*, *56*, 475–488.

Fuguitt, G. V., Brown, D. L., & Beale, C. L. (1989). *Rural and small town America*. New York: Russell Sage.

Galpin, C. J. (1915). *The social anatomy of an agricultural community*. Madison: University of Wisconsin Agricultural Experiment Station Research Bulletin 34.

Gans, H. J. (1962). Urbanism and suburbanism as ways of life a reevaluation of definitions. In: A. M. Rose (Ed.), *Human Behavior and Social Process* (pp. 625–648). Boston: Houghton Mifflin.

Garkovich, L., & Bell, A. M. (1995). Charting trends in rural sociology 1986–1995. *Rural Sociology*, *60*, 571–584.

Gieryn, T. F. (2000). A space for place in sociology. *Annual Review of Sociology*, *26*, 463–496.

Haggett, P. (1965). *Locational analysis in human geography*. London: Edward Arnold.

Haller, A. O. (1957). The influence of planning to enter farming on plans to attend college. *Rural Sociology*, *22*, 137–141.

Haller, A. O. (1958). Research problems on the occupational achievement levels of farm-reared people. *Rural Sociology*, *22*, 137–141.

Haller, A. O. (1982). A socioeconomic regionalization of Brazil. *Geographical Review*, *72*, 450–464.

Hamrick, K. (Ed.) (2003). *Rural America at a glance*. Washington, DC: U.S. Department of Agriculture, Economic Research Service, Rural Development Research Report RRDR97–1, September. http://www.ers.usda.gov/publications/RRDR97-1.

Hauser, P. M., & Duncan, O. D. (1959). *The study of population: An inventory and appraisal*. Chicago: University of Chicago Press.

Hofferth, S. L., & Iceland, J. (1998). Social capital in rural and urban communities. *Rural Sociology*, *63*, 574–598.

Horan, P. M., & Tolbert II, C. M. (1984). *The organization of work in rural and urban labor markets*. Boulder: Westview Press.

LaGory, M. (1993). Spatial structure and the urban experience: Ecology and the new urban sociology. Urban sociology in transition. *Research in Urban Sociology*, *3*, 111–125.

Lewis, O. (1952). Urbanism without breakdown. *Scientific Monthly*, *75*, 31–41.

Lichter, D. T. (1992). Migration, population redistribution, and the new spatial inequality. In: D. L. Brown, D. R. Field & J. J. Zuiches (Eds), *The Demography of Rural Life: Current Knowledge and Future Directions for Research* (pp. 19–46). University Park, PA: North Central Regional Center for Rural Development.

Lobao, L. M. (1994). The place of space in current sociological research. *Environment and Planning A*, *26*, 665–668.

Lobao, L. M. (1996). A sociology of the periphery vs. a peripheral sociology: Rural sociology and the dimension of space. *Rural Sociology*, *61*, 77–102.

Lobao, L. M., & Saenz, R. (2002). Spatial inequality and diversity as an emergent research area. *Rural Sociology*, *67*, 497–511.

Long, L. (1988). *Migration and residential mobility in the United States*. New York: Russell Sage.

Miller, M. K., Stokes, C. S., & Clifford, W. B. (1987). A comparison of the rural-urban mortality differential for deaths from all causes, cardiovascular disease and cancer. *Journal of Rural Health*, *3*, 23–33.

Morton, L. W. (2004). Spatial patterns of rural mortality. In: N. Glasgow, L. W. Morton & N. E. Johnson (Eds), *Critical Issues in Rural Health* (Chap. 4). Ames, IA: Iowa State Press.

Newby, H. (1983). The sociology of agriculture – Toward a new rural sociology. *Annual Review of Sociology*, *9*, 67–81.

Pahl, R. E. (1966). The rural-urban continuum. *Sociologia Ruralis*, *6*, 299–327.

Richards, R. O. (1978). Urbanization of rural areas. In: D. Street and Associates (Eds), *Handbook of Contemporary Urban Life*. San Francisco: Jossey-Bass.

Ross, E. A. (1916). Folk depletion as a cause of rural decline. *Papers and Proceedings Eleventh Annual Meeting American Sociological Society, Columbus, Ohio* (Vol. 11). Chicago: University of Chicago Press.

Ryder, N. B. (1964). Notes on the concept of a population. *American Journal of Sociology*, *47*, 447–463.

Salamon, S. (2003). *Newcomers to old towns: Suburbanization of the heartland* (with collaboration of K. Davis-Brown). Chicago: University of Chicago Press.

Salant, P., Carley, L. R., & Dillman, D. A. (1997). Lone eagles among Washington's in-migrants: Who are they and are they moving to rural places. *Northwest Journal of Business and Economics*.

Sewell, W. H., & Haller, A. O. (1965). Educational and occupational perspectives of farm and rural youth. In: L. G. Burchinal (Ed.), *Rural Youth in Crisis: Facts, Myths and Social Change* (pp. 149–169). Washington, DC: U.S. Department of Health, Education, and Welfare, Government Printing Office.

Silber, I. F. (1995). Space, fields, boundaries: The rise of spatial metaphors in contemporary sociological theory. *Social Research*, *62*, 323–337.

Singleman, J., & Deseran, F. A. (1993). *Inequalities in labor market areas*. Boulder, CO: Westview Press.

Slesinger, D. P. (1974). The relationship of fertility of measures of metopolitan dominance: A new look. *Rural Sociology*, *39*, 350–361.

Slesinger, D. P. (1991). Health care in rural America. In: C. B. Flora & J. A. Christianson (Eds), *Rural Policies for the 1990s* (pp. 140–150). Boulder, CO: Westview Press.

Sorokin, P. A., & Zimmerman, C. C. (1929). *Principles of rural-urban sociology*. New York: Henry Holt.

Summers, G. F., Horton, F., & Gringeri, C. (1990). Rural labour market changes in the United States. In: T. Marsden, P. Lowe & S. Whatmore (Eds), *Rural Restructuring: Global Processes and Their Responses*. London: David Fulton.

Sweet, J. A., & Bumpass, L. L. (1992). Disruption of marital and cohabitation relationships: A social-demographic perspective. In: T. Orbuch (Ed.), *Close Relationship Loss: Theoretical Approaches* (pp.67–89). New York: Springer-Verlag.

Tickamyer, A. R. (2000). Space matters: Spatial inequality in future sociology. *Contemporary Sociology*, *29*, 805–813.

Tisdale, H. (1942). The process of urbanization. *Social Forces*, *20*, 311–316.

Tittle, C. R., & Stafford, M. C. (1992). Urban theory, urbanism, and suburban residence. *Social Forces*, *70*, 725–744.

U.S. Census Bureau (2003). About metropolitan and micropolitan statistical areas. www.census.gov/population/www/estimates/aboutmetro.htm.

Vining, D. R., & Straus, A. (1977). A demonstration that the current deconcentration of population in the United States is a clean break with the past. *Environment and Planning*, *9*, 751–758.

Voss, P. (2003). Rambling about space: The role of space in rural sociological thought & practice. Lecture given at the Department of Rural Sociology, University of Wisconsin-Madison.

Warner, W. K., England, J. L., & Ward, C. (1992). Scale of organizing: The case of rural school consolidation. Paper presented at the Eighth World Congress for Rural Sociology, State College, Pennsylvania.

White, K. J. C., & Guest, A. M. (2003). Community lost or transformed? Urbanization and social ties. *City & Community*, *2*, 239–259.

Wilson, D., & Moss, P. (1997). Spatiality studies in urban geography. *Research in Urban Sociology*, *4*, 1–24.

Wilson, T. C. (1993). Urbanism and kinship bonds. *Social Forces*, *71*, 703–712.

Wirth, L. (1938). Urbanism as a way of life. *American Journal of Sociology*, *44*, 1–24.

Zelinsky, W. (1991). Is the rural-urban dichotomy withering away? Paper presented at the 14th Congress of the European Society for Rural Sociology, Geissen, Germany, July 19.

APPENDIX

Residence Definitions

Operationalizing a concept always requires a certain degree of arbitrariness, and this certainly is true of the distinction between rural and urban. The evolution of the U.S. Census definition over the past century is illustrative, and has reflected a continuing effort to approach a measure that reflects population concentration, while at the same time being doable in the constraints of a national census, but with due respect for the need to provide comparability across censuses.

Between 1910 and 1940 rural constituted all people living outside incorporated places having 2,500 or more population, with those living in such places constituting the urban sector. I have never seen an explanation for why the number 2,500 was chosen. (Note that villages of less than 2,500 along with those outside any place were included as rural). Similar definitions adapted by other countries have used many different threshold sizes. Political incorporation needed to be included in this early definition because incorporated places have clearly defined boundaries for purposes of governance and were units of census enumeration. Thus, unincorporated places above 2,500 had to be considered rural. In 1950 this definition was modified to reflect the increasing importance of deconcentrated settlement that does not necessarily respect political boundaries. From 1950 on unincorporated places over 1000 population were delimited by the Bureau and those over 2,500 were considered to be urban. In addition, *Urbanized Areas* (UAs) were delimited to include the thickly settled territory, whether incorporated or unincorporated, around cities of 50,000 or more. The fringe added in this way became part of the urban population, and so was deducted from the rural population. In later censuses the UA became thickly settled territory having a total population of 50,000, removing the requirement that it include an incorporated place with more than 50,000. Finally, in the most recent census of 2000, advances in computation and more elaborate geographic identification made it possible to automate the Urbanized Area delineation. Consequently, the Bureau expanded the delineation of thickly settled areas to include all those having at least 2,500 people. Those areas having between 2,500 and 50,000 people are called *Urban Clusters*, and this further removed territory formerly considered rural, but makes the urban distinction consistent in including all thickly settled areas over the urban threshold in size (Bureau of the Census, 2001).

Concern about monitoring the increasing interdependence of rural and urban settlement, especially around large cities, led to the establishment of Standard Metropolitan Areas in 1950 (Fuguitt et al., 1989). These were county-based units, but there was also an alternative town-based version in New England. These units include major cities (later simply UAs) along with any adjacent counties closely related to the center. After 1960 a major criterion for including adjacent counties was based on patterns of commuting revealed through census data. After 1980 the name of the unit was changed to Metropolitan Statistical Area (MSAs). Again, this system of areas was modified for the Census of 2000, by adding *Micropolitan Areas*. Each metropolitan area must include at least one urbanized area of more than 50,000 population, and each micropolitan area must include at least one urban cluster with at least 10,000 but no more than 50,000 population. Both metropolitan and micropolitan areas may include adjacent counties having sufficient commuting ties to the central county (U.S. Census Bureau, 2003). Note that people residing in metropolitan areas make up the metropolitan population, whereas the nonmetropolitan population includes those living in micropolitan areas and those outside either metropolitan or micropolitan areas.

What can we conclude from this rather detailed account? First even though we are committed to employing the simplest definition based on population conconcentration, without grappling with the inclusion of sociocultural or occupational variables, making definitions based on census data is necessarily complex and must be arbitrary. Second, changes over time in census procedures may make comparable research across censuses difficult. Third, technological changes have enabled the Bureau to come closer to the goal of measuring urban or metropolitan as consistent indices of level of concentration.

Finally, both urban and metropolitan play a role in understanding the settlement fabric, but they are definitely not synonymous. Those interested in studying issues of residence find the metro-nonmetro distinction very attractive. Based on county units that rarely change substantially across censuses, they provide a wider array of data and an easier basis for doing comparisons over time. Yet the two delineations are far from the same, and have become more different with successive censuses. In 1980, 60% of the rural population was nonmetropolitan so 40% was found in metropolitan areas. At the same time 13% of the urban population was nonmetropolitan. By the census of 2000 more than one-half of the rural population (51%) was found in metropolitan areas, and the urban population proportion in nonmetropolitan areas had dropped to nine.

Analysis has reaffirmed the importance of distinguishing between the micropolitan (or, previously counties having smaller cities) and other parts of the nonmetropolitan territory. Recent work has shown similarities between the micropolitan and the metropolitan groupings (Cromartie & Brown, 2003). This does not mean, however, that "real" nonmetropolitan is limited to the areas outside of micropolitan areas. Both need to be taken into account to understand the portion of the country that is beyond the immediate reach of large urban centers.

PART II: APPLICATIONS IN U.S. SOCIETY

ASSIMILATION IN AMERICAN SOCIETY: OCCUPATIONAL ACHIEVEMENT AND EARNINGS FOR ETHNIC MINORITIES IN THE UNITED STATES, 1970 TO 1990

C. Matthew Snipp and Charles Hirschman

ABSTRACT

We examine changes in the socioeconomic position of racial and ethnic minorities for the period of 1970–1990. In particular we use data from the decennial censuses to estimate models showing the direct and indirect affects of minority status on occupational status and income. The period of 1970–1990 is significant because it covers an approximate generation following the introduction of affirmative action and other measures designed to improve the opportunities for ethnic minorities in the workplace. We find that for most of the groups we examine, there have been small improvements over time. However, ethnic minority status remains a strong and persistent disadvantage for most of these groups, and especially for Black and Latino workers.

The Shape of Social Inequality: Stratification and Ethnicity in Comparative Perspective
Research in Social Stratification and Mobility, Volume 22, 93–117
© 2005 Published by Elsevier Ltd.
ISSN: 0276-5624/doi:10.1016/S0276-5624(04)22004-2

INTRODUCTION

For most of the last century, social scientists and the public at-large harbored a view of American society which was homogeneous and culturally monolithic. The immigrant experience through the late 19th and early 20th centuries lent credence to this vision. Newcomers to this nation embraced the ideals of Anglo-American society, and discarded as quickly as possible the evidence of their ethnic heritage. There was of course, a considerable economic incentive offered for these immigrants. Assimilation, and especially conformity to Anglo-American ideals was essential to achieve the upward social mobility that America promised these immigrants.

At the beginning of the 21st century, we begin trying to imagine American society in the next one hundred years, the so-called "American Dream" of upward mobility in exchange for cultural conformity seems an increasingly tattered and yellowing proposition. New waves of immigrants originating from non-European locations have been reluctant to surrender their culture, language, and traditions. Claims to political power are increasingly organized on the basis of personal identity, especially racial and gender identities. Sensing this shift in political power, other groups such as American Indians have made a concerted effort to assert the importance of their ethnic ancestry.

This paper examines the social and economic inequality that exists among the major racial and ethnic groups in American society. However, it is not enough to simply document the inequalities that exist across groups. Equally, if not more important is knowledge about how these inequalities have changed over time, and how racial and ethnic inequality may evolve in the near future. For this reason, this paper also will examine the changes that have taken place in the most recent generation of American history, 1970–1990. These years are especially important because they mark the decades following the introduction of public policy, Affirmative Action and Equal Employment Opportunity Programs, explicitly designed to enhance the socioeconomic standing of historically disadvantaged minorities, without regard to their conformity to traditional Anglo-American ideals.

WHY RACE AND ETHNICITY MATTER

Why race and ethnicity matter, and especially why the social and economic inequality connected with racial and ethnic characteristics are a concern, are issues that can be addressed from at least two perspectives. One is that the distribution of scarce resources, and ultimately the determination of life chances, on the basis of ascribed racial characteristics contradicts values deeply in-grained within the core of American society. This belief stipulates that all Americans have an equal opportunity to compete for the wealth of this nation, and that the distribution of resources is determined by a fair process governed only by personal merit and achievement. Hard work and ingenuity is rewarded above everything else. This idea is not only an article of faith, it is also a guiding principle for a great deal of public policy.

From another more sociological perspective, racial and ethnic characteristics serve as a fundamental organizing principle for American society. In particular, racial and ethnic characteristics are ascribed traits which can be used to study the relative openness of a society with respect to social mobility. In theory, economic resources and personal well-being in highly rigid and closed societies are determined entirely by ascribed characteristics such as race or family background. By the same logic, economic well-being in fluid open societies is determined by effort and ability in a competition among equal actors (Lenski, 1966). As a result of this thinking, sociologists for decades have devoted extensive amounts of time and effort studying social mobility and status attainment. The purpose of this research has been to assay the extent to which American society is becoming more or less rigidified vis-à-vis other nations and over time. The data we report below can be seen as part of this on-going effort.

Likewise, the stratification literature in sociology and especially studies of status attainment have traditionally viewed changes in economic differentials as one important gauge of economic assimilation (Hirschman, 1983). That is, for example, the presence of black-white differentials in education, income, occupational status, or other measures of economic well-being are regarded as *prima facie* evidence that economic assimilation has not occurred. Indeed, data from the General Social Survey suggest that racial differentials have grown smaller in recent years but a significant gap continues to exist (Grusky & DiPrete, 1990).

ANTECEDENT LITERATURE

Assimilation Theory

An early, if not the first comprehensive treatment of assimilation was published over 75 years ago by Park and Burgess (1921) in their classic textbook. Park and Burgess defined assimilation as "a process of interpenetration and fusion in which persons and groups . . . are incorporated with them into a common cultural life" (Park & Burgess, 1969, p. 360). In a later work, Park elaborated his thinking about assimilation as a process consisting of four distinct phases:

contact, competition, accommodation, assimilation (Park, 1950, p. 150). Park's model (1950) of assimilation was meant to describe a process in which culturally distinct groups came into contact and eventually fused into a unified cultural whole.

Knowing the context of Park's thinking helps to clarify the sources of this perspective. In particular, Park and his colleagues were situated at the University of Chicago and writing in the 1920s when the city of Chicago was teeming with recent European immigrants. The process of assimilation and Park's description of it seemed to capture well the experience of these European newcomers. It is difficult to underestimate the influence of Park's influence on thinking about assimilation. The Park and Burgess textbook in particular dominated the field for nearly 20 years (Hirschman, 1983). Since the publication of Park's ideas, sociologists have tended to think about assimilation as an evolutionary process taking place in more or less discrete periods or phases.

The next major milestone in assimilation theory appeared with the publication of Milton Gordon's book *Assimilation in American Life* (1964). Gordon's work went beyond past models of assimilation by articulating seven different types of assimilation: cultural (acculturation), structural, marital (amalgamation), identificational, attitude receptional (absence of discrimination), and civic (absence of value and power conflict). Gordon also posited that while these different types of assimilation were connected to one another, one did not necessarily follow from the other. He contended, for example, that while African-Americans had been culturally assimilated, they had not been given large-scale entrance into cliques, clubs, and institutions of the host society, i.e. structurally assimilated. For sociologists concerned with racially induced socioeconomic disparities, it is this type of assimilation that matters most.

In the years since the publication of Gordon's book, assimilation theorists have proposed a variety of schemes to describe the incorporation of ethnic minorities into the dominant society. These theories in one way or another attempted to display a growing awareness of the complexity of assimilation and to account for the fact that for some groups, especially blacks, assimilation was not a prospect for the foreseeable future. For example, Greeley (1974) suggested that assimilation is neither linear, nor unidirectional. Another revision of Gordon (1964) argued that the subprocesses of assimilation could be found working independently of one another (Yinger, 1985).

Over the years, assimilation theory has endured a hailstorm of criticism. Some of these criticisms were anticipated by Horace Kallen who in 1915, argued that it was not reasonable to expect immigrants to surrender their culture and identity as a condition for participating in American society. More recently, assimilation theory has been most vigorously challenged by the literature on ethnic resurgence. These critics counter that ethnicity plays a central role in the lives of even the most acculturated groups (Glazer & Moynihan, 1970). Even among groups once considered destined for extinction, such as American Indians, there has been a remarkable revival of ethnic awareness (Cornell, 1988; Nagel, 1996).

In response, others have argued that the content of modern ethnic culture, and that the role of ethnicity in modern America is sharply different from its earlier incarnations. For many groups in American society, ethnicity is largely a symbolic construction that has a relatively minor role in their lives (Gans, 1979). Alba (1986) for example, writes about the "twilight of ethnicity" among Americans of Italian descent.

Assimilation Research

There is a very large body of empirical research that deals with assimilation from a variety of perspectives. For convenience, this literature can be organized in terms of socioeconomic assimilation, segregation (especially with regard to schools and housing), intermarriage, and prejudice (Hirschman, 1983). Socioeconomic assimilation is a key indicator of what Gordon (1964) called structural assimilation. Editorial limits prevent a complete review of the assimilation literature but it is worthwhile to re-iterate what is known about socioeconomic assimilation in anticipation of the data presented below.

For European immigrants, it is not unfair to characterize their experience with assimilation as a "rags to riches" story, exemplifying American cultural beliefs about the reward for hard work and ingenuity. Indeed, most of these groups enjoyed considerable upward mobility throughout the 20th century (Greeley, 1978; Lieberson, 1980). This is not to say that groups such as Jews or Catholics were unaffected by bigotry and discrimination. On the contrary, there is a long tradition of anti-Semitism and anti-Catholic sentiments in this country (Baltzell, 1964). Nevertheless, the large socioeconomic differences that existed among these European ethnics in 1900 had virtually disappeared by the 1980s (Lieberson & Waters, 1988).

Compared with the immigrants form Europe, other groups such as American Indians, African-Americans, and non-European immigrants have not fared as well. The American Indian population is small, but from the standpoint of assimilation studies, they are very interesting. Unlike other groups, they were placed under intense pressure from the federal government to assimilate into American society. From the 1880s to the 1970s, the federal government instituted a variety of measures – policy directives and programs – designed to encourage and sometimes force assimilation (Fixico, 1988; Hoxie, 1984). Despite these efforts, many if not most American Indians remain outside the urban mainstream of the modern U.S.

economy. In fact, efforts to assimilate American Indians by helping them move to cities has done little to improve their social and economic status (Gundlach & Roberts, 1978; Snipp & Sandefur, 1989). In terms of education, occupation, and income, American Indians are about the same or below the attainments of African-Americans (Snipp, 1989).

Unlike American Indians, African-Americans never have been the targets of assimilationist measures, so it is perhaps less surprising that they possess a disproportionately small share of the nation's wealth. Compared with European immigrants, it is especially clear that they have not participated fully in the national economy. Comparing blacks with European immigrants at the turn of the century, Lieberson (1980) found that black migrants settling in northern cities had about the same amount of schooling as did European immigrants. He rules out differences in family structure and cultural orientations as reasons for why European immigrants and their descendants out paced African-Americans in upward mobility, especially gains in education. Instead, Lieberson (1980) argues that in the period between World War I and World War II, racial attitudes in the North hardened and discrimination against blacks intensified, at the same time discrimination against European ethnics declined.

Educational differentials as well as other key indicators of socioeconomic assimilation persist but in recent years, the gap separating African-Americans from other groups, especially whites, has declined noticeably. With respect to education, for example, the gap between blacks and whites began to narrow in the 1950s (Featherman & Hauser, 1978). By 1975, the percentage of black high school graduates attending college was about the same as the percentage of white high school graduates enrolled in college. Since 1975, the black-white differential in college attendance has increased significantly. Similarly, black-white differences in earnings declined in the 1970s but increased in the 1980s (Farley & Allen, 1987). Finally, in an important new study, Oliver and Shapiro show that the greatest inequality between blacks and whites is with respect to wealth holding – the assets of whites significantly exceed the assets of blacks with similar education and occupations.

Although Latinos, especially Mexicans, began moving to the United States in the early part of the 20th century, they are relatively recent immigrants. During the 1960s and 1970s they were the largest and fastest growing immigrant group. It is difficult to discuss intelligently the assimilation of Latinos because of the heterogeneity of this population, which includes Cubans, Mexicans, Puerto Ricans, and immigrants from central and South America. The diversity of immigrant experiences within these populations defies simple description. Even with a small group such as Cubans, there are substantial differences with respect to the circumstances of their immigration and the extent to which they have become assimilated (Portes & Bach, 1985).

Bean and Tienda (1987) argue persuasively that the similarities and differences among these various groups of Latinos can be traced to the circumstances of their immigration, especially migration and settlement patterns, and to their ensuing labor market experiences. Residential and occupational concentration are two conditions which have helped establish and maintain their Latino ethnicity and have made them especially resistant to assimilation (Bean & Tienda, 1987, p. 12). The extent to which Latinos are becoming residentially and occupationally assimilated will foretell their entry into the dominant culture.

Bean and Tienda (1987, p. 13) also note that "Cubans and other Latin American immigrants are seldom identified as a minority group, but Mexicans and Puerto Ricans usually are." This distinction is revealing because minority groups can be thought of as populations subject to prejudice and discrimination, while ethnic groups are cultural collectivities which are not stigmatized to the same extent (Vincent, 1974; see also Bean & Tienda, 1987, p. 14). Not surprisingly, examining the socioeconomic characteristics of these groups, one finds a hierarchy reflecting different degrees of structural assimilation among these groups. Predictably, Cubans and other Latin Americans have somewhat higher levels of educational and occupational attainments than Mexicans or Puerto Ricans. An equally unsurprising finding is that native born Latinos in general enjoy a higher socioeconomic status than the foreign born (Bean & Tienda, 1987). It has been suggested that Latinos are gradually becoming assimilated in American society more quickly than other groups, particularly African-Americans. This point is difficult to substantiate but between 1980 and 1990, the economic standing of Hispanics (as a group) improved steadily throughout this decade while the situation of African-Americans deteriorated significantly.

Although there is considerable diversity in the Latinos population, there are a few commonalties such as language. But compared to Latinos, the diversity in the so-called Asian and Pacific Islander population is spectacular and except for the continent of origin, there are few visible similarities. Among others, this group includes Hmong, Japanese, Cambodians, Asian Indians, Laotians, Native Hawaiians, Samoans, Maori, Vietnamese, and ethnic populations within larger groups such as Chinese. Many of these groups are exceedingly small in number, making it exceedingly difficult to obtain data about them. However, in the 1980s, Asian and Pacific Islanders were, as a whole, the fastest growing segment of the American population (O'Hare, 1992). This rapid increase came about because of significant change in U.S. immigration laws in 1965, and again in 1986 (Chiswick & Sullivan, 1995). Prior to these changes, especially before 1965, most immigration originated from Europe or Canada. Since then, the majority of immigrants have come from Mexico, Latin America, and Asia.

Of course, this most recent wave of Asian immigration is not the first. Chinese men were recruited for railroad construction in mid-nineteenth century. Japanese and Asian Indian laborers were recruited for farm work by western agricultural interest in the late nineteenth century (Barringer et al., 1955). However, this episode of immigration was curtailed in the early twentieth century with a series of measures adopted by the federal government first to limit Chinese immigrants, and later the Japanese.

Asians, especially those of Japanese and Chinese descent pose an anomaly for assimilationist thinking about racial and ethnic inequality. Particularly surprising is that despite a long history of discrimination and persecution, some Asian groups, notable Japanese and Chinese, have attained a higher level of social and economic well-being than the white population. Only part of this success can be linked to their high levels of schooling and concentration in urban areas (Hirschman & Wong, 1984). This has caused some observers to describe Asians as "model minorities" because their success is attributed to their hard work and ingenuity, and in some respects, they are an anomaly in assimilation theory.

Given the diversity of the Asian and Pacific Islander population, the evidence of their assimilation into American society is very mixed but not very surprising – especially with regard to their socioeconomic success equal to, or in some cases higher than the white population. Some of this success is due to the fact that these groups have been in American society longer than other groups. The same can be said about Asian Indians. Groups that have recently immigrated, especially those who were forced to flee in the aftermath of the Vietnam war typically are not as well-assimilated as other Asian groups as measured by their socioeconomic standing. Whether these disadvantages will persist into the future is an open question. Immigration is not a singular handicap insofar as other Asians, notably Filipinos, are not well-integrated into the economic mainstream (Barringer et al., 1993; Hirschman & Wong, 1984).

Finally, it should be noted in closing this discussion that for all of these groups – especially African-Americans, Latinos, American Indians, and Asians – there is compelling evidence to suggest that there is growing internal diversity with respect to their assimilation into the economy. This has been written about most extensively in regard to African-Americans. At the same time segments of the African American population have enjoyed greater opportunities and have attained a middle-class lifestyle, others have been alarmed by the creation of an apparent underclass destined for poverty and hardship (Landry, 1987; Wilson, 1987).

ANOTHER LOOK AT SOCIOECONOMIC ASSIMILATION

In light of the existing research dealing with assimilation it is useful to periodically review current data for the purpose of monitoring potentially important changes in the socioeconomic standing of racial and ethnic minorities. In the pages that follow, we will focus on the socioeconomic characteristics of whites, African-Americans, Latinos, Japanese, Chinese, Filipinos, and American Indians. The selection of these groups is partially dictated so it will be possible to take advantage of earlier work by Hirschman and Wong (1984). Specifically, our intent is to examine the current status of these populations as well as to note any significant changes that have happened to them over the years from 1970 to 1990.

Data and Methods

The data reported here have been obtained from the public use files for each decennial census since 1970. For relatively large populations such as African-Americans, there are a number of other data sources that we could have employed. However, for smaller populations such as American Indians, the decennial census is unmatched in terms of its sample size and the amount of data that it provides. Most data sources simply do not have enough cases to permit a detailed analysis.

Because we are focusing on indicators of economic well-being, namely socioeconomic status and personal earning, as well as changes in these indicators since 1970, we have imposed several restrictions on the data we will present. In particular, this analysis is limited to a sample of men between the ages of 25–64, we readily concede that this limits the generalizability of our findings but we also believe that there are several good reasons for restricting our data to this subset of the population.

First, gender differences and their interaction with respect to assimilation over time are sufficiently complex to require a separate study of their own. Women's roles and particularly their participation in the economy have changed dramatically since 1970. This, confounded with changes in racial and ethnic assimilation clearly deserves a separate study that exceeds the scope of this work. Second, the age range we use id dictated by the fact that we are most interested in persons who are economically active. Persons younger than 25 years are likely still in school or newly entering the workforce. Persons over 64 years are in most cases about to enter retirement. Third, our analysis of earnings is based on men who reported positive incomes and who responded positively to questions about weeks worked in the prior year and hours worked in the week preceding the census. Finally, these restrictions facilitate comparisons with other studies of assimilation, Hirschman and Wong (1984) in particular. Although these restrictions limit our results to a select segment of the population, other work leads us to believe that our estimates of interethnic differences will be conservative, and will thus lend confidence to any conclusions we might reach (Hirschman & Wong, 1984). In particularly, the interethnic differences we report below would in all likelihood be larger if we included persons who, at the time the census was taken, were not active members of the workforce.

Our estimates of interethnic differences are net of other factors known to affect occupational status and earnings. That is, we have estimated models using OLS regression that take into account a number of factors including age, education, residence, and recency of immigration. A complete list and description of these variables is presented in Table 1. However, the coding of these variables deserves further discussion.

Table 1. Definition and Measurement of Demographic and Socioeconomic Characteristics.

Measure	Definition
Ethnicity	Self-reported racial identification. Hispanic is self-identified in a separate item in the 1980 and 1990 censuses, and based on a composite measure in the 1970 census. Whites are non-Hispanic whites; black Hispanics are coded black.
Age	Age at last birthday: 25–34; 35–44; 45–64.
Birthplace/length of U.S. residence	A composite measure based on country or birth and place of residence five years ago. Coded as: native born; foreign born, in U.S. 5 years ago; foreign born not in U.S. 5 years ago.
Place of residence	State or region of residence April 1, 1970, 1980, 1990. Coded for: California; New York; Hawaii; South, metropolitan area; South, non-metropolitan area; rest of U.S. metropolitan; rest of U.S. non-metropolitan.
Years of schooling	Number of years of formal schooling completed. Coded as: 0–8; 9–11; 12; 13–15; 16 or more.
Occupational SEI	Duncan's Socioeconomic Index, updated for 1980 and 1990.
Sector	Composite variable based on class of worker and industry classification. Coded as: self-employed; government employed; retail trade, not self employed; other periphery, not self-employed, not government employed, not in retail trade; other core, not self-employed or government employed.
Weeks worked last year	Number of weeks worked in 1969, 1979, and 1989. Coded as: less than 50 weeks, 50 or more weeks.
Hours worked last week	Hours worked during the week prior to the census (April 1). Coded as: less than 40, 40, more than 40.
Earnings	Total income received from wages and salaries, self-employment income from farm and non-farm sources. Earners with zero or negative incomes were excluded from each sample.

Our dependent variables, occupational status and earnings, are relatively straightforward. The measure of occupational status is the Duncan Socioeconomic Index which has been updated to reflect changes in the occupational structure, especially in the 1990 data (see Hauser & Warren, 1995). Earnings are defined as wage and salary income as well as any income received from farm or non-farm self-employment. Estimates of income have been adjusted for inflation to constant 1990 dollars.

Most of our independent variables are conventionally scaled, such as education which is measured in years of completed schooling. However, there are several others which merit clarification. The ethnic categories we use are based on the self-reported responses to the so-called "race question" on the census form. To these categories, we have added an additional category for "Hispanic." In 1970, this item was a composite based on Spanish surname, Puerto Rican birthplace or parentage, and Spanish Language. For 1980 and 1990, Hispanic was a category of self-identification and separate from the race question. Needless to say, this vitiates comparisons of 1970 with later years but does not render them altogether meaningless (see Bean & Tienda, 1987). Note also that the white category includes only persons who did *not* report positively to the Hispanic identifiers, i.e. they are non-Hispanic whites.

We have combined the variables for place of birth and "place of residence 5 years ago" to construct a measure of immigration. Foreign born persons who were not living in the United States five years prior to the decennial census (1965, 1975, and 1985) are presumed to be recent immigrants. Place of residence is measured in terms of region, state, and metropolitan location. This set of contrasts is based on the analysis of Hirschman and Wong (1984) who attempted to identify locations approximate to areas and labor markets associated with ethnic economic success, particularly for Asians. Finally, the economic sector variable is one which has become commonplace in studies of economic inequality. Again, the coding of this variable follow the scheme employed by Hirschman and Wong (1984) which attempts to incorporate ethnic concentrations in self-employment and retail trade with a more conventional classification for core and periphery industries. The measures for residence and sector are admittedly crude approximations but we contend they are the best that can be constructed from existing data.

There are two other matters which deserve comment before we present our findings. One is the issue of heterogeneity in our measure of Hispanic. We are well aware of the diverse composition of this population, as we have noted earlier. However, small numbers for groups such as Cubans and Puerto Ricans as well as the problems of disaggregating these groups in the 1970 data makes it virtually impossible to handle each of these groups separately. The alternative – would be to exclude the Hispanic population from this analysis, and this seemed an even less desirable strategy. An equally, if not thornier problem is the matter of compositional changes due to immigration. This problem is perhaps most serious for Hispanics but not exclusive to them. One solution would be to limit this analysis to native born ethnic populations but in many respects, we believe this an overly restrictive measure for recency of immigration and this, along with an interpretation of the findings mindful of immigration induced change, should be adequate to prevent us from overstating our conclusions.

Ethnic Differences in Socioeconomic Assimilation

The numbers of whites and blacks in the decennial census public use files are very large. For this reason, we have selected a representative sample of these groups for our analysis. In contrast, other groups such as American Indians or Filipinos are relatively small in number. Consequently, for other groups besides blacks and whites we have included in our analysis all of the respondents available in the public use files. The numbers in Table 2 show the sample sizes for each group in each year of our data.

The statistics in Table 2 also show the means and standard decisions for the dependent variables, SEI and earnings, for each group in the years 1970, 1980 and 1990. In nearly all respects, the results presented in this table are fully consistent with findings of other studies. The SEI and earnings of Japanese workers, and the SEI of the Chinese exceed those of the white sample. On the other hand, the earnings and occupational status of Hispanics, Blacks, American Indians, and Filipinos to a lesser degree, are well below the earnings and occupational status of Whites.

Over time, these results are remarkably stable insofar as the socioeconomic hierarchy of these groups is virtually unchanged between 1970 and 1990. In terms of occupational status, all of these groups experienced a small amount of upgrading between 1970 and 1990. Most of these gains were in the 4.0–6.0 range except for Japanese workers who gained 9.0 points. Likewise, there was little change in the earnings hierarchy but the results are somewhat less stable than for occupational status. In constant dollars, white earnings have been stagnant since 1970. Blacks, on the other hand, enjoyed modest gains in each decade since 1970; the same can be said for Japanese workers. The other groups experienced modest gains in one decade and declines in another. The source of this instability is not easily accounted for but it might very likely be the result of compositional differences due to immigration, changing racial self-identification (in the case of American Indians), or reporting errors in the earnings data.

Table 2. Sample Sizes, Means, and Standard Deviations[a] for SEI and Earnings[b] by Year and Ethnicity, 1970–1990.

Ethnicity	1970			1980			1990		
	N	SEI	Earnings	*N*	SEI	Earnings	*N*	SEI	Earnings
White	3139	36.5 (20)	35125 (23726)	3063	39.4 (20.5)	35278 (23042)	3013	40.1 (20.9)	35523 (31477)
Black	2726	24 (13.9)	20436 (12500)	2540	28.3 (16.4)	23061 (14898)	2482	30.2 (17.7)	23172 (18127)
American Indian	939	27.4 (16.4)	22159 (16054)	2272	31.9 (18.3)	25925 (19334)	2585	31.2 (17.6)	21511 (17293)
Japanese	1130	38.4 (21.9)	35429 (22189)	1647	44.7 (21.3)	37205 (23245)	3037	47.4 (20.9)	42750 (33909)
Chinese	911	43 (24.1)	32243 (22959)	1859	46.7 (23.5)	31424 (24307)	2977	46.1 (23.2)	32017 (30035)
Filipino	611	33.7 (23.2)	25044 (17473)	1379	39.1 (22.7)	30463 (23807)	3029	38.2 (21.2)	28945 (26332)
Hispanic	1394	28.6 (17.5)	27304 (19561)	2231	29.7 (17.5)	23797 (17163)	2841	32.7 (18.8)	25760 (23803)

Sources: 1970, 1980, 1990 Public-Use files, U.S. Bureau of the Census.
[a] Standard deviations in parentheses.
[b] Earnings are for the years 1969, 1979, and 1989, computed in constant 1989 dollars.

Moving on to the independent variables, the numbers in Table 3 show the marginal percentage distributions for these measures. It is important to keep in mind that these percentage distributions do not represent profiles of the population at-large. Readers should remember first the restrictions imposed on the sample, namely that these data represent employed men in the civilian labor force. Furthermore, smaller sub-samples for black and white men were extracted from the master files to facilitate data processing, while all respondents were used for smaller groups such as Filipinos and American Indians. It is not true, for example, that white men comprised 15.1% of the workforce in 1990, but it is true that they represent about 15% of the respondents in our data.

Nonetheless, it is possible to discern several broad social trends that are important in the context of the economic assimilation of ethnic minorities in the United States. First, the age distributions of these workers reflect the aging of the "baby-boom" generation workers at the same that they indicate an increasingly youthful workforce. The percentage of men aged 35–44 increased by 16% between 1970 and 1990, while in the same period, the percentage of younger men ages 25–34 increased by 18%. An influx of younger immigrant workers was no doubt an agent in these changing age distributions.

The percentages in Table 3 also show that the proportion of immigrant workers in these data more than doubled, from 16.9 to 35.5%, in the decades between 1970 and 1990. During these decades, Hispanics and Asians accounted for a large part of the immigrant population. And not surprisingly, the percentage of workers living in California, where many Hispanic and Asian immigrants settled, also increased sharply between 1970 and 1990 from 19 to 29%, a gain of 53%. Other patterns in the residential distribution of workers, immigrant or native, are not as pronounced.

A growing concentration of youthful workers, as older, less educated workers leave the labor force, also accounts for rising levels of education between 1970 and 1990. The percentage of workers with less than a high school education declined from 47.2% in 1970 to 14.9% in 1990. Likewise, the percentage of persons with four years or more of post-secondary education doubled, from 14.8% in 1970 to 30.6 in 1990.

In terms of labor force participation, the data in Table 3 reveal fewer dramatic changes over time among these groups of male workers. The kinds of work these men engaged in changed slightly, most likely because of the re-structuring of the U.S. economy in the 1970s and 1980s. There was a somewhat smaller percentage of men working in the so-called core industries in 1990 than in 1970. Likewise, the percentage of men employed in retail and other periphery industries increased over these decades. With regard to the amount of time these men worked, there are even fewer clear patterns. For reasons that are not altogether clear, the only

Table 3. Percentage Distributions for Selected Determinants of Occupational Status and Personal Earnings.

Variable	1970	1980	1990
Ethnicity			
Black	24.9	16.2	12.6
Chinese	8.3	14.1	14.7
Filipino	5.6	10.9	15.0
Hispanic	12.7	13.8	14.1
American Indian	8.6	14.1	13.2
Japanese	10.3	12.6	15.2
White	29.6	18.3	15.1
Age (years)			
25–34	30.0	39.3	35.4
35–44	28.3	27.6	32.7
45–64	41.7	33.1	31.9
Immigration status[a]			
Greater than 5 yrs	12.8	22.3	27.0
Less than 5 years	4.1	8.9	8.5
Native	83.1	68.8	64.5
Residence			
California	19.2	26.3	29.2
Hawaii	6.9	7.3	7.4
New York	9.4	8.5	7.0
Other U.S. – metro[b]	24.5	24.4	13.4
Other U.S. – non-metro	13.4	8.5	17.7
South – metro	13.8	17.1	13.1
South – non-metro	12.8	7.9	12.2
Education[c] (in years)			
0–8	28.2	14.4	7.3
9–11	19.0	13.5	7.6
12	27.2	27.6	27.8
13–15	10.8	19.7	26.8
16 and over	14.8	24.8	30.6
Employment sector[d]			
Government	17.9	18.1	16.9
Core	42.8	39.5	38.1
Periphery	19.3	20.0	22.0
Retail	9.4	11.6	11.8
Self-employed	10.6	10.8	11.2
Weeks worked			
50 or more	74.1	67.4	71.0
Less than 50	25.9	32.6	29.0

Table 3. *(Continued)*

Variable	1970	1980	1990
Hours worked			
40	49.4	50.9	48.2
Less than 40	19.0	22.0	19.0
More than 40	31.6	27.1	32.8

Source: U.S. Bureau of the Census, 1970, 1980, and 1990 Public-Use Microdata samples.
[a] See text for definition of Immigration Status.
[b] Other U.S. is defined as all states except California, Hawaii, New York, and states in the South. Metro areas are defined as Standard Metropolitan Statistical Areas in 1970 and 1980, and are Statistical Metropolitan Areas in 1990. The Census Bureau provides extensive details about how these areas are defined.
[c] Defined in terms of years of completed schooling.
[d] See text for definition of Employment Sector.

distinct pattern is that hours and weeks worked were relatively low in 1980, but more or less comparable in 1970 and 1990.

Models of Ethnic Stratification, 1970–1990

Although the preceding data reveal much about the relative status of ethnic minorities in American society, they are nonetheless gross estimates. In particular, they do not take into account the fact that some of the observed differentials across these groups are due to differences in factors such as education, place of residence, or labor force participation and not strictly the result of assimilation per se. For example, some groups may have higher incomes because they are more heavily concentrated in higher paying urban labor markets and not because they are more assimilated into the economy.

To explore how other factors in addition to socioeconomic assimilation a may affect the economic well-being of ethnic minorities, we have estimated a series of OLS regression equations. These equations allow us to accomplish two tasks. First, they allow us to estimate the net effects of ethnicity on occupational attainment, holding constant group differences in education, residence, and sector employment. Likewise, we have estimated the net effects of ethnicity on earnings, again holding constant group differences in places of residence, education, sector of employment, and occupation and labor force participation in addition. The second task of this analysis is to estimate the relative importance of these factors vis-a-vis their direct and indirect effects on occupational and earnings attainment.

In the tables that follow, the effects of ethnicity are expressed as deviations from the white level. The gross effects are the level of ethnic inequality between whites and ethnic minorities as measured in Duncan SEI points in constant 1989 dollar. The other coefficients are a decomposition which allows us to examine the direct and indirect effects of ethnicity on attainment. Total effects represent the sum of direct and indirect effects net of age and birthplace. Direct effects are the effects of ethnicity on occupational and earnings attainment, net of the aforementioned factors such as residence, education, and industrial sector. Indirect effects are the effects of ethnicity on occupational and earnings attainment as mediated through the variables held constant to obtain the direct effects. The indirect effects are obtained by successive regression equations and subtracting the ethnic regression coefficients with the intervening variable from the ethnic coefficients in the preceding equation without the intervening variable (Hauser & Alwin, 1975).

Occupational Attainment

The coefficients in Table 4 show the gross, total, direct, and indirect effects of ethnicity on occupational attainment for the years of 1970, 1980, and 1990. The gross effects show that on average, the statuses of occupations held by blacks, American Indians, and Hispanics are between 7 and 13 points lower than the occupational statuses of whites in each decade between 1970 and 1990.

Holding constant age and birthplace has little effect on these estimates insofar as the "total" effects and virtually identical to the unadjusted gross effects. However, decomposing the total effects into their constituent direct and indirect components is revealing. This shows that, for blacks, American Indians, and Hispanics about half of their lower occupational status (between 4 and 7 points) is due to deficits in human capital, i.e. schooling. Place of residence and employment sector has little explanatory power in accounting for the deficit in occupational status that exists for these groups relative to white male workers.

In addition to schooling, the remaining deficit in occupational status for blacks, American Indians, and Hispanics relative to whites is manifest in the "direct effects" of ethnicity. These effects are net of residence, schooling, employment sector, and represent the residual of unmeasured variables not in the equation, such as family background, as well as the outcomes of racial discrimination. Quite clearly, educational deficits and a variety of other unobserved factors such as family background and racial discrimination, account for the lower occupational statuses of employed black, Hispanic, and American Indian men relative to employed white men.

Compared to employed black, Hispanic, and American Indian men, the experiences of Asian men are somewhat different. Japanese and Chinese men in particular enjoy a modestly higher gross level of occupational status than white men, and this lead has increased slightly since 1970. In contrast, Filipino men have

Table 4. Effects of Ethnicity on Occupational Attainment of Men Aged 25–64, in the Labor Force 1970, 1980, 1990.

Effects/Year	Black	American Indian	Japanese	Chinese	Filipino	Hispanic
Gross						
1970	−13	−9	2	6	−3	−8
1980	−11	−8	5	7	0	−10
1990	−10	−9	7	6	−2	−7
Total						
1970	−13	−10	2	5	−4	−9
1980	−11	−8	6	8	0	−10
1990	−10	−9	8	7	−1	−7
Indirect via						
Residence						
1970	0	−1	−2	−1	−3	−1
1980	1	−1	−3	−1	−2	−1
1990	0	−1	−6	−4	−5	−1
Schooling						
1970	−7	−6	6	5	2	−6
1980	−6	−4	7	6	5	−7
1990	−5	−5	10	7	7	−5
Sector						
1970	1	0	0	0	0	0
1980	0	0	0	0	0	0
1990	0	0	0	0	0	0
Direct						
1970	−7	−3	−2	1	−3	−2
1980	−6	−3	2	3	−3	−2
1990	−5	−3	4	4	−3	−1

a slightly lower level of occupational status. Interestingly, unlike other minorities, Asian men residing in areas with large populations of co-ethnics, namely California and Hawaii, have occupational statuses which are slightly lower than Asian men living elsewhere. In the absence of this liability, the occupational statuses of Japanese and Chinese men in California and Hawaii would be an average of 1 to 6 points higher.

Looking at the direct and indirect effects, it becomes clear that higher levels of education play a central role in elevating the occupational status of Asian men. Furthermore, the effects of education are larger in 1990 than in 1970. The residual direct effects of ethnicity are also different for Chinese and Japanese men. For these men, their ethnic heritage exerts a small but positive influence on their occupational status; suggesting that unmeasured characteristics such as family background may provide a modest advantage for them in the labor force. However, Filipino men do not appear to enjoy the same advantages.

In sum, it is clear that Japanese and Chinese men have advantages in the labor market that men from other minority groups do not possess. Primarily, these advantages accrue from higher levels on educational attainment which are reflected in occupational statuses that exceed even those of white men. And while Japanese and Chinese men are adversely affected by their place of residence, unlike other minorities, their ethnic background does not exert a direct negative influence on their occupational standing, as it does for blacks, Filipinos, Hispanics, and American Indians. It is also worth noting parenthetically that employment sector has no effect whatsoever on the occupational success of any of these groups.

Earnings Attainment

Table 5 shows the estimates from multivariate models of earnings attainment in 1969, 1979, and 1989. Recall that earnings are based on labor performed in the year preceding the census (e.g. 1969 for the 1970 census), and that these estimates are presented in constant 1989 dollars. The results in Table 5 show a striking shortfall in the gross earnings of all groups relative to whites, with the exception of Japanese men who have enjoyed earnings that have steadily exceeded those of white men.

The patterns of change over time are complex, however. The earnings of American Indians and blacks lag farthest behind those of white workers, about $13,000 to $15,000 in 1969. This gap diminished for both groups during the 1970s but was virtually unchanged for blacks during the 1980s, a time when earnings were stagnant for most of the nation (Levy, 1995), and increased for American Indians. Hispanic earnings also lagged far behind white earnings (between $8,000 and $11,000 over the two decades) but this gap grew during the 1970s, perhaps due to an influx of immigrants, then declined in the 1980s. Employed Chinese men also displayed a similar pattern of increase and decline, but their earnings lagged behind white men by less than $4,000 over the twenty year period.

In terms of the factors that mediate the earnings gap between minority and white workers, the results in this analysis parallel but do not exactly replicate the results for occupational status. For employed non-Asian men, place of residence has a negligible effect. This was also true for Asian men in the 1970s, but during the 1980s, place of residence was associated with lower earnings for these workers, by as much as $3,900 for Japanese men in 1989. At the same time, all groups of Asian men profited from higher levels of schooling. In 1989, for example, education elevated the earnings of Filipino men by $4,700 and the earnings of Japanese men by $6,800.

Table 5. Effects of Ethnicity on the Earnings[a] Attainment of Men Aged 25–64, in the Labor Force 1969, 1979, 1989.

Effects/Year	Black	American Indian	Japanese	Chinese	Filipino	Hispanic
Gross						
1969	−14691	−12965	301	−2885	−10083	−7820
1979	−12216	−9351	1930	−3852	−4815	−11481
1989	−12351	−14012	7227	−3507	−6578	−9763
Total						
1969	−14664	−12772	547	−1858	−8232	−7398
1979	−11934	−9017	2524	−2582	−3339	−10665
1989	−12062	−13326	7634	−3502	−7138	−9329
Indirect via						
Residence						
1969	−115	−1307	2466	1682	2017	−1307
1979	67	−488	−266	12	5	174
1989	−465	−843	−3911	−2451	−3398	−561
Schooling						
1969	−4084	−3476	3138	2689	166	−4165
1979	−3360	−2571	3758	3020	2726	−4516
1989	−3276	−3777	6822	5092	4742	−3518
Sector						
1969	−1000	−1118	456	220	−1118	−730
1979	−1081	−849	−145	−921	−541	−629
1989	−1355	−912	−327	−1023	−1178	−539
Occupation						
1969	−2023	−1030	−571	348	−959	−763
1979	−1638	−835	601	953	−798	−673
1989	−1635	−992	1405	1361	−877	−421
Weeks and hours worked						
1969	−980	−1118	20	−318	−652	−763
1979	−1353	−1344	−46	−514	−1071	−721
1989	−1618	−2011	162	−594	−1114	−747

[a] In constant 1989 dollars.

By the same token, deficits in human capital explain a substantial amount, more than any other single factor, the lower earnings of black, Hispanic, and American Indian workers relative to white workers. In fact, the mediating effect of education on earnings was at least twice as large as the effects of employment sector, occupation, or labor force participation (weeks and hours worked) among black men, and even larger for Hispanics and American Indians. For instance, holding constant age and birthplace, employed Hispanic men earned $9,300 less than white men in 1989 – education accounted for $3,500 of the total earnings gap, and second most important were lower numbers of weeks and hours worked, which accounted for $750 of the difference.

Overall, the effects of mediating factors such as employment sector, occupation, and weeks and hours worked pale in comparison to the influence of schooling in accounting for the differences in earnings between white male workers and ethnic minorities. They are smallest and most erratic among Asian and Hispanic workers, though somewhat larger and persistently negative for blacks and American Indians.

Turning to the estimates of the "direct" effects, it is abundantly clear that ethnicity and whatever other unmeasured variables are embedded in these parameters have a pronounced influence on earnings and are mostly negative. For blacks, Hispanics, Filipinos, and American Indians, the adverse effects of their ethnic heritage on their earnings exceeds even the disadvantages accruing from their lack of education. The liability associated with ethnic background has declined since 1969, perhaps reflecting diminishing levels of employment discrimination over the past twenty years. However, the "cost" associated with being anything but Japanese ranged from $3,500 to $5,900 in 1989. And even for Japanese men, the advantages conferred by their ethnicity was $3,500 and most likely reflected unmeasured benefits from their family background.

In closing, the results presented here leave no doubt that groups such as American Indians, blacks, Hispanics, and Filipinos earn substantially less than white workers. The earnings gaps for these workers relative to employed white men can be partially explained by their occupation, sector, and labor force participation. However, like the occupational status, education represents the largest mediating factor accounting for the earnings deficits between white and minority men. Yet, even after these factors are considered, a very large residual remains to be explained. This residual, of course, includes factors omitted from our models, but it is troubling because its sheer magnitude suggests that even if other factors such as family background were included, it is doubtful that they would be adequate to fully explain the earnings deficit directly associated with the ethnic background of minority workers. This leaves few other conclusions other than the fact that ethnic background, most likely through discriminatory practices by employers, has a significant direct effect on the earnings of minority workers, and on the life chances for them and their families.

CONCLUDING COMMENTS

It is abundantly clear that American society has not realized the egalitarian promises so deeply embedded within the prevailing ideologies of the past

100 years. The ideals of a society in which rewards are distributed solely on the basis of merit, without regard to ascriptive characteristics such as race and ethnicity are simply just that: unfulfilled promises. The findings reported in this paper leave little doubt about the fact that racial and ethnic minority groups which have a long history of economic hardships in this nation continue to experience significant disadvantages in the labor market.

These findings are especially significant because they represent two decades of an important social experiment initiated by the federal government in the late 1960s. Responding to a period of protest, a variety of programs and policies were instituted for the purpose of addressing persistent socioeconomic disadvantages which affect ethnic and racial minorities in the schools and the labor market. These measures have come to be known as "Affirmative Action."

At this moment, these policies are as controversial today as the day they were introduced. Perhaps even more so today, as politicians and assorted interest groups mobilize to abolish Affirmative Action programs. Proposition 209 passed in 1996 by California voters explicitly forbids any kind of preferential considerations for racial and ethnic minorities. The arguments against Affirmative Action are based on a widespread perception that racial and ethnic minorities enjoy an unfair and disproportionate advantage in the competition for education and employment. There is an ancillary idea that these programs are not only unfair, they are no longer necessary for remedying past injustices; indeed they are last hurdles to be overcome in making America a truly colorblind society. The data reported in this paper raise serious doubts about the veracity of these claims.

Quite to the contrary, the analyses reported here show very clearly that after 20 years, America is still a very long way from the ideal of a colorblind society. In fact, these results render a complicated picture of racial and ethnic inequality in American society. Focusing on occupational status as a measure of inequality, the findings are varied and not terribly dramatic. Black, American Indian, and Hispanic men lag behind white men by about 10 SEI points. Asian American men enjoy about the same occupational status as white men.

On the other hand, earnings inequality is a much more serious problem for racial and ethnic minorities in America. Black, Hispanic, and American Indian working men have earnings that are about $10,000 less than the earnings of employed white men. Furthermore, there are few signs of progress in the period under study; the earnings gap is about as large in 1990 as it was in 1970. Chinese and Filipino men also have lower earnings than white men, though the gap is smaller than for Black, Hispanic, and American Indian men. Japanese men are the only workers to have earnings higher than white men.

Three hypotheses can be advanced to explain the persistence of social and economic inequalities among groups such as Blacks, Hispanics, and American

Indians. One is the persistence of human capital deficits in these groups. Blacks, American Indians, and Hispanics all have less schooling than Whites, and less than the schooling of Asian Americans. The results of this study suggest that Black, Hispanic and American Indian men would experience only modest educational disadvantages if they possessed the same level of education as White men. For example, there would still be a 5 SEI point deficit for black men, but this is half of their current handicap. In contrast, Asian American men have higher levels of occupational attainment simply by virtue of their higher levels of schooling. Earnings inequality is a more serious problem, and less amenable to remedy by additional years of schooling. More education among the most disadvantaged groups of workers would close the earnings gap but only by about 25–33%. Chinese and Filipino men are better educated but still have lower earnings than white men.

A second hypothesis attributes these inequalities by to the structure of local labor markets, regional economic conditions, and macro-economic fluctuations. There is some evidence here to suggest that at least Asian American men are disadvantaged by their geographic concentrations. However, this handicap is offset somewhat by their higher levels of schooling. For other disadvantaged groups, factors such as industrial sector and residence have a much smaller if not altogether negligible impact.

A third possible explanation focuses on the inequality which cannot be accounted for by the differences among these groups, education for example. That is, after all of the differences in worker qualifications are taken into account, there is a substantial amount of occupational and earnings inequality that cannot be explained. This residual no doubt includes a large number of unmeasured variables absent in our data. One such variable that is virtually impossible to measure is employer discrimination. The persistence of substantial inequalities across racial and ethnic groups points to discriminatory practices in the labor market. To be sure, such practices are virtually impossible to detect. However, in slack labor markets where employers have opportunities to select and reward workers, it is likely they will favor those who most closely fit employer "tastes" in worker appearance and behavior.

In closing, it also should be noted that the empirical results presented in this research are not well-suited to address a large number of hypotheses that might explain the persistence of the occupational and earnings inequalities. In particular, we have no information about the social networks and informal ties – the opportunity structures – that broker the careers of these workers. Likewise, that we have no information about the family history of these workers, no information about family background for example, is an especially significant omission. We hope that in the future, the data needed to address neo-classical theories of inequality for racial and ethnic minorities will become available. Without such information, social scientists will be left to puzzle and merely speculate about why racial and ethnic inequality remains an enduring feature of American society.

ACKNOWLEDGMENTS

This paper was prepared for an edited volume honoring the retirement of Archibald O. Haller, Professor Emeritus in the Department of Rural Sociology at the University of Wisconsin-Madison. The authors wish to acknowledge the financial support provided to the senior author by the Vilas Trust at the University of Wisconsin-Madison which was used for the processing and analysis of the data presented herein. Some of the data presented in this paper also appear in another report presented at the conference "American Society: Diversity and Consensus, A Cornell Symposium Honoring Robin Williams, Jr." held October 20–21, 1996, Ithaca, NY.

REFERENCES

Alba, R. (1986). The twilight of ethnicity among Americans of European ancestry: The case of Italians. *Ethnic and Racial Studies*, *8*(1), 134–158.

Baltzell, E. D. (1964). *The protestant establishment: Aristocracy and caste in America*. New York: Vintage Books.

Barringer, H., Gardner, R. W., & Levin, M. (1993). *Asians and Pacific Islanders in the United States*. New York: Russell Sage.

Bean, F., & Tienda, M. (1987). *The Hispanic population of the United States*. New York: Russell Sage.

Chiswick, B., & Sullivan, T. (1995). The new immigrants. In: R. Farley (Ed.), *State of the Union: America in the 1990s* (Vol. 2, pp. 211–270). New York: Russell Sage.

Cornell, S. (1988). *The return of the native*. New York: Oxford.

Farley, R., & Allen, W. R. (1987). *The color line and the quality of life in America*. New York: Russell Sage.

Featherman, D., & Hauser, R. (1978). *Opportunity and change*. New York: Academic Press.

Glazer, N., & Moynihan, D. P. (1970). *Beyond the melting pot* (2nd ed.). Cambridge, MA: MIT Press.

Gordon, M. (1964). *Assimilation in American life*. New York: Oxford University Press.

Greeley, A. (1974). *Ethnicity in the United States: A preliminary reconnaissance*. New York: Wiley.

Greeley, A. (1978). *Ethnicity, denomination, and inequality*. Beverly Hills: Sage.

Grusky, D., & DiPrete, T. (1990). Recent trends in the process of stratification. *Demography*, *27*, 617–637.

Gundlach, J., & Roberts, A. (1978). Native American Indian migration and relocation: Success or failure. *Pacific Sociological Review*, *12*, 117–128.

Hirschman, C. (1983). The melting pot reconsidered. *Annual Review of Sociology*, *9*, 397–423.

Hirschman, C., & Wong, M. G. (1984). Socioeconomic gains of Asian Americans, Blacks and Hispanics: 1960–1976. *American Journal of Sociology*, *90*(3), 584–607.

Landry, B. (1987). *The new black middle class*. Berkeley: University of California Press.

Levy, F. (1995). Incomes and income inequality. In: R. Farley (Ed.), *State of the Union. America in the 1990s* (Vol. 1, pp. 1–58). New York: Russell Sage.

Lieberson, S. (1980). *A piece of the pie: Black and white immigrants since 1880*. Berkeley: University of California Press.

Lieberson, S., & Waters, M. (1988). *From many strands: Ethnic and racial groups in contemporary America*. New York: Russell Sage.

Lenski, G. (1966). *Power and privilege: A theory of social stratification*. New York: McGraw-Hill.

Nagel, J. (1996). *American Indian ethnic renewal*. New York: Oxford.

Park, R. (1950). *Race and culture*. Glencoe: Free Press.

Park, R., & Burgess, E. (1921). *Introduction to the science of sociology*. Chicago: University of Chicago Press.

Snipp, C. M. (1989). *American Indians: The first of the this land*. New York: Russell Sage.

Snipp, C. M., & Sandefur, G. (1989). Earnings of American Indians and Alaska natives: The effects of residence and migration. *Social Forces*, *66*, 994–1008.

Wilson, W. J. (1987). *The truly disadvantaged, the inner city, the underclass, and public policy*. Chicago: University of Chicago Press.

CHANGES IN THE STRUCTURE OF STATUS SYSTEMS: EMPLOYMENT SHIFTS IN THE WAKE OF DEINDUSTRIALIZATION

William J. Haller

ABSTRACT

This paper reviews, adapts, and applies the general conceptual framework for analyzing changes in status systems developed by Haller (1970) to socioeconomic change in older industrial regions using deindustrialization in the Pittsburgh area during the early 1980s as an example. Although the general framework was formulated to be applicable to any human society it requires slight modification to make it useful for analyzing social structural change in societies with modern institutional structures. Its application to assessing some of the contours and consequences of socioeconomic change resulting from deindustrialization in the Pittsburgh region qualifies the observation that shifts in the social structures produced by modern market economies occur at a "glacial" rate. Social structural shifts may also occur suddenly, even in the absence of political forces such as are typically associated with rapid social structural change. The consequences of such shifts can be devastating for communities situated in the wrong place at the wrong time, but the economic

The Shape of Social Inequality: Stratification and Ethnicity in Comparative Perspective
Research in Social Stratification and Mobility, Volume 22, 119–147
© 2005 Published by Elsevier Ltd.
ISSN: 0276-5624/doi:10.1016/S0276-5624(04)22005-4

and human costs of such social structural shifts are not inevitable. In the case of deindustrialization in the Pittsburgh region, these costs could have been mitigated with greater attention to early warning signals in the market and better communication and coordination between private and public sectors and labor to anticipate and plan for a smoother regional transition.

INTRODUCTION

"Changes in the Structure of Status Systems," Haller's (1970) presidential address to the Rural Sociological Society, outlined the general conceptual issues for measuring social change in terms of wealth, power and prestige, the universal content dimensions of status from classical sociology and contemporary stratification research. Nearly 30 years later I focused on some of the major social structural shifts, and their consequences, that accompanied the restructuring of older industrial regions the midwestern and northeastern United States. This research used the Pittsburgh region, with the closure of its large integrated steel plants, as an example. Specifically, census tract data permitted a test of the relationship between the contraction of manufacturing employment and the spread of poverty among the region's communities to determine whether deindustrialization may be a cause of growth in the urban underclass (as posited, for example, by Massey & Denton, 1993; Wacquant & Wilson, 1993; Wilson, 1990, 1996). The idea that it is has been in the literature for many years, but mainly as an article of faith. Empirical corroboration has been quite another matter. Because a generalized conceptual framework for analyzing changes in the structure of status systems should be universally applicable, this chapter provides: (1) a summary of Haller (1970) on how to conceptualize and measure the various dimensions of status systems; (2) an adaptation of the generalized framework tailored for stratification research on societies with modern institutional structures; and (3) use of this revised framework to help organize and interpret my findings on the ramifications of industrial restructuring for the status system of the Pittsburgh region during the 1970s and 1980s. Haller (1970) clarifies the limits faced by researchers seeking to measure and assess all the important dimensions of a status system. Likewise, the research presented and discussed below is limited only to certain aspects of change in the socioeconomic structure of a single metropolitan region during a specific period. Nevertheless, within these limits it is still possible to assess change in a status system and some of the broader implications for its population.

GENERALIZED CONTENT AND STRUCTURE DIMENSIONS OF STRATIFICATION

Haller (1970) rooted his framework for analyzing structural changes in status systems in the Weberian revision of the basic Marxian model, emphasizing "relationships among social units whose incumbents are unequal in wealth, power, or prestige . . . [variables that] constitute the minimum set of hierarchical inequalities which apparently discriminate among all peoples." He described these as *content* dimensions of status, and noted that they are used both by scholars who view "the constancy of complex social systems" as the fundamental problem requiring sociological explanation in addition to those concerned with understanding change. Among the former, wealth and prestige are emphasized; among the latter, "power is that by which some control others for their own benefit; wealth is the main benefit sought or protected; and prestige (if considered at all) seems to be a non-coercive inducement used by the powerful to insure compliance at low cost" (1970, p. 470).

Other candidates for the basic content dimensions of status were discussed. These included socioeconomic status (SES), education, and color (henceforth called race).[1] SES was dismissed because by 1970 it had become a buzzword. Among its various meanings, as an index of household consumption it is really a measure of wealth. In another definition of SES, consistent with an early SES index developed by Sewell (1940), it encompasses "material possessions, cultural possessions, and social participation" (Haller & Saraiva, 1973, pp. 2, 8). This definition of SES works as a summary of all three content dimensions of status, "accounting for practically all the common variance of indicators of wealth, prestige and power" (Haller, 1970, p. 471). Thus, SES is not an additional content variable.

Concerning formal education as a general status dimension (e.g. Svalastoga, 1965, p. 16), Haller points to its lack of generality when compared to the continuity of wealth, power, and prestige inequalities across all known human societies, indicating his "preference is to consider advanced education as a common but not indispensable precondition for higher status in societies with exceedingly elaborate occupational structures" (1970, p. 472).

Race, too, is dismissed as a general status variable in Haller (1970). He points out, however, that race can and does become intertwined systematically with inequalities in wealth, power and prestige because of the insidious effects of discrimination. "When color and the three content dimensions become correlated, people base their cues for interaction, and often their laws governing it, on color. In this way color almost takes on an independent existence as a status variable.

Clearly, color lacks universality, and its status effects may easily be derived from the three universal status content dimensions" (p. 472). This is consistent with the position that prejudice based on race is widespread and affects the probabilities for acquiring wealth, power, and prestige because of the stigmatizing and labeling in which people so commonly indulge. If wealth, power, and prestige are withheld perpetually from members of a minority group it becomes easy for members of the majority to misdirect responsibility for their own exclusionary practices on those whom they disadvantage. Thus, the three content dimensions of status which appear to be universal are wealth, power and prestige, as consistent with the early writings on stratification by classical sociologists (Marx & Weber, in particular).

Of these three content dimensions, wealth may be the easiest to understand but it is not the easiest to measure. Haller uses a standard definition of wealth as access to goods and services. But, "the concept 'wealth' has several referents" some of which may be non-monetary, from various perquisites to "differences in access to food [which] discriminate among the penniless" (1970, p. 476).

Power has proven the most difficult of the status dimensions either to understand or to measure. Weber's definition of power, "the probability that one actor within a social relationship will be in a position to carry out his own will despite resistance, regardless of the basis on which this probability rests" (1947, p. 152) is perhaps the one most widely accepted. Haller (1970) acknowledged the tradition of developing and testing theory on social power in experimental small-group settings (e.g. Berger et al., 1966; Gamson, 1968) but noted how little has been done to measure power empirically in the wider society. An important exception was reported in Saraiva (1969) and Haller and Saraiva (1970). The central idea was that "a measure of political influence applicable to any citizen can be formulated by asking each respondent to indicate the highest level at which he had ever succeeded in attempts to gain an objective by working through the incumbent of an office" (Haller, 1970, p. 477). Prestige, by contrast, has been much more thoroughly researched. "Occupational prestige – the average evaluation of specific occupational titles by adult members of a society – is almost universally conceded to be the key prestige variable." Additionally, the occupational prestige hierarchies of urbanized areas vary little from country to country and, in the United States, have "not changed notably since 1925" (ibid.). But occupational prestige hierarchies are based merely on how people subjectively rank occupations, and may therefore remain stable even if the shares of employment among the occupations within them shift.

Besides the content dimensions of status, there are its structural dimensions: central tendency, dispersion, skewness, stratifigraphy, flux, and crystallization (Haller, 1970, p. 478). These refer to the distributions of the phenomena described by the content dimensions of status across a human population. Central tendency, dispersion, skewness and stratigraphy each reflect simple concepts from descriptive statistics. Haller (1970, p. 479) indicates that central tendency should usually be taken to refer to the mean. Dispersion is best described in statistical terms as the variance; skewness, as in statistics, is the degree of asymmetry between the tails of a distribution. Stratigraphy refers to the order and relative position of social strata – the unequal positions along one of the content dimensions of status where the population is unevenly clustered.

Flux refers to the extent that peoples' present positions are determined by their past positions. A society with no flux would be one where humble origins proscribe access to wealth, power, or prestige while high-status origins guarantee it, and all intermediate outcomes are likewise predetermined. All five of these structural dimensions are applicable to any of the content dimensions of status which, taken together, describe a status system.

The sixth structural dimension is status crystallization, typically portrayed as a societal-level dimension of stratification. It is the extent to which wealth, power and prestige are correlated in a society or smaller social unit. However, the possible degree of status crystallization can and does vary according to a person's position in the overall status system. In particular, those in the underclass of advanced industrialized societies are caught in an extreme situation of status crystallization because they have minimal resources in terms of wealth, power or prestige such that their position at the bottom is virtually inescapable.

At higher levels in the stratification systems of advanced industrialized societies such tight relationships among the content dimensions of stratification begin to loosen, reducing the degree of status crystallization. This becomes apparent when considering the relative freedom characterizing the dominant classes where "we find individuals whose wealth liberates them from the need to sell their labor time for a living" (Portes, 2000, p. 261). Some wealthy people may prefer to enjoy life with work, occupying themselves in lower-status positions because they find intrinsic pleasure in the required tasks and in enjoying the company of others involved with their responsibilities. Such people contribute to a reduction in the degree of status crystallization when they participate in the labor force also because their wealth does not necessarily provide them with power or authority in the workplace.

ADAPTATION OF THE GENERALIZED FRAMEWORK ON STATUS DIMENSIONS

Some dimensions of status are more easily measured than others. An additional benefit of such a general conceptual framework is that it permits us to understand the connections among the various streams of stratification research, which have become highly specialized. However, in order to apply the general framework to any specific set of findings some generality must be sacrificed to purchase tractability within the historically and geographically specific institutional contexts where the mechanisms governing stratification and mobility are embedded. Formal education is such an institution (rather than a general content dimension of stratification, as already discussed) and it offers some purchase on predicting mobility outcomes. This is mainly because educational credentials offer considerable convenience in calibrating individuals' capabilities against the range of tasks valued to highly unequal degrees in advanced societies. As indicated by Haller (1970, p. 472):

> . . . formal education lacks the . . . generality of the other three . . . as complexity [in the division of labor] increased there was an increased demand for people steeped in certain symbolic skills (language, logic, and mathematics) . . .. Existing systems such as the family were obviously unable to fulfill the need, and formal educational systems developed. They are hierarchical because some kinds of knowledge are preconditions for others. Thus, on average, the higher the level of general knowledge and the greater the refinement of specific knowledge, the higher the prestige, power, and wealth.

Likewise, the labor market is another institution in which primary mechanisms governing the distribution of wealth, power, and prestige are embedded. It, too, serves as a convenient mechanism for calibrating individual abilities against unequally valued societal needs because what one has done lately is typically the best indicator of what one is prepared to do next. Of course, the accuracy of these institutional mechanisms in predicting individuals' abilities is no better than those who exercise this authority within them. Neglecting the fallibility of bureaucratic actors reifies the functioning of institutional mechanisms in determining who gets what, how and why.

Viewed from this perspective, it is obvious that research on many critical issues in contemporary stratification systems, such as educational and occupational attainments and labor market segmentation, do not represent major breakthroughs in the generalized understanding of stratification but rather contribute to understanding specific social structures within identifiable social and institutional contexts of particular places and times. Much contemporary stratification research provides cross-sectional snapshots of some of the parameters of stratification resulting from the organizational patterns of educational institutions and labor markets. There is as of yet very limited understanding of the ramifications of change in one content dimension of status on another, either in any abstract sense or within specific institutional settings. But despite limitations in understanding the parameters of social structures, and the effects of changes in one content dimension on another, the advancement of relatively certain knowledge of contemporary stratification systems, and illumination of the likely implications of the choices which continuously confront policymakers, lies in concentrating on the distributions of (and tradeoffs between) wealth, power and prestige within the institutions which house and manage inequality.

Most relevant are labor markets, educational systems and the linkages between them. Obviously, occupational attainment automatically confers a certain degree of status, and places one in a position within the occupational hierarchy which is embedded within the system of power relations which manage productivity while simultaneously providing earnings and, hence, a certain degree of wealth. (As far as prestige is concerned, some measure of it comes with the job depending on its rank in the occupational prestige hierarchy.) Specialized training and educational attainment is a prerequisite for entering most occupations. However, occupational attainment is possible only with respect to the current structure of occupations and job openings. Thus, a major shift in the structure of occupations supported by an economy's labor market is very consequential for the overall status system in which it is embedded: "If we are to understand status attainment we must ultimately understand the moving status system with respect to which status attainment behaviors occur" (Haller, 1970, p. 475). When specific segments of the occupational structure grow faster than the ability of the educational system to provide suitable aspirants to meet the demand, employers must import the required labor, export the work, or make do with suboptimal help. On the other hand, entire segments of a local occupational structure sometimes contract very quickly, expelling the incumbents of previously supported positions and restricting the occupational choices of those whose hopes and needs were tied to these positions and their future availability. Inadequate preparation to adapt to such changes forces some to accept less desirable positions and sidelines others, barring their access to earnings and any prestige or authority conferred through formal employment. The following analysis and discussion of the structural changes in the labor market of the Pittsburgh region resulting from the economic restructuring of the steel industry and its relation to growth in the urban underclass provides an illustrative example.

PLANT CLOSURES AND MANUFACTURING EMPLOYMENT LOSSES IN THE PITTSBURGH METROPOLITAN AREA

Traditionally, employment in the steel industry was concentrated in a limited number of major production regions where large-scale integrated plants developed.[2] The onset of economic restructuring as a primary trend in many manufacturing sectors since the mid-1970s intensified the corporate quest to minimize production costs. Because steel is a basic material for many manufactured goods demand for steel declined during the 1970s as energy costs rose and lighter, cheaper substitute materials were introduced in many products. With increasing imports and advancements in production technology competitive pressures in the industry increased while the market for steel worldwide was shrinking. This resulted in major shifts away from Fordist models of industrial organization towards flexible production, which both undercut the power of organized labor and permitted diversification of many companies' product lines. In the United States, the regional concentration of steel production meant that employment declines in the industry at the national level weighed disproportionately on specific communities within a relatively small number of major metropolitan areas and smaller company towns. Of course, such regionally concentrated industrial restructuring was not limited to steel or to the Pittsburgh region. Many cities and towns that became famous for the manufactures they specialized in suffered similarly, if not necessarily on such a scale. The consequences of such job losses were particularly severe because the job searches of workers in basic manufacturing are typically restricted to local labor markets, and because the resources of people in disadvantaged situations to carry out effective job searches are even more limited. Thus, the disproportionate impacts of deindustrialization in specific regions compounded the redundancy suffered by laid-off workers in these places.

To provide some sense of the magnitude of the manufacturing employment losses in the Pittsburgh region during this time, data from the U.S. Bureau of Labor Statistics Regional Economic Information System (REIS) permits comparison of manufacturing and non-manufacturing employment trends between the Pittsburgh region and the United States as a whole for the period from 1976 to 1986 and from 1986 to 1996. These are shown in Table 1.

There was a slight overall decline in manufacturing employment for the United States as a whole from 1976 to 1996, this decline represented a 0.6% increase from 1976 to 1986 and approximately a 1.3% decrease from 1986 to 1996. The Pittsburgh region alone, however, lost nearly 115,000 manufacturing jobs from 1976 to 1986, and more than 18,000 additional manufacturing jobs during the subsequent ten year period. The Pittsburgh region's job losses from 1976 to 1986 were so severe that total employment actually declined during this period. Particularly noticeable are the differences in the decline of employment in manufacturing between the Pittsburgh region alone and the entire United States. While from 1976 to 1996, employment in manufacturing declined by 0.7% for the United States as a whole, the Pittsburgh region suffered a 52.7% drop. The concentration

Table 1. Manufacturing and Non-Manufacturing Employment in the United States and in the Pittsburgh Region, 1976–1996.

	Year			Percent Change		
	1976	1986	1996	1976–1986	1986–1996	1976–1996
Panel A: United States						
Manufacturing	19,372,300	19,489,700	19,231,500	0.6	−1.3	−0.7
Non-manufacturing	78,252,900	104,116,500	130,148,400	33.1	25.0	66.3
Total[a]	97,625,200	123,606,200	149,379,900	26.6	20.9	53.0
Panel B: Pittsburgh Region						
Manufacturing	251,771	137,306	119,073	−45.5	−13.3	−52.7
Non-manufacturing	792,803	888,254	1,042,792	12.0	17.4	31.5
Total[a]	1,044,574	1,025,560	1,161,865	−1.8	13.3	11.2

Source: Regional Economic Information System (U.S. BLS, 1996; U.S. Bureau of Economic Analysis, 1997).
[a] Totals refer to non-farm employment.

Table 2. Integrated Steel Plant Closures in the Pittsburgh Region Since 1982.

Company, Plant Name, and Location	Year
USS Carries Furnaces, Rankin	1982
USS Duquesne Works, Duquesne	1984
USS Clairton Works, Clairton[a]	1984
J&L Pittsburgh South Side, Pittsburgh	1985
J&L Hazelwood Works, Pittsburgh[a]	1985
USS Homestead Works, Homestead	1986
Wheeling-Pittsburgh Monesson Works, Monesson	1986
USS National Works, McKeesport	1987
USS Christy Park Works, McKeesport[b]	
LTV Aliquippa Works, Aliquippa[c]	
USS Edgar Thompson Works, Braddock[c]	

Notes: The J&L plants in Pittsburgh were bought by LTV prior to closure. USS National Works in Mckeesport were previously owned by Republic Steel.

Sources: Hoerr (1988), Guilherme Heraclito de Lima (1991), Hall (1997).

[a] Coke production was maintained after the indicated year.

[b] Sold in 1986.

[c] Production capacity sharply curtailed during mid-1980's, LTV Aliquippa's ore-based capacity was eliminated by 1991.

of manufacturing employment represented by the steel industry in Table 1 is, of course, greater for the Pittsburgh region than for the entire United States (and was also greater in 1976 than in 1996). Nevertheless, the other manufacturing sectors in the Pittsburgh region were tied closely to steel and were therefore also vulnerable. Decline in the region's steel industry had upstream impacts within the broader local manufacturing economy. As Hoerr (1988, p. 570) explains, "Each 1,000 jobs lost in the primary metals industry forces the loss of an additional 130 jobs at firms that supply that industry." The sharp declines in the manufacturing base of the Pittsburgh region's economy occurred mainly during the early 1980s. Table 2 shows the sequence of closures among the large integrated steel plants in the Pittsburgh region during this period.

Deindustrialization and the Underclass

The social and economic implications of the employment losses resulting from these massive shifts in the regional industrial employment base associated with the global restructuring of the steel industry warrant examination. To emphasize that such shifts in the occupational structure carry pervasive effects across the content and structural dimensions of status, the question of whether these industrial shifts led to the growth of an underclass is addressed. The growth of an underclass within a region due to shifts in the structure of employment resulting from corporate reaction to the vicissitudes of unanticipated market conditions is viewed here as part of a major realignment of wealth, power and prestige within the regional population. Additionally, it implies a crystallization at the bottom of the social hierarchy among these content dimensions of status and a reduction in flux because, once trapped in the underclass, a person's chances of reentry into the mainstream are greatly reduced.

Singh (1991, p. 506) indicated that the underclass in older industrial regions of the United States has had "readily observable spatial, economic, and demographic components as well as behavioral dimensions." Examples include weak labor force attachment among adults, extreme poverty, high dropout rates among teenagers, unwed parenting, and welfare dependency. These phenomena are commonly associated with the kind of community distress which perpetuates poverty and isolation. As research on the underclass accumulated, a number of empirical indicators came into use to estimate its scope and size. Ricketts and Sawhill (1988) proposed one set of indicators to identify underclass areas as census tracts with scores of at least one standard deviation above the national mean for the number of: (1) males sixteen years of age and older, unemployed or not in the labor force; (2) households headed by women with at least one child and no spouse present; (3) households receiving public assistance; and (4) youths aged sixteen to nineteen years not in enrolled in school and not high school graduates. According to this criterion all four indicators must intersect at these levels. Such places are very uncommon in the U.S. and most likely are found only deep within the most isolated inner-city ghettoes.

An alternative criterion for identifying underclass areas is extreme poverty. Specifically, census tracts where the poverty rate is 40% or higher are taken to identify underclass areas (Massey & Denton, 1993; Wilson, 1993).[3] This latter criterion is met much more frequently than the former because Ricketts and Sawhill's (1988) criterion for identifying underclass areas is exceedingly stringent. To meet the Ricketts and Sawhill criterion based on the 1980 census, for example, a tract must have: (1) 56% of the males sixteen years of age and older, unemployed or not in the labor force; (2) 60% of the households headed by women with at least one child and no spouse present; (3) 34% of the households receiving public assistance; and (4) 36% of the youths aged sixteen to nineteen years not in enrolled in school and not high school graduates. Such places are extremely rare, even during the worst of times. On the other hand, census tracts fitting the extreme poverty criterion for the underclass could be found in practically every major metropolitan area in the U.S. since 1980.

DATA AND METHODS

The data used are the tract-level data from the Decennial Censuses of 1970, 1980, and 1990 for the Pittsburgh Standard Metropolitan Statistical Area (SMSA), which includes Allegheny, Beaver, Washington and Westmoreland Counties. These were published by the U.S. Census Bureau as the Population Counts 4A (Counts 4A) for 1970, and the Summary Tape Files 3A (STF3A) for 1980 and 1990. The tract-level data provides estimates for the entire population within tracts based on the 20% sample from the long form of the Decennial Census. Because during this twenty-year period there were changes to the variables provided by the Census Bureau and to the numbering and boundaries of many of the census tracts in the Pittsburgh SMSA comparability issues arose between the variables (particularly with the 1970 data) and between the census tracts themselves. The adjustments used to correct for these comparability issues are described below. Even though it is only available at ten-year intervals, the decennial census offers certain advantages. Primarily, it provides estimates for complete coverage of local populations for detailed geographical units rather than attempting overall representativeness for large areas. Because of the spatial isolation of the underclass within the most dilapidated areas of metropolitan regions the sampling designs for data collection aimed at achieving a general representation of the population do not capture a sufficient proportion of this population to show its extent. By the same token the Public Use Microsample from the decennial census does not provide geographic information in sufficient detail to identify which cases are residents of underclass areas and which are merely outliers who happen to reside elsewhere. The comparability issues between the variables in the data provided by Census Bureau for each decade and the measures taken to ensure comparability between the observations across time using the Census Tract Comparability Charts (U.S. Bureau of the Census, 1983, 1993) are described below.

THE VARIABLES AND THEIR COMPARABILITY OVER TIME

The Summary Tape File 3A (STF3A) for 1990 was designed to be comparable with the STF3A for 1980, so comparability issues arise mainly between the Counts 4A for 1970 and the STF3A for 1980 and 1990. These variables are measures of employment structure, the composition of employment by industry and occupation of the populations of each tract; and underclass indicators (the proportion of the population in each census tract measured by any of the five variables used to define and measure the underclass).

Employment Structure Measures

The efforts of the Census Bureau to maintain comparability between the 1980 and 1990 censuses meant that the categories for the Standard Industrial Classifications (SIC) and the Occupational Titles provided in the STF3A were almost identical for these two years.[4] The only difference was in employment counts for the industry "Personal Services" which, in 1980, also included "Entertainment and Recreation Services." However, these industry groups were aggregated in the 1980 STF3A so they were also aggregated for the 1970 and 1990 data sets to make the employment counts for "Personal Services" comparable for all three panels.

The industrial and occupational classifications for the Counts 4A for 1970 were considerably more detailed than the classifications in the STF3A for 1980 and 1990. Therefore, these were similarly aggregated to make the industrial and occupational employment counts from 1970 comparable with those from 1980 and 1990.

Underclass Indicators

The underclass indicators as derived from the 1980 and 1990 STF3A provided by the Census Bureau are fully comparable. However, the files from the 1970 Census (Population Counts 4A) are not. Specifically, the questions for households headed by women with children (WHEAD) and public assistance (ASSIST) were restricted to families, rather than extended to households in general. This may pose less of a problem than it appears because household living arrangements coincided with nuclear families to a greater extent in 1970 than in 1980 or 1990. Thus, the proportions for WHEAD and ASSIST may be underestimated for 1970, but not to a great extent. Additionally, the school enrollment and employment questions which provide the counts for youths who dropped out of high school (DROPOUT) covered the sixteen to twenty-one year age group, rather than those aged sixteen to nineteen. Therefore, the DROPOUT indicator for 1970 is somewhat inflated. There are some minor technicalities that interfere with the precise comparability of poverty rates between the 1970 census and the censuses of 1980 and 1990. However, "these changes resulted in a minimal increase in the number of poor at the national level" (U.S. Bureau of the Census, 1992, Appendix B, p. 54). Also, problems arising from the census undercount of 1990 are disregarded because the bias it introduces is conservative, leading to lower estimates of the size of the underclass.

The Units of Analysis and Their Comparability Over Time

Census tracts are the units of analysis in this research. According to the Census Bureau's Geographical Areas Reference Manual (U.S. Bureau of the Census, 1994, pp. 10, 11), "Census tracts are small, relatively permanent geographic entities within counties (or the statistical equivalents of counties) delineated by a committee of local data users." Nevertheless, census tracts are still sometimes subject to change from decade to decade. Because census tract boundaries were changed in several instances in the Pittsburgh SMSA from 1970 to 1980 and from 1980 to 1990, the Census Tract Comparability Charts for the 1980 and 1990 Censuses were used as a guide to trace the changes in their tract boundaries and their labeling between decades. Aggregating tracts for comparability produced a data set with 615 comparable observations for each panel. The labeling conventions used by the Census Bureau to indicate the presence and types of tract changes serve as a guide to the kinds of changes that were implemented. The few tracts with the suffix, "0.99" were removed from the data set because that suffix indicates residence on ships or boats. Tracts split after 1970 were aggregated together again because comparability with the 1970 tracts took precedence over attempting to maintain roughly homogeneous population sizes.

Methods

Two methods are used to examine the relationship between regional restructuring and underclass growth in the Pittsburgh SMSA. The first is a simple tabular presentation of the relevant descriptive statistics; the second is a series of regression models to test whether employment change in durable manufacturing industries (and also in blue-collar occupations aggregating workers in precision production occupational categories together with machine operators) are significant determinants of joblessness among working-age males and poverty, two key outcomes associated with the underclass.

In the regression models the tract-level changes between 1970 and 1980 and between 1980 and 1990 are represented as change variables, constructed using the differences between the relevant population proportions at the tract level from the earlier and later time points. Change variables are thus derived for employment in durable manufacturing industries, blue-collar occupations, males age sixteen and older not in the labor force or unemployed, and persons below the poverty line. Change variables constructed from two underclass indicators, the proportions of jobless working-age males and persons below the poverty line, are used as dependent variables in separate regression equations to test for significant relationships between regional restructuring and underclass growth.

Race is expected to be an important variable, but using tract-level changes in racial composition does not reflect the hypotheses derived from the literature on restructuring and the underclass. The argument to be tested by including race is that the black population was hit harder by the structural shifts in the economy in the way these shifts impacted neighborhoods and communities, generally. In other words, the theoretical perspective on which this work is based (e.g. Wilson, 1990) concerns the differential impacts of restructuring by race within residential areas, not change in the racial composition of residential areas per se. Because there is no theoretical justification for using the proportion of the black population at either the earlier or later time point, the average between the two is used. The independent variables are applied consistently in a set of models covering the 1970–1980 period and the 1980–1990 period, with either proportional change in the male population aged sixteen and older not in the labor force or unemployed, or proportional change in the population below the poverty line as dependent variables.

Industrial and Occupational Shifts and Changes in the Underclass Indicators in the Pittsburgh Metropolitan Area

Because the specific form of an occupational structure depends on the mix of industries and the production technologies used in those industries, the impact of deindustrialization on overall employment, and employment by industry and occupation require explicit consideration. Specifically, deindustrialization reduced employment in some industries and occupations, particularly basic manufacturing industries and blue-collar occupations. This *absolute* reduction in employment for specific industries and occupations resulted in: (1) a jump in the unemployment rate, which was much more severe in some communities than others; (2) a shift in the *proportions* of the total employment accounted for by the rest of the region's industrial sectors; and (3) a corresponding shift in the regional occupational structure (because the structure of occupations in an economy depends on its mix of industries). And, of course, as a region's occupational structure shifts so, too, does the structure of its status system.

Shifts in the proportion of employment by industry for the Pittsburgh region as a whole are presented in Table 3. The most pronounced trend was in durable

Table 3. Employment by Industry in 1970, 1980 and 1990, Pittsburgh SMSA.

	Percent of Total			Percent Change		
	1970	1980	1990	1970–1980	1980–1990	1970–1990
Extractive	1.6	1.8	1.4	0.2	−0.4	−0.2
Construction	5.3	5.3	5.9	0.0	0.6	0.6
Durable manufacturing	25.9	20.6	10.2	−5.3	−10.4	−15.7
Non-durable manufacturing	5.8	5.0	4.3	−0.8	−0.7	−1.5
Transportation	4.2	3.9	5.3	−0.3	1.4	1.1
Communication	2.9	2.8	2.8	−0.1	0.0	−0.1
Wholesale	4.0	4.2	4.6	0.2	0.4	0.6
Retail	16.4	17.1	20.0	0.7	2.9	3.6
Financial services	4.5	5.4	7.1	0.9	1.7	2.6
Business services	3.3	4.8	4.8	1.5	0.0	1.5
Personal services	4.3	3.5	4.0	−0.8	0.5	−0.3
Health services	6.2	8.7	11.6	2.5	2.9	5.4
Education	7.4	8.1	8.5	0.7	0.4	1.1
Other professional services	4.2	4.4	7.4	0.2	3.0	3.2
Public administration	4.1	3.4	3.1	−0.7	−0.3	−1.0
Total employed	870,902	938,432	923,049			

Sources: U.S. Bureau of the Census, 1970, 1980, and 1992.

manufacturing industries, which includes all subsectors of steel production. In 1970 more than one quarter of the region's employment was concentrated in durable manufacturing industries. From 1970 to 1980 the share of the region's employment in durable manufacturing shrank by more than 5%. From 1980 to 1990, however, the share of employment in the region accounted for by durable manufacturing shrank by more than 10%. Another important difference regarding employment changes during these decades in the region can be seen in total employment. The employment decline in durable manufacturing during the 1970s took place within the context of a growing job base. As can be seen in Table 3, the employment decline in durable manufacturing industries during the 1980s was so large that the gross number of jobs in the region was reduced because of it.

Employment trends by occupational category for the Pittsburgh SMSA are presented in Table 4. Declines in blue-collar production jobs, as expected with declines in durable manufacturing industries, can be seen in two of the broad occupational categories, "Machine Operators, Assemblers and Inspectors" and "Precision Production, Craft and Repair Occupations." Other industrial sectors also employed workers whose jobs fell under these categories so their decline does not fully mirror the decline of durable manufacturing. Nevertheless, the drop

Table 4. Employment by Occupation in 1970, 1980 and 1990, Pittsburgh SMSA.

	Percent of Total			Percent Change		
	1970	1980	1990	1970–1980	1980–1990	1970–1990
Executive	7.3	9.5	12.2	2.2	2.7	4.9
Professional	12.2	12.8	15.6	0.6	2.8	3.4
Technical	3.6	3.4	4.4	−0.2	1.0	0.8
Sales	7.9	10.2	12.5	2.3	2.3	4.6
Administrative support	18.1	17.6	17.3	−0.5	−0.3	−0.8
Private household	1.0	0.4	0.3	−0.6	−0.1	−0.7
Protective service	1.4	1.6	1.6	0.2	0.0	0.2
Other service	10.4	11.5	12.3	1.1	0.8	1.9
Farming, forestry and fishing	0.5	0.6	0.9	0.1	0.3	0.4
Precision production	15.5	13.0	10.4	−2.5	−2.6	−5.1
Machine operators	12.8	8.7	4.6	−4.1	−4.1	−8.2
Transportation	4.0	5.1	3.9	1.1	−1.2	−0.1
Handlers and laborers	5.4	5.5	4.0	0.1	−1.5	−1.4
Total employed	870,902	938,432	923,049			

Sources: U.S. Bureau of the Census, 1970, 1980, and 1992.

in blue-collar employment at the regional level was substantial: among machine operators nearly 30,000 jobs were lost from the Pittsburgh SMSA from 1970 to 1980 and more than another 39,000 were lost from 1980 to 1990.[5]

Regarding the processes linking deindustrialization to increase in the underclass indicators, it must be pointed out that the relationship between the industrial and occupational changes associated with deindustrialization and the underclass indicators are not always related through the underclass itself. Deindustrialization increases unemployment and simultaneously reduces the share of employment accounted for by manufacturing, but this does not mean that workers dislocated by plant closures simply drop into the underclass. Loss of income due to job losses in manufacturing increases local poverty rates, particularly because the educational requirements for the basic manufacturing jobs which were lost were insufficient to provide transferable credentials or skills, restricting the ability of most production workers to move into jobs with equivalent or better pay. This permits employers greater selectivity in choosing among workers or applicants for the remaining jobs. In the Pittsburgh region during this period,

> ...for both men and women, leaving most jobs in the goods-producing sector and entering jobs in the service-producing sector was costly... That leavers from goods producing sectors could not maintain their high earnings suggests they lacked transferable skills, which is hardly surprising. On average, well-paid blue-collar workers have high school educations, while the

> high-paid service workers are managers, doctors, lawyers, and teachers whose jobs require much more education (Jacobson, 1987, pp. 442, 443).

Thus, less-educated workers did not move into higher-paying jobs and minority youths from poorer neighborhoods with no college and scant work experience were extraordinarily disadvantaged in competition with experienced white males and their spouses for available vacancies in unskilled jobs.

The summary statistics for changes in the underclass indicators from 1970 to 1980 and from 1980 to 1990 for the Pittsburgh SMSA as a whole are given in Table 5. For the region as a whole, the percentage of males aged sixteen and older not in the labor force or unemployed decreased from nearly 28% to under 17%. But by 1990 this rate rose to nearly 40%. Furthermore, this shows a persistence of weak labor force attachment resulting from deindustrialization because this was more than five years after the greatest number of steel plant closures in the region's history. The population below the poverty line stood at the same rate in 1980 as in 1970. This, however, cannot be interpreted as a sign of stability in the poverty rate, which probably fluctuated with the turbulent national economic conditions of the 1970s. By 1990 the regional poverty rate was nearly 4% higher.

The percentages given for youths not attending high school and not high school graduates are comparable for 1980 and 1990, but the coverage for 1970 includes a wider age group and is therefore somewhat inflated. Specifically, the number of

Table 5. Average Percentages for Underclass Indicators in 1970, 1980 and 1990, Pittsburgh SMSA.

	Percent of Total			Percent Change		
	1970	1980	1990	1970–1980	1980–1990	1970–1990
Males 16 and older not in labor force or unemployed	27.9	16.4	39.2	−11.5	22.8	11.3
Population 16 and older below poverty[a]	10.7	10.7	14.6	0	3.9	3.9
Youths not attending school and not high school graduates[b]	9.4	5.6	8.5	−3.8	2.9	−0.9
Households on public assistance income[c]	5.8	9.6	9.7	3.8	0.1	3.9
Households headed by women with at least one child, no spouse present[c]	5.4	5.1	6.0	−0.3	0.9	0.6

Sources: U.S. Bureau of the Census, 1970, 1980, 1992.

[a] Poverty figures are for previous year.

[b] 1970 covers 16–21 year olds, 1980 and 1990 cover 16–19 year olds.

[c] 1970 covers families, not households.

dropouts reported for 1970 may include youths who dropped out five years earlier, as opposed to only three years earlier. The overall dropout rate for the region was nearly 3% greater in 1990 than it was in 1980.

The welfare system was designed to provide temporary assistance for dislocated workers and the poor. During a period of industrial decline it seems likely that the number of people turning to the welfare system for support should increase because of the loss of jobs and incomes by those lacking transferable skills. Nevertheless, the percentage of households on public assistance did not stand at a much higher rate in 1990 than in 1980 for the region as a whole. The lower rate in 1970 is, in part, due to lack of comparability of the tract-level data from the 1970 census because the Census Bureau reported welfare receipt for families rather than households in 1970, as explained above. The percentage of households headed by women with at least one child and no spouse present was lower in 1980 than in 1970, and increased by only about 1% from 1980 to 1990. For the region as a whole, during the period from 1980 to 1990 all the underclass indicators exhibited some increase. However, for households on public assistance the rate was only approximately 0.1% higher in 1990 than in 1980. The changes from 1980 to 1990 in the indicators of the behavioral aspects of the underclass appear very weak in comparison to those associated with shifts in the regional employment structure.

In summary, deindustrialization in the Pittsburgh region during the 1980s had a profound effect on joblessness among males and a substantial effect on the poverty rate. The other underclass indicators, associated more with behavioral phenomena (dropping out of school, welfare receipt, and unwed parenting) did not exhibit such increases. In order to reinforce the causal argument between deindustrialization and increased status crystallization due to underclass growth a series of regression models for the Pittsburgh are presented below.

The Impacts of Employment Shifts

This section addresses the association between the economic restructuring and increases in the underclass in the Pittsburgh region. First, changes in the proportions of employment among residents in durable manufacturing industries and in blue-collar occupations, residence in the central city, and the proportion of black residents are regressed in separate models on change in male joblessness and change in the poverty rate from 1970 to 1980 and then from 1980 to 1990. Because the period of accelerated deindustrialization occurred during the early 1980s, the models for the latter period address the theoretical expectation that regional restructuring via deindustrialization increases the size of the underclass. The regression models for the 1970 to 1980 period are provided mainly for the sake of comparison with those for the 1980 to 1990 period. Second, to examine the possibility that the proportion of the population below the poverty line and joblessness among males are mutually reinforcing, change in male joblessness is added as an independent variable in the model to explain change in localized poverty and, conversely, change in localized poverty is added as to the model to explain change in male joblessness.

The first set of results is shown in Table 6. As can be seen in Panel A, from 1980 to 1990 declines in durable manufacturing employment, as expected, are significantly related to increases in male joblessness, with a slope of −0.29. Likewise, declines in blue-collar employment are significantly related to increases in male joblessness, and with a more gradual slope (−0.11). The proportion of the black population was also significant with a slope of 0.21. Central city residence addresses the spatial mismatch argument offered by Kasarda (1988). Because of the historical concentration of basic manufacturing outside the City, particularly in the Mon

Table 6. Unstandardized Coefficients for Regional Restructuring Estimates on Change in Proportion of Jobless Males and Population in Poverty, Pittsburgh SMSA.

Independent Variables	1970–1980	1980–1990
Panel A: Change in males, 16 and older not in the labor force or unemployed		
Durable manufacturing	0.18** (0.047)	−0.29*** (0.046)
Blue collar occupations	−0.09* (0.047)	−0.11* (0.063)
Black population	−0.04*** (0.009)	0.21*** (0.015)
Central city	−0.02*** (0.006)	0.01 (0.008)
Constant	−0.10*** (0.005)	0.16*** (0.007)
R^2	0.08	0.32
n	612	611
Type of test	Two-tailed	One-tailed
Panel B: Change in population below the poverty line		
Durable manufacturing	−0.14*** (0.036)	−0.11** (0.039)
Blue collar occupations	0.07* (0.036)	0.00 (0.053)
Black population	−0.02** (0.008)	0.08*** (0.013)
Central city	0.03*** (0.005)	−0.00 (0.007)
Constant	−0.01** (0.004)	0.02** (0.006)
R^2	0.09	0.08
n	612	611
Type of test	Two-tailed	One-tailed

Note: Standard errors given in parentheses.
Sources: U.S. Bureau of the Census, 1970, 1980, 1992.
*$p < 0.05$.
**$p < 0.01$.
***$p < 0.001$.

Valley, the Pittsburgh region did not conform closely to the spatial patterns of employment typical of so many American cities (and it is not significant in any of the models for the period from 1980 to 1990). The model for these determinants of male joblessness from 1980 to 1990 in the Pittsburgh region has an R^2 of 0.32.

Regarding the corresponding model for 1970–1980, because there is no theoretical investment in the directions of the coefficients for these variables, significance levels for the two-tailed tests are reported. From 1970 to 1980, changes in the number of working-age males not in the labor force or unemployed were positively associated with changes in durable manufacturing. Before the period of accelerated deindustrialization, it was not uncommon for working-class men in the Pittsburgh region to voluntarily exit and reenter the labor force and live on savings between jobs, and many could do so without placing their future earning ability in jeopardy. Additionally, during the period from 1970 to 1980, residence in predominantly black areas was negatively associated with male joblessness. This is likely because federal discrimination lawsuits over unfair employment and promotion practices in the steel industry placed employers in Southwestern Pennsylvania under close federal scrutiny over their dismal record regarding racial inclusion. This is evidence in support of what Dickerson (1986) described as a Pyrrhic victory for blacks in Southwestern Pennsylvania because the decades-long legal struggle, resulting a small measure of advancement towards parity with whites, was quickly rendered irrelevant by the plant closures of the 1980s.

As can be seen in Panel B, the same set of independent variables do not predict poverty growth under conditions of deindustrialization as well as they predict increases in male joblessness. Declines in durable manufacturing employment are significantly related to poverty growth but changes in blue-collar employment are not. Additionally, the proportion of the black population was significantly related to poverty growth from 1980 to 1990, reflecting the harsher impacts of deindustrialization on predominantly black communities.

The corresponding estimates of the model for poverty growth covering the period from 1970 to 1980 show declines in durable manufacturing as a significant determinant of poverty for this period, even stronger than for the period of 1980 to 1990. As can be seen in Tables 3 through 6, discussed above, deindustrialization was not unknown in the region before 1980.

Additionally, there is a significant positive association between change in blue-collar employment and poverty from 1970 to 1980. This anomaly may be due to the fact that blue-collar workers were concentrated residentially while manufacturing supported both blue- and white-collar jobs.

The underclass concept suggests that some of its key features may become mutually reinforcing. This idea is examined in Table 7 by adding joblessness among working-age males to the model used to explain increases in the population below

Table 7. Unstandardized Coefficients for Regional Restructuring Estimates and Relationships Between Structural Underclass Indicators, Pittsburgh SMSA.

Independent Variables	1970–1980	1980–1990
Panel A: Change in males 16 and older not in the labor force or unemployed		
Population below poverty line	−0.03 (0.053)	0.48*** (0.044)
Durable manufacturing	0.18*** (0.048)	−0.24*** (0.043)
Blue collar occupations	−0.08* (0.048)	−0.12* (0.057)
Black population	−0.04*** (0.010)	0.18*** (0.014)
Central city	−0.02*** (0.006)	0.01 (0.008)
Constant	−0.10*** (0.005)	0.15*** (0.006)
R^2	0.08	0.44
n	612	611
Type of test	Two-tailed	One-tailed
Panel B: Change in population below the poverty line		
Jobless Males	−0.02 (0.031)	0.34*** (0.031)
Durable Manufacturing	−0.14*** (0.037)	−0.01 (0.037)
Blue Collar Occupations	0.07* (0.037)	0.04 (0.048)
Black Population	−0.02** (0.007)	0.01 (0.014)
Central City	0.03*** (0.005)	−0.00 (0.007)
Constant	−0.01* (0.005)	−0.04** (0.007)
R^2	0.09	0.23
n	612	611
Type of test	Two-tailed	One-tailed

Notes: Standard errors given in parentheses.
Sources: U.S. Bureau of the Census, 1970, 1980, 1992.
*$p < 0.05$.
**$p < 0.01$.
***$p < 0.001$.

the poverty line, and increases in the population below the poverty line to explain joblessness among working-age males. As can be seen in Panel A of Table 7, change in the population below the poverty line appears to be a significant determinant joblessness among males with a relatively steep positive slope. Although this may seem to make little sense compared to the idea of a causal link from joblessness to poverty, it underlines the necessity of a minimum package of resources needed to carry out a successful job hunt. Note that the R^2 jumps from 0.32 to 0.44 with this sole modification to the model. Likewise, when change in the proportion of jobless males is added to the model explaining change in the proportion of the population below the poverty line it is significant with a steep slope in the expected direction. Furthermore, all the other previously significant independent variables are no longer significant and R^2 jumps from 0.08 to 0.23, suggesting that poverty increases due to deindustrialization occurred via the resulting joblessness among males. It appears that under conditions of deindustrialization, poverty and joblessness (at least among males) become mutually reinforcing, creating a highly problematic situation of perennial exclusion from the work force combined with persistent poverty within the neighborhood.

The major shifts in the structure of the region's employment base by industry and occupation that occurred as a consequence of the closure of so many of the large

Table 8. Average Percentages of Households at Various Income Levels by Race, 1969, 1979 and 1989, Pittsburgh SMSA.

Rough Adjusted Income Strata (100 = 1980–1982 Dollars)	Average Percentages		
	1969	1979	1989
Panel A: All neighborhoods			
<$3,000	2.0	4.0	7.3
$3,000–$7,000	6.6	9.7	12.9
$7,000–$10,000	4.5	8.3	5.8
$10,000–$15,000	9.9	15.6	15.2
$15,000–$21,000	15.0	14.1	13.8
$21,000–$30,000	31.0	20.0	17.8
$30,000–$40,000	13.5	16.0	9.4
$40,000–$60,000	13.5	8.3	11.4
>$60,000	4.0	4.0	6.4
Panel B: Predominantly white neighborhoods			
<$3,000	1.7	3.5	6.1
$3,000–$7,000	6.2	8.9	12.1
$7,000–$10,000	4.3	8.0	5.6
$10,000–$15,000	9.7	15.3	15.2
$15,000–$21,000	14.9	14.3	14.0
$21,000–$30,000	31.6	20.6	18.4
$30,000–$40,000	13.8	16.6	9.8
$40,000–$60,000	13.8	8.7	12.0
>$60,000	4.0	4.2	6.9
Panel C: Predominantly black neighborhoods			
<$3,000	5.6	10.7	19.5
$3,000–$7,000	13.9	21.0	22.4
$7,000–$10,000	7.6	12.3	7.4
$10,000–$15,000	14.0	18.9	15.2
$15,000–$21,000	16.1	12.2	11.6
$21,000–$30,000	21.3	12.0	11.7
$30,000–$40,000	8.6	8.3	4.9
$40,000–$60,000	9.2	3.4	5.3
>$60,000	3.6	1.0	1.9

Sources: U.S. Bureau of the Census, 1970, 1980, 1992; Consumer Price Index.

integrated steel plants during the early 1980s are shown to have been significant determinants of the key phenomena associated with the underclass. If industrial restructuring had such profound implications for the regional economy and, in particular, for its traditional employment base in the steel industry it is reasonable to expect to see a shift in the stratigraphy of income for the regional population as a whole. Table 8 therefore shows shifts in the shape of the income distribution in constant dollars. The increases in the proportions of the population caught within the two lowest income strata over this twenty-year period are apparent. However, it is the expected declines in the middle-income strata, particularly within the $30,000 to $40,000 range during the 1980s, which show the consequences of the displacement of the unionized blue-collar manufacturing economy by an economy dominated by the constellation of various jobs defined within the service sector (particularly its large, unskilled segment).

It is important, however, to emphasize the point that none of the forces involved with the economic restructuring in the Pittsburgh region during this period, or their effects in the redistribution of wealth, power and prestige, were color blind. To underscore this, Table 8 also shows the shift in the stratigraphy of wealth separately for predominantly white and black neighborhoods in the Pittsburgh region. The overall shifts shown in Panel A of Table 8 are largely reproduced in Panel B because it reflects the changes for the majority. Within predominantly black neighborhoods, however, the proportion of the population within the lowest income stratum doubled during the 1970s and then doubled again during the 1980s, as reflected in Panel C. The income declines within predominantly black neighborhoods were not, however, isolated to the poor. All the income strata from $15,000 and up in these neighborhoods declined substantially during the 1970s and 1980s with merely a few percentage points rise in the strata from $40,000 and above during the 1980s following the increases of income among the wealthiest during this period.

Conclusions

Haller (1970) pointed out that most scholarly attention to changes in the structures of status systems emphasized sudden or profound changes, either by addressing specific periods of crisis or as a result of juxtaposing periods from disparate points in history. He also raised another possibility, that changes in the structures of status systems are going on continuously, but incrementally so that hardly anyone ever notices: "almost no one seems to consider seriously the possibility that changes in status may be going on all the time-though at a 'glacial' rate" (Haller, 1970, pp. 472, 473).

The research presented in this chapter, however, addresses a major shift in social structure more in keeping with the common tradition and shows that a slow rate of change is not the only characteristic of glacial movement exhibited by status systems. Both may develop large areas of conflicting pressure, producing fault lines that sometimes result in sharp and sudden fractures. Haller (1970, p. 483) points to the gratuitous complexity that would be involved with any attempt at a full, comprehensive analysis of all three content dimensions within their respective structural dimensions simultaneously and, of course, the work presented here makes no such attempt. Nevertheless, it is certainly possible to speculate on some additional aspects of the changes in the structure of the local status system of the Pittsburgh region during the 1970s and 1980s. The shifts in the regional employment base certainly had implications for the changes in the distribution of income, and therefore wealth. Power inequalities and dynamics were not addressed explicitly but remained backstage in this analysis.

Although it may be possible to show the changing contours of occupational prestige in the region concomitant with its changing structure of employment such an exercise seems of secondary importance. Although the levels of occupational prestige within the region most likely declined merely as a result of the shift in employment structure (with the main thrust from unionized manufacturing to low-level services), it is also possible that the local prestige rankings among occupations may have shifted slightly. What was shown here were the rough shifts in the local structure of income distribution, which had a much more direct bearing on the well being of the majority of the population. Most importantly, such profound shifts in the regional employment base, which were brought about by the broader economic forces and decisions of corporate leaders within the steel industry, were implicated as significant causes of increases in the numbers of jobless men specifically and also for the number of people living below the poverty line. Shifts in the structure of social status systems are ubiquitous and may normally occur continuously and slowly. However, as the preceding analysis of the Pittsburgh region during the 1970s and 1980s has shown they can also occur as sharp discontinuities, disrupting lives and routines of entire strata in specific places very suddenly.

In addition to these social scientific points which address the three primary dimensions of social stratification and the rates of continuous versus abrupt socioeconomic change, there is the matter of the human costs of abrupt socioeconomic change which cannot be measured merely in terms of losses in jobs and income. The consequences of deindustrialization in the Pittsburgh region for which there is relatively good statistical coverage also came with profound human costs in the communities most affected. The following reports from the Mon Valley illustrate these. One area mental health agency noted a heightened incidence of self-referrals: "The clients suffer from personality disorders, chronic anxiety, chronic depression and acute depression. This depression and frustration has often led to severe marital conflicts, sometimes resulting in separation and divorce. Domestic violence is on the rise, and many afflicted families struggle for survival . . ." (Snyder, 1986, p. 52). The Turtle Creek Valley MH/MR Unemployment Project Newsletter (July/August 1985) reported on the heightened sensitivity and confusion which disrupted the normal emotional balance of some families during this period, "[Two women] spoke of hating their sons, 13-years-old . . . who they experienced as maliciously eating them out of house and home . . . at a time when they [could] barely afford to put food on the table . . . both had husbands who had been laid off . . . both could not understand how they had gone from [being] loving mothers to hateful mothers" (Snyder, 1986, p. 52). Hoerr (1988, p. 11) states that by 1987:

> The mill towns, once so alive with the heavy throb of industry, now gave off the weak pulse of welfare and retirement communities. The degree of suffering caused by lost jobs, mortgage foreclosures, suicides, broken marriages, and alcoholism was beyond calculation. Many people, especially the young, had left the valley, but middle-aged and older workers, unable or unwilling to migrate from the only home they had known, went through the anguish of trying to start new careers.

Stress related to unemployment and depressed wages also increased racial tensions in these communities. Cunningham (1986, p. 91) reported the "Worsening poverty of black people in a climate of latent racial hostility." Margolis, Burtt and McLaughlin (1986, p. 30) indicated that, "Interviews revealed some white residents and leaders already attributed the growth of the economic problems to the influx of blacks . . . if the economic base continues to decline, race relations could worsen." This misperception prevailed even in areas where the size of the black population remained essentially constant since the 1950s. Such decimation to the manufacturing base of the Pittsburgh region and the consequent rending of the social fabric of many of its communities suggests that researchers who study social structural change need not forego all interest they may have in abrupt structural shifts and social dislocations as outmoded intellectual fashion left over from the nineteenth century, even while bearing in mind that the typical course of structural change in advanced industrial societies is relatively gradual and not usually so sensitive to the changing fortunes of specific economic sectors. Whatever conscious or unconscious power alignments ultimately pushed such rapid change in the Pittsburgh region with such deleterious consequences for so many among the local population, those who would have preferred the indefinite perpetuation of the status quo were powerless to stop it. Although the earlier status quo based on the region's traditional manufacturing economy was hardly egalitarian, for decades it afforded hundreds of thousands of people a higher standard of living than what they might otherwise have had. Local government officials and union leaders attempting to preserve the status quo were not as powerful as the market forces and corporate actors aligned against them. Nevertheless, the economic and human costs of restructuring might have been reduced with closer attention to early warning signs in the relevant market trends, and communication and coordination between public and private sectors and labor to enable a smoother regional economic transition.

NOTES

1. Additionally, gender-based inequalities are nearly universal, but gender as a general content dimension of stratification was not discussed in Haller (1970). He was mainly considering households as the primary units of analysis during a time when households headed by women were more scarce than today.

2. Access to water transport and the local availability of important factors of production such as coking coal were among the reasons why steel producing regions developed in some places and not others.

3. The criterion of 40% below the poverty line for extreme poverty corresponds very closely to ghetto areas based on observation of housing conditions and the judgments of city and local census bureau officials (Wilson, 1993, p. 13n).

4. The SIC codes in the STF3A are at the two-digit level, with fourteen industry categories for 1980 and fifteen industry categories for 1990, while the major occupational groupings consisted of thirteen major occupational titles for both census decades.

5. Although handlers and laborers can also be considered blue-collar, this group was not included with the blue-collar category here because it represents relatively low-paid work in comparison to machine operators and precision production workers so it is not expected to have had a large impact on the economic base of the communities where these workers lived.

REFERENCES

Berger, J., Cohen, B. P., & Zeldich, M., Jr. (1966). Status characteristics and expectation states. In: J. Berger, M. Zeldich, Jr. & B. Anderson (Eds), *Sociological Theories in Progress* (pp. 29–46). Boston: Houghton Mifflin.

Cunningham, J. (1986). Conclusions: The issues that emerge. In: J. Cunningham & P. Martz (Eds), *Steel People: Survival and Resilience in Pittsburgh's Mon Valley* (pp. 87–97). Pittsburgh: River Communities Project School of Social Work, University of Pittsburgh.

Dickerson, D. C. (1986). Out of the crucible: Black steelworkers in western Pennsylvania, 1875–1980. In: J. Howard & R. Smith (Eds), *SUNY Series in Afro-American Studies*. Albany: State University of New York Press.

Gamson, W. A. (1968). *Power and discontent*. Homewood, IL: Dorsey Press.

Hall, C. G. (1997). *Steel phoenix: The fall and rise of the U.S. steel industry*. New York: St. Martin's Press.

Haller, A. O. (1970, December). Changes in the structure of status systems. *Rural Sociology*, *35*(4), 469–487.

Haller, A. O., & Saraiva, H. U. (1970). *Status measurement and the variable discrimination hypothesis in an isolated Brazilian region*. Unpublished paper.

Haller, A. O., & Saraiva, H. U. (1973). *The variable discrimination hypothesis in stratification*. Unpublished manuscript.

Heraclito de Lima, J. G. (1991). *Restructuring the U.S. steel industry: Semi finished steel imports, international integration, and U.S. adaptation*. Boulder: Westview Press.

Hoerr, J. P. (1988). *And the wolf finally came: The decline of the American steel industry*. Pittsburgh: University of Pittsburgh Press.

Jacobson, L. (1987). Labor mobility and structural change in Pittsburgh. *Journal of the American Planning Association*, *53*(4), 438–448.

Kasarda, J. D. (1988). Jobs, migration, and emerging urban mismatches. In: M. G. H. McGreary & L. E. Lynn, Jr. (Eds), *Urban Change and Poverty* (pp. 148–198). Washington, DC: National Academy Press.

Margolis, M., Burtt, R. E., & McLaughlin, J. (1986). Impact of industrial decline: Braddock, North Braddock and Rankin. In: J. Cunningham & P. Martz (Eds), *Steel People: Survival and Resilience in Pittsburgh's Mon Valley* (pp. 9–32). Pittsburgh: River Communities Project School of Social Work, University of Pittsburgh.

Massey, D., & Denton, N. A. (1993). *American apartheid: Segregation and the making of the underclass*. Cambridge: Harvard University Press.

Portes, A. (2000). The resilient importance of class: A nominalist interpretation. *Political Power and Social Theory*, *14*, 249–284.

Ricketts, E. R., & Sawhill, I. V. (1988). Defining and measuring the underclass. *Journal of Policy Analysis and Management*, *7*(2), 316–325.

Saraiva, H. (1969). *The 'variable discrimination' hypothesis and the measurement of socioeconomic status in an isolated Brazilian region*. Unpublished dissertation. Madison: University of Wisconsin.

Sewell, W. H. (1940). *The construction and standardization of a scale for the measurement of socio-economic status of Oklahoma farm families*. Technical Bulletin 9. Stillwater: Oklahoma Agricultural Experiment Station.

Singh, V. P. (1991, November). The underclass in the United States: Some correlates of economic change. *Sociological Inquiry*, *61*(4), 505–521.

Snyder, K. (1986). The valley's people: Effects on family and elderly. In: J. Cunningham & P. Martz (Eds), *Trouble in Electric Valley: Local Leaders Assess the Difficult Future of East Pittsburgh and Turtle Creek* (pp. 48–59). Pittsburgh: River Communities Project, School of Social Work, University of Pittsburgh.

Svalastoga, K. (1965). Social differentiation. In: *McKay Social Science Series*. New York: David McKay Company.

Turtle Creek Valley MH/MR (1985, July/August). *The unemployment project newsletter*. Turtle Creek: Turtle Creek Valley Mental Health/Mental Retardation.

U.S. Bureau of Economic Analysis (1997). *Regional economic information system: 1969–1996*. CD-ROM. Washington, DC: U.S. Department of Commerce.

U.S. Bureau of Labor Statistics (1996). *Consumer price index*. Washington, DC: Bureau of Labor Statistics.

U.S. Bureau of the Census (1970). Census of population and housing 1970, fourth count summary tape, file 4A. Princeton University Data Archives, Study Number 808. Washington, DC: Bureau of the Census.

U.S. Bureau of the Census (1980). Census of population and housing 1980, summary tape file 3A [computer file]. In: *Inter-University Consortium for Political and Social Research, Ann Arbor [Distributor]*. Washington, DC: Bureau of the Census [Producer].

U.S. Bureau of the Census (1983). Census tracts, Pittsburgh PA: Standard metropolitan statistical area. In: *1980 Census of Population and Housing*. Washington, DC: U.S. Department of Commerce.

U.S. Bureau of the Census (1992). Census of population and housing 1990, Summary tape file 3 on CD-ROM. Washington, DC: Bureau of the Census.

U.S. Bureau of the Census (1993). Population and housing characteristics for census tracts and block numbering areas, Pittsburgh PA PMSA. In: *1990 Census of Population and Housing*. Washington, DC: U.S. Department of Commerce.

U.S. Bureau of the Census (1994). *Geographical areas reference manual*. Washington, DC: U.S. Department of Commerce.

Wacquant, L. J. D., & Wilson, W. J. (1993). The cost of racial and class exclusion in the inner city. In: W. J. Wilson (Ed.), *The Ghetto Underclass: Social Science Perspectives* (pp. 25–42). Newbury Park: Sage.

Weber, M. (1947). *Max Weber: The theory of social and economic organization* (first Free Press paperback edition 1964). New York: Oxford University Press.

Wilson, W. J. (1990). *The truly disadvantaged: The inner city, the underclass, and public policy* (2nd ed.). Chicago: University of Chicago Press.

Wilson, W. J. (1993). The underclass: Issues, perspectives, and public policy. In: W. J. Wilson (Ed.), *The Ghetto Underclass: Social Science Perspectives* (pp. 1–24). Newbury Park: Sage.

Wilson, W. J. (1996). *When work disappears: The world of the new urban poor*. New York: Alfred A. Knopf.

PHYSICAL AND MENTAL HEALTH STATUS OF ADOLESCENT GIRLS: A COMPARATIVE ETHNIC PERSPECTIVE

Meredith Kleykamp and Marta Tienda

Adolescence is a time of physical and emotional transition, and a crucial period for identity formation as sex roles become differentiated and individual identities, including ethnicity, crystallize. The onset of puberty brings physical changes that require emotional adjustment and impose behavioral challenges for youth, who begin experimenting with adult behaviors even as they have little appreciation for how some actions can affect their health status in later years. Experimentation with adult roles often places adolescents at risk of health-compromising behavior, particularly when drugs, alcohol or tobacco are involved. For girls, unprotected sexual activity not only increases the risk of contracting sexually transmitted diseases, but also the likelihood of unintended pregnancy. Eating disorders and early initiation into sexual activity also jeopardize long-term health prospects of youth. Experiences with physical or sexual abuse lead to high levels of stress, low self-esteem and suicidal ideation – all indicators of poor mental health. To the extent that self-esteem is tied to social or cultural identity, minority adolescents are particularly vulnerable to poor self-images, especially if their differences are made conspicuous by language difficulties or phenotypic markers.

The Shape of Social Inequality: Stratification and Ethnicity in Comparative Perspective
Research in Social Stratification and Mobility, Volume 22, 149–185
© 2005 Published by Elsevier Ltd.
ISSN: 0276-5624/doi:10.1016/S0276-5624(04)22006-6

Parental absence during formative years also places adolescents at risk. Single parents are less able to monitor their children's activities, and youth from parent-absent homes are more likely to engage in risky health behaviors such as alcohol and drug use. Residing in a broken home exposes adolescents to greater risk of physical and sexual abuse, especially adolescent girls (Finkelhor, 1994). Single-parent households also face greater difficulty gaining access to health insurance and health care due to limited financial resources. Poverty and poor economic prospects force many adolescents, especially racial and ethnic minorities, to live in dangerous neighborhoods, placing them at risk of physical harm and violence.

Social class differences between minority and non-minority youth complicate the task of disentangling differences in health status and behavior due to group membership from that due to social and economic circumstances. Therefore, it is important not only to document the range of variation in the health status of minority adolescents, but also to consider whether such differences persist when they are compared to whites with similar characteristics that affect health status and care-seeking behaviors. Race and ethnic differences in the risks of health-compromising behavior emerge during adolescence not only because minority teens are more likely to be poor, which implies more limited access to preventive health care and greater exposure to risky social environments, but possibly also because of group-specific differences in tolerance for physical abuse, early sexual activity, and trust in the medical system.

In this chapter we examine the physical and mental health status of adolescent girls from a comparative ethnic perspective. Because there are relatively few studies of Hispanic girls' health, it is difficult to know whether and which aspects of health status or risk-taking and health-seeking behavior are unique to them. Therefore, we compare Hispanic adolescent girls to black and white adolescent girls and to Hispanic boys on various indicators of physical and mental health status. Comparisons with black and white adolescent girls help identify possible cultural differences; comparisons with Hispanic boys isolate sex differences within a common culture. Using the *Commonwealth Fund Survey of the Health of Adolescent Girls*, we consider two dimensions of health status, namely physical and mental well-being, and several behavioral indicators that either compromise or enhance health status. To characterize the mental health status of Hispanic adolescent girls we examine measures of self-esteem, depression, stress, and suicidal ideation. Most of these psychiatric disorders are inter-correlated, hence youth who fare poorly on one outcome will most likely fare poorly on one or more of the others. Indicators of physical well-being examined include exposure to physical and sexual abuse, experiences with violence, and perceived safety. Consideration of risky and health-compromising behavior by adolescents, namely the prevalence and correlates of substance abuse, eating disorders, and health-seeking behavior provides some leverage for making recommendations about possible policy interventions.

Overall, we find that adolescent girls are particularly vulnerable during adolescence when they reside in parent-absent homes. Girls who live with a single parent or neither parent are highly susceptible to poor mental health, abuse and violence, substance abuse and are often unable to obtain adequate medical care. Family structure appears to have stronger influence on the well-being of girls than on boys. Low socioeconomic status also places youth at higher risk of poor mental health, as well as abuse or violence, and it has a particularly strong effect on adolescent girls' use of illegal drugs. Much of the differences between white, black and Hispanic adolescents' well-being can be explained by family structure and socioeconomic status.

Our approach to characterizing the health status of adolescent girls is largely descriptive, but with due attention to the statistical significance of differences among groups compared. These differences are assessed using bivariate contrasts between boys and girls, and comparisons among white, black and Hispanic adolescent girls. Because demographic groups differ in socioeconomic and other characteristics that are systematically related to health status, physical well-being, and propensity to seek preventive care, we also use multivariate techniques to determine whether, in what ways, and to what extent minority adolescents differ from their white age counterparts.

The chapter is organized along the three domains of health status examined plus a section on health-seeking behavior. Each section is prefaced by a selective review of previous studies that situate results from the *Commonwealth Fund Survey* against a backdrop of existing empirical evidence. Before proceeding with the empirical results, the following section briefly describes the measurement of the core constructs analyzed. The concluding section provides a general appraisal of minority girls' health status, highlighting whether and in what ways they differ from white adolescents, and identifying how policy intervention may enhance prevention.

DATA AND MEASUREMENT

The *Commonwealth Fund Survey of the Health of Adolescent Girls* is a nationally representative sample of adolescent girls ($N = 3{,}586$) and boys ($N = 3{,}162$) enrolled in grades five through twelve during the 1996–1997 school year (Commonwealth Fund, 1997). Adolescents were selected from (and interviewed in) classrooms in 265 schools drawn from a nationally representative cross-section of public, private and parochial schools. Approximately half of the respondents were enrolled in middle school and the other half in high school. Because the sampling involved a stratified design that ensured representation of black, white and Hispanic adolescents from urban, suburban and rural schools, weights inverse to the probability of selection are applied to reach national representativeness. We restrict our analysis to 2,833 girls (353 Hispanic; 1,947 white; and 533 black) and 2,353 boys (269 Hispanic; 1,668 white; and 416 black) who provided valid data about their race and ethnic origin. Appendix Table A.1 reports weighted and unweighted sample sizes for the adolescent surveys.

Operational Definitions

The dependent variables examined fall into three general categories, namely mental health, physical well-being, and risk-taking behaviors that compromise health status and various indicators of health-seeking behavior, including differences in access to care. We also discuss key independent variables, particularly socioeconomic status, family structure, and race and ethnic origin.

Mental Health Status. We analyze four indicators of mental health status using the adolescent surveys. These include: self-esteem, depression, reports of suicidal thoughts, and stress. In the adolescent survey, *self-esteem* is measured using Rosenberg's 10-item self-esteem scale (Rosenberg, 1965). Each item is coded on a Likert-type scale indicating how strongly respondents agree or disagree with specific items. Low scale values represent low self-esteem on a scale ranging from a minimum value of 10 to a high of 40. For tabular analyses, values are coded into an ordinal scale where scores under 25 represent low esteem, scores from 25 to 34 represent moderate esteem, and values in excess of 34 represent high esteem. Cases missing data for one item are assigned the mean of the other 9 items, but those missing responses on more than one item are excluded.

Depression is based on the Children's Depression Inventory (Kovacs & Beck, 1977) using 14 of the 27 items in the original scale. Item-specific values range from 1 to 3, but responses are recoded with values ranging from 0 (low) to 2 (high), and summed. The minimum and maximum values are 0 and 28, representing low and high depression, respectively. We coded a maximum of two missing responses, substituting the mean of other values for the non-responses. In the few instances where multiple answers were provided, the most conservative is used. For tabular analyses, we constructed an ordinal scale where scores of 8 or less represented low or no depression, scores of 9–12 moderate depression, and in excess of 12 high depression.

Respondents who reported they had thought of suicide (both those who think about suicide but would not do it, and those who claim they wanted to kill themselves) are classified as *suicidal*. To measure *stress*, we sum a 17-item scale with item values from 0 to 2 where 2 indicates high stress and 0 no stress. Adolescents with composite average scores of 1.2 or greater, or who had at least six items with a score of 2, are classified as *high stress*; those with average composite scores ranging from 0.6 to 1.2 are classified as *moderate stress*; and those with average scores under 0.6 and no single item with a score of 2 are grouped into the *low stress* category.

Finally, adolescents were asked about the number of stressful life events experienced in the last year. These include: moving to a new home; a new family member; a new school; a serious family illness; parental separation/divorce; parental job loss; death of a close family member; death of a close friend; parental legal difficulties; or unspecified stressful events. Of these events, four are considered the most *negative* stressful life events: parental separation/divorce; parental job loss; death of a close friend; and family experiencing legal difficulties. For this index, based on a simple tally of negative events, values range from a minimum of 0 to a maximum of 4.

Health Status and Physical Well-Being. We analyze several indicators of overall health status and physical well-being. The most common self-reported measure of health status asks respondents to describe their own health status as excellent, good, fair or poor. To measure physical and sexual abuse, in separate questions, adolescents reported whether they were ever physically or sexually abused. For those who responded affirmatively, several follow-up questions inquired about the location and perpetrator of the abuse. From these two items, we create a composite variable with four categories, namely: no abuse; both sexual and physical abuse; sexual abuse only; and physical abuse only. These items feature very little missing data (about 3%). Because safety is a fundamental aspect of physical as well as emotional well-being, and several recent studies have shown that domestic violence is far more prevalent than previously acknowledged, we examine whether domestic violence was a serious problem for adolescents. Adolescents reported whether violence in the home, or the threat of domestic violence, ever made them want to leave home, which we use as a measure of severe violence. This variable contains two categories indicating affirmative (nearly 25% of adolescent girls responded yes) or negative responses.

Risky and Unhealthy Behaviors. Behaviors that compromise future health status and emotional well-being include the incidence of eating disorders and use of alcohol, tobacco and drugs. To identify eating disorders, adolescents reported binging and purging behavior, and the frequency of their cigarette use. Regular cigarette smokers are youth who smoked several cigarettes the week before the survey, and include self-designated smokers who did not specify a frequency of use. Similarly, for alcohol consumption, respondents were asked to describe their use of alcohol, based on several possibilities, including: never used; tried once or twice; occasional drinker; monthly drinker; weekly drinker or drinker (frequency unspecified). Regular drinkers are those who reported drinking monthly, including drinkers of unspecified frequency. Drug use refers to use during the past month.

Health Seeking Behavior and Access to Care. Most indicators reported above characterize health status or behaviors that compromise physical or mental well-being. Because previous studies document race and ethnic differences health care behavior, we also investigate whether minority girls have difficulty accessing the health care system. Adolescents reported if they had a regular health care provider and the usual source of care (i.e. physician, clinic, school nurse). Responses are collapsed into five categories: physician, clinic (non-school based), school nurse or clinic, emergency room and other. The "other" category includes a variety of sources, such as: parents, pharmacy, hospital, military medical service provider, home and any other provider.

Girls also reported the frequency they saw a physician during the past 12 months; whether there was a time that they needed care but had not received it; if they had *ever* gone without needed health care; and whether some topics are either too embarrassing or uncomfortable to discuss with their health care providers. The latter question helps to evaluate claims that cultural factors are responsible group differences in health-seeking behavior.

Independent Variables

Our main interest is in the health status of adolescent girls, but comparisons with boys help identify gender-specific differences. To compare minority and non-minority girls, we use self-reported race or ethnic background which is based on a set of pre-specified categories.[1]

Only 4–5% of girls and boys, respectively, reported that they did not know their race or ethnic background, but this low non-response rate is based on the subset of youth who actually answered the question. Because this item was placed at the end of the adolescent survey, a non-trivial share of students did not answer, most likely because they ran out of time. This is consistent with diagnostic analyses showing that students who did not respond to the race/ethnic item also failed to provide answers to several items located at the end of the survey instrument. Our interest in ethno-racial comparisons requires valid data on the race and ethnic identifier. We also excluded adolescents who reported Asian or Native American origin.[2]

Socioeconomic Status. In general, it is difficult for youth to provide accurate information about their parents' income and socioeconomic status. In the *Commonwealth Fund Survey* socioeconomic status is assessed with responses to questions about mother's education and a qualitative evaluation of income adequacy. Respondents were asked to select among four responses indicating family financial status: family has a hard time getting enough money for food, clothing, and basic living costs; family has just enough money for food, clothing and basic living costs; family has few problems buying what your family needs; family has no problems buying what your family needs and is able to buy special things. Adolescents who reported that their families had a difficult time meeting basic living costs are classified as low socioeconomic status, as were youth whose mother had a high school education or less and could barely meet basic needs. Youth from families with college-educated mothers who could just meet basic needs are classified as lower middle class, while those who experienced no difficulties meeting basic needs represent the middle class, irrespective of parental educational status. Finally, youth whose families experienced no financial difficulties and were able to purchase extras are the upper middle class if their mother had high school or some college education. Youth whose parents are college graduates and who experienced no difficulty meeting needs or providing for special things are considered affluent.

Sample Characteristics

Table 1 reports the age composition and family characteristics of adolescent boys and girls. There are no sex differences in the age composition of the sample, although Hispanic girls are slightly more highly represented at the lower ages than their black and white counterparts.[3] In general, the socioeconomic profiles of adolescent boys resemble those of adolescent girls, with the noteworthy exception that white and Hispanic boys are more likely than girls to be classified in the upper socioeconomic group. Also, whereas 8% of Hispanic girls failed to answer questions about their parents' education and/or their income shortfalls, only 4% of Hispanic boys did so. These discrepancies in non-response may contribute to sex differences in socioeconomic status, but the impact is not likely to be large.

The largest race and ethnic differences in background characteristics obtain for family structure. Whites are most likely to reside with two parents – 80% and 84% for girls and boys, respectively. Consistent with national data, black youth are more than twice as likely as whites to live with one parent. Among Hispanics, slightly higher shares of adolescent girls compared to boys lived with a single parent – 24% vs. 21%, respectively – and girls are slightly more likely than boys to live with neither parent – 5.1% vs. 3.9%, respectively. Family structure is important for health outcomes because it is associated with risks of violence and abuse, with stress, and lower levels of parental supervision that in turn influence the likelihood

Table 1. Social and Family Characteristics of Adolescent Girls and Boys by Race and Hispanic Origin (Means or Percents)[a].

	Adolescent Girls				Adolescent Boys			
	Hispanic	White	Black	Total	Hispanic	White	Black	Total
Age								
9–10	11.3	7.3	5.5	7.5	12.8	5.9	6.0	6.7
11–12	24.5	23.3	32.3	25.0	27.0	21.5	31.7	23.7
13	15.4	12.7	13.6	13.2	12.6	13.3	10.9	12.9
14	13.8	13.8	11.5	13.4	12.8	14.8	13.8	14.4
15	9.3	14.6	11.7	13.5	7.8	15.0	11.1	13.5
16	15.0	12.7	13.7	13.1	12.9	12.8	11.8	12.6
17	7.2	10.6	8.1	9.8	6.5	10.2	8.3	9.5
18+	3.4	5.0	3.6	4.6	7.7	6.6	6.4	6.7
Total	100.0	100.0	100.0	100.0	100.0	100.0	100.0	100.0
Mean	13.5	13.9	13.6	13.8	13.5	14.1	13.7	14.0
N	312	1,827	427	2,566	303	1,853	374	2,530
% missing	3.2	1.3	1.0	1.5	3.3	1.9	4.0	2.4
Socioeconomic status								
Low	16.4	11.5	13.4	12.4	18.2	8.2	15.1	10.4
Lower middle	15.1	10.8	20.0	12.9	12.7	8.2	14.0	9.6
Middle	25.8	26.2	22.1	25.4	31.5	29.2	21.0	28.3
Upper middle	26.9	26.1	20.4	25.3	17.0	25.1	25.0	24.1
Upper	15.8	25.3	24.1	24.0	20.6	29.3	25.1	27.6
Total	100.0	100.0	100.0	100.0	100.0	100.0	100.0	100.0
N	297	1,801	419	2,516	301	1,824	365	2,490
% missing	8.0	2.7	2.9	3.4	4.0	3.5	6.4	4.0
	χ^2 (8) = 90.78, p = 0.000				χ^2 (8) = 144.59, p = 0.000			
Family structure								
2 parent	70.5	79.6	46.5	73.0	75.5	83.7	48.3	77.5
1 parent	24.4	18.1	41.4	22.7	20.6	14.8	44.2	19.8
Other	5.1	2.3	12.1	4.3	3.9	1.5	7.6	2.7
Total	100.0	100.0	100.0	100.0	100.0	100.0	100.0	100.0
N	313	1,829	425	2,567	300	1,846	373	2,519
% missing	2.8	1.2	1.3	1.4	4.1	2.3	4.4	2.9
	χ^2 (4) = 421.70, p = 0.000				χ^2 (4) = 461.89, p = 0.000			
Sample share	12.4	71.1	16.5	50.1	12.1	72.9	15.0	49.9
Sample *N*	322	1,852	431	2,605	313	1,890	390	2,593

[a] χ^2 degrees of freedom in parenthesis.

that youth will engage in health-compromising behaviors. The key message from Table 1 is that race and ethnic differences in health status must consider group variation in socioeconomic status and family structure in order to draw inferences about group-specific behavior and outcomes.

MENTAL HEALTH STATUS OF HISPANIC ADOLESCENTS

During their teen years, minority youth are subjected to the difficulties and pressures associated with the cognitive, social and emotional changes that accompany puberty, as well as the compounding effects of minority group status. Not surprisingly, between the ages of 14 and 18, the co-incidence of psychiatric disorders, such as depression, low self-esteem, stress and suicidal thoughts peaks (Institute of Medicine, 1989; Millstein & Litt, 1990). These include issues of social identity, cultural integration and discrimination – all of which have mental health implications that are particularly salient among Hispanics (Porter & Washington, 1993).

Peers exert a powerful influence on adolescents' self-esteem, especially among girls (Brown, 1990, p. 191). Self-esteem influences (and is influenced by) identity formation, which in turn shapes adult aspirations. The importance of self-esteem to general mental health stems from its high association with depression, suicidal behavior and a myriad of conduct disorders (Harter, 1990; Knight, 1994). Feelings of worthlessness, an aspect of low self-esteem, are also a symptom of depression (Lennon, 1996). Thus, promoting high self-esteem among adolescents is an important goal to prevent low mental health status among adults.

There is some evidence that childhood sexual abuse influences adult self-esteem (Beitchman et al., 1992; Green, 1993; McCauley et al., 1997). Walitzer and Sher (1996) and Geller et al. (1998) trace low self-esteem to alcohol problems, psychiatric disorders and eating disorders, but their analysis does not address adequately issues of causal order, namely whether low esteem leads to alcohol, psychiatric, and eating disorders, or whether the latter produce low esteem. Most likely the relationship is reciprocal and self-reinforcing, but the key triggers may differ over the life cycle and among population subgroups.

Some research suggests that native- and foreign-born youth manage stress differently (Burnam et al., 1987; Cervantes & Castro, 1985). Native-born Mexican Americans have higher rates of mental disorders and substance use disorders than do their foreign-born peers, even though the foreign-born have likely experienced greater stress from immigration and adaptation to a new culture (Burnam et al., 1987). Most research about Hispanics takes for granted that language differences, cultural differences and migration experiences produce high levels of stress. Therefore, many analysts assume that group differences capture cultural differences and exposure to stressful experiences (Burnam et al., 1987; Vega et al., 1984). Such tautological reasoning ignores the logically prior question, namely whether Hispanic adolescents actually experience a higher level of stress than their non-Hispanic counterparts, all things being equal. Our analyses provide relatively weak evidence.

Several studies show that depression is more prevalent in girls compared to boys (Brooks-Gunn & Reiter, 1990; Millstein & Litt, 1990). Although the highest rates of depression are found among women aged 25–44, Horton (1992) claims that the peak age of onset appears to be declining. Also, a recent study by Bifulco et al. (1998) reveals that depressive episodes during adolescence predict depression in mature women. Specifically, she finds that adult women who endured a period of depression before age 20 are highly likely to experience another.

Table 2 compares mental health of black, white and Hispanic boys and girls using reported self-esteem, depressive symptoms, stress levels and experiences of negative life events. Because social class and family structure are associated with poor mental health and because Hispanics and Blacks are more highly represented among lower socioeconomic strata than their white age-mates, we evaluate differentials in mental health status after adjusting for socioeconomic and family structure characteristics.

Tabular results show a strong association between self-esteem and minority group status for adolescent girls.[4] Hispanic and white teens are similar in their level of self-esteem, and both groups report lower self-esteem compared to their black age mates. Over half of black adolescent girls report high self-esteem, but only two in five white or Hispanic girls do so.

Although the association between group membership and depression is not statistically significant, the bivariate results are consistent with prior studies showing that adolescent Hispanics experience a higher incidence of depression than their white and black age mates. Specifically, 11% of Hispanic teens reported a high level of depressive symptoms compared to 6% of black girls.

Not surprisingly, girls with low self-esteem and high levels of depression are more likely to think about or attempt suicide than their counterparts with high esteem. Among adolescent women, the highest levels of suicidal ideation correspond to Hispanics and the lowest to blacks. In fact, the mental health status of Hispanic adolescents is worse than that of black girls on all three indicators. Low self-esteem is the key marker of poor mental health status for Hispanic adolescents.

Comparisons with Hispanic adolescent boys revealed additional dimensions of disadvantage in Hispanic girls' mental health status. Relative to their male counterparts, Hispanic girls are significantly more likely to report low self-esteem,

Table 2. Mental Health Indicators of Adolescent Girls and Boys by Race and Hispanic Origin (Means or Percents)[a].

	Adolescent Girls				Adolescent Boys			
	Hispanic	White	Black	Total	Hispanic	White	Black	Total
Self-esteem								
Low	11.0	11.8	7.2	11.0	9.0	6.5	4.1	6.4
Moderate	48.2	49.0	40.6	47.6	43.6	38.8	40.9	39.6
High	40.8	39.2	52.2	41.4	47.5	54.8	55.1	54.0
Total	100.0	100.0	100.0	100.0	100.0	100.0	100.0	100.0
Mean	32.2	32.0	33.7	32.3	32.9	33.9	34.0	33.8
N	289	1,749	371	2,409	255	1,693	325	2,274
% missing	10.4	5.5	13.8	7.5	18.5	10.4	16.6	12.3
	χ^2 (4) $= 45.41, p = 0.001$				χ^2 (4) $= 18.85, p = 0.160$			
Depression								
Low	73.8	78.1	82.7	78.3	81.4	86.0	86.5	85.5
Moderate	15.4	12.8	11.5	12.9	12.4	8.0	7.3	8.4
High	10.8	9.2	5.8	8.8	6.2	6.0	6.2	6.0
Total	100.0	100.0	100.0	100.0	100.0	100.0	100.0	100.0
Mean	5.7	5.2	4.6	5.2	4.7	4.0	3.8	4.0
N	311	1,785	391	2,487	288	1,767	339	2,394
% missing	3.4	3.6	9.2	4.5	7.9	6.5	13.1	7.7
	χ^2 (4) $= 18.924, p = 0.064$				χ^2 (4) $= 14.54, p = 0.257$			
Stress								
None/low	10.7	9.5	13.6	10.3	15.1	17.0	14.1	16.4
Moderate	42.4	48.4	47.4	47.5	51.3	55.8	47.0	54.1
High	46.9	42.1	39.1	42.2	33.5	27.2	38.9	29.5
Total	100.0	100.0	100.0	100.0	100.0	100.0	100.0	100.0
Mean	1.0	1.0	0.9	0.9	0.8	0.8	0.9	0.8
N	284	1,693	367	2,344	259	1,668	310	2,237
% missing	11.8	8.6	14.8	10.0	17.4	11.7	20.6	13.7
	χ^2 (4) $= 18.15, p = 0.101$				χ^2 (4) $= 14.87, p = 0.005$			
Negative life events								
0	53.6	59.1	50.1	56.9	62.5	63.1	45.6	60.4
1	29.7	29.1	33.2	29.8	24.3	25.4	32.0	26.2
2	11.2	8.9	11.3	9.6	8.7	7.5	16.7	9.0
3	4.2	2.5	4.4	3.0	2.8	2.6	4.4	2.9
4	1.4	0.5	0.9	0.7	1.7	1.4	1.3	1.4
Total	100.0	100.0	100.0	100.0	100.0	100.0	100.0	100.0
Mean	0.7	0.6	0.7	0.6	0.6	0.5	0.8	0.6
N	318	1,834	426	2,578	307	1,824	374	2,505
% missing	1.4	1.0	1.0	1.0	1.9	3.5	4.1	3.4
	χ^2 (8) $= 40.16, p = 0.015$				χ^2 (8) $= 111.4, p = 0.000$			

[a] χ^2 degrees of freedom in parenthesis.

higher levels of depressive symptoms, and higher stress levels. Hispanic girls also report more negative life events than Hispanic boys. However, the association between minority group status and each of the mental health status indicators is not uniformly significant.[5] While there appear to be large sex differences in the sources of poor mental health status, differences between Hispanics and whites are modest to negligible. This conclusion is bolstered by a multivariate analysis that simultaneously considers various correlates of mental health status in addition to minority group status, including social class, family structure, urban residence and school type.

Socioeconomic correlates of mental health status are well documented – if poorly understood. Urban residence, particularly in poor inner city neighborhoods, is associated with exposure to unsafe and stressful environments (Furstenberg et al., 1999; Kotlowitz, 1991). Growing up in such circumstances is highly stressful for adolescents, who often must navigate numerous daily challenges in their personal and social lives. Finally, school type may be related to levels of stress and general well-being because of differences in expectations for success. Especially important in this connection is the difference between public and private schools, but often these contrasts are wiped out by social class variation of the respective student bodies.

Multivariate results reported in Table 3 indicate that most of the observed differences between Hispanic and white adolescent girls in reported levels of self-esteem, depression and stress stem from other factors, but particularly socioeconomic status, that are systematically correlated both with group membership and mental health status.[6] Although the predictive power of the models is tiny, mainly because the variance in the response outcomes is also small, the results basically corroborate inferences based on the bivariate tabulations. That is, black adolescent girls are *more* likely than whites to report high self-esteem and *less likely* to admit high levels of depression, but Hispanic girls are as likely as whites to report low self-esteem and high depression. That the race differences persist among girls with similar socioeconomic backgrounds, family structure, urban residence and who attend similar types of schools attests to their robustness. However, these results challenge prior claims that Hispanic adolescents have worse mental health than their white age counterparts. If they do, it is because they are more likely to be poor, to live in cities, and to live with a single parent or no parent. For both girls and boys, the main influence on mental health status is socioeconomic status and the relationship is in the expected direction. Low status predicts high stress and depression, and low self-esteem.

Auxiliary analyses based on combined analyses of boys and girls (not reported) confirm that girls have significantly lower self-esteem than boys, and they reaffirm the higher reported self-esteem of blacks. The difference in self-esteem is driven by

Table 3. Correlates of Mental Health of Adolescent Girls and Boys (Coefficients)[a].

	Self-Esteem		Depression		Stress	
	Girls	Boys	Girls	Boys	Girls	Boys
Race/ethnicity						
Hispanic	0.25 (0.36)	−0.31 (0.54)	0.23 (0.32)	0.43 (0.38)	0.02 (0.03)	−0.02 (0.04)
Black	1.56*** (0.45)	0.53 (0.35)	−0.83** (0.32)	−0.55* (0.26)	−0.07** (0.03)	0.02 (0.03)
Socioeconomic status						
Low middle	2.05*** (0.55)	0.42 (0.61)	−1.53*** (0.47)	−1.24** (0.47)	−0.05 (0.04)	−0.10 (0.06)
Middle	1.93*** (0.48)	0.91 (0.50)	−1.67*** (0.38)	−1.07** (0.41)	−0.05 (0.03)	−0.15*** (0.04)
High middle	2.90*** (0.46)	2.92*** (0.44)	−2.39*** (0.37)	−2.33*** (0.36)	−0.11*** (0.03)	−0.24*** (0.04)
High	30.87*** (0.48)	30.00*** (0.48)	−30.26*** (0.36)	−2.68*** (0.38)	−0.16*** (0.03)	−0.24*** (0.04)
Family structure						
Single parent	−0.01 (0.32)	−0.48 (0.36)	0.09 (0.24)	0.57* (0.28)	0.00 (0.02)	0.02 (0.03)
Non−family living	−0.21 (0.69)	−1.31 (1.02)	1.00 (0.54)	1.31* (0.58)	0.03 (0.05)	0.10 (0.08)
Residence						
Suburban	−0.27 (0.38)	−0.42 (0.32)	−0.12 (0.30)	0.06 (0.28)	0.04 (0.03)	0.00 (0.03)
Rural	−0.32 (0.37)	−0.44 (0.33)	−0.24 (0.26)	0.17 (0.26)	0.05* (0.03)	0.00 (0.02)
School type						
Private	−1.31 (0.78)	−0.36 (0.50)	0.32 (0.50)	0.07 (0.36)	0.01 (0.05)	0.03 (0.06)
Catholic	−0.41 (0.60)	−0.30 (0.63)	0.32 (0.40)	0.45 (0.44)	0.12* (0.06)	0.03 (0.05)
R^2	0.06	0.06	0.07	0.06	0.03	0.04
N	2,298	2,140	2,378	2,240	2,241	22

[a] Includes controls for age modeled in one- or two- year ago categories.
$^{*}p < 0.05$.
$^{**}p < 0.01$.
$^{***}p < 0.001$.

the positive self-images of black girls because there are no significant differences in reported self-esteem among adolescent boys. These results also confirm the monotonic association between socioeconomic status and self-esteem, with the highest levels corresponding to the most affluent youth.

In summary, the analyses reported in Table 3 lend no support to claims that Hispanic adolescents experience poorer mental health than whites, but there is consistent evidence that black youth are in better mental health than their white counterparts. The mental health status differences between boys and girls appear to be larger than those among race and ethnic groups. Because adolescents, particularly the very young, are less reliable observers of their subjective health status than mature adults, the reported mental health status indicators may be subject to high levels of reporting error. Therefore, our tentative conclusion about weak to trivial differences in mental health status between Hispanic and white adolescent girls requires further empirical scrutiny using objective measures of mental health. If our suppositions about adolescents' limited ability to represent accurately their emotional well-being are correct, then we would expect more pronounced race and ethnic differences in objective indicators of health status – namely, physical well-being. We turn to this subject next.

RACE AND ETHNIC DIFFERENCES IN ADOLESCENT PHYSICAL WELL-BEING

Compared to other population groups, such as the elderly, farmworkers or adults working in hazardous occupations, adolescents are generally in good physical health. Therefore, teenagers have less contact with health care providers, on average, than either young children or most adults. Nevertheless, as a developmental period, adolescence poses formidable challenges for youth, particularly those from disadvantaged backgrounds. These challenges often are related to teens' health-compromising behavior. Impoverished social and physical environments pose serious health risks for adolescents by exposing them to violence, abuse, and varied opportunities for transgressive behavior. We consider various indicators of physical well-being to address whether, and in what ways, the health status of minority teens differs from that of whites and whether girls differ from teenage boys in physical well-being.

Table 4 reveals no race and ethnic differences in self-reported health status among adolescent girls (p-value = 0.377). However, teenage boys report better health status than girls, as 36% claim to be in excellent health compared to only 23% of girls. By contrast to girls, boys' self-reported health status differs along race and ethnic lines. Hispanic boys are significantly less likely to report excellent

Table 4. Indicators of Physical Well Being of Adolescent Girls and Boys by Race and Hispanic Origin (Percents)[a].

	Adolescent Girls				Adolescent Boys			
	Hispanic	White	Black	Total	Hispanic	White	Black	Total
Health status								
Excellent	23.3	22.9	24.5	23.2	29.1	36.1	39.2	35.8
Good	61.6	61.7	56.2	60.8	56.9	51.3	45.6	51.1
Fair	14.9	14.5	18.2	15.1	11.5	12.1	14.6	12.4
Poor	0.3	0.9	1.1	0.9	2.6	0.5	0.5	0.8
Total	100.0	100.0	100.0	100.0	100.0	100.0	100.0	100.0
N	310	1,800	404	2,514	285	1,819	375	2,478
% missing	3.8	2.8	66.2	3.5	9.1	3.7	3.9	4.4
	χ^2 (6) = 13.28, $p = 0.377$				χ^2 (6) = 50.36, $p = 0.004$			
Abuse								
Sexual only	5.9	4.7	4.3	4.8	2.9	1.4	0.6	1.5
Physical only	9.6	8.2	5.1	7.9	8.8	6.7	8.4	7.2
Sexual and physical	4.7	5.1	5.6	5.1	4.4	1.4	1.9	1.8
No abuse	79.8	82.0	84.9	82.2	83.9	90.5	89.1	89.5
Total	100.0	100.0	100.0	100.0	100.0	100.0	100.0	100.0
N	304	1,741	404	2,450	284	1,766	365	2,415
% missing	5.7	6.0	6.1	6.0	9.2	6.6	6.4	6.9
	χ^2 (6) = 14.42, $p = 0.338$				χ^2 (6) = 43.61, $p = 0.051$			
Violence at home made you want to leave								
Yes	31.5	26.5	25.6	26.9	32.0	21.1	22.7	22.6
No	68.5	73.5	74.4	79.1	68.0	78.9	77.3	77.4
Total	100.0	100.0	100.0	106.0	100.0	100.0	100.0	100.0
N	290	1,712	387	2,388	274	1,724	341	2,338
% missing	10.2	7.6	10.1	8.3	12.6	8.8	12.7	9.8
	χ^2 (2) = 7.16, $p = 0.210$				χ^2 (2) = 32.65, $p = 0.003$			
Percent rarely or never feel safe								
At home	2.8	2.1	2.8	2.3	2.4	1.4	2.1	1.6
N	8.7	36.8	11.3	56.8	7.0	24.2	7.7	38.9
At school	5.1	2.6	8.0	3.8	8.7	3.3	12.2	5.3
N	15.9	46.9	33.3	96.1	26.0	60.1	45.2	131.3
In neighborhood	5.4	3.4	7.7	4.3	6.9	3.0	9.5	4.4
N	16.5	61.3	31.5	109.3	20.1	54.6	34.8	109.4

[a] χ^2 degrees of freedom in parenthesis.

health compared to white and black boys, yet approximately similar shares of Hispanic boys and girls report only fair to good health. As we show below, these sex and ethno-racial differences in self-reported physical well-being can be traced partly to low rates of insurance, restricted access to health care, residence in high-risk environments and exposure to violence and abuse during adolescence.

Abuse and Violence

Violence and abuse pose very serious physical and mental health risks for women in general, and adolescents in particular. Several studies demonstrate that sexual abuse is associated with numerous psychological problems, including depression, suicidal behavior, low self-esteem and anxiety, eating disorders, and substance abuse (Coble et al., 1993; Flisher et al., 1997; Mennen, 1994; Nelson et al., 1995; Silverman et al., 1996). There is also emerging consensus that childhood and adolescent sexual abuse has lasting consequences for emotional and physical well-being. Adolescents are at greater risk of physical abuse compared to young children, partly because their cognitive development places them in confrontational situations more frequently than younger children, and partly because their physical development makes them more attractive targets (Coble et al., 1993; Straus, 1994). Moreover, female children are more likely to be abused than males (Silverman et al., 1996; Straus, 1994), and black children are more often victims of severe physical abuse than whites (Hampton & Gelles, 1991). Lindholm and Willey (1986) shows that black boys and girls are equally likely to suffer abuse, but that Hispanic and Anglo females are more likely than their male peers to be victims of physical or sexual abuse.

Table 4 provides suggestive evidence that Hispanic adolescents experience higher rates of abuse than white or black girls (20% vs. 18% and 15%, respectively), but the overall association between group membership and reported experiences of abuse is not statistically significant. Girls experience a higher incidence of sexual and physical abuse compared to boys. Approximately 10% of adolescent boys reported experiencing physical or sexual abuse (mainly physical abuse), compared to nearly 18% of young girls. Moreover, sexual abuse is the most frequent form of abuse reported by girls. Among boys, Hispanics are significantly more likely than whites or blacks to experience both physical and sexual abuse – 84% vs. 90%, respectively.

Violence is also common among urban youth living in impoverished environments, and its constant threat places many at risk of psychological problems (Kotlowitz, 1991). Boys experience higher risk of injury and death from firearms than do girls. According to Schwab-Stone et al. (1995), over 40% of urban youths witnessed a stabbing or shooting in the past year, three out of four urban adolescents report feeling unsafe in one or more of their common social environments. The Children's Defense Fund (1997) reports that 1.6 million adolescents ages 12–17 years old were victims of violent crime in 1994. Although Schwab-Stone et al. (1995) finds trivial race and ethnic differences in domestic violence among young girls, Hispanic teenage boys report significantly higher levels of domestic violence than their black and white peers.

Because minorities, and especially blacks, are disproportionately concentrated in poor urban neighborhoods, race and ethnic differences in abuse may simply reflect group differences in social environments. This suggests that black and Hispanic children are more likely than whites be victims of violence because they are poor and live in dangerous neighborhoods, and not because of group-specific proclivities to engage in violent behavior (American Psychological Association, 1993; Children's Defense Fund, 1997; Hammond & Yung, 1993; Strauss, 1994). To investigate this possibility, we computed several logistic regressions for each of the physical health status measures using the same covariates used to predict mental health status. *Good health* includes youth who reported excellent or good health. *Abuse* denotes youth who were ever physically or sexually abused, and *violence* refers to experiences with domestic violence severe enough to make youth want to leave home.

Table 5, which reports the multivariate results for boys and girls separately, indicates that self-reported health status of adolescent girls does *not* differ among blacks, whites or Hispanics, as suggested by the tabular analyses. Results from an analysis that combines boy and girls (not reported) shows that girls are significantly *less likely* than boys to report good to excellent health even after taking into account variation in socioeconomic status and family structure. The largest differences in self-reported health status correspond to socioeconomic status. Girls from high status families are 2.6 times as likely as those from low status families to report excellent to good health, and the comparable odds ratio for boys is 3.2. Similarly, youth from middle status families are approximately 1.7 times as likely as those from low status families to see themselves as in excellent to good health. Unequal access to health care services may undergird the observed socioeconomic inequities in health status – a point addressed in the final section.

The multivariate analyses confirm significantly higher odds of abuse among girls compared to boys (pooled results, not reported), with girls almost two times (odds ratio = 1.91) as likely as boys to report having ever been physically or sexually abused. However, the sex-specific analyses reported in Table 5 reveal trivial race and ethnic differences in experiences of physical and sexual abuse among girls or boys of comparable socioeconomic status. It is conceivable that understandings of what constitutes abuse differ among black, white and Hispanic youth. If so, then the

Table 5. Correlates of Physical Well-Being of Adolescent Girls and Boys (Odds Ratios)[a].

	Good Health		Ever Abused		Experienced Violoence	
	Girls	Boys	Girls	Boys	Girls	Boys
Race/ethnicity						
Hispanic	1.02	0.90	1.09	1.51	1.14	1.43*
Black	0.79	1.01	0.71	0.88	0.80	0.83
Socioeconomic status						
Low middle	1.65*	1.26	0.50***	0.58	0.62**	0.51**
Middle	1.68**	1.77**	0.60**	0.63	0.58***	0.46***
High middle	2.24***	2.92***	0.50***	0.41**	0.48***	0.30***
High	2.64***	3.23***	0.41***	0.22***	0.26***	0.25***
Family structure						
Single parent	1.00	0.74	1.41**	1.40	1.36*	1.13
Non-family living	0.97	1.56	2.20***	2.71*	2.20***	1.64
Residence						
Suburban	1.07	1.15	0.89	1.00	1.00	0.67*
Rural	0.93	0.90	0.85	0.86	0.93	0.78
School type						
Private	0.74	0.64	0.74	1.03	0.51*	0.67
Catholic	1.24	0.48***	0.54*	0.39	0.73	0.70
N	2,394	2,296	2,329	2,270	2,273	2,194

[a] Includes controls for age modeled in one- or two- year categories.
*$p < 0.05$.
**$p < 0.01$.
***$p < 0.001$.

apparent absence of ethno-racial differences merely reflects unequal conceptions of and tolerance for physical abuse and sexual abuse (Hampton & Gelles, 1991).[7] Unfortunately, there is no straightforward way of addressing this issue with the available data, and an equally plausible hypothesis – one supported by results reported in Table 5 – is that the race and ethnic differences observed in Table 4 reflect *group* variation in socioeconomic circumstances that are conducive to abuse. In fact, the statistical results show large socioeconomic differences in the odds of abuse, with lower status girls at significantly higher risk of experiencing abuse compared to girls from higher status families.

Race and ethnic differences in girls' experiences with physical and sexual abuse also stem from the weaker protections they receive in disrupted families. Teenage girls residing in single parent homes are 1.4 times as likely, and girls who live with neither parent 2.2 times as likely, to experience physical and/or sexual abuse compared to their counterparts living with both parents. Also, girls attending Catholic schools are only half as likely as public school students to report one or more episodes of abuse. Thus, to the extent that minority girls are more likely to reside in poor, parent-absent families and attend public schools, the risk of experiencing physical or sexual abuse is greatly compounded. These circumstances, rather than minority group status *per se*, largely explain why minority girls experience higher average rates of abuse. For boys, socioeconomic variation in experiences of abuse are weaker, as only two of the four socioeconomic coefficients reach statistical significance. However, residence in non-family arrangements places adolescent boys at particularly high risk of abuse.

The descriptive tabulations reported in Table 4 indicate that white and black girls are more likely than their male counterparts to report experiences of violence in the home, but that white and Hispanic youth are equally likely to experience violence. Statistical analyses based boys and girls combined (not reported) confirm that girls are 20% more likely than boys to report experiences of domestic violence. The sex-specific analyses reported in Table 5 show that Hispanic boys (but not girls) report more domestic violence than their black and white age peers. Whereas Hispanic boys were 1.4 times as likely as their white counterparts to report having experienced severe violence at home, blacks were no more likely than whites to do so. That family structure differences in the likelihood of domestic violence obtain for girls but not boys attests to their greater vulnerability when one or both parents are absent. Adolescent girls residing in parent absent families were 1.4 times as likely, and those residing in non-family living arrangements were over twice as likely to witness extreme violence at home as their counterparts living with both parents. This points to another deleterious consequence of the rise of single-parent families – one that has received less attention than teenage parenting, poor educational outcomes and deviance (McLanahan & Sandefur, 1994). For both boys and girls domestic violence is also significantly associated with low socioeconomic status. Teenage girls from high status families are only 40% as likely as those from low status families to witness domestic violence, but for boys, the risk is half as great.

In summary, the two factors that place girls at highest risk of domestic violence are parent absence and low socioeconomic status. To the extent that minority girls are more likely than whites to reside in low-income families with only one or no parent, their odds of witnessing domestic violence are more than doubled. Although adolescents seldom have much control over their family arrangements and virtually no control over their socioeconomic status, these circumstances compromise their health in myriad ways. Not only do they expose youth to risky environments, but they are also conducive to risky behaviors, such as substance abuse. The next section considers whether and in what ways adolescent girls may differ from each other and from boys in these respects.

RISKY AND UNHEALTHY BEHAVIOR AMONG HISPANIC ADOLESCENTS

Although the previous sections illustrate how social and environmental circumstances place teenagers at risk of unhealthy outcomes, most studies on adolescent health focus on their high-risk and problem behaviors, such as unprotected sex, reckless driving, and substance abuse (Office of Research on Women's Health, 1998). In this section we discuss race and ethnic differences in substance abuse to determine whether and how minority girls differ from whites and from boys in their propensity to use alcohol, cigarettes or drugs. We also consider the prevalence of unhealthy eating practices as a weight control strategy.

Unhealthy Eating Behavior

Rew (1998) claims that teen Hispanic girls are less satisfied with their bodies than whites, which puts them at higher risk of eating disorders. Bulimia, a disorder characterized by binge eating and vomiting or using laxatives for weight control, generally begins in late adolescence, but it also presents among junior high students (Herzog & Copeland, 1985; Horton, 1992). Low self-esteem and depression, typically more common among Hispanics than blacks or whites, are both a cause and consequence of bulimia. However, the causal direction is difficult to establish because of its high co-morbidity with other psychiatric conditions (Herzog et al., 1992).

As a group, Hispanics are as likely as whites but more likely than blacks to report attempting weight loss, and they are most likely to use a variety of methods to do so. According to the Centers for Disease Control (1998), 61–62% of Hispanic and white adolescent girls reported current attempts at weight loss compared to 51% of blacks. Among boys, the comparable shares are 33% for Hispanics vs. 22% for whites and 20% for blacks. Hispanic teens also report the highest rates of bulimic behavior (10%), and the highest use of diet pills to lose weight. Both are unhealthy and dangerous methods of weight loss (CDC, 1998).

Table 6 confirms the CDC findings in that significantly higher shares of Hispanic adolescent women report bingeing and purging behavior compared to blacks and whites. Among teenage girls, approximately one-in-five Hispanics admitted to bingeing and purging behavior compared to 15% of whites and 13% of black girls.

Table 6. Self-Reported Unhealthy Behavior of Adolescent Girls and Boys by Race and Hispanic Origin (Percents)[a].

	Adolescent Girls				Adolescent Boys			
	Hispanic	White	Black	Total	Hispanic	White	Black	Total
Binge/purge								
Yes	20.7	15.1	12.6	15.4	13.9	4.6	14.8	7.2
No	79.3	84.9	87.5	84.6	86.1	95.5	85.2	92.8
Total	100.0	100.0	100.0	100.0	100.0	100.0	100.0	100.0
N	298	1,691	392	2,381	295	1,813	365	2,474
% missing	7.6	8.7	9.0	8.6	5.7	4.0	6.4	4.6
	χ^2 (2) = 18.35, $p = 0.020$				χ^2 (2) = 139.64, $p = 0.000$			
Regular alcohol use								
Yes	8.0	12.4	6.7	11.0	19.1	14.6	13.6	71.1
No	92.0	87.6	93.3	89.0	80.9	85.4	86.4	28.9
Total	100.0	100.0	100.0	100.0	100.0	100.0	100.0	100.0
N	312	1,814	414	2,540	304	1,828	364	2,495
% missing	3.2	2.0	3.9	2.5	2.8	3.3	6.8	3.8
	χ^2 (2) = 28.78, $p = 0.002$				χ^2 (2) = 9.70, $p = 0.170$			
Regular cigarette smoking								
Yes	11.0	13.2	5.8	11.7	14.0	12.2	11.6	12.3
No	89.0	86.8	94.2	88.3	86.0	87.9	88.4	87.7
Total	100.0	100.0	100.0	100.0	100.0	100.0	100.0	100.0
N	315	1,824	425	2,564	306	1,861	373	2,540
% missing	2.2	1.5	1.4	1.6	2.3	1.5	4.3	2.0
	χ^2 (2) = 36.00, $p = 0.000$				χ^2 (2) = 2.14, $p = 0.733$			
Drug use-past month								
Yes	12.8	15.3	8.8	13.9	25.0	14.4	15.6	15.9
No	87.2	84.8	91.2	86.1	75.0	85.6	84.4	84.2
Total	100.0	100.0	100.0	100.0	100.0	100.0	100.0	100.0
N	305	1,796	405	2,505	303	1,790	350	2,443
% missing	5.5	3.0	6.1	3.8	3.4	5.3	10.2	5.8
	χ^2 (2) = 23.48, $p = 0.008$				χ^2 (2) = 44.42, $p = 0.001$			
Exercise								
Never	2.8	1.3	1.9	1.6	1.9	1.0	1.3	1.1
<1-2/Wk	9.2	7.8	9.9	8.3	4.2	4.0	3.2	3.9
1-2/Wk	15.8	13.0	11.8	13.1	7.2	7.7	7.3	7.6
3/Wk	19.8	20.5	15.3	19.5	11.9	14.1	11.9	13.5
Almost daily	52.4	57.5	61.1	57.5	74.8	73.3	76.3	73.9
Total	100.0	100.0	100.0	100.0	100.0	100.0	100.0	100.0
N	300	1,802	415	2,516	305	18,28	375	2,508
% missing	7.1	2.7	3.7	3.4	2.5	3.3	3.9	3.3
	χ^2 (8) = 31.14, $p = 0.085$				χ^2 (8) = 9.76, $p = 0.869$			

[a] χ^2 degrees of freedom in parenthesis.

Minority boys also exhibit a higher rate of bingeing and purging behavior than their white counterparts. A multivariate analysis based on a model that combines girls and boys confirms that girls are 2.5 times as likely as boys to binge and purge as a means of controlling weight, and that Hispanics are significantly more likely to do so than either whites or blacks (results not reported). The sex-specific analyses (Table 7) reaffirm the persistence of race and ethnic differences in the likelihood of bingeing and purging. Among boys, both blacks and Hispanics are more likely than whites to report bingeing and purging behavior. For them, participation in competitive sports based on weight classes are probably the main reason for extreme eating behavior. However, self-images that place a high value on slenderness and are bundled with self-esteem are largely responsible for girls' dangerous eating practices. Race and ethnic differences in bulimic behavior are

Table 7. Correlates of Health Compromising Behaviors of Adolescent Girls and Boys (Odds Ratios)[a].

	Bingeing		Regular Alcohol Use		Regular Cigarette Use		Drug Use	
	Girls	Boys	Girls	Boys	Girls	Boys	Girls	Boys
Race/ethnicity								
Hispanic	1.48*	3.49***	0.52**	1.56*	0.73	1.36	0.67	2.18***
Black	0.83	2.63**	0.39***	0.76	0.34***	0.74	0.41***	0.87
Socioeconomic status								
Low middle	0.93	0.38*	0.82	0.92	0.75	0.77	0.69	0.70
Middle	0.74	0.53	0.89	1.23	1.07	0.95	0.62**	0.84
High middle	0.68	0.59	0.78	1.31	0.95	1.18	0.62**	0.93
High	0.61*	0.73	1.12	1.22	0.67	0.71	0.54**	0.90
Family structure								
Single parent	0.97	1.14	1.92***	1.13	1.25	1.40*	1.70***	1.69**
Non-family living	1.43	2.66**	2.56**	2.76***	1.46	2.97**	1.83*	3.04**
Residence								
Suburban	0.97	0.63	0.92	0.74	0.71	0.66	0.84	0.90
Rural	1.20	1.25	1.01	1.25	0.98	1.06	0.81	1.07
School type								
Private	0.76	0.68	0.67	1.44	0.68	1.10	0.54	1.42
Catholic	0.95	1.12	0.94	0.92	1.38	0.80	0.67	0.47
N	2,272	2,303	2,417	2,321	2,440	2,358	2,394	2,286

[a] Includes controls for age modeled in one- or two- year age categories.
*Significant at $p \leq 0.05$.
**Significant at $p \leq 0.01$.
***Significant at $p \leq 0.001$.

more pronounced for boys compared to girls, as Hispanic males are 3.5 times as likely as whites, and blacks 2.6 times as likely, to binge and purge for weight control. Hispanic girls are approximately 1.5 times as likely as their white peers to engage in unhealthy behavior to control their weight, but there are no significant race differences.

Alcohol, Cigarette and Drug Use

Substance use by teenage girls is widespread, but there is some disagreement about whether use of specific substances is rising or falling, and whether use is higher among boys or girls (see CDC, 1998; Commonwealth Fund, 1999; Office of Women's Health, 1998, p. 25). In the main, these disagreements revolve around measurement issues. Because adolescence is a time of experimentation and risk-taking, most teens report that they have tried alcohol, cigarettes, and drugs. However, only a small subset become regular users (The Commonwealth Fund, 1999). Measures of ever use neither discriminate the problematic aspects of substance abuse, nor indicate whether race and ethnic differences in consumption of tobacco, alcohol and drugs carry over into problematic use patterns.

There is also a striking lack of consensus about race and ethnic differences in substance abuse, however. Horton (1992) reports that about one in four teenage girls (ages 12–17 years) acknowledge using alcohol in the past month. The Office of Research on Women's Health (1998) shows that alcohol use is higher among white compared to minority teenage girls, despite the stresses associated with pervasive minority poverty. Based on a 1997 sample of high schools, the Centers for Disease Control (1998) report that 83% of Hispanics ever used alcohol and girls were about as likely (82%) as boys (84%) to do so. A recent study finds that alcohol use among Hispanic teens is highly correlated with stress, anxiety and depression (Alva, 1995). According to the CDC, Hispanics report the highest levels of lifetime alcohol use, which suggests that alcohol consumption during adolescence may eventuate into problem behavior during adulthood.

Although it is not possible to establish which adolescents who experiment early with drugs, tobacco and alcohol will become lifetime substance abusers, the tabulations in Table 6 show similar race and ethnic patterns of regular alcohol use among adolescent girls. These differences are both substantively and statistically significant. Consistent with the assessment of the Office of Research in Women's Health (1998) white teens are more likely than minority adolescents to report regular alcohol use – 12% compared to about 7% to 8%. In contrast to the CDC survey, which shows about equal use of alcohol by teenage boys and girls, the *Commonwealth Fund Survey* reveals appreciable sex differences in regular alcohol use, with the boys using alcohol regularly nearly twice as much as minority girls. A statistical test (not shown) confirms that girls are only 0.74 times as likely as boys with similar characteristics to use alcohol on a regular basis.

Table 7 reports the multivariate analyses of alcohol use based on regular use, which helps discriminate the problematic aspects of adolescent drinking from experimentation. For this analysis students who drink at least once a month, at least weekly, or an unspecified regular frequency are designated as regular drinkers. The odds of regular alcohol use among Hispanic girls are half (0.52) those of white adolescents of comparable family background, and for black girls the comparable odds are approximately one-third as high (0.39). There are no race differences in regular drinking for boys, but Hispanic boys are 1.5 times as likely as whites to report regular alcohol use.[8]

Not surprisingly, regular teen alcohol use is associated with lack of parental supervision, but there is some indication that this effect is stronger for girls. Only when boys reside with neither parent is their alcohol consumption problematic. In contrast, girls who reside in single-parent families are about 2 times as likely, and those in non-family arrangements 2.6 times as likely to drink regularly compared to girls from two parent families. That there are no significant socioeconomic status differences in regular alcohol use attests to the prevalence of this behavior among adolescents, which appears to be independent of income.

Like alcohol, tobacco use differs among race and ethnic groups, but not uniformly for boys and girls. For adolescents, regular users smoke several cigarettes or more per week. About 11–13% of Hispanic and white girls report regular tobacco use compared to approximately 6% of black teenagers. In contrast to Horton (1992), who claims that girls are more likely than boys to smoke, we find no sex differences in either ever use or regular use of tobacco. Because of the widespread use of tobacco and its relatively easy access, there are no socioeconomic differences in the propensity of youth to become regular smokers during adolescence. However, black girls are only 0.34 times as likely as whites to become regular teen smokers, which is consistent with the descriptive tabulations reported in Table 6. Hispanic girls are as likely as whites to smoke during adolescence.

Although girls' smoking behavior is not influenced by their family structure, that of boys is highly responsive to the amount of parental supervision. Teenage boys who live with one parent are 1.4 times as likely to smoke as their age counterparts with two parents present. Moreover, adolescent boys who live with neither parent are almost three times as likely to use tobacco regularly compared to boys reared in an intact family. Like most prior research, we find no differences in regular tobacco use by place of residence or type of school attended – probably because peers are more decisive than adults in shaping this behavior.

Finally, illegal drug use shows some differentiation along race and ethnic lines, but differentially for boys and girls. The descriptive tabulations reported in Table 6 show that white girls report the highest rate of illegal drug use (15%), blacks the lowest rate (9%), with Hispanics between the extremes at 13%, but closer to whites. A different pattern obtains for boys, as Hispanics report the highest rates of drug use, nearly 10% points above black and white youth. Hispanic boys use illegal drugs at twice the rate of Hispanic girls, black boys do so at just under twice the rate of black girls, but there are only trivial sex differences for white teens.

The multivariate analysis reported in Table 7 confirms the lower rates of drug use among black girls compared to whites of comparable socioeconomic status, but no significant differences obtain between Hispanics and whites. Black teenage girls are only 0.4 times as likely as white teens to have used illegal drugs in the month before the survey. Unlike tobacco and alcohol use, which is relatively pervasive among adolescents and therefore less influenced by social class, girls' drug use is highly variable by socioeconomic status, but (surprisingly) not for boys. Girls from affluent families are only 0.5–0.6 times as likely as those from poor families to use drugs, and their risk of drug use is highly sensitive to parental supervision. Specifically, adolescent girls residing with only one parent are 1.7 times as likely as those with two parents to report using illegal drugs in the previous month, and those with no parent present are even more likely to do so. Even stronger family structure effects obtain for adolescent boys inasmuch as those who reside with neither parent are about three times as likely as teenage boys with two parents present to use illegal drugs, and those with one parent are 1.7 times as likely to use drugs compared to boys from intact families.

An important sex difference in drug use is that Hispanic boys are twice as likely as white boys to use illegal drugs, while there were no ethnic differences among girls. Rather, parental supervision is the most decisive influence on this and other health compromising behaviors for girls. In fact, a common theme based on the analyses of risk-taking behavior is that parent absence rather than membership in an ethno-racial group, renders adolescents vulnerable to participation in activities that compromise their health. And as we show below, parental absence also influences health-seeking behavior in deleterious ways.

HEALTH SEEKING BEHAVIOR AND ACCESS TO CARE

A recent study by the American College of Physicians-American Society of Internal Medicine (ACP-ASIM) reports that Hispanics are highly vulnerable to poor health outcomes because a large share of the population lacks health insurance, which lowers the likelihood that they will have regular health care providers and receive timely screening and diagnosis of serious illnesses (Current Topics, 2000). Although we find few statistically significant differences in health status and risk-taking behaviors among Hispanic, black and white teenagers of comparable socioeconomic status and environmental circumstances, Schoen and associates (1997) claim that the neediest girls, namely those who suffer from abuse, who lack health insurance, and who are poor are least likely to receive needed medical care.

Because minority youth are more likely to be poor than whites, observed ethno-racial differences in access to health services probably reflect socioeconomic differences rather than group-specific variation in health-seeking behavior. However, Flores et al. (1999) argue that race and ethnic differences in health-seeking behavior persist among groups of comparable socioeconomic status, particularly when pan-ethnic groups are separated into their national subgroups. They conclude that non-financial factors – such as cultural beliefs, language differences, and provider practices – are responsible for the disparities in health status and use of medical services among similarly situated groups. Given the diversity of opinion regarding ethno-racial differences in teen's health-seeking behavior, in this section we examine whether minority youth are less likely than whites to access needed medical services, and evaluate whether observed inequities reflect group-specific behavior or underlying differences in financial circumstances that affect health care utilization.

For these analyses, we examined several measures representing both access to services and health-seeking behavior. Teenagers were asked whether they had a regular health care provider and the usual source of care (i.e. physician, clinic, school nurse). Responses for the adolescents are collapsed into five categories: physician, clinic (non-school based), school nurse or clinic, emergency room and other. The latter category represents various sources that include parents, pharmacy, hospital (ER), and others, such as traditional healers.

In the *Commonwealth Fund Survey*, teens reported the number of times they had seen a physician in the past 12 months. This information is used to designate two groups – those who reported at least one visit in the prior year and those who had not. Adolescents who reported having a check-up during the prior 12 months are also classified as having a doctor visit within the previous year. Teenagers also answered a general question about health insurance coverage. To assess group differences in constraints to medical care, adolescents were asked about unmet need for medical services, and specifically whether they had ever needed medical care and not received it.

Table 8 presents tabular data comparing the health-care seeking behavior of Hispanic, white and black teenagers. Consistent with numerous prior studies, minority teens are less likely to have a regular health care provider. Approximately

Table 8. Help-seeking Behavior and Access to Health Care of Adolescent Girls and Boys by Race and Hispanic Origin (Percents)[a].

	Adolescent Girls				Adolescent Boys			
	Hispanic	White	Black	Total	Hispanic	White	Black	Total
Regular provider								
Yes	78.8	86.0	78.3	83.8	70.8	82.1	73.9	79.5
No	21.2	14.0	21.7	16.2	29.2	17.9	26.1	20.5
Total	100.0	100.0	100.0	100.0	100.0	100.0	100.0	100.0
N	302	1,763	402	2,467	290	1,759	367	2,417
% missing	6.4	4.8	6.6	5.3	7.4	6.9	5.8	6.8
	χ^2 (2) = 41.25, $p = 0.0002$				χ^2 (2) = 56.56, $p = 0.0003$			
Any doctor visits in past 12 mos.								
Yes	75.1	76.8	68.4	75.2	70.6	74.7	69.4	73.4
No	24.9	23.2	31.6	24.8	29.4	25.3	30.6	26.6
Total	100.0	100.0	100.0	100.0	100.0	100.0	100.0	100.0
N	319	1,840	428	2,587	313	1,885	385	2,583
% missing	0.9	0.6	0.7	0.7	0.0	0.2	1.3	0.4
	χ^2 (2) = 26.29, $p = 0.0046$				χ^2 (2) = 12.25, $p = 0.0983$			
Where is usual source of care								
Doctor's office	50.3	72.5	50.5	66.2	47.1	67.9	45.4	62.1
Clinic (non-school)	34.2	19.8	31.5	23.5	35.0	20.7	27.7	23.4
ER	4.6	3.4	7.4	4.2	5.7	4.9	11.4	6.0
School clinic or nurse	0.8	0.8	2.6	1.1	1.7	0.9	2.8	1.3
Other	12.2	6.3	15.9	8.6	11.6	7.8	16.3	9.5
Total	102.1	102.8	107.9	103.5	101.1	102.2	103.6	102.3

Table 8. *(Continued)*

	Adolescent Girls				Adolescent Boys			
	Hispanic	White	Black	Total	Hispanic	White	Black	Total
N	304	1,785	406	2,495	281	1,783	371	2,435
% missing	5.8	3.6	5.6	4.2	10.3	5.7	4.8	6.1
		χ^2 (2) = 13.05, $p = 0.0419$				χ^2 (2) = 32.26, $p = 0.0042$		
Have insurance								
Yes	84.0	91.4	90.5	90.4	78.5	93.3	86.5	15.9
No	16.0	8.6	9.5	9.6	21.5	6.7	13.5	84.2
Total	100.0	100.0	100.0	100.0	100.0	100.0	100.0	100.0
N	232	1,484	331	2,047	237	1,448	281	1,966
% missing	28.1	19.9	23.1	21.4	24.5	23.4	27.9	24.2
		χ^2 (2) = 25.38, $p = 0.0025$				χ^2 (2) = 120.47, $p = 0.000$		
Ever needed care but not gotten it								
Yes	29.8	26.8	34.7	28.5	27.2	19.7	29.3	22.0
No	70.2	73.2	65.3	71.6	72.8	80.3	70.7	78.1
Total	100.0	100.0	100.0	100.0	100.0	100.0	100.0	100.1
N	297	1,683	390	2,371	265	1,687	345	2,296
% missing	7.8	9.1	9.5	9.0	15.3	10.8	11.6	11.4
		χ^2 (2) = 20.16, $p = 0.0212$				χ^2 (2) = 41.60, $p = 0.0002$		
Ever too embarassed to talk to MD about problem								
Yes	41.5	35.7	41.7	37.4	30.9	19.9	23.0	21.7
No	58.5	64.3	58.3	62.6	69.1	80.1	77.0	78.3
Total	100.0	100.0	100.0	100.0	100.0	100.0	100.0	100.0
N	264	1,530	348	2,143	252	1,501	296	2,049
% missing	18.1	17.3	19.2	17.7	19.4	20.6	24.0	21.0
		χ^2 (2) = 13.05, $p = 0.0419$				χ^2 (2) = 32.26, $p = 0.0042$		

[a] χ^2 degrees of freedom in parenthesis.

one-in-five minority adolescent girls lack a regular health care provider, and for boys, these shares are higher still – almost 30% for Hispanic boys and 26% for blacks. As shown in Table 9, the Hispanic-white gap in the absence of a regular provider persists once controls are introduced for socioeconomic status, family structure, and social environment (i.e. residence and school type), but the racial gap disappears once these compositional differences are taken into account. Specifically, Hispanic girls are only 0.67 times as likely as whites to have a regular health care provider, but black teens are as likely as whites to do so if they share the same family background and family structure. Minority teenage boys are only half as likely as white teens to have a regular provider. Adolescent girls from high status families are 1.8–2.2 times as likely as those from low status families to have a regular health care provider. Moreover, those who reside with neither parent are only half as likely as their counterparts living in two-parent families to report having a regular provider – again revealing the vulnerability of youth to parent absence.

Despite the fact that black teenage girls are as likely as whites to have a regular provider, the multivariate results reported in Table 9 reveal that they are only 0.71 times as likely as whites to have made a doctor visit during the 12 months preceding the survey. Hispanic teen girls are less likely than whites to report having a regular provider, but they are as likely as white teens of similar socioeconomic circumstances to visit a physician. On balance, these findings suggest that socioeconomic circumstances and family structure are more decisive than group membership in shaping health-seeking behavior, although race differences persist for teenage girls. Adolescents who live with neither parent are especially vulnerable to receive inadequate care.

Teenage girls also differ in the sources of their health care. That minority teens are significantly less likely than white adolescents to receive their health care in a physician's office indicates their lower reliance on the private medical system. Only half of all Hispanic and black girls report that they usually receive their health care in a doctor's office, compared to 72% of whites, and the shares of teenage boys is lower still. Nearly one-third of minority teenage girls receive their health care in a general clinic, as do approximately similar shares of teen minority boys, compared to only 20% of white girls.[9] Between 3% and 5% of minority teenage girls receive their health care in the emergency room, which is an expensive fallback for the absence of a regular provider. School clinics provide medical services for less than 1% of white and Hispanic girls, but nearly 3% of black teens. Results reported in Table 9 indicate that ethno-racial differences in sources of care do not simply mirror class differences in ability to pay. Minority youth are only 0.4–0.5 times as likely as comparably situated whites to receive care in a private physician's office. Also, teen girls who reside with neither parent are only half as likely as girls who live with both parents to receive care in a physician's office.

Table 9. Correlates of Help-Seeking Behaviors of Adolescent Girls and Boys (Odds Ratios)[a].

	Regular Provider		Doctor Visit in Past Year		Usual Source is MD Office		Has Insurance		Needed but not Received Care	
	Girls	Boys	Girls	Boys	Girls	Boys	Girls	Boys	Girls	Boys
Race/ethnicity										
Hispanic	0.67*	0.49***	0.94	0.84	0.44***	0.41***	0.53**	0.34***	1.19	1.34
Black	0.71	0.53**	0.71**	0.91	0.50***	0.44***	1.17	0.72	1.22	1.46*
Socioeconomic status										
Low middle	1.66*	0.97	1.36	1.08	1.44*	1.21	2.16**	2.62**	0.85	0.51**
Middle	1.42*	1.54*	1.31	1.50*	1.40*	1.71**	2.57***	2.23**	0.72*	0.50***
High middle	1.87***	1.54*	1.58**	1.54*	1.77***	1.57*	4.48***	3.89***	0.46***	0.40***
High	2.17***	1.86**	1.63**	1.78**	1.92***	1.98***	8.55***	5.17***	0.40***	0.29***
Family structure										
Single parent	0.82	0.99	0.80	0.83	0.82	0.78	0.78	0.78	1.28*	1.16
Non-family living	0.48**	0.75	0.55**	0.65	0.47***	0.80	0.40**	0.62	1.45	1.96*
Residence										
Suburban	1.28	0.85	0.94	1.03	1.22	1.23	1.26	1.14	0.89	0.81
Rural	1.06	0.88	1.09	0.99	0.91	0.74	0.96	1.19	1.04	0.96
School type										
Private	1.14	1.05	0.84	0.67	1.80*	2.07**	1.66	1.73	0.77	1.10
Catholic	1.06	0.69	0.88	0.74	1.10	0.88	1.78	1.10	0.68*	1.10
N	2,352	2,244	2,461	2,393	2,368	2,253	1,964	1,833	2,257	2,133

[a] Includes controls for age modeled in one- or two- year categories.
*Significant at $p \leq 0.05$.
**Significant at $p \leq 0.01$.
***Significant at $p \leq 0.001$.

The availability of a regular health care provider and the freedom to seek medical services from private providers often is tied to availability of health insurance. Hispanic girls are about twice as likely as whites to be uninsured: 16% vs. 8.6% for black and white girls, respectively. Similar ethno-racial differentials in access to health insurance obtain among teenage boys, except that the inequities are more pronounced. Table 9 reveals that Hispanic youth are only 0.53 (girls) to 0.34 (boys) times as likely as white youth to have access to health insurance even when their socioeconomic and social circumstances are approximately similar. However, the racial differences in access to health insurance observed in Table 8 derive largely from black-white differences in social class, family structure and social environments.

Given these differentials in health insurance coverage, it is unsurprising that minority women are more likely to report not having received needed health care on at least one occasion. Group differences in unmet medical service needs are both substantially higher and statistically significant among minority teens: 35% for black girls and 30% for Hispanics, compared to 27% for white teenagers. However, the multivariate analyses indicate that the ethno-racial differences in unmet medical needs are due largely to differences in social class and family structure rather than differences in propensity to seek and receive needed medical care. The only exception is black teenage boys, who are 46% more likely than whites to go without necessary medical care.

In summary, these findings on access to health care and health care seeking behavior reaffirm minority youth's unequal access to the medical system because they are significantly more likely to be uninsured, to lack a regular provider, and to receive their medical care outside of the private health care delivery system. Because access to health insurance limits access to health care services, these differences portend ill for the ability of young girls, but especially Hispanics, to obtain needed preventive services. Differences in access to a regular provider suggest that Hispanic girls and minority boys are least likely to receive adequate preventive care of all groups considered. To the extent that teenagers' access to health care shapes lifelong physical and mental well-being, the inequities in access among young women portend poorer health for these groups in the future.

CONCLUSIONS AND IMPLICATIONS FOR FURTHER RESEARCH

Our main objectives in this chapter are to document the physical and mental health status of Hispanic teenage girls, including behavior that compromises or enhances their health status, and to ascertain whether these differences are "real," that is, whether they represent group-specific differences in mental and physical well-being, or whether they proxy for variation in social class, family structure and social environments conductive to poorer health.

On balance, we find relatively few differences between Hispanic and white teenagers on various indicators of mental health once comparisons are restricted to girls of similar socioeconomic status and family structure. However, racial differences in self-esteem, depression and stress reveal that white rather than black teens experience worse mental health outcomes. As important, we show that adolescents who reside either with a single parent or with no parent are especially vulnerable to poor mental health. That both black and Hispanic girls are appreciably more likely than white adolescents to live with one or neither parent places them at very high risk of poor mental health, as measured by self-esteem, depressive symptoms, episodes of stress, and suicidal ideation. Comparable mental health consequences of family structure do not obtain for teenage boys.

Although indicators of teen girls' average physical well-being reveal that Hispanics and blacks are more disadvantaged than their white peers, particularly in their experiences with abuse and violence, these differences are non-existent among girls of comparable social class, family structure and social environments. That is, the multivariate analyses indicate trivial race and ethnic differences in health status, physical and/or sexual abuse, and experiences with domestic violence among girls with similar social circumstances. While informative about ultimate causes, the fact remains that minority and non-minority girls do not share similar social environments, which is why both black and Hispanic teenagers experience higher rates of abuse and violence than their white counterparts, on average. This is particularly so for those who do not live with either parent, who are over twice as likely to experience physical or sexual abuse compared to their age-mates who reside with two parents. Family structure emerges as an important protective factor for adolescent girls, yet secular trends indicate that the share of Hispanic girls who reside with a single parent (or no parent) is increasing, thereby exposing increasing numbers to the risk of violence and abuse. This insight has important policy implications and challenges for youth advocates to devise surrogate protections for young girls whose living arrangements impair their safety.

We also find limited evidence that Hispanic girls are more likely than white teens to engage in behaviors that compromise their health status, except for bingeing as a weight control strategy. Once comparisons are restricted to girls of comparable socioeconomic status and family structure, Hispanic girls are no more or less likely than other teenage girls (or boys) to use alcohol regularly, to smoke regularly, or to use drugs. However, we find very strong evidence that lack of parental supervision increases adolescents' propensity to engage in substance abuse. In fact, family structure effects on alcohol, tobacco and drug use were generally larger than those of social class, which reveals the vulnerability of youth to transgressive behavior when parental supervision is weak or absent. The only noteworthy exception is drug use, which is more prevalent among lower status girls (but not boys). These findings reinforce the conclusion about the need for policy alternatives to protect girls reared in parent-absent homes, particularly those where neither parent is present.

Our analyses of access to health care do show some differences among Hispanic, white and black teens of comparable social class and family structure. Especially noteworthy is the lower health insurance rate of Hispanic teens relative to their black and white peers, which limits their access to the health care system. In fact, Hispanic youth (both boys and girls) as well as black teens are significantly less able to access the private health care system, as indicated by the significantly lower shares of minority adolescents who usually receive their health care in a physician's office. Equally striking are the strong family structure effects on teenage girls ability to access private medical care. Specifically, adolescents who do not reside with both parents – disproportionately minority youth – are least likely to receive health care in a physician's office. More importantly, these girls are also much more likely to report that they have not visited a physician in the past year and even have gone without needed care. Again, these results underscore the high vulnerability of adolescents who reside with single parents, or no parents.

In conclusion, we offer two policy recommendations for enhancing the health status of Hispanic girls, and protecting their physical and mental well-being as adults. First, expansion of health care insurance must become a national priority. Although the Children's Health Insurance Program technically extends health care insurance to economically disadvantaged youth, either lack of information, the complexities of enrollment, and/or the legal status of parents have limited the shares of Hispanics who avail themselves to these benefits. Equally important are the pronounced effects on various health status outcomes and behaviors produced by the weak or absent parental supervision available to teenagers who reside in "broken homes." Conceivably community strategies can be devised to reduce the vulnerability of teenagers raised by single parents, but especially those not living with either parent. The current national emphasis on mentoring programs is an important stride in that direction, but organized institutional strategies are also warranted.

NOTES

1. These categories are: White (not Hispanic); Black or African-American (not Hispanic), Hispanic/Latino – white, Hispanic/Latino – black; Hispanic/Latino – Unspecified; Asian, Asian Indian or Pacific Islander; Native American or Alaskan Native;

and Other. We exclude the latter three because they are comparatively smaller and to produce a more manageable set of comparisons.

2. We have conducted extensive sensitivity analyses of the missing responses on the race and ethnic item. These results showed that students who failed to provide their race and ethnic origin were in lower grades (7th grade or below), of low socioeconomic status and attended urban public schools compared to their counterparts whose race and ethnic status was provided.

3. Because Hispanic girls (and boys) also had higher rates of non-response to all items reported in this table, it is conceivable that the observed age differences partly reflect the non-random character of missing data.

4. The test of independence that is displayed by default is based on the usual Pearson χ^2 statistic for two-way tables. To account for the survey design, the statistic is turned into an f statistic with non-integer degrees of freedom using a second order Rao and Scott correction. Although the theory behind Rao Scott is complicated, the p-value for the corrected F-statistic can be interpreted in the same way as a p-value for the Pearson χ^2 for "ordinary" data. These statistics take into account the sample design, which is why some large χ^2 statistics are not significant statistically, i.e. have low p-values.

5. That is, for girls, the association between self-esteem and group membership is statistically significant, but not for adolescent boys, while the obverse is true for stress. Group differences in depression levels and suicidal ideation are not statistically significant.

6. We use the interval composite scores of self-esteem, depression and stress as dependent variables for each case rather than the categories described in Table 2, which are derived from the continuous measure. In such cases, OLS regression techniques are appropriate. We use the SVYREG command in the Stata statistical package to execute the analysis, which is the linear regression command for stratified samples requiring weighting of observations. This command produces the appropriate standard errors.

7. This interpretation is consistent with focus groups conducted with adolescents at a multi-ethnic high school in the Midwest where white, black and Hispanic students alike reported that physical discipline was far more common among Hispanics and blacks, especially the latter, compared to whites. However, experiences of sexual abuse were not discussed. See Hampton and Gelles (1991).

8. An analysis based on ever use showed no significant differences between Hispanic and white girls, indicating that the former are as likely as whites to experiment with alcohol, but they are less likely to become regular users during adolescence.

9. It is conceivable that poor minority girls receive their health care in community clinics, many of which offer services based on ability to pay. However, the survey did not ask whether services were obtained.

ACKNOWLEDGMENTS

We acknowledge institutional support from the Office of Population Research at Princeton University and technical assistance from Pamela Bye-Erts and Adair Iacono. Direct all correspondence to Marta Tienda at the address indicated above.

REFERENCES

Alva, S. A. (1995). Psychological distress and alcohol use in hispanic adolescents. *Journal of Youth and Adolescence*, *24*(4), 481–498.

American Psychological Association (1993). *Commission on youth and violence summary report. Violence and youth: Psychology's response* (Vol. 1). Washington, DC: American Psychological Association.

Beitchman, J. H., Zucker, K. J., Hood, J. E., DaCosta, G. A., Akman, D., & Cassavia, E. (1992). A review of the long term effects of child sexual abuse. *Child Abuse and Neglect*, *16*, 101–118.

Bifulco, A., Brown, G. W., Moran, P., Ball, C., & Campbell, C. (1998). Predicting depression in women: The role of past and present vulnerability. *Psychological Medicine*, *28*(1), 39–50.

Brooks-Gunn, J., & Reiter, E. O. (1990). The role of pubertal processes. In: S. S. Feldman & G. R. Elliot (Eds), *At the Threshold: The Developing Adolescent* (pp. 16–53). Cambridge, MA: Harvard University Press.

Brown, B. B. (1990). Peer groups and peer cultures. In: S. S. Feldman & G. R. Elliott (Eds), *At the Threshold: The Developing Adolescent* (pp. 171–196). Cambridge, MA: Harvard University Press.

Burnam, M. A., Hough, R. L., Karno, M., Escobar, J. I., & Telles, C. A. (1987). Acculturation and lifetime prevalence of psychiatric disorders among Mexican immigrants. *Journal of Health and Social Behavior*, *28*(1), 89–102.

Centers for Disease Control and Prevention (1998). CDC surveillance summaries. *Morbidity and Mortality Weekly Reports*, *47*(SS-3). Atlanta, GA: Centers for Disease Control.

Cervantes, R. C., & Castro, F. G. (1985). Stress, coping and Mexican American mental health: A systematic review. *Hispanic Journal of Behavioral Sciences*, *7*(1), 1–73.

Children's Defense Fund (1997). *The state of America's children: Yearbook 1997*. Washington, DC: Children's Defense Fund.

Coble, Y. D., Estes, E. H., Head, C. A., Karlan, M. S., Kennedy, W. R., Numann, P. J., Scheider, K. A., Scott, W. C., Skelton, W. D., Steinhilber, R. M., Strong, J. P., Wagner, H. N., Loeb, J. M., Rinaldi, R. C., Stewart, B., & Voegtle, K. (1993). Adolescents as victims of family violence. *JAMA – Journal of the American Medical Association*, *270*(15), 1850–1856.

Commonwealth Fund, The (1997). *The commonwealth fund survey of the health of adolescent girls*. Rochester, NY: Louis Harris & Associates.

Commonwealth Fund, The (1999). *Improving the health of adolescent girls*. New York: Commonwealth Fund.

Current Topics (2000). *Pan American Journal of Public Health*, *7*(4), 275–277.

Finkelhor, D. (1994). The international epidemiology of child sexual abuse. *Child Abuse & Neglect*, *18*(5), 409–417.

Flisher, A. J., Kramer, R. A., Hoven, C. W., Greenwald, S., Alegria, M., Bird, H. R., Canino, G., Connell, R., & Moore, R. E. (1997). Psychosocial characteristics of physically abused children and adolescents. *Journal of the American Academy of Child and Adolescent Psychiatry*, *36*(1), 123–131.

Flores, G., Bauchner, H., Feinstein, A. R., & Nguyen, U.-S. D. T. (1999). The impact of ethnicity, family income, and parental education on children's health and use of health services. *American Journal of Public Health*, *89*(7), 1066–1071.

Furstenberg, F. F., Cook, T. D., Eccles, J., & Sameroff, A. J. (Eds) (1999). *Managing to make it: Urban families and adolescent success*. Chicago: University of Chicago Press.

Geller, J., Johnston, C., Madsen, K., Goldner, E. M., Remick, R. A., & Birmingham, C. L. (1998). Shape- and weight-based self-esteem and the eating disorders. *International Journal of Eating Disorders*, *24*(3), 285–298.

Green, A. H. (1993). Child sexual abuse: Immediate and long-term effects and intervention. *Journal of the American Academy of Adolescent Psychiatry*, *32*, 890–902.

Hammond, R. W., & Yung, B. (1993). Psychology's role in the public health response to assaultive violence among young African-American men. *American Psychologist*, *48*(2), 142–154.

Hampton, R. L., & Gelles, R. J. (1991). A profile of violence toward black children. In: R. L. Hampton (Ed.), *Black Family Violence: Current Research and Theory* (pp. 21–34). Lexington, MA: Lexington Books.

Harter, S. (1990). Self identity and development. In: S. S. Feldman & G. R. Elliott (Eds), *At the Threshold: The Developing Adolescent* (pp. 352–387). Cambridge, MA: Harvard University Press.

Herzog, D. B., & Copeland, P. M. (1985). Eating disorders. *New England Journal of Medicine*, *318*, 295–303.

Herzog, D. B., Keller, M. B., Sacks, N. R., Yeh, C. J., & Lavori, P. W. (1992). Psychiatric comorbidity in treatment-seeking anorexics and bulimics. *Journal of the American Academy of Child and Adolescent Psychiatry*, *31*(5), 810–818.

Horton, J. A. (1992). *The women's health data book*. Elsevier.

Institute of Medicine: Division of Mental Health and Behavioral Medicine (1989). *Research on children and adolescents with mental, behavioral, and developmental disorders: Mobilizing a national initiative*. Washington, DC: National Academy Press.

Knight, G. P. (1994). An examination of the cross ethnic equivalence of measures of negative life events and mental health among Hispanic and Anglo-American children. *American Journal of Community Psychology*, *22*(6), 767–784.

Kotlowitz, A. (1991). *There are no children here: The story of two boys growing up in the other America* (1st ed.). New York: Doubleday.

Kovacs, M., & Beck, A. T. (1977). An empirical-clinical approach toward a definition of childhood depression. In: J. G. Schulterbrandt & A. Raskin (Eds), *Depression in Childhood: Diagnosis, Treatment, and Conceptual Models* (pp. 1–25). Raven.

Lennon, M. C. (1996). Depression and self-esteem among women. In: M. M. Falk & K. Scott Collins (Eds), *Women's Health: The Commonwealth Fund Survey* (pp. 207–236). Baltimore, MD: Johns Hopkins University Press.

Lindholm, K., & Willey, R. (1986). Ethnic differences in child abuse and sexual abuse. *Hispanic Journal of Behavioral Sciences*, *8*(2), 111–125.

McCauley, J., Kern, D. E., Kolodner, K., Dill, L., Schroeder, A. F., DeChant, H. K., Ryden, J., Derogatis, L. R., & Bass, E. B. (1997). Clinical characteristics of women with a history of child abuse-unhealed wounds. *JAMA – Journal of the American Medical Association*, *277*(17), 1362–1368.

McLanahan, S., & Sandefur, G. (1994). *Growing up with a single parent: What hurts, what helps*. Cambridge, MA: Harvard University Press.

Mennen, F. E. (1994). Sexual abuse in Latina girls: Their functioning and a comparison with white and African American girls. *Hispanic Journal of Behavioral Sciences*, *16*(4), 475–486.

Millstein, S. G., & Litt, I. F. (1990). Adolescent health. In: S. S. Feldman & G. R. Elliott (Eds), *At the Threshold: The Developing Adolescent* (pp. 431–456). Cambridge, MA: Harvard University Press.

Nelson, D. E., Higginson, G. K., & Grantworley, J. A. (1995). Physical abuse among high-school students – Prevalence and correlation with other health behaviors. *Archives of Pediatrics and Adolescent Medicine*, *149*(11), 1254–1258.

Office of Research on Women's Health (1998). *Women of color health data book: Adolescents to seniors*. Bethesda, MD: National Institutes of Health.

Porter, J. R., & Washington, R. E. (1993). Minority identity and self-esteem. *Annual Review of Sociology*, *19*, 139–162.

Rew, L. (1998). Access to health care for Latina adolescents. *Journal of Adolescent Health*, *23*, 194–204.

Rosenberg, M. (1965). *Society and the adolescent self-image*. Princeton: Princeton University Press.

Schoen, C., Davis, K., Scott Collins, K., Greenberg, L., Des Roches, C., & Abrams, M. (1997). *The commonwealth fund survey of the health of adolescent girls*. New York: Commonwealth Fund.

Schwab-Stone, M. E., Ayers, T. S., Kasprow, W., Voyce, C., Barone, C., Shriver, T., & Weissberg, R. P. (1995). No safe haven: A study of violence exposure in an urban community. *Journal of the American Academy of Child and Adolescent Psychiatry*, *34*(10), 1343–1352.

Silverman, A. B., Reinherz, H. Z., & Giaconia, R. M. (1996). The long-term sequelae of child and adolescent abuse: A longitudinal community study. *Child Abuse and Neglect*, *20*(8), 709–723.

Straus, M. B. (1994). *Violence in the lives of adolescents*. New York: W. W. Norton and Co.

Vega, W., Warheit, G., Buhl-Auth, J., & Meinhardt, K. (1984). The prevalence of depressive symptoms among Mexican Americans and Anglos. *American Journal of Epidemiology*, *120*(4), 592–607.

Walitzer, K. S., & Sher, K. J. (1996). A prospective study of self-esteem and alcohol use disorders in early adulthood: Evidence for gender differences. *Alcoholism-Clinical and Experimental Research*, *20*(6), 1118–1124.

APPENDIX

Table A.1. Comparison of Weighted and Unweighted *N*'s and Sample Shares.

Adolescent Girls				Adolescent Boys			
Hispanic	White	Black	Total	Hispanic	White	Black	Total
Weighted							
12.4	71.1	16.5	50.1	12.1	72.9	15	49.9
322	1,852	431	2,605	313	1,890	390	2,593
Unweighted							
12.5	68.7	18.8	54.63	11.4	70.9	17.7	45.37
353	1,947	533	2,833	269	1,668	416	2,353

Source: Commonwealth Fund Survey of Adolescent Girls and Boys (1997).

THE BLACK-WHITE ACHIEVEMENT GAP IN THE FIRST COLLEGE YEAR: EVIDENCE FROM A NEW LONGITUDINAL CASE STUDY

Kenneth I. Spenner, Claudia Buchmann and Lawrence R. Landerman

ABSTRACT

In the United States, an achievement gap between whites and blacks persists at all levels of schooling from elementary school to higher education. Definitive reasons and remedies for minority underperformance remain unclear. This study examines how students acquire and utilize "collegiate capital" which, in turn, relates to their academic achievement in the first year of college. Results indicate that significant black-white differences in academic achievement emerge as early as the first semester of students' first year in college. Controls for family background, parental involvement, prior ability, cultural capital acquired during the middle- and high-school years, and other factors produce a moderate reduction in the achievement gap, but over half of the gap remains unexplained. The study is part of a larger research project that involves a longitudinal study of two cohorts – the graduating classes of 2005 and 2006 – at a major private university. Through the assessment of pre-college differences and extensive data collected via student surveys and

The Shape of Social Inequality: Stratification and Ethnicity in Comparative Perspective
Research in Social Stratification and Mobility, Volume 22, 187–216
© 2005 Published by Elsevier Ltd.
ISSN: 0276-5624/doi:10.1016/S0276-5624(04)22007-8

academic records during the college years, the goal of the larger project is to illuminate the factors underlying raced-based variations on a range of academic outcomes such as educational performance and attainment, but also several new measures of collegiate intellectual development such as students' ecological integration, perceptions of other groups, and satisfaction with college.

Conventional scholarly wisdom on stratification patterns for blacks and whites in the United States suggests a pattern of gradual convergence in the post-World War II years (Featherman & Hauser, 1978; Hirschman & Snipp, 1999; Hout, 1988). Further, educational achievement has been one of the key factors in the convergence (Kuo & Hauser, 1995). However, a growing literature in the U.S. finds a persistent gap in academic performance between whites and blacks at all levels of schooling from elementary school to higher education (Jencks & Phillips, 1998). African Americans also continue to have higher rates of drop out and lower educational attainment than whites. In one of the more comprehensive studies to date, Bowen and Bok (1998) postulate several reasons for the minority achievement gap, including poorer academic preparation, dis-identification with achievement in response to academic hardships, and racial distrust. Status attainment and human capital perspectives maintain that students' educational outcomes are a function of their family background, cognitive abilities, and achievement orientations. Yet prior research that accounts for these factors provides only modest reductions in the black-white performance gap. Thus, the precise reasons and remedies for minority underperformance remain unclear. This paper and the larger research project of which it is part apply this question to the arena of higher education: What are the causes of the black-white achievement gap and what might reasonably be done to ameliorate it?

In the sections that follow, we review the evidence and explanations that have been offered for the achievement gap. Then we discuss the *Campus Life and Learning Project*, a new longitudinal study of racio-ethnic differentials in educational performance in higher education. The research project features a prospective panel study of two cohorts, the graduating classes of 2005 and 2006, at Duke University, a major private university in the southeastern United States. We will survey about 1,500 students annually through college exit or graduation and at least two years thereafter. A number of recent or ongoing studies use samples of multiple institutions or large heterogeneous samples of individuals in order to investigate racial differences in educational outcomes. These include general surveys of children and youth, such as the National Educational Longitudinal Surveys, Children of the National Longitudinal Survey, and High School and Beyond. Other studies by Bowen and Bok (1998) and Massey et al. (2003) are also designed to investigate racial achievement gaps in higher education. The *Campus Life and Learning Project* is distinctive in that it entails a panel study of a single institution with the goal of providing a more detailed, in-depth assessment of the wide range of explanations for race-based performance differences than can be achieved with larger, multiple institution studies.

After discussing the design of the larger research project and situating Duke University within the population of four-year colleges and universities in the United States, we utilize data from two survey waves for the first cohort in the study to examine the determinants of academic performance, measured by grade point average (GPA), at the end of the first semester of the first college year. One goal of this analysis is to establish whether an achievement gap between minority students and whites in our sample emerges as early as the end of the first semester of college and to compare the size of this gap, if found, to achievement gaps found in other studies. A second goal is to estimate a "net" achievement gap, controlling for a wider range of status attainment, human capital and cultural capital factors than in prior studies of the achievement gap. Results indicate that an achievement gap emerges early in the college career. A range of pre-college factors is responsible for ameliorating or exacerbating the achievement gap in first semester grades. We conclude the paper with a discussion of the main implications of the findings and our plans for future research.

THE MINORITY ACHIEVEMENT GAP: EVIDENCE AND EXPLANATIONS

A long tradition of research has found substantial differences in college grades, persistence in college, and graduation rates between African-Americans and whites in the United States (Cleary, 1968; Crouse & Trusheim, 1988; Nettles, 1988; Nettles et al., 1986; Ramist et al., 1994). Kane (1998) used data from the High School and Beyond (HSB) study to investigate differences in the educational performance of minority and white students. The HSB data refer to a sample of students in the graduating class of 1982 from over 1,000 public and private high schools in the United States, who were periodically resurveyed over the next 10 years. Using a sub-sample of 2,912 students, Kane reports that black and Hispanic students scored from 0.3 to 0.4 of a letter grade lower in college GPA (on a four-point scale), compared with white students. Generally, the size of the achievement gap between Hispanic students and whites was about half of the gap for blacks. Adjustments for family background (parental education and income), and other controls (gender, SAT score, and high school GPA) reduced the gap by about half for both groups.

Vars and Bowen (1998) provide a similar estimate of the gap. They used the College and Beyond dataset, which contains a sample that is more selective than HSB, as the data refer to more than 10,500 students entering six private universities and five selective liberal arts colleges in 1989. Vars and Bowen report a black-white gap in college GPA of 0.5 of one letter grade. Controls for family socioeconomic background, attendance at private versus public high school, and prior achievement reduced the gap by about half. More generally, using data for two cohorts from the College and Beyond data, Bowen and Bok (1998) show that while the black-white gap in SAT test scores has narrowed, the gap in college performance is nearly as large for the college cohort entering in 1989 as it was for the 1976 cohort. Moreover, the gap may be largest among students in the highest echelons of the SAT distribution and at more selective institutions.

What factors can explain these persistent differences in academic performance? A variety of explanations have been offered to account for the minority achievement gap. Most arguments focus on differences in various forms of capital for students of different racio-ethnic groups. Other arguments maintain that a combination of institutional factors are to blame for these achievement differences. Below we summarize the main factors emphasized by prior research attempting to explain minority achievement gaps. They include status attainment variations, social and cultural capital differentials, negative stereotype threat, and the racial climate of college classrooms and campuses.

Status Attainment and Human Capital

Scholars of education have long recognized that an individual's experiences during childhood, and the financial resources and socioeconomic standing of one's family are very important for later educational attainment and achievement. At the same time, it is important to note that prior research suggests that status attainment and human capital variables *do not* fully explain the gap; at most, they account for one-half of the gross differential in grades and test scores (for summary, see Bowen & Bok, 1998, pp. 53–90; Jencks & Phillips, 1998, pp. 1–51). The most frequently studied measures include family socioeconomic background (family income, parents' occupation and education levels), family size and structure, rural background, cognitive ability measured by prior test scores, and the quality of prior schooling (including class size, teacher characteristics, school type, etc.).

Cultural Capital

A growing literature finds that cultural capital, conceptualized as high status cultural knowledge (Bourdieu, 1977; DiMaggio, 1982) or cognitive and linguistics skills (DeGraaf et al., 2000), shapes students' educational outcomes in primary and secondary school. Using two cohorts of data from the National Educational Longitudinal Survey, Roscigno and Ainsworth-Darnell (1999) find that cultural capital ("high brow" cultural trips and classes) and household educational resources explain a modest portion of the racial gap in high school grades and mediate some of the effects of family background. Notably, little research has applied these ideas to the study of higher education. Cultural capital factors may have modest independent and mediating effects on initial college grades, and may explain a small part of the racio-ethnic achievement gap. It is also possible that cultural capital matters more for college enrollment than for academic performance in college. Moreover, once students are enrolled in college, their cultural capital may require rebuilding or redirection in order for it to exert effects on educational achievement or other college-level outcomes. This latter issue raises questions about the amount of "carry-over" of cultural capital acquired by students prior to college to their experiences once in college. Thus far research has not examined such questions.

Social Capital

Primary proponents of the concept of social capital, Bourdieu (1977) and Coleman (1988), see social capital as inherent in the character of social relations among people (versus human capital such as skills that reside within individuals). Accordingly, relationships have varying levels of trust, obligations and normative expectations (Coleman, 1988). Some initial research shows promise for the idea that high-school students with well-developed social capital have higher educational outcomes (e.g. Stanton-Salazar & Dornbusch, 1995), but, as in the case of cultural capital, the role of social capital in the realm of higher education represents largely uncharted waters. Several aspects of social capital may be related to students' college outcomes. For example, how do students use social networks (family, peer, professional and academic ties) as sources of support and information? How do these networks change in terms of their diversity and frequency over the college years?

Homophily is a well-established tendency in human association and networks (McPherson et al., 2001). Some studies have found considerable diversity in the degree of racial homophily in the social networks of college students

(Smith & Moore, 2000), which may also have implications for academic performance. For example, Bowen and Bok (1998) speculate that black students are disadvantaged by racially-homophilous peer networks, because the black distribution of ability, as measured by test scores, is lower on average than the white distribution. Finally, resource provision, information and social support in associational ties are not only a matter of personal networks and individual action, but also of structured opportunities provided by institutional life. Students at many colleges and universities live in dormitories with randomly-assigned roommates in their first year. Differences in the composition of students' social networks, which are partly determined by such institutional regulations, may be related to student perceptions, aspirations, and performance differentials.

Negative Stereotype Threat

Psychologists Claude Steele, Joshua Aronson and colleagues (Aronson et al., 1999, 2002; Steele, 1997, 1999; Steele & Aronson, 1995, 1998) have advanced a novel explanation for the achievement gap. Negative stereotype threat begins with cultural beliefs about how members of different social groups perform in different situations, but especially those involving competitive performance. Negative stereotype threat is activated (i.e. perceived by ego) when such beliefs are made salient and referenced in a situation and, in turn, have affective and motivational consequences, such as fear and anxiety. These responses interfere with the efficiency of information processing and eventually reduce performance on evaluated tasks such as exams. Over the longer term these experiences may lead to dis-identification with being a "good" student and devaluing of academic performance.

Empirical evidence supporting the negative stereotype threat hypothesis has been reported for African American students relative to their white counterparts in standardized test performance (Aronson et al., 2002; Steele & Aronson, 1995), for women's performance at mathematics relative to male's performance (Croizet et al., 2001), white men's performance at mathematics relative to their Asian counterparts (Aronson et al., 1999), and students of different socioeconomic status/social class levels and their performance on intellectual tasks (Croizet & Claire, 1998). Nearly all of the empirical support for the theory has been based on experimental or quasi-experimental designs, or on elementary or secondary school populations (Voelkl, 1997). In the typical experimental or quasi-experimental design, the salience of the stereotype is manipulated and subsequent test performance is the measured dependent variable.

Micro- and Institutional Climates

Perspectives emphasizing the importance of institutional climate for minority underperformance involve several classes of explanations. "Climate" refers to the extent to which prospective interactional ties of a focal environment (i.e. classroom, dormitory, social or institutional group) are perceived as welcoming and integrative, on the one hand, or hostile and exclusive, on the other (Astin, 1993; Hurtado, 1992; Hurtado et al., 1998; Pascarella et al., 1986; Pfieffer, 1976). These researchers maintain that some minority students perceive various environments as hostile for their ascriptive social group which, in turn, leads to: (a) reduced levels of ecological integration or a sense of belonging; (b) increased stress that can impede academic performance; (c) different patterns of social and human capital acquisition and help-seeking behavior in academic settings.

Other research in this genre focuses on the "micro-politics" of classrooms such as differential treatment by teachers, school personnel and peers (Roscigno & Ainsworth-Darnell, 1999), black students' fears of acting white (Cook & Ludwig, 1998), and oppositional cultures of minority groups (Ainsworth-Darnell & Downey, 1998; Farkas et al., 2002; Ogbu, 1974, 1986).

RESEARCH DESIGN AND ANALYSIS

The Larger Project

In order to investigate the salience of these various perspectives in determining racial differences in academic achievement, we have designed a major research effort that improves upon prior studies in several important ways. The *Campus Life and Learning Project* entails a multi-year, prospective panel study of two consecutive cohorts of students admitted to Duke University, the incoming classes of 2001 and 2002 (graduating classes of 2005 and 2006). Duke is a private research university located in Durham, North Carolina with an undergraduate enrollment of about 6,000 students from the United States and several foreign countries. In contrast to other studies that investigate multiple institutions, the study is designed to capture the rich details of students' experiences and the structure of a single institution of higher education. It includes several key types of data and audit points that permit comparison to other institutions of higher education, particularly private elite colleges and universities.[1]

First, each cohort is surveyed via mail in the summer preceding their enrollment.[2] This survey gathers data on factors predicted by status attainment and human capital arguments to be important for academic achievement: measures of the quality of prior schooling and pre-college achievement orientations, (including self-esteem, self-perceptions of ability in different domains, and expectations for future performance). It provides a more detailed assessment of family socioeconomic background than prior research because we have access to data on students' financial aid and other aspects of family wealth.

The pre-college survey also allows a detailed assessment of students' cultural capital during their middle- and high-school years. Questions regarding students' exposure to high culture, interaction with parents over curricular and non-curricular matters, parental participation in school activities, and household educational resources will provide valuable data to address questions regarding the role of prior cultural capital in academic achievement in college.

Second, surveys administered during the college years contain a core set of questions, supplemented with questions regarding students' social networks, social and cultural capital, performance attributions patterns, and the like. Additional modules will include questions on time-use, choice of major, residential life, campus climate, advising, support networks, finances, and faculty-student interaction. In order to examine the role of social capital, the in-college surveys will obtain replicate panels of assessments of students' strong-tie and weak-tie networks over time, along with selected characteristics (race, gender and location) of these ties. We will examine how peer networks of black and white students vary and how these variations relate to a range of college experiences and outcomes. Moreover, we can investigate how various aspects of institutional support for specific student populations, such as scholarship athletes or students participating in highly-integrated, small-group curriculum programs, are related to differential academic outcomes and satisfaction with the college experience, net of individual factors. Additionally, as a first step toward testing negative stereotype threat arguments in a survey design framework with college students, we will measure the affective, motivational and behavioral components of this sequence for the most challenging class a student takes each semester and track students' academic performance and several identity components in each survey wave. One challenge for the negative stereotype threat explanation is to move beyond the laboratory and to demonstrate external validity in broader real world contexts of actual classrooms and students' academic careers. The study will measure students' perceptions of climate in academic (i.e, classroom), residential, social, and extracurricular arenas at multiple time points during the college years in order to investigate the salience of these perspectives for explaining the relationship between campus and classroom climate and students' college performance and experiences.

The study also features several assessments of climate at local (living group, social group), university, and community levels. Archival methods will capture salient issues and events in the local media and student culture, while focus groups interviews will capture elements of the racial climate that are not easily measured in surveys. Also, for each respondent, we anticipate building a full temporal, spatial segregation profile that maps the racio-ethnic composition of his or her classes, living groups, and social networks. This type of information should illuminate the developmental patterns of integration and segregation as they evolve over the undergraduate career.

Finally, students will be surveyed after they leave Duke University, whether by graduation or early exit. Figure 1 summarizes the design and data collection points, and illustrative information that we expect to gather each year.[3] The sampling design randomly selects about 350 whites in each cohort and all black and Hispanic students, and about two-thirds of the Asian students in each cohort. Thus, the full design across both cohorts will have about 700 whites and 800 non-whites.[4]

The Current Study

While the larger Campus Life and Learning project is in still in progress, the current paper utilizes the pre-college and first year surveys for the incoming class of 2001 with two immediate goals. First, we seek to establish the size of the achievement gap between minority students and whites in our sample and compare it to other studies. The dependent variable for these analyses is first semester GPA. Second, we estimate a "net" achievement gap, controlling for status attainment and human capital factors and an extensive set of cultural capital measures taken prior to students' arrival at college. Prior research finds an achievement gap in the range of one-half of one letter grade between black and white students, and about half of this amount between Hispanic and white students. In both cases, roughly half of the gap is explained by pre-college differentials in family background, prior schooling, test scores, and cultural capital.

Note that the sample for this study was not designed to be representative of the U.S. population of college and university students. Rather, it is more representative of highly selective institutions of higher education in the United States. In their sample of the cohort entering college in 1989, Bowen and Bok (1998, p. 337) define their top tier of selective institutions as those with combined average SAT scores (verbal and mathematics) of 1300 or higher. Their sample included institutions like Bryn Mawr, Swarthmore, and Williams colleges, and Princeton, Duke, Rice, Stanford and Yale universities. The average SAT score of Duke's 2001 entering cohort was 1385. SAT scores have been rising over time so some upward shift since 1989 is to be expected. Further, Bowen and Bok (1998) reported a performance gap of .5 letter grade in overall college GPA between white and black sample members

Pre-Collegiate Variables

Pre-College Survey
Demographic
- Racio-Ethnic Identity
- Parental Racio-Ethnic Identity
- Citizenship
- Religious Affiliation

Family Structure
Family Capital
Cultural Capital
Schooling Experiences
Diversity Exposure
Performance Expectations
Performance Attributions
Non-Cognitive Resources
Identity Encapsulation
Social Support Network
Gender Roles
College Expectations
Admissions Resources
SES Constellation
Occupational Aspirations
Psychological Stressors

SISS
High School Curriculum
Test Scores (SAT, ACT, etc.)
GPA
Reader Rating Scores
High School Extracurricular
Financial Aid and Support

Collegiate Wave 1 Variables

Survey conducted during the sample's first year

Academic
- Record
- Course difficulty
- University academic climate and diversity*
- Classroom climate general and diversity
- Integration**
- Proposed major

Social/interpersonal networks and support
Residential Life
- Climate and diversity
- Integration

Extracurricular
Durham community
College development
- Stressful events and coping flexibility
- Stereotype threat

First Year Specialized Modules-Survey and qualitative research
- Transitions to college
- Pre-major advising/academic risk assessment
- Scholarship recipients
- Student athletes
- FOCUS Program

* Diversity includes breadth of network and experiences of discrimination
** Integration is the degree to which a student is strongly affiliated with a given domain, resources and opportunities available in that domain.

Post-College Variables

Graduation
Educational Attainment
Occupational Attainment
Income Attainment
Life and Job Satisfaction
Satisfaction with Duke

=> Time Line of Measures of Institutional Structure and Climate

	Application to Duke	2001	'02	'03	'04	'05	'06	'07	'08
Design Timeline ➔									
Data Collection									
Class of 2005 (Cohort 1)		Wave - 1	W2	W3	W4	W5		W6	
Class of 2006 (Cohort 2)			W1	W2	W3	W4	W5		W6

Fig. 1. Summary of Major Design Components.

Table 1. Percentage Enrollment by Racio-Ethnic Category for U.S. Four-Year Public and Private Higher Education Institutions and Duke University (1999 data).

Racio-Ethnic Category	Public Four-Year	Private Four-Year	Duke University
White, non-hispanic	74.9	75.8	69.7
Black, non-hispanic	10.7	11.4	8.0
Hispanic	6.9	6.2	4.2
Asian	6.5	6.0	14.2
Other	1.0	0.6	3.9

Notes: "Other" for public and private four-year institutions includes those for whom racio-ethnic category is unknown. For Duke this category includes racio-ethnic category unknown and a category for "Bi- or Multi-racial." Data sources: For public and private four-year institutions: U.S. Department of Education, National Center for Education Statistics, 2002. *Digest of Education Statistics, 2001.* NCES 2000-130, by Thomas D. Snyder. Washington, D.C. For Duke University, Office of the Registrar (unpublished data).

for the class of 1989 from 28 sampled selective institutions. Their broader sample includes the likes of Columbia University, Northwestern University, Wellesley College, the University of Michigan, and the University of North Carolina at Chapel Hill.

Table 1 provides a further comparison of the racio-ethnic composition of the Duke student body compared with all U.S. public and private higher education institutions in 1999 (the last year for which the latter data are available). Duke is fairly comparable to other universities with the exception that Duke University has about twice the percentage of Asian students (similar to other private elite institutions) and somewhat more students in the "Other" category. The latter difference is likely because the Duke admissions form includes a category in which students can describe themselves as "Bi- or Multi-racial."[5]

As noted above, the pre-college survey contains questions on a wide range of issues including students' social and economic background, past schooling experiences, social and cultural capital, social psychological characteristics, and expectations for college life. Procedures for all surveys follow Dillman's (1978) Total Design Method. Table 2 provides the response rates by racio-ethnic group for the pre-college survey of the incoming Class of 2001. Response rates varied by racio-ethnic groups, ranging from 75 to 86%, and only about 2.5% of the sample refused to participate. The overall response rate was high; 80% of the sample completed the pre-college survey.

The analysis refers to members of the 2001 entering cohort who had relatively complete data ($N = 673$, 80% of total sample) and who authorized release of their Duke student records (91% of the above respondents).

Table 2. Population, Sample, and Response Rates: Pre-college Survey, Incoming Class of 2001.

	Total	Asian	African-American	Hispanic	Other[a]	White
Population	1631	238	182	123	114	974
Sampled (sampling fraction)	837 (0.51)	147 (0.62)	178 (0.98)[b]	120 (0.98)[b]	36 (0.32)	356 (0.37)
Completed (response rate)	673 (0.80)	114 (0.78)	137 (0.77)	103 (0.86)	27 (0.75)	292 (0.82)
Refusals (n)	21	2	1	3	3	12
Other nonresponse	143	31	40	14	6	52

[a] "Other" includes Native American, multiracial ethnic identification, no ethnic identification.
[b] Sampling fractions for African-Americans and Hispanics are not 100% because of late changes in intention to matriculate.

This sub-sample is comprised of 610 respondents (73% of original sample members).[6]

The dependent variable in all analyses is GPA (on a 4 point scale) at the end of the first semester of the first college year. Independent variables include a range of individual, family background, and other factors. Students' race is gathered from multiple U.S. Census type questions that separately measure whether the respondent is Hispanic, and then asks for racial identification (White, Black, American Indian or Pacific Islander, Asian, Biracial or Multi-racial, or Some Other Race). The respective categories include both foreign and native born (i.e. Asian includes both foreign born and Americans of Asian descent). We combined Bi-/Multi-racial with the Other category. Virtually all of respondents who self-identified as Hispanic listed racial categories other than Black; accordingly, we assigned these cases to the Hispanic group. Following prior studies (Bowen & Bok, 1998; Roscigno & Ainsworth-Darnell, 1999) we include controls for sex, citizenship, parental education, occupation, income, labor force status, family structure, number of siblings, and the type of high school (private or public) attended. In most cases, students are asked to provide information on background measures for the time period of their senior year in high school; in other cases the more general period of high school years was used. The student's score on the verbal and math areas of the Scholastic Aptitude Test, typically taken in the fall of the senior year of high school was also included as an independent variable. These controls are more comprehensive than those of prior studies, since most datasets do not include all of the above measures. For example, the only other published study to examine minority achievement gaps in the first semester of college (Massey et al., 2003), does not control for SAT scores or most of the family background measures used here.

Other independent variables capture pre-college levels of human, cultural and social capital. While there is debate about the exact boundaries of human, social and cultural capital (Farkas, 1996), here we try to specify pre-college capital as completely as possible, regardless of whether the exact components fall in one domain or another. Measures of pre-college human capital include: whether the student applied for financial aid during the first year of college; whether the student was enrolled in the engineering college during the first year (usually indicating more rigorous preparation in natural science and mathematics fields in high school); average hours studied per week in high school (Rau & Durand, 2000); a measure of the importance of being a good student for one's self identity; and self-ratings of ability in challenging math/science and literature courses (Spenner & Featherman, 1978).

For measures of cultural and social capital, we drew heavily on prior research (DeGraaf et al., 2000; DiMaggio, 1982; Roscigno & Ainsworth-Darnell, 1999;

Stanton-Salazar & Dornbusch, 1995). Based on these studies, the instrumentation included a pool of approximately 40 items that captured various aspects of high culture activities, popular culture activities, types of interactions with parents, parent's involvement in school and other domains of the respondent's life, and educational resources available to the respondent in the home and elsewhere. These were measured for the high school years and, for 15 of the items, the middle school years. We factor analyzed these items, for middle and high school separately, under a variety of model conditions.[7] Four factors emerged that were fairly comparable across the middle and high school years, although there were some small variations in variables that loaded high. The high-culture factor is defined by the respondent visiting a museum, art gallery, zoo, or science center, or attending opera, ballet and the theater, either alone or with parents. The popular culture factor involves going to popular musical concerts, sporting events (high school), or parents talking with friends (middle school). A "parents-school-activity" factor is defined by parental participation in school-based Parent Teacher Association activities or other types of school activities. Finally, a parent homework factor involves parents regularly checking to see if homework is completed or assisting with homework.

Results

Table 3 provides descriptions, means, and standard deviations for all variables for each racio-ethnic group. Data for the entering cohort of 2001 shows a black-white performance gap in first semester grades of 0.39, slightly smaller than that found by Bowen and Bok (1998), but slightly larger than that reported by Massey et al. (2003). Recall, this is but one semester of grades and the gap typically grows some over the undergraduate years. In general, whites are advantaged on most socioeconomic measures compared with black and Hispanic students. For example, average family income for white students ranges from $100,000 to $150,000 (USD) per year, for Hispanic and Asian students from $75,000 to $100,000, and for black students from $50,000 to $75,000 per year. A few other racio-ethnic differences are noteworthy: three out of four black students are female;[8] Asian students are less likely to be U.S. citizens; only four out of ten white students had mothers who worked full time in the labor force while they were in high school, compared with more than seven out of ten black students. Finally, black students and, to a lesser extent, Hispanic and Other students are less likely to have lived in an intact family and more likely to have experienced the divorce of their parents during their high school years (see Massey et al., 2003, for similar findings).

Table 3. Measures and Descriptive Statistics by Racio-Ethnic Status, Pre-College Survey, Class of 2005.

Variable	Metric/Notes	Mean (Standard Deviation) by Racio-Ethnic Group				
		White	Black	Hispanic	Asian	Other
Race	Dummy variable for groups; left out category = white; U.S. Census questions	285	104	93	97	31
Sex	0 = Male	0.49	0.74	0.46	0.48	0.61
	1 = Female	(0.59)	(0.33)	(0.36)	(0.47)	(0.47)
Citizenship	0 = Other	0.98	0.93	0.91	0.70	0.89
	1 = U.S. Citizen	(0.17)	(0.19)	(0.21)	(0.43)	(0.30)
Father's education[1]	1 = Less than high school graduate	4.97	3.95	4.43	5.04	4.98
	2 = High school graduate	(1.28)	(1.13)	(1.05)	(1.13)	(0.97)
	3 = Some college/vocational school					
	4 = College graduate					
	5 = Some graduate school or Master's Degree					
	6 = Higher professional degree (Ph.D., J.D., M.D.)					
Mother's education[1]	1 = Less than high school graduate	4.41	3.77	4.09	4.11	4.29
	2 = High school graduate	(1.16)	(1.02)	(0.89)	(1.02)	(0.85)
	3 = Some college/vocational school					
	4 = College graduate					
	5 = Some graduate school or Master's Degree					
	6 = Higher professional degree (Ph.D., J.D., M.D.)					
Father's occupation	Duncan Socioeconomic Status Score assigned to 1990 census 3-digit occupation	57.72 (19.92)	49.13 (13.35)	53.76 (13.51)	58.77 (15.50)	58.32 (14.54)
Mother's occupation	Duncan Socioeconomic Status Score assigned to 1990 census 3-digit occupation	52.41 (16.23)	50.41 (11.11)	50.91 (11.71)	50.94 (16.59)	54.67 (11.57)
Mother working	0 = No	0.39	0.72	0.51	0.43	0.44
	1 = Yes, mother employed full-time in labor force during respondent's senior year of high school	(0.57)	(0.34)	(0.36)	(0.46)	(0.48)

Table 3. *(Continued)*

Variable	Metric/Notes	Mean (Standard Deviation) by Racio-Ethnic Group				
		White	Black	Hispanic	Asian	Other
Parent's income	1 = less than $1,000	8.33	6.44	7.25	7.48	8.32
	2 = $1,000 to $9,999	(2.42)	(1.64)	(1.71)	(2.07)	(1.79)
	3 = $10,000 to $19,999					
	4 = $20,000 to $29,999					
	5 = $30,000 to $49,999					
	6 = $50,000 to $74,999					
	7 = $75,000 to $99,999					
	8 = $100,000 to $149,999					
	9 = $150,000 to $199,999					
	10 = $200,000 to $499,999					
	11 = $500,000 or more					
Intact family	0 = Not Intact	0.88	0.53	0.79	0.88	0.79
(senior year)	1 = Intact	(0.38)	(0.38)	(0.29)	(0.30)	(0.39)
Experienced	0 = No	0.06	0.12	0.10	0.02	0.10
parent's divorce during high school	1 = Yes	(0.28)	(0.24)	(0.22)	(0.13)	(0.29)
Number of	Number (including step and half siblings)	1.72	2.17	1.87	1.42	1.69
siblings		(1.30)	(1.28)	(1.31)	(0.97)	(1.31)
Financial Aid-	0 = No	0.53	0.63	0.54	0.56	0.58
Applicant	1 = Yes (applied for aid packages from univ.)	(0.59)	(0.36)	(0.36)	(0.46)	(0.48)
Public high school	0 = No	0.66	0.70	0.52	0.79	0.62
	1 = Yes (major school attended)	(0.56)	(0.35)	(0.36)	(0.38)	(0.47)
Middle school	Sum of variables that load high on factor					
cultural capital[2]	Scales range from 1 = Never to 5 = Very Often					
Factor 1 - high	Respondent visits museum, art gallery	12.15	9.68	11.20	10.75	12.78
culture	Respondent attends opera, ballet, etc.	(3.86)	(2.38)	(2.43)	(2.88)	(3.13)

	Respondent visits zoo, science center, etc.					
	Respondents visit museum/art gallery w/parents					
	Respondent attends opera/ballet/etc. w/parents					
Factor 2 - pop	Parents talk with friends	11.84	9.76	11.29	9.36	11.61
culture	Respondent goes to movies	(2.30)	(1.93)	(1.46)	(2.35)	(2.23)
	Respondent attends pop. music concerts					
	Respondent attends sporting event					
Factor 3 -	Parents participate in parent-school organization	6.23	4.86	5.57	4.78	5.72
parent-school	(e.g., PTA) Parents participate in other school activities	(2.13)	(1.57)	(1.55)	(1.88)	(1.79)
Factor 4 -	Parents check homework completion	5.99	5.86	5.73	5.35	5.48
parent-homewk	Parents help with homework	(1.81)	(1.30)	(1.32)	(1.61)	(1.93)
High school cultural	Sum of variables that load high on factor					
capital	Scales range from 1 = Never to 5 = Very Often					
Factor 1 - parent	Parents participate in PTA	5.75	4.57	5.03	4.65	5.11
school	Parents participate in other school activities	(2.28)	(1.46)	(1.55)	(1.92)	(1.81)
Factor 2 - high	Respondent visits museum, art gallery	6.97	6.15	6.84	7.06	7.35
culture	Respondent attends opera, ballet, etc.	(2.54)	(1.53)	(1.40)	(1.96)	(2.16)
	Respondent visits zoo, science center, etc.					
Factor 3 - pop	Respondent attends pop music concerts	5.93	4.98	5.62	4.76	5.93
culture	Respondent attends sporting events	(1.64)	(1.03)	(0.96)	(1.57)	(1.24)
Factor 4 -	Parents check if homework is done	4.46	4.34	4.07	3.85	4.26
parent-homewk	Parents help with homework	(1.96)	(1.25)	(1.24)	(1.57)	(1.56)
Test score - verbal	Scholastic aptitude test[3] (max = 800)	696.5	644.61	661.8	701.4	695.9
		(79.43)	(44.25)	(42.79)	(68.25)	(60.06)
Test score -	Scholastic aptitude test[3] (max = 800)	720.5	633.3	677.2	750.4	711.9
mathematics		(61.38)	(40.19)	(36.27)	(44.99)	(65.28)
Scholarship	0 = No	0.03	0.01	0.02	0	0.03
athlete	1 = Yes	(0.21)	(0.07)	(0.10)	(0)	(0.16)

Table 3. *(Continued)*

Engineering	0 = Arts & sciences college	0.22	0	0.08	0.30	0.15
	1 = Engineering college	(0.48)	(0)	(0.19)	(0.43)	(0.34)
Hours studied -	Average number of hours spent studying or doing	13.22	14.99	13.78	15.86	15.90
high school	homework per week in high school	(9.83)	(6.68)	(6.09)	(8.06)	(8.84)
Student identity	Importance of "being a good student" to overall identity	4.31	4.74	4.58	4.34	4.30
	(from 5 = extremely important to 1 = not at all important)	(0.93)	(0.40)	(0.56)	(0.82)	(0.84)
Self- rated ability	Sum of two questions: For recent (a) most challenging	8.25	7.93	8.27	8.29	8.09
	math/natural science class and (b) challenging literature	(1.38)	(0.87)	(0.82)	(1.11)	(1.06)
	class (from 5 = very much above average to 1 = very much					
	below average)					
Relative at Duke	Family member or relative attended Duke University					
	0 = No	0.26	0.10	0.09	0.10	0.11
	1 = Yes	(0.51)	(0.23)	(0.20)	(0.28)	(0.30)
Grade point	0.0 = F, 1.0 = D, 2.0 = C, 3.0 = B, 4.0 = A	3.27	2.8	3.12	3.32	3.24
average		(0.64)	(0.36)	(0.34)	(0.47)	(0.44)

[1]The temporal referent for family background measures was senior year in high school unless noted otherwise.

[2]Factors were identified through analysis of covariance matrices under a promax solution. The mean loading was 0.73; the minimum loading was 0.43. The solution chosen was one that provided a clearly interpretable pattern separation with items loading on one and only one factor.

[3]Most students had SAT scores in their records. A small number of students had ACT scores, which we transformed to an SAT analog score by regression inputation.

There are few racio-ethnic differences in the measures of cultural capital and school involvement. Black students report lower levels of participation in high culture and popular culture activities (along with Asian students) in middle and high school. Also of note, Asian students have the lowest levels of parental participation in homework activities compared with any racio-ethnic group, yet have the highest first semester GPA's of any group. Massey et al. (2003) reported a similar finding for Asian students, wherein Asian parents were less involved in monitoring and assisting in homework.

For SAT scores measured during the senior year of high school, Asians score the highest, followed by white students, students of bi- or multi-racial identification, Hispanics and blacks. The black-white difference is a sizeable 52 points on the verbal score (over one standard deviation in the black distribution, and two-thirds of a standard deviation in the white distribution). The mathematics test score difference is even larger at 87 points (more than two standard deviations in the black distribution and 1.4 standard deviations in the white distribution). These sizable differences are in line with those found in several of other studies. Also, note that no black student respondents are in the School of Engineering, compared with 22% of whites and 30% of Asian students. This finding, too, reflects differences found at the national level. Finally, one out of four white students had a relative or family member who attended Duke; in contrast, only about 10% of members of other racio-ethnic groups reported having a family member or relative who had attended Duke. This distinction is potentially important because some portion of these respondents could be categorized as "legacy" cases. Legacy cases are of several types. One group involves student applicants who may be given some special consideration in the admissions process because their parents or relatives attended the same institution. Another group involves applicants from wealthy families who receive special consideration in the admission process because their family makes or promises to make a sizeable donation to the college or university. Legacy cases have been discussed in the literature (for example, Bowen & Bok, 1998) and have been a contentious issue in the media recently, but to the best of our knowledge there is no empirical research based on systematic data on this issue.

Table 4 reports parameter estimates for nested regression models of racio-ethnic group and several sets of control variables on first semester GPA.[9] These are unstandardized coefficients with standard errors in parentheses. Whites comprise the excluded category relative to dummy variables for other racio-ethnic groups.

It is striking that already in this first semester of college, for students at a highly selective university who have taken only four college courses, a sizable gap emerges. Blacks score 0.39 of one letter grade lower than whites in the first semester. The academic performance of other racio-ethnic groups is not significantly different from that of whites, although the Hispanic-white gap is

Table 4. Nested Regression Models of Racio-Ethnic Group and Sets of Control Variables on First-Semester Grade Point Average (Ordinary Least Squares).

Independent Variable	Model					
	1	2	3	4	5	6
	b (s.e.)	b (s.e.)	b (s.e.)	b (s.e.)	b (s.e.)	b (s.e.)
Racio-ethnic group[a]						
Black	−0.39**	−0.36**	−0.35**	−0.39**	−0.19**	−0.23**
	(0.07)	(0.08)	(0.08)	(0.08)	(0.08)	(0.08)
Hispanic	−0.14	−0.12	−0.13	−0.15	−0.03	−0.07
	(0.08)	(0.08)	(0.08)	(0.08)	(0.08)	(0.08)
Asian	0.04	0.04	0.01	−0.04	−0.09	−0.11
	(0.06)	(0.06)	(0.06)	(0.07)	(0.06)	(0.06)
Other	−0.03	−0.04	−0.06	−0.06	−0.05	−0.06
	(0.10)	(0.10)	(0.10)	(0.10)	(0.10)	(0.10)
Demographics						
Female		−0.00	−0.01	−0.01	0.05	0.04
		(0.04)	(0.04)	(0.04)	(0.04)	(0.04)
Citizenship		b	b	b	b	b
Father's education		0.02	0.03	0.03	0.01	0.02
		(0.02)	(0.02)	(0.02)	(0.02)	(0.02)
Mother's education		0.03	0.03	0.04	0.03	0.03
		(0.02)	(0.02)	(0.02)	(0.02)	(0.02)
Father's SEI		b	b	b	b	b
Mother's SEI		b	b	b	b	b
Mother working		b	b	b	b	b
Parents' income		b	b	b	b	b
Intact family		−0.05	−0.02	−0.05	−0.06	−0.04
		(0.06)	(0.06)	(0.06)	(0.06)	(−0.06)
Parent divorce		b	b	b	b	b
Number of siblings		−0.03	−0.03	−0.03	−0.03	−0.03
		(0.02)	(0.02)	(0.02)	(0.02)	(0.02)
Financial aid applicant		b	b	b	b	b
Public high school		b	b	b	b	b
Middle school capital						
High culture			b	b	b	b
Pop culture			b	b	b	b
Parent-school			b	b	b	b
Parent-homework			−0.05**	−0.05**	−0.04**	−0.04**
			(0.01)	(0.01)	(0.01)	(0.01)
High school capital						
Parent-school				b	b	b
High culture				b	b	b
Pop culture				−0.05**	−0.04*	−0.03*
				(0.02)	(0.02)	(0.01)

Table 4. *(Continued)*

	Model					
Independent Variable	1	2	3	4	5	6
Parent-homework				b	b	b
Test scores						
SAT-verbal					0.001^{**}	0.001^{**}
					(0.0003)	(0.0003)
SAT-math					0.002^{**}	0.002^{**}
					(0.0004)	(0.0004)
Other capital measures						
Scholarship athlete					b	
Engineering					b	
Hours studied					b	
Student identity						0.08
						(0.03)
Self-rated ability						b
Relative at Duke						-0.14^{*}
						(0.05)
Regression summary						
R^2	0.051	0.063	0.088	0.102	0.172	0.197

[a]Excluded category: white.
[b]Variable trimmed from model.
*indicates $p \leq 0.05$.
**Indicates $p \leq 0.01$.

one-third of the black-white gap, and the difference between Hispanics and whites may become significant with added data from a second cohort or with passing semesters and accumulation of grades.

Notably, model 6, which includes all pre-college control variables, reduces the black-white achievement gap by 41%, leaving 59% of the gap unexplained by measured pre-college differentials. Most other studies report reductions in the neighborhood of 40% of the gross gap, once pre-college factors are controlled (Bowen & Bok, 1998; Jencks & Phillips, 1998; Massey et al., 2003). Hence, this key finding for the Campus Life and Learning population is similar to that found in other studies, even though our models include more extensive control variables than prior research. These results strongly indicate that there are important within-college processes that differentiate students by racio-ethnic group, even very early in their collegiate careers.

In estimating the nested models, we tested all of the independent variables listed in the left-hand column of Table 4. The final model reported in each column reflects a trimmed equation that includes predictors that maintain their significance through to the full model (model 6), and predictors found to be important in other research. Predictors designated as "trimmed" were tested both individually and as a group for each model. Gender and parent's education, found to be important in some other studies, are not significant predictors of first semester grades at conventional levels of probability. Students from families with larger sib-ships bear a small but statistically nonsignificant disadvantage in grades. This coefficient approaches significance in some equations. In contrast to the findings of some prior studies, neither intact family nor parental divorce has a significant net effect on first semester grades for this group of students. The means in Table 3 demonstrate that black and, to some extent, Hispanic students were more likely to come from non-intact families. In our results, it appears the control for racio-ethnic status removes any disadvantage for students from nonintact families.

The middle and high school capital variables do not make much difference in first semester grades, once racio-ethnic group and demographic factors have been controlled. During the middle school years, parental involvement in homework signals a small disadvantage in college grades; we suspect this reflects a process whereby parents respond to their child's lack of self-direction or underachievement, relative to the larger group of students with credentials to secure admission to an elite institution. Also, active participation in sporting events and popular music concerts in high school has a significant negative effect on first semester grades, net of other factors. Overall, the pre-college level of cultural capital makes little difference, at least at this early point in the college career. As noted above, it is possible that cultural capital is expended in the college admissions process or needs readjustment and re-accumulation in the college environment. We will explore these possibilities with additional waves of data in the future.

Test scores are strong predictors of first semester grades. Alone, they explain 7% of the total variance and about one third of the explained variance (relative to model 6) in first semester grades. With combined SAT verbal and mathematics scores controlled, the achievement gap is reduced from 0.39 to 0.19 of a letter grade of GPA. A one-hundred point increment in verbal test scores, with other factors statistically controlled, translates into one-tenth of letter grade of GPA; a one-hundred point increment in mathematics scores translates into a one-fifth of a letter grade of GPA. However, a key portion of the achievement gap remains unexplained even after adjusting for test scores. Controls for other pre-college factors increase the net unexplained gap.

Among other capital measures, enrollment in the school of engineering, hours studied in high school, and self-rated ability at challenging high school classes do not significantly affect first semester college grades. Unfortunately, there are not enough intercollegiate athletes in this first cohort to be able to assess the effect of athlete status. A student's academic identity, measured as the extent to which a student views "being a good student" as an important part of his or her overall identity prior to college, significantly affects first semester grades. This finding is important for several reasons. First, there is precedent for this finding of identity effects on behavior in the literature (Gecas & Burke, 1995) and we plan to explore it further with future waves of data. Second, this finding is consistent with evidence that domain identification (i.e. valuing the self highly in particular performance arenas) mediates the relationship between negative stereotype threat and performance (Aronson et al., 1999). Those who identify closely with a particular arena, in this case academic achievement, may be more subject to the effects of negative stereotype threat on academic performance. Finally, such an identity configuration might predict persistence in courses and areas of study in the face of academic difficulty or failure.

A final interesting finding relates to the fact that students whose relative or family member previously attended Duke scored significantly lower in first semester grades (by about one-seventh of a letter grade, $b = -0.14$) than other students. It is likely that not all of these students constitute true legacy cases, for example some students responding yes to this question may have had a sibling attend Duke; nonetheless legacy cases are subsumed within this category of students. This gap of one-seventh of a letter grade is about one-third of the gross black-white gap, and about two-thirds the size of the net black-white gap (model 6). At first glance the negative effect may seem like an anomaly; one might expect relatives and family members with experience at the student's education institution to provide a *positive* form of social capital for the student. On further consideration of legacy cases in U.S. higher education, and elite institutions in particular, there are several plausible explanations for the significant achievement gap between this group of students and others. If higher education institutions bend their admission standards to accommodate the children and relatives of alumni and alumnae (particularly potential donors), these students may be less prepared and have lower ability to endure the demands of college than their counterparts who have no familial ties to the university. Alternatively, legacy cases on balance, once admitted, may "coast" or pursue their studies less vigorously compared with non-legacy cases. They may believe that their legacy status buffers them from the negative outcomes of poor academic performance in college. We will explore these possibilities in greater detail with future waves of data.

CONCLUSION

The black-white achievement gap has drawn significant attention of both the scholarly and policy communities in the United States in recent years. This is the case at all levels of education, but particularly in higher education, and perhaps even more so at elite institutions. The achievement gap is sizable and virtually all efforts to explain it with traditional background and capital factors leave a substantial portion of the gap unexplained. Why is this the case? This paper begins to address this question by examining racio-ethnic differentials in GPA at the end of the first semester of the first college year at one highly-selective U.S. university. In an attempt to examine some of the major explanations for the minority achievement gap, including those of status attainment and cultural and social capital differences, the analyses control for a wider range of factors than prior studies. The empirical analyses establish the emergence of a substantial and significant black-white achievement gap after as little as four months into the college career. This gap is nearly as large as that reported in other national studies for the entire college career. Other minority groups demonstrate no significant differences at this early stage of college, though it is possible that significant achievement gaps for other minority groups emerge over the college career.

Controlling for a range of pre-college factors – in terms of family background, middle- and high-school cultural capital, parental involvement, test scores and other capital measures – indicates that about 40% of gap in first-semester GPA can be explained by differences in the socioeconomic background and academic preparation between black and white students. These findings accord well with theoretical perspectives that emphasize capital deficits as a central explanation for the lagging educational achievement of minority groups. When students arrive at college, they bring with them widely-varying amounts of pre-college human, social, and cultural capital that predict a portion of the variation in academic achievement in the first college semester. The importance of a student's pre-college academic identity for first-semester achievement is also interesting, as it suggests a form of psychic capital that might serve to buffer students as they make the transition from high school to college. Students who report a strong identification with "being a good student" may respond differently to educational challenges and successes than other students. As we follow these students through college we can examine whether those who identify strongly with the academic domain benefit from this identity or, alternatively, as the work of Steele and Aronson (Aronson et al., 1999; Steele, 1997; Steele & Aronson, 1995) suggests, whether they are more susceptible to the detrimental effects of negative stereotype threat. At the same time it is striking that such a large portion, about 60%, of the gap remains unexplained.

Overall, our results raise several implications for future research as well as for policymakers hoping to find ways to increase the achievement of all students. First, scholars seeking to understand achievement gaps in college performance would do well to examine students' academic experiences at even earlier stages in the life course, from early childhood to young adulthood, as some of the sources of achievement differences in college are likely to be found in the educational experiences and family backgrounds of children well before they reach college. For example, some research indicates that minority achievement differences emerge as early as preschool and kindergarten (Barbarin et al., 2004). A greater dialogue between researchers focused on college students and scholars examining similar issues at earlier stages of childhood and youth could yield very fruitful results.

In light of the above comments, it may be tempting to conclude that college administrators and policy makers can do little to enhance the achievement of minority students or that elementary and secondary school practitioners are best suited to address the problems of the minority achievement gap. While early interventions are certainly important, institutions of higher education should not relinquish their critical role in addressing the minority achievement gap. The knowledge that achievement gaps emerge so early in the college career is powerful in that it can be utilized by college faculty and administrators to craft effective programs to identify high-risk students early and to find effective ways to help these students acclimate to college and strengthen their academic skills. Such efforts might entail better support services and counseling as well as tutoring and academic skills training. By no means is this an easy task since universities, especially elite universities, and students, especially minorities students, may be resistant to programs that emphasize academic-skills training and tutoring and are seen as remedial. But absent such interventions, students who are struggling academically are more likely to drop out or experience other academic episodes detrimental to their college performance and completion. As we follow two cohorts of Duke Students over their college careers we will assess how students utilize a range of university academic and counseling services and the efficacy of such programs for improving their academic achievement. These investigations should yield more detailed knowledge for those interested in implementing successful programs to address achievement gaps in college.

Finally, this study is the first to document an achievement gap between students with familial ties to their university (of whom whites comprise the largest category) and students with no such familial ties. This finding is especially interesting in light of the current heated debates regarding affirmative action in U.S. higher education. It underscores the point that universities consider a wide range of factors beyond academic achievement in their admission processes. While the national conversation has disproportionately focused on issues of race, it should be broadened to include legacy students as well. As we continue with this research, we hope to illuminate the mechanisms behind the achievement gaps of both minority and legacy students and to see how such differences in academic performance change over the college years.

NOTES

1. The stream of measurements, assessments, and data merges proceeds over the college years of each cohort. The design provides full access to Duke University's Student Information Support System (SISS) database for those students who give signed release. Information from admissions files, registrar files (including grades, courses taken, honors, special programs, and study-abroad activities each semester), financial aid files, and residential housing files will be merged and analyzed.

2. The pre-college survey is administered and coded by an external data collection subcontractor, Research Triangle Institute. RTI is well-established national survey research organization. We will employ two strategies to deal with potential sample selection bias resulting from the possibility that those who matriculate at Duke are not representative of those admitted. As an alternative to conventional selection modeling techniques, which can provide misleading results (Stolzenberg & Relles, 1990), Land and McCall (1993) present a Bayesian 'mixture-modeling' technique for estimating the sensitivity of sample parameter estimates to various assumptions about sample selection bias. Given an observed response and covariate data, this technique provides a subjective probability interval within which a sample statistic would lie if non-respondents had been included. Necessary covariate information (demographic, pre-college grades, test scores, etc.) will be available for both matriculates and non-matriculates. In the fourth year of the project we plan to obtain academic performance data from a sub-sample of those who were admitted to Duke but matriculated elsewhere. While attempting to track and survey a parallel sample of non-matriculates over four years would be prohibitively costly and time-consuming, obtaining minimal performance data at year four (via postcard and/or 5-minute telephone interview) is feasible. It will allow us to compare matriculates and non-matriculates on key variables, and empirically examine the assumptions used in the (Land-McCall) sensitivity analysis.

3. Wave 1 (pre-college) and wave 2 (first year) instrumentation can be viewed at: http://www.soc.duke.edu/dept/faculty/kspen.html.

4. We have conducted extended analyses of the power in our sampling design under a variety of assumptions, including different analysis models and with tabular data. The key results lead us to be confident that we have sufficient statistical power in the research design (i.e. given the sample of approximately 1500 students over two cohorts). The statistical power for analyses of a single cohort is marginal but still acceptable, including with the response rates reported here.

5. We also made comparisons of Duke to other so-called "Elite" (Harvard, Princeton, Yale, Dartmouth, the University of Pennsylvania, Brown, Stanford and Columbia) and "Top 50" (based on SAT scores) universities. In general Duke is identical to or slightly below the elite institutions and clearly above the Top 50 institutions. For example, Duke's first year retention rate is 97% compared with 97% for the elite and 90% for the Top 50. The 2000 graduation rate is 93% of those matriculating, compared with 93% for the elite, and 80% for the Top 50 institutions. The student-faculty ratio is 9.0:1, versus 8.22:1 for the elite and

10.69:1 for the Top 50 institutions. Finally, the 25th and 75th percentile of SAT scores for Duke are 1300 and 1500; for elite institutions, 1334 and 1522; for Top 50 institutions, 1234 and 1424. These comparisons help to situate Duke in the national distribution.

6. We compared those in the analysis sub-sample to non-respondents in the sample using pre-college admissions data (test scores and demographics). The only significant difference ($p < 0.05$) involved SAT mathematics scores, with respondents scoring about 15 points higher than non-respondents. All other differences were not significant (SAT verbal, high school rank, overall admissions rating, mother's and father's college graduation, financial aid applicant, and public-private high school).

7. This included input correlation and covariance matrices, varimax and promax rotations, and 3, 4, 5, and 6 factor solutions.

8. This is the case in the population of Duke students and not a non-response differential.

9. We handled missing data as follows. For continuous independent variables with fewer than 5% missing, we used mean imputation as described in Cohen and Cohen (1975). For dichotomous variables with fewer than 5% missing we used stochastic imputation for binary outcomes (Little & Rubin, 1987) with cut points set to the mean of a dummy variable. Variables with more than 5% missing included verbal SAT (15.6%), mathematics SAT (18.4%), and family income (7.7%). For these variables, we imputed a regression-predicted score using other variables in the design (ACT scores plus high school record for SAT; demographic variables plus financial aid information for family income). Prediction equations explained more than 60% of the variance in each regression imputed outcome, an indication that minimal bias will be present in substantive models using these imputed variables (Landerman et al., 1997).

ACKNOWLEDGMENTS

An earlier version of this paper was presented at the XV World Congress of Sociology of the International Sociological Association, July 2002, Brisbane, Australia. The authors gratefully acknowledge support for this research provided by grants from the Andrew W. Mellon Foundation and Duke University. The authors bear sole responsibility for the contents of the paper. We thank A. Y. Bryant, Kara Bonneau, Will Tyson, Tammy McLamb, and Megan Hay for their assistance.

REFERENCES

Ainsworth-Darnell, J. W., & Downey, D. B. (1998). Assessing the oppositional culture explanation for racial/ethnic differences in school performance. *American Sociological Review*, *63*, 536–553.

Aronson, J., Fried, C. B., & Good, C. (2002). Reducing the effects of stereotype threat on African American college students by shaping theories about intelligence. *Journal of Experimental Social Psychology*, *38*, 113–125.

Aronson, J., Lustina, M. J., Good, C., Keough, K., & Steele, C. (1999). When white men can't do math: Necessary and sufficient factors in stereotype threat. *Journal of Experimental Social Psychology*, *35*, 29–46.

Astin, A. W. (1993). *What matters in college: Four critical years revisited*. San Francisco: Jossey Bass.

Barbarin, O. A., McCandies, T., Coleman, C., & Atkins, T. (2004). Ethnicity and culture. In: P. Allen-Meares & M. W. Fraser (Eds), *Intervention with Children and Adolescents: An Interdisciplinary Perspective* (pp. 27–53). New York: Allyn & Bacon.

Bourdieu, P. (1977). Cultural reproduction and social reproduction. In: J. Karabel & A. H. Halsey (Eds), *Power and Ideology in Education* (pp. 487–511). New York: Oxford.

Bowen, W. G., & Bok, D. (1998). *The shape of the river: Long term consequences of considering race in college and university admissions*. Princeton, NJ: Princeton University Press.

Cleary, T. A. (1968). Test bias: Prediction of grades of negro and whites students in integrated colleges. *Journal of Educational Measurement*, *5*, 115–124.

Cohen, J., & Cohen, P. (1975). *Applied multiple regression/correlation analysis for the behavioral sciences*. New York: Wiley.

Coleman, J. S. (1988). Social capital in the creation of human capital. *American Journal of Sociology*, *94*(Suppl.), S95–S120.

Cook, P. J., & Ludwig, J. (1998). The burden of 'acting white': Do black adolescents disparage academic achievement? In: C. Jencks & M. Phillips (Eds), *The Black-White Test Score Gap* (pp. 375–400). Washington, DC: Brookings Institution Press.

Croizet, J.-C., & Claire, T. (1998). Extending the concept of stereotype threat to social class: The intellectual underperformance of students from low socioeconomic backgrounds. *Personality and Social Psychology Bulletin*, *24*, 588–594.

Croizet, J.-C., Desert, M., Dutrevis, M., & Leyens, J. (2001). Stereotype threat, social class, gender, and academic under-achievement: When our reputation catches up to us and takes over. *Social Psychology of Education*, *4*, 295–310.

Crouse, J., & Trusheim, D. (1988). *The case against the SAT*. Chicago: University of Chicago Press.

DeGraaf, N. D., DeGraaf, P. M., & Kraaykamp, G. (2000). Parental cultural capital and educational attainment in the Netherlands: A refinement of the cultural capital perspective. *Sociology of Education*, *73*, 92–111.

Dillman, D. A. (1978). *Mail and telephone surveys: The total design method*. New York: Wiley-Interscience.

DiMaggio, P. (1982). Cultural capital and school success: The impact of status culture participation on the grades of U.S. high school students. *American Sociological Review*, *47*, 189–201.

Farkas, G. (1996). *Human capital or cultural capital? Ethnicity and poverty groups in an urban school district*. New York: Aldine De Gruyter.

Farkas, G., Lleras, C., & Maczuga, S. (2002). Does oppositional culture exist in minority and poverty peer groups. *American Sociological Review*, *67*, 148–155.

Featherman, D. L., & Hauser, R. M. (1978). *Opportunity and change*. New York: Academic Press.

Gecas, V., & Burke, P. J. (1995). Self and identity. (pp. 41–67) In: K. S. Cook, G. A. Fine & J. S. House (Eds), *Sociological Perspectives on Social Psychology*. Needham Heights, MA: Allyn & Bacon.

Hirschman, C., & Snipp, C. M. (1999). The state of the American dream: Race and ethnic socioeconomic inequality in the United States, 1970–1990. In: P. Moen, D. Dempster-McClain & H. A. Walker (Eds), *A Nation Divided: Diversity, Inequality, and Community in American Society* (pp. 889–1105). Ithaca, NY: Cornell University Press.

Hout, M. (1988). More universalism, less structural mobility: The American occupational structure in the 1980s. *American Journal of Sociology*, *93*, 1358–1400.

Hurtado, S. (1992). The campus racial climate: Contexts of conflict. *Journal of Higher Education*, *63*, 539–569.

Hurtado, S., Milem, J. F., Clayton-Pederson, A. R., & Allen, W. (1998). Enhancing campus climates for racial/ethnic diversity: Educational policy and practice. *The Review of Higher Education*, *21*, 279–302.

Jencks, C., & Phillips, M. (Eds). (1998). *The black-white test score gap*. Washington, DC: Brookings Institution Press.

Kane, T. J. (1998). Racial and ethnic preferences in college admissions. In: C. Jencks & M. Phillips (Eds), *The Black-White Test Score Gap* (pp. 431–456). Washington, DC: Brookings Institution Press.

Kuo, H.-H. D., & Hauser, R. M. (1995). Trends in family effects on the education of black and white brothers. *Sociology of Education*, *68*, 136–160.

Land, K. C., & McCall, P. L. (1993). Estimating the effect of nonignorable nonresponse in sample surveys: An application of Rubin's Bayesian method to the estimation of community standards for obscenity. *Sociological Methods and Research*, *221*, 291–316.

Landerman, L. R., Land, K. C., & Pieper, C. F. (1997). An empirical evaluation of predictive mean matching method for imputing missing values. *Sociological Methods and Research*, *26*, 3–33.

Little, R. J. A., & Rubin, D. B. (1987). *Statistical analysis with missing data*. New York: Wiley.

Massey, D. A., Charles, C. Z., Lundy, G. F., & Fischer, M. J. (2003). *The source of the river: The social origins of freshmen at America's selective colleges and universities*. Princeton, NJ: Princeton University Press.

McPherson, M., Smith-Lovin, L., & Cook, J. M. (2001). Birds of a feather: Homophily in social networks. In: K. S. Cook & J. Hagen (Eds), *Annual Review of Sociology* (Vol. 27, pp. 415–444). Palo Alto: Annual Reviews.

Nettles, M. T. (1988). *Toward black undergraduate student equality in American higher education*. New York: Greenwood Press.

Nettles, M. T., Theony, A. R., & Grosman, E. (1986). Comparative and predictive analyses of black and white students' college achievement and experiences. *Journal of Higher Education*, *57*, 289–318.

Ogbu, J. (1974). *The next generation: An ethnography of education in an urban neighborhood*. New York: Academic Press.

Ogbu, J. (1986). The consequences of the American caste system. In: U. Neisser (Ed.), *The School Achievement of Minority Children* (pp. 19–56). Hillsdale, NJ: Lawrence Erlbaum.

Pascarella, E. T., Edison, M., Nora, A., Hagedorn, L., & Terenzini, P. (1986). Influences on students' openness to diversity and challenge in the first year of college. *Journal of Higher Education*, *67*, 174–195.

Pfieffer, C. M. (1976). Relationship between scholastic aptitude, perception of university climate, and college success for black and white students. *Journal of Applied Psychology*, *9*, 341–347.

Ramist, L., Lewis, C., & McCamley-Jenkins, L. (1994). *Student group differences in predicting college grades: Sex, language, and ethnic groups*. New York: College Entrance Examination Board.

Rau, W., & Durand, A. (2000). The academic ethic and college grades: Does hard work help students to 'make the grade'?" *Sociology of Education*, *73*, 19–38.

Roscigno, V., & Ainsworth-Darnell, J. (1999). Race, cultural capital and educational resources: Persistent inequalities and achievement returns. *Sociology of Education*, *72*, 158–178.

Smith, S. S., & Moore, M. R. (2000). Intraracial diversity and relations among African-Americans: Closeness among black students at a predominantly white university. *American Journal of Sociology*, *106*, 1–39.

Spenner, K. I., & Featherman, D. L. (1978). Achievement ambitions. In: R. H. Turner, J. Coleman & R. C. Fox (Eds), the *Annual Review of Sociology* (Vol. 4, pp. 373–420). Palo Alto, CA: Annual Reviews.

Stanton-Salazar, R. D., & Dornbusch, S. M. (1995). Social capital and the reproduction of inequality: Information networks among Mexican-origin high school students. *Sociology of Education*, *68*, 116–135.

Steele, C. M. (1997). A threat in the air: How stereotypes shape the intellectual identities of women and African-Americans. *American Psychologist*, *52*, 613–629.

Steele, C. M. (1999). Thin ice: Stereotype threat and black college students. *Atlantic Monthly* (August), 44–54.

Steele, C. M., & Aronson, J. (1995). Stereotype threat and the intellectual test performance of African Americans. *Journal of Personality and Social Psychology*, *69*, 797–811.

Steele, C. M., & Aronson, J. (1998). Stereotype threat and the test performance of academically successful African Americans. In: C. Jencks & M. Phillips (Eds), *The Black-White Test Score Gap*. Washington, DC: Brookings Institution.

Stolzenberg, R. M., & Relles, D. A. (1990). Theory testing in a world of constrained research design: The significance of Heckman's censored sampling bias correction for nonexperimental research. *Sociological Methods and Research*, *18*, 395–415.

U.S. Department of Education, National Center for Education Statistics (2002). *Digest of Education Statistics, 2001*. NCES 2000-130, by T. D. Snyder. Washington, DC: U.S. Department of Education.

Vars, F. E., & Bowen, W. G. (1998). Racial and ethnic preferences in college admissions. (pp. 431–456). In: C. Jencks & M. Phillips (Eds), *The Black-White Test Score Gap*. Washington, DC: Brookings Institution Press.

Voelkl, K. (1997). Identification with school. *American Journal of Education*, *105*, 294–318.

PART III: COMPARATIVE APPLICATIONS

STATUS ALLOCATION IN VILLAGE INDIA

Bam Dev Sharda

ABSTRACT

Village India's status allocation process is largely ascriptive. The most frequently held generalization maintains that India's system of stratification is unique: "closed" and "non-permeable." The preferred model of status allocation, therefore, has been that of caste. In this paper, I test the caste determination hypothesis for village India, and an alternative socioeconomic ascription hypothesis. There are two forms of the latter hypothesis: (1) an agrarian mode of production hypothesis that maintains that landed wealth and socioeconomic origins (e.g. father's occupation) determine current occupation; and (2) an agrarian modernization hypothesis that states that in industrializing agrarian economy, landed wealth and education determine a person's occupational status. My analyses of the data from village India for 1962 and 1977 do not support either the caste determination hypothesis (caste effects on current occupation are insignificant) or the modernization hypothesis (the effects of education remain near zero, even after green revolution and India's progressive legislation). Consistent support is, however, found for the agrarian mode of production hypothesis.

The Shape of Social Inequality: Stratification and Ethnicity in Comparative Perspective
Research in Social Stratification and Mobility, Volume 22, 219–254
© 2005 Published by Elsevier Ltd.
ISSN: 0276-5624/doi:10.1016/S0276-5624(04)22008-X

STATUS ALLOCATION IN VILLAGE INDIA

This paper proposes a socioeconomic model for the analysis of status allocation in village India. It thus differs sharply from the main line of thought on Indian stratification system (Dumont, 1970), a view that has endured since the seventeenth century when the term "caste" was first incorporated into the English language to describe society in India (see Dumont, 1970, p. 21). This paper shows the near irrelevance of caste and importance of agricultural land and occupational status origins for stratification of village India.

For village India, the preferred model so far has been that of caste (Dumont, 1970; Ghurey, 1961; Hutton, 1969; Ibbetson, 1916; Maine, 1881; Mayer, 1968; Nesfield, 1885; Srinivas, 1966; Weber, 1958). This classical view attributes the processes of ascription of inter-generational status transmission to caste (locally known as *jati*). In light of recent research on stratification it appears that the effects of caste on occupational status transmission have been over stated (Elder, 1996; Gartrell, 1981; Sharda, 1977; Singh, 1976). Current reflections on status allocation in India have been stimulated by the emergence of modern status allocation models and their application to developing nations.

Status allocation models arose in the 1960s following the convergence of several essential advances in research methods. These include dependable methods for eliciting data on human subjects, which procedures are applicable anywhere in the world; sampling methods permitting generalizations from small samples to large populations; statistical methods capable of summarizing and analyzing masses of data, especially those permitting tests of causal network hypotheses; and rapid data processing by means of computers. To these were added sociological concepts of transmitted and seemingly non-transmitted elements of status allocation processes (commonly called "ascription and achievement") (Linton, 1936). It was the coalescing of these intellectual developments that laid the groundwork for today's models of status allocation phenomena. The first such models were applied to population of the United States (Blau & Duncan, 1967; Sewell et al., 1969). Tests on populations of developing nations followed shortly afterwards (Hansen & Haller, 1973). Researchers concerned with India's supposedly unique stratification structure, "caste," soon reasoned that perhaps status allocation in that nation, too, might be less unique than the previous thinkers had held; indeed, that models developed elsewhere might well provide a more realistic picture of such processes – albeit, with India-specific parameter values (Elder, 1996; Gartrell, 1981; Sharda, 1977; Singh, 1976).

THEORY

Status Allocation Process

Status allocation refers to the process whereby individual members of the society are distributed into hierarchically ordered positions in stratification structures. The "allocation" terminology came into being as a critique of the "status attainment" models, to make it clear that the process does not assume free will (Haller, 1982; Kerchkoff, 1976). However, the components and empirical referents of the two models remain unchanged. Inter-generational status transmission processes incorporate both ascription and achievement (see e.g. Haller & Saraiva, 1991). Models of the process are, in principle, fully deterministic, even though they incorporate social psychological variables as aspirations of youth and the expectations others hold for them (Haller, 1982; Kerchkoff, 1976).

The pioneering work of Blau and Duncan (1967) described the process of status attainment in American society. Their findings have been replicated periodically: (1) that occupational attainment in the United States is influenced by educational attainment; and (2) that a sizeable proportion of the effects on educational attainment is due to status origins. The Wisconsin model replicates these effects (see Haller & Portes, 1973) and elaborates the link between SES and educational attainment through socialization processes, especially the influence of significant others.

Status Allocation in Village India

The most frequently held generalization maintains that India's status allocation process is purely ascriptive because India's system of stratification is "closed" and "non-permeable." The generalization further states that an individual cannot change one's position during his life because of his caste status. One of the prominent exponents of the Indian caste system is Louis Dumont. According to Dumont (1970, p. 21, emphasis in original):

> . . . caste system divides the whole society into a large number of hereditary groups, distinguished from one another and connected together by three characteristics: *separation* in matters of marriage and contact, whether direct or indirect (food); *division* of labour, each group having, in theory or by tradition, a profession from which their members can depart only within certain limits; and finally *hierarchy*, which ranks the groups as relatively superior or inferior to one another.

Elsewhere (p. 67) he adds that the Indian hierarchy consists of five levels – four varnas plus the "untouchable," demarcated by "a series of successive dichotomies or inclusions" – the four varnas being the well-known "castes" of Brahmans, Kshatryas, Vaishyas and Shudras (see Appendix C, and discussion in methodology section).

The emphasis on caste in Indian stratification literature does not always deny the possible importance of such variables as occupational prestige, income and education (Elder, 1996). However, when other socio-economic variables are considered at all by modern scholars, they are treated as "subordinate to [ritual] hierarchy" (Dumont, 1970, p. 251). Some scholars maintain that when economic or political gains are made by individuals, these gains are used to enhance the ritual rank of their entire caste (Bailey, 1957; Gough, 1960; Marriott & Inden, 1974). Lack of systematic evidence to support these assertions suggests both the need for careful testing of the Caste hypothesis, and if it was found to be untenable, for an alternative model of status allocation. As indicated, the Blau and Duncan (1967, Chap. 5) model of status attainment (allocation) is frequently used to describe the stratification processes of societies, including those of developing societies (Bills & Haller, 1984; Bills et al., 1985; Haller & Saraiva, 1991; Hansen & Haller, 1973; Holsinger, 1975; Sharda, 1981; among others).

A truncated version of the Blau and Duncan basic model was tested in village India for 1962 (Sharda, 1977). No information was available for father's education in the 1962 data and hence this variable was dropped from the model. The results were later compared with the 1962 Occupational Changes in a Generation (OCGI) survey data from the rural United States (Sharda, 1981). Whereas the rural and overall model of the United States were essentially the same, village India showed a marked contrast with the rural United States. Similar findings were reported for the Wisconsin model for Costa Rica (Hansen & Haller, 1973) and rural Brazil (Richards & Hansen, 1982) which showed marked differences from the patterns observed in the United States.

The alternative socioecomic model or models, I present two similar ones, of status allocation that I propose argues that occupations of village India are tied to the agrarian mode of production. The way children are socialized and trained for village occupations has very little to do either with the caste rank of their families; most either follow agricultural occupations or aspire to be in one. Neither does it have much to do with education, the key variable in the U.S. and other countries with capitalist-industrial mode of production. The status allocation regime remains largely ascriptive due mainly to the structure of village India economy, but this itself has little or nothing to do with the caste system. Contemporary scholars (see e.g. Gupta, 1984, p. 7) have suggested that the "Indian village community has undergone progressive change . . . during the [last] thirty years" due to industrialization, urbanization, education and progressive legislation and that the influence of caste on occupational allocation is decreasing and the role of education increasing. Land reform legislation, Free and Compulsory Act of 1961 and "green revolution" of late 1960s are often cited in support of these claims.

In this paper I extend and compare models over time for 1962 and 1977 data. The extended models incorporate family land ownership (an indicator of rural wealth) and "caste" status as exogenous variables. The comparison of these models will allow me to assess the claims of various positions. One is the caste determination hypothesis, the other two are agrarian society hypotheses. One of these holds that landed wealth and occupational status origins alone determine one's occupational status in village India. The other, industrializing agrarian society hypothesis holds that in such societies education has a strong influence on occupational status. This should also be the case with village India.

METHODS

Village India: The Population

Some 578,000 villages are home to nearly three quarters of the one billion people of India. Visaria and Visaria (1995, pp. 4, 5) point out that "the diversity of language and culture among India's 26 states and 6 union territories is unique among the nations of the world . . . There are 18 official languages and at least 50 regional tongues." Though there are no officially recognized religions in India, 83% people belong to the Hindu faith. The 1991 census enumerated 1,091 scheduled castes (formerly called untouchable) which constitute 17% of the total population. Another 8% belong to some 500 tribes. Although information is not solicited by the census, the Anthropological Survey of India (1991) reported that there are "4,384 communities" or castes (locally known as *jatis*) and tribes. Drawing a representative sample of village India, therefore, is a task of monumental proportions.

Scholars have often selected a "typical village" for their fieldwork for their study of village India. Charles Metcalf (1832, p. 331) described village life to the Select Committee of the British House of Commons in 1832 in the following way:

> The village communities are little republics, having nearly everything they want within themselves, and almost independent of any foreign relations. They seem to last where nothing else lasts. Dynasty after dynasty tumbles down; revolution succeeds revolution . . ., but the village community remains the same . . . This little state in itself, has, I conceive, contributed more than any other cause to the preservation of the people of India, through all the revolutions and changes which they have suffered, and is in a high degree conducive to their happiness, and to the enjoyment of . . . freedom and independence.

This image of village "republics" was more or less retained by scholars throughout the colonial period and formed the basis of "village studies" approach "focusing on the totality of a village social system" (Epstein, 1978, p. 127) in post-independence India as well. The studies of Indian villages in 1950s and 1960s were done with a sense of urgency to capture village social organization before changes destroyed their self sufficiency and independence. Two famous anthologies: *India's Villages* edited by M. N. Srinivas, and *Village India* edited by McKim Marriott (Marriott, 1961; Srinivas, 1966) presented fieldwork from "typical villages." However, as has been argued elsewhere, this image of village India as isolated and self sufficient communities is unwarranted. It was further argued (Sharda, 1977, p. 23) that: "studies of single villages, though useful in certain respects, might be too confined to provide an understanding of rural India's inter-village social structure and social processes."

Few villages are really isolated from others. They tend to form networks of inter-village marriage, trade and job connections. Gough (1960, 1966), for example, reported that a south Indian village she studied was linked to 18 other villages through marital alliances. Similarly, villages are not economically self sufficient now (and probably never were) and were traditionally inter-linked through trade and labor exchange relations known as the "jajmani system." The villages are of different sizes and have different levels of isolation from cities and roads. It is, therefore, difficult to capture this diversity by studying a single village. There are no "typical" villages. Therefore, a sample containing a diverse group of villages is essential to capture the representativeness of village India and to understand the status allocation process.

The Sample

The 1961 Census of India introduced an innovation by including an additional study of 583 villages. Fieldwork in these villages was undertaken in 1962–1963. This sample was carefully selected in order to "construct a map of village India's social structure" (Census of India, 1961, p. 56). However, only about half of the village studies were actually published before funding ran out. I carefully selected eleven villages from these 583 census villages so as to represent a broad range of cultural and dialectic-linguistic variations within two adjacent states of India, Punjab and Haryana. The villages differed in terms of their jati (endogamous caste) composition – some were single jati villages while others were multi-jati villages. While all of these villages were located in an important wheat-producing region, the villages did illustrate different levels of economic development and productive resources, different sizes, and different degrees of isolation from communication centers (roads and cities). I was allowed to code data from the "Head of Household Schedule" pertaining to these villages on the premises of the Census office in New Delhi, India. All individual sample members are employed male head of households.

This region underwent rapid agricultural transformation, described as the green revolution, beginning in the late 1960s. Frankel (1978, p. 317; see also 1971) describes the heartland of the *green revolution* as: "Three states, Punjab, Haryana and (western) Uttar Pradesh. By 1969, Mexican seeds covered almost the entire irrigated acreage and spurred a spectacular increase in production." In 1977, I resurveyed the same 11 villages and data from this survey are also reported herein. For the analysis of status allocation, the data in both years were limited to all male heads of households between the ages of 20 and 64. They numbered 761 in 1962 and 823 in 1977.

This is of importance to the present researcher because the economic changes thus wrought would be expected to reduce the presumed effects of caste on occupational status and to have increased the occupational effects of education presume to result from India's industrialization (Gupta, 1984).

Changes in the distribution of occupations by major divisions, for the two time periods, are reported in Table 1. It is clear that the index of dissimilarity (26.6) is sizeable and indicate a significant shift away from farming (nearly 20%) and away from the mainly "jajmani" occupations (classified mainly as "service").

Major increases, on the other hand, have been in "modern" skilled and unskilled or manual occupations, and "professional, administrative and technical" category. These are the effects of green revolution (see e.g. Frankel, 1971; Sharda, 1986). Did these changes alter the status allocation process? The answer is clearly

Table 1. Distribution of Respondents According to Major Occupational divisions, 1962 and 1977.

Nos.	Occupational Division	1962	1977	Percent Change
0–2	Professional, Technical, and Administrative	1.8	9.4	7.6
3	Clerical and related services	2.4	3.2	0.4
4	Sales	4.6	6.1	1.5
5	Service (mainly jamani)	10.2	3.4	6.8
6	Farmers, Fisherman, Hunters, and Loggers	72.9	52.6	19.7
7–9	Skilled and Unskilled Workers	8.1	25.3	17.2
Total		100.0	100.0	53.2

Note: 1. Index of dissimilarity is defined as one half of the sum of absolute differences between to distributions.

Index of Dissimilarity = 26.6.

no, not yet, as the evidence will show. I now describe the model and its components for the status allocation process of village India – the model by which I shall test the effect of caste, education and other variables on occupational status.

The Model, Variables and their Measurement

The basic Blau and Duncan (1967) model includes socioeconomic origins as exogenous variables measured through father's education and father's occupation. Father's education was excluded from the status allocation model of village India because of the lack of data. The revised village India model, however, included family landholding and caste rank as additional exogenous variables. Endogenous variables of the model were education, 1st job, and the dependent variable was current occupation of the respondent. To be consistent in the measurement of variables, higher scores were assigned for higher values and similarly lower scores reflect lower values of status dimensions. These variables were measured as follows:

Occupational Status

Occupational status variables in both 1962 and 1977 – father's occupation, 1st job and current occupation – were measured by means of the Rural India Occupational Prestige Scale or RIOPS (Appendix A). In reviewing literature on occupational prestige in India, substantial rural-urban differences were found, particularly for agricultural occupations (Sharda, 1979). Since the purpose of this paper is to describe the status allocation of village India and changes therein, it was decided to use the prestige scale developed in that setting as well. This is not unique about village India either. For example, Haller, Holsinger and Saraiva (1972) argued that status scales for rural agrarian populations (e.g. of Brazil), are bound to be different than standard scales developed in industrialized nations. Hansen and Converse (1976) also caution against indiscriminate application of alleged universal scales of occupational prestige to sub-populations of Third World nations.

Family Landholding

Landed wealth has always been considered an important resource for rural stratification systems (Lenski, 1966, pp. 226–230). The extent to which size of family's landholdings affect a person's occupational status needs to be assessed empirically. Bi-variate correlation among the two is only moderate, 0.297 in 1962 and 0.294 in 1977 (Appendix B). Family landholding is an exogenous variable in the model and was measured by the number of acres owned by the parents' household when the respondents were growing up.

Education of the Respondent

The education of the respondent was measured in this study in terms of the approximate number of school years completed. There was, however, one limitation of this measure in 1962 but not in 1977. In 1977, the exact years completed were given. The question asked by the census in 1962 was in terms of "educational standards" met by the respondent. And this is to be understood in terms of various levels of education such as primary school (5 years of schooling), middle school (8 years of schooling), high school (10 years of schooling) and college (14 years of schooling). So the approximate numbers of years successfully completed was substituted for the level ("Standards") reported for the head of the households. There was another category of "read and write," but with no standards. This was assigned a score equivalent to one year of education, a conservative estimate. This is warranted given that a large proportion of students drop out of elementary schools. Of course, illiterates who could neither read nor write were assigned a score of zero.

Caste Status

Castes – jatis – were assigned levels of ritual status in the hierarchy. Scholars agree that the Hindu conceptions of purity and pollution is the basic principle according to which jatis are placed in the hierarchy (Dumont, 1970; Marriott, 1959; Srinivas, 1968; Stevenson, 1954). However, agreements break down after that. The two major issues that pertain to caste ranking are: (1) what units should be ranked; and (2) by what criteria.

Units of the Caste system. The term "caste" was introduced by the Portuguese and has been accepted into the English language. However, there are at least two levels on which scholars ranked "castes": varna and jati. Four varnas are the categories described in the *Rig Veda*, considered to be the oldest religious literature of the Hindus, going back perhaps 7,000 years ago. A hymn of the *Rig Veda* describes the origin of the universe through the sacrifice of *Purusha*, the Cosmic Being. As part of this origin (de Barry, 1958, pp. 14, 15):

> when they divided *Purusha* . . . His mouth became the brahman; His two arms were made into rajanys; his two thighs the vaishyas; from his two feet the shudras were born.

Thus in the *Manu Smriti* the four varnas were later ranked in that order: brahmans at the top, followed by kshatrya (rajanya of the *Rig Veda* or Rajputs of today), followed by Vaishyas and Shudras. The upper three varnas are eligible to undergo the *upanayana* ceremony (donning of the sacred thread) through which they become twice born. Anthropologists argue that the nearly five thousand jatis and "tribes" that we find in India today are not the same as varnas; that they are not even due to their fission or fusion over the millennia. With the exception of Brahmans, who seem to be everywhere, other jatis are not common to all regions but are found locally (and regionally). However, jatis, based on their local standing can be classified into "jati clusters" on the basis of commensality and can be placed rather well within the varna scheme. Moreover, multiple jati clusters can fit within a single varna. Rank ordering of jati units, whether defined as varnas, jatis or jati-clusters becomes important since within certain varnas, jati-clusters are also ranked.

How Castes are ranked?. Three different approaches have been used for the ranking of "castes": by attribution, by interaction and by reputation. Different theories have been advanced for each approach.

Attributional theories (Kolenda, 1963; Stevenson, 1954) argue that "castes" could be ranked by virtue of their attributes. For example, the *upanayana* ceremony or the donning of sacred thread effectively classifies castes into higher (twice born) and lower (non-twice born). Other attributes are vegetarianism, claiming purity of blood, prohibition of divorce and widow remarriage, etc. Nineteenth century census takers did provide rich details of "attributes" of different jatis (see e.g. Ibbetson, 1916). However, as Gough (1966) reported, "in spite of the higher ritual value normally accorded to vegetarianism, we find that some meat eaters [e.g. Kshatrya/Rajput] in fact outrank some vegetarians [e.g. Bania]. Similarly, in spite of the value placed on female chastity and widow celibacy, we find in Kerala [a south Indian state] that very high ranking Kshatrya queens have in the past practiced divorce and widow remarriage. And castes with ritually clean occupations, such as basket makers, often rank very low." Thus one cannot rely on the attributes of local jatis alone for ranking.

McKim Marriott (1959) suggested an alternative theory known as the interactional theory of jati ranks. Ritual jati rank in a village, he argues, can be established through a matrix of prestations and commensality of water and food, and sharing of tobacco pipe, cots and buildings. However, as Dumont (1970, pp. 289–291) points out, this approach is limited to a single village and is, therefore, of limited utility.

A third approach, suggested by Freed (1963), called the reputational approach, is similar to the one used in community studies in the United States. In the reputational approach, "knowledgeable" persons are asked to rank jatis on some scale. This too has its limitations. One needs to have a complete list of groups (jatis) before undertaking the study. No such ranking was undertaken in 1962.

For the 1962 data, I was limited to constructing a scale based on attributes reported by the British census takers and village studies. However, I had the thirty plus jatis in my sample ranked by "knowledgeable" scholars (see Appendix C). In the 1977 study, I was able to validate this ranking by "knowledgeable" people from these villages.

Whereas the overall rankings of jati-clusters were remarkably similar, there were some interesting differences in the two rankings. Part of the difference is due to sanskritization and part involves ambiguities in specific ritual behaviors. Srinivas (1968) described a process whereby low castes (jatis), after becoming economically successful locally, start emulating the life styles (manners, dress) of Brahmans and other high castes. These jatis adopt brahmanical rituals in order to improve their caste rank in the region. He labelled this process: "Sanskritization." Sanskritization, seen as the "caste mobility" of a jati over time, changes the ritual level of a jati in some villages. Such changes would appear large to the participants, but in fact be quite small within the structure as a whole.

On the other hand Jats, Gujjars and Kamboj did not don the sacred thread, the major brahmanic ritual for entry into the "twice born" category. So they technically did not qualify to be twice born. However, all other jatis agreed that they should be classified as Vaishyas. Jats, however, argued that they should be classed as part of the Kshatrya varna (see also Banerjee, 1970). Since no other jati agreed with their claim, I left them in the Vaishya category and placed them below other Vaishyas (e.g. Bania) who are eligible to don the sacred thread. Jats also did not like having Gujjars and Kambojh classified alongside of them since the jats labelled them "cattle herders" or "gardeners" and not farmers. Furthermore, certain names of formerly untouchable jatis, e.g. Rai, Arya and Mahajan, are also the titles of Vaishya jatis in the region. I was informed that the Arya Smaj (a Hindu reform movement) had a lot to do with name changes of former untouchable jatis. During the pre-independence decades, the Arya Smaj carried on a campaign to eradicate untouchability and encouraged untouchable jatis in selected districts of Punjab to don the sacred thread. The Arya Smaj also assigned these families the names of new jatis, segmented from their traditional jati. All of their jati names are honorific titles claimed by various Vaishya jatis. Some jatis chose the name "Harijan" as their jati name under the influence of Mahatma Gandhi. However, today these new "segmented jatis" do not don the sacred thread anymore. They now accept their low status relative to other jatis. Despite these "caste mobility" movements in this region, when jatis were grouped into "jati-clusters" and rank ordered, the rankings of the two time periods 1962 and 1977 remained remarkably similar, suggesting no changes in the overall caste hierarchy.

Table 2. Standardized and Unstandardized Regression Coefficients for Three Equation Status Allocation Model, Village Indian Male Heads of Households, Aged 20–64 in 1962 ($n = 761$) and 1977 ($n = 823$).

Dependent Variable	Independent Variables											
	CASTE		FOCC		FLAND		EDU		FJOB		R^2	
	1962	1977	1962	1977	1962	1977	1962	1977	1962	1977	1962	1977
Standardized coefficients												
EDU	0.256*	0.296*	−0.092*	−0.039	−0.075*	−0.009					0.05	0.08
FJOB	0.041	0.241*	0.599*	0.418*	0.052	0.113*	−0.042	−0.162*			0.41	0.32
OCC	0.001	0.048	0.355*	0.120*	−0.025	0.112*	0.025	−0.032	0.497*	0.575*	0.62	0.49
Unstandardized coefficients												
EDU	0.291	0.549	−0.010	−0.005	−0.007	−0.001						
	(0.045)	(0.065)	(0.005)	(0.004)	(0.003)	(0.008)						
FJOB	0.469	3.130	0.682	0.364	0.050	0.184	−0.430	−1.123				
	(0.374)	(0.410)	(0.037)	(0.026)	(0.029)	(0.049)	(0.292)	(0.210)				
OCC	0.006	0.627	0.397	0.104	−0.247	0.181	−0.247	−0.224	0.489	0.573		
	(0.296)	(0.367)	(0.035)	(0.025)	(0.231)	(0.042)	(0.230)	(0.185)	(0.029)	(0.030)		

Notes: a = Figures in parenthesis refer to standard errors of estimates.
b = See Appendix A.
*Significant metric coefficient is at lest-twice the standard error of estimates.

Data Analysis

The data were analysed using the standard methods used in status allocation analyses. Coefficients of bivariate correlations, their means and standard deviations are reported in Appendix B. Table 2 presents the results of multiple regression. Both the standardized regression coefficients (B) and unstandardized regression coefficients (b) are reported, along with the standard error of estimates. Using the techniques of path analysis, indirect and direct effects were calculated and presented in summary form for various models in Table 4 (Duncan, 1971; Finney, 1972; Pastore et al., 1975). Other data are reported in the rest of the tables in simple percentage form to provide detail or support a point in the discussion.

FINDINGS AND DISCUSSION

The results of village India analyses are reported in Table 2 below. The major findings of the comparison are that: (1) the effect of the family's caste on the occupational status of the subjects was negligible as indicated by the zero-order coefficients of determination: 1962 – $r^2 = 0.096$, 1977 – $r^2 = 0.081$. (2) In village India, occupational status was determined more by inheritance than in the United States or other developed industrialized nations. (3) Educational attainment was unrelated to occupational attainment. Even in 1977, after the green revolution, the effect of education on occupational attainment remained near zero. (4) Furthermore, both in 1962 and 1977, the only significant predictor of educational attainment was a person's caste (jati) rank. (5) Both family land and father's occupation had small but significant negative effects on educational attainment in 1962. In 1977, however, their influence remained negative but was not statistically significant.

Based on this evidence, what kind of status allocation model fits village India today? I start with the description of the well known "classical" caste model of allocation. I then review the results I have found for this model. I also examine the implications of caste status being the only significant variable for educational attainment in view of Singh's (1976, p. 109) assertion that: "the emerging stratification system [is] certainly more ordered and probably more rigid than the previous one." Furthermore, I test the hypothesis that the process may be changing due to industrialization.

When we compare systems of stratification in different societies, we often refer to two criteria of allocation: one is the relative openness of the system. This aspect is usually associated with the volume of mobility, and the manner in which individuals are sorted or sifted in stratified roles. The other is the degrees of ascription or achievement, the later usually measured through the effects of education. In contrast to the developed Western societies, village India, because of the "caste" system, is usually considered to have a "closed" system of stratification. Similarly, the pattern of status allocation is considered to be entirely "ascriptive" rather than based upon "achievement," the dominant method of allocation in developed Western societies.

Status Allocation in Village India: The Caste Model

In village India, according to the traditional view of scholars, status is allocated by "caste ascription." Every individual is born into a caste, a *Jati*, and since individuals cannot change their jati, their status (to the extent it is based solely on their caste) is considered to be fixed from birth. This model, however, is based on the misconception that every caste is assigned a "traditional" or "caste occupation." Any deviation from the traditional model, therefore, is viewed as a threat to the system itself. Hindu caste society has created rigid rules that all members must follow. It may be possible to escape from deviations in urban areas where neighbors do not know each other and where social relations are of the "secondary group" type. However, deviations are less tolerated in village India where everyone knows everyone else and where social relations are of the "primary group" type. Inter-generational social mobility in such a system is, therefore, considered to be extremely low, if any exists at all. Individuals cannot choose their social status. Children are socialized into the roles prescribed by caste rules rather than by families.

In describing the Indian system of stratification as closed and non-permeable, scholars have viewed the Hindu caste system wherein:

> religious values and ideas were the sole determinants of attitudes toward and chances for social mobility; in which little if any such mobility actually occurred; in which there were no discrepancies or incongruities between an individual's position in the "caste" dimensions; in which there was practically no social change and resultant consequences for the alteration of positions of individuals in society; and in which, in sum, a state of near-perfect integration, stability and individual immobility prevailed for endless centuries, even millennia (Barber, 1968, p. 18).

Although the full history will never be known, whatever "ancient" (Sinder, 1958; Stein, 1968) and "contemporary" historical and field research is available casts doubt on the assertion that an ideal-typical "closed" or "non-permeable" system of stratification ever prevailed in India (see Silverberg, 1968).

There are three reasons for this "older" misconception of the Hindu caste system (Barber, 1968; Sharda, 1977). First, the older picture was derived from a selective reading of religious and ideological literature of the Hindus – the Dharamshatras – not from objective historical accounts of the past. The official religious and ideological literature of the Hindus was written by the priestly literati who claimed their social and moral superiority over other social classes and conferred upon themselves the right to uphold the caste observances. Scholars continue to draw on these religious-ideological sources to support their theories of the caste system. A prominent example (Marriott & Inden, 1974) claims that reported social mobility is contained within the individual varnas of the four-varna scheme mentioned in the ancient Dharamshastras, and there has always been a *maximum* overlap between ritual and secular hierarchies of these varnas. This theoretical orthodoxy has been slow to change (Elder, 1996).

Secondly, many Western scholars describe the Indian stratification system in this way in order to contrast the West with *unique features* of the Indian stratification system (see, e.g. Dumont, 1970). Barber (1968, p. 28) points out that Weber "purposely chose India as a case to compare with the West; [because he] . . . wanted to contrast its relative social immobility and its immobilizing religious values and ideologies with the relative social mobility and mobilizing values and ideologies of the west" (see also Weber, 1958). Weber and other scholars (for example, Sorokin, 1927) highlighted the differences but ignored the similarities in their comparative descriptions of the stratification systems of India and the West.

The third reason for the persistence of the older view (Barber, 1968, p. 27) is that "by describing conditions of radical inequality and complete lack of mobility in India, westerners could, in the light of their own values, be asserting a moral superiority to the Indians which helped to justify their imperialistic policy in India. Colonialism could be justified as a means of bringing better ways to an immoral stratification system." Thus, historical inaccuracies, an emphasis on uniqueness, and political considerations have all contributed to a mistaken view of India's stratification system. For a recent critique and re-evaluation of the older view, see Milner (1994).

Caste, the Jajmani System and Ascription

The link between caste and occupation is usually established through the Jajmani System in village India, where different castes are locked into a ritual-economic exchange known as the *jajmani system*.

The jajmani system was first described in detail by William Wiser (1936) in a village in Uttar Pradesh province, the heartland of northern India. This has been the subject of considerable discussion since (see for example: Beidelman, 1959; Elder, 1970; Gould, 1964; Gupta, 1984; Harper, 1959; Kolenda, 1963; Rowe, 1963; among others). Despite differences concerning the domain of the jajmani system, researchers all emphasize how permanently the ritual exchanges are institutionalized. For example, a barber family, like other *kamin* (literally meaning low status worker) families; e.g. washerman, carpenter, sweeper, etc. provide regular services to a high caste farming family – the *jajmans* – throughout the year and receives fixed remuneration in kind at the harvest time. They also receive gifts of clothing, money and grains on ceremonial occasions. Such remuneration was differential on the basis of the caste status of *kamins*: high for high castes and low for low castes, and it is not purely an economic reward for their services. The term kamin does not cover the Brahmin family although they are integral part of the jajmani system. The Brahmin priests are called *purohits* and *padhas* instead. Although the kamin terminology is common knowledge in the area, the kamins are frequently referred to as *laagis* (literally meaning attached (to a family)). However, the mechanization of agriculture unleashed mainly by green revolution technology and the resultant expansion of commerce have greatly affected these relationships, especially in recent decades. Similarly, improvements in communications have led to the out-migration of the poor and the educated. I, therefore, argue that the influence of caste is not strong enough to make the occupational structure non-permeable.

The proportion of kamins (i.e. ritual-exchange workers) however, never exceeded more than 10–15% of the village labor force (Sharda, 1970). Some lower castes have also made caste-wide efforts to dissociate them from their traditional occupations (Burman, 1970). Table 3 reports the proportion of people following "occupations fitting the jati name," and those following "agricultural" and "*other* non-agricultural pursuits that do not fit the jati name."

Those castes numbered one (i.e. Brahmans), 10 through 16 among the Shudra *varna* category, and numbered 20 through 23 and 25 from the former-Untouchable category, are expected to provide "services" under the jajmani system arrangements. It will be noted that their numbers are small in the total labor force and that very rarely are more than 50% of the caste members in "occupations fitting their jati-name." In addition, the basis of the jajmani system have changed considerably since the 1960s. Agricultural households now prefer to pay cash for services, which often are seasonal or piecemeal, rather than providing a fixed amount of grain at harvest time. The jajmani system is becoming extinct very fast. Hence, whatever small number of "traditional occupations" were linked to *kamin* castes are being disconnected.

Is Agriculture a "Traditional" Caste Profession?

In the older view, there is often a tendency to link agriculture to certain castes as their "traditional occupation." The view also asserts that some castes (e.g.

Table 3. Distribution of Village Indian Males, Ages 20–64, According to Occupations Fitting Jati Names, and Agriculture or other Non-Agricultural Occupations, 1962.

Varna Rank	Jati Name (RANK)	Jati (Traditional)	Occupation Fitting Jati Name (%)	Agricultural Pursuits (%)	Other Non-Agricultural Pursuits (%)	Total %	Total *N*
Twice born							
I.	Brahmans						
	Brahmans	Priestly	2.6	76.9	20.5	100	39
Total all brahmans			2.6	76.9	20.5	100	39
II.	Kshatrya						
	Rajputs/Kshatrya	Warriors	0.0	75.0	25.0	100	16
Total all kshatryas			0.0	75.0	25.0	100	16
III.	Vaishya–I						
	Khatri/Sood/Mahajan	Trade/Finance	17.4	34.8	47.8	100	23
	Khatri-Sikh	-do-	0.0	50.0	50.0	100	2
	Bania/Gupta/Argawal	-do-	66.7	16.7	16.7	100	6
	Arora/Arora-Sikh	-do-	0.0	83.3	16.7	100	6
Non-twice born							
	Vaishya-II						
	(i) Jat/Jat-Sikh	Agriculture	(90.3)	90.3	9.7	100	144
	Vaishya-II						
	(ii) Gujjar	Dairy Farming	9.2	90.8	0.0	100	65
	Kamboj/Kamboh/Saini	Gardner/Agriculture	(75.5)	75.0	25.0	100	8
Total all vaishyas			0.5	83.7	11.8	100	254
IV.	Shudra Jatis						

Table 3. *(Continued)*

Varna Rank	Jati Name (RANK)	Jati (Traditional)	Occupation Fitting Jati Name (%)	Agricultural Pursuits (%)	Other Non-Agricultural Pursuits (%)	Total %	Total N
A.	Non-scheduled castes						
	1. Sat shudra						
	Sunar	Goldsmith	50.0	0.0	50.0	100	2
	Tarkhan/Ramgarhia/Khati/Barhai/Dhaman	Carpenter	50.0	33.3	16.7	100	30
	Lohar	Blacksmith	0.0	0.0	100.0	100	1
	Chhimba/Chimmai	Tailor	50.0	50.0	0.0	100	8
	2. Asat (impure) Shudra: not untouchable						
	Nai	Barber	60.0	0.0	40.0	100	5
	Kumhar/Ghumar/Parjapt	Potter	38.5	53.8	7.7	100	13
	Jhiwar	Water-career	0.0	18.8	81.2	100	16
	Julaha/Dhanuk	Weaver	0.0	0.0	100.0	100	8
	Teli	Oil Crusher	0.0	0.0	100.0	100	1
Total non-scheduled caste jatis			33.3	28.6	38.1	100	84
B.	Scheduled castes (formerly untouchable castes)						
	1. Asat (impure) Shudra: formerly untouchable						
	Jogi	Begger	0.0	83.3	16.7	100	6
	Mirasi	Drummer	0.0	0.0	100.0	100	1
	Chamar/Ramdasia/Harijan	Leather Worker	21.2	30.8	48.0	100	52
	Arya/Mahajan	-do-	0.0	0.0	100.0	100	3
	Chuhra/Balmiki	Sweeper	0.0	100.0	0.0	100	6
	Mazhbi-Sikh	Scavenger	0.0	76.5	23.5	100	

	2. Segmented jatis: Formerly untouchable						
	Rai/Rai-Sikh	Leather Worker	0.0	96.9	3.1	100	32
	Bawaria	Hunter	0.0	0.0	100.0	100	2
	Majhili	Fisherman	0.0	100.0	0.0	100	1
Total scheduled caste jatis			8.0	62.8	29.2	100	137
V.	Former "criminal tribes"						
	Bazighar	Acrobat	0.0	93.7	6.3	100	63
	Ghadhila	Vagrant tribe	0.0	75.0	25.0	100	4
	Sansi	Pig-farmer	8.0	44.0	48.0	100	25
Total former "tribes"			2.2	79.3	18.5	100	92
O.	Others (caste not given)						
	Christian		–	15.4	84.6	100	13
	Muslims		–	100.00	0.0	100	1
	Don't Know		–	100.00	0.0	100	1
Total others			–	26.6	73.4	100	15

Note: Figures in ( )s are not counted in the totals. Those jatis claim agriculture as their caste occupation. However, agriculture is not considered to be a caste profession (see Banerjee, 1970 and discussion in the paper).

Brahmins) are "prohibited" from engaging in agricultural pursuits. This view is wrong on both counts. In a very informative paper, Banerjee (1970, p. 241) investigated this question historically and found that:

> Inspite of the emphasis given to agriculture as a vocation, we do not find any mention of agricultural castes in the list of castes already formed during the later Vedic times. In the *Sabhapravan*, *Aranyakaparvan*, and *Shantiparavan* of the *Mahabharta* and in the *Ayodhyakanda* of the Ramayna we see [an] emphasis on *Varta*. *Varta* is constituted by agriculture, trade and commerce, mining and animal husbandry. Thus in the Epics, also, we do not come across any agricultural caste, neither do we find this occupation prohibited for any caste or *varna*.

Banerjee notes that *Baudhanya* "frowns upon [but never prohibits] agriculture as an occupation for Brahmins," arguing that "agriculture destroys the study of [the] Vedas and [the] Vedas destroy agriculture." *Harita* similarly allows agriculture as an occupation for Brahmins in distress. *Manu*, however, is one of the few sources that does prohibit agriculture (being a *pramita* occupation, i.e. an occupation that result in loss of life) for both the Brahmin and the Kshatria *varnas*. Thus, there is overwhelming support in the Dharamshtras for agriculture, the main occupation of people in village India, to remain open to all castes. This is also the conclusion to be drawn from data in Table 3.

Concerning the question whether there are some castes claiming agriculture as their traditional occupation, Banerjee makes a distinction between "castes practicing agriculture" and "occupational castes of agriculturists." After surveying the literature, Banerjee (1970, p. 243) concludes that "we find that in the Indo-Aryan speaking part of India there is no caste that fully satisfies the definition of a *genuine agricultural caste* in having agriculture as their caste [traditional] occupation and with a mythological association with this particular occupation" (emphasis added). Banerjee does, however, state that the "Jats of Punjab [part of my sample] have almost become an agriculturalist caste; they are practically identified with agriculture in the region. This transition from a group of people, or a tribe, practicing agriculture to a [group claiming] hereditary agriculture is not yet complete. It is still primarily a land-owning caste aspiring to be classed as Kshatriya" (Banerjee, 1970, p. 242). Thus it is a mistake to classify agriculture as a traditional occupation of any single caste and consider other castes as non-agriculturist castes, (e.g. artisans or former-untouchable or even Brahmins). It is clear from Table 3 that landholding, and hence becoming an agriculturist, is a major avenue for status mobility in village India by *all* castes. Agriculture is the most likely occupation of those not in the occupations of their "caste." Landholding and employment in agricultural occupations are certainly important avenues of status allocation in village India as they are in other parts of the rural world. *A Critique of the Caste Model of Status Allocation.*

Since the 1960s, this older orthodox view has been increasingly challenged on empirical grounds as well (Elder, 1996; Gartrell, 1977; Sharda, 1977; Silverberg, 1968; Singh, 1976). The very title of Silverberg's (1968) edited collection; *Social Mobility in the Caste System* indicates that the older view is incorrect.

The results reported in Appendix B show that father-to-son correlations between caste rank and occupational attainment variables for both 1962 and 1977 are only moderate and not nearly perfect or even high, as predicted by the caste model. For example, in 1962 the correlation between caste rank and current occupation was 0.310 whereas the correlation between caste rank and 1st Job was 0.303 (Appendix B). Moreover, caste effects on both of these variables (1st job and current occupation) became near zero in the multiple regression model (Table 2). These analyses show, therefore, that whatever influence caste may have on occupational attainment is mediated by father's occupation since the correlation between caste and father's occupation is 0.426 (Appendix B). In 1977, the effect of caste on 1st job was significant but for the current occupation it was not. It is, therefore, clear that green revolution has not altered the status allocation process in any significant manner and that in any case it had little or nothing to do with caste. The significant effect of caste on education in both years ($\beta = 0.256$ and $b = 0.291$ in 1962 and $\beta = 0.296$ and $b = 0.549$ in 1977) might possibly suggest the indirect influence of caste in getting more prestigious "modern" jobs (the new jobs) by higher caste people outside of village India. This future research has to decide. But its direct effect on current occupation was essentially zero.

People are born in various castes and the ritual status that is ascribed to the caste of birth usually remains little changed throughout one's life-time, irrespective of socioeconomic status. Ritual status is an added dimension of status in India; a dimension that is practically orthogonal to occupational status, that is, the extent to which ritual status influences one's occupation has been subject to theoretical debate and now to empirical test, that it has been found wanting. The older orthodox view makes a deductive fallacy in assuming that all other statuses are derived from ritual status (e.g. Dumont, 1970). Nothing could be farther from the truth. In modern times, many castes are abandoning their "traditional caste occupations." One of these is the Brahmins, or I should say the Brahmins are leading this defection.

In conclusion, it is argued that the Jajmani system, which linked castes with occupations, is crumbling under the weight of modern developments in agriculture, and that agriculture is not really a caste occupation. By providing newly emerging opportunities for steady work, agricultural occupations remain the source of upward mobility for members of all castes. Therefore, the influence of caste on occupational status allocation in village India. The total effects of caste is rather small. The total effects of caste were only 0.010 in 1962 and 0.149, mostly indirect effects, through first job, and were far less than the total effects of father's

Table 4. Summary of Effects.

Variable	Direct Effects		Indirect Effects							
	1962	1977	Via EDU		Via FJOB		Via EDU & FJOB		1962	1977
			1962	1977	1962	1977	1962	1977		
Status allocation models of village india										
EDU	−0.025	−0.032			−0.021	0.093			−0.046	−0.032
CASTE	0.001	0.048	−0.006	−0.009	0.020	0.139	−0.005	−0.029	0.010	0.149
FLAND	−0.025	0.112	0.002	0.000	0.026	0.065	0.001	0.001	0.004	0.178
FOCC	0.355	0.120	0.001	0.001	0.298	0.241	0.002	0.004	0.656	0.366

Notes: Calculated From Table 2.
Definition of Variables: EDU = Education of respondent; CASTE = Caste rank; FLAND = Family land; FOCC = Father's occupation.

occupation, 0.656 for 1962 and 0.366 for 1977 (Table 4). How, then, does one explain the high level of ascription in village India?

After all, father-to-son correlations for father's occupation and son's occupation are: 0.684 in 1962 and 0.439 in 1977 (Appendix B). Here I introduce a non-caste alternative socioeconomic model of ascription.

Socioeconomic Ascription: the Non-caste Aspects of Status Allocation

Even in its non-caste aspects, status allocation in village India is largely ascriptive. This is evident from the summary of effects table (Table 4). The direct effects of father's occupation are significant in both years (Table 2), the only origin variable that shows consistent results. Table 4 further demonstrate the strong total effects of father's occupation in both years (0.656 in 1962 and 0.366 in 1977). This high degree of status inheritance is largely a function of the virtual absence of industrialization in the countryside, except for the green revolution itself, and of a low level of diversity in the rural economy. Among those of rural origins, most upward mobility may take place through out-migration to industrial cities or through service in the armed forces and other bureaucratic agencies outside the village, although tests of this hypothesis remain to be done. In order to understand the role of high ascription in the status allocation of village India, one needs to understand the socioeconomic processes related to the organization of agriculture rather than to the role of caste.

The essential point is that in 1962 agriculture was at a near-subsistence level and the economy was "underdeveloped," farms were very small. These small farms supported a limited division of labor with a limited degree of inequality. Similarly,

Table 5. Migration Status of Families, Village India, 1977.

No	Category	*N*	%
I	Non-Migrants	554	59
II	Refugee Families (1947–50)	255	27
III	Other Immigrants		
	(1951–60)	31	3
	(1961–67)	13	1
	(1968–77)	89	9
Total		942	99*

Note: ( ) = Period of migration.
*Does not equal 100 due to rounding errors.

having few if any imports or exports, there were few service occupations. This, too, restricted economic inequality. In addition, there was no manufacturing and there were very few handicraft industries in the villages.

With the green revolution, beginning in the late sixties, some mechanization of farms has taken place. The economy is being differentiated with new occupations that relate to repair and production of farm implements (Table 1). In addition, the state has provided community development services to the villagers in the 1950s and 1960s. There is, therefore, some new in-migration of skilled manual workers and of low level professionals into these villages. In my 1977 survey (see Table 5), I found that whereas only 3% of the respondents were reported to be in-migrants during the 1951–1960 decade, this proportion was 11% during 1961–1977. Only 1% indicated that they were moving into the village to buy land. The rest cited other reasons, mainly "non-agricultural" jobs.

Although their numbers remain low, in-migration of non-agricultural skilled and educated labor apparently started to take place in the 1970s. This trend seems to be continuing. As the proportion of labor force in agriculture is declining, the proportion of the labor force in non-agricultural pursuits is increasing. This should have impact on the status allocation process. It has.

The direct effect of father's occupation on current occupation, for example, declined sharply from $b = 0.397$ ($\beta = 0.355$) in 1962 to $b = 0.104$ ($\beta = 0.120$) in 1977 (Table 2; also see Table 4). (Both were statistically significant indicating their relative importance in the model.) The decline in the size of regression coefficient is due to the decline of agricultural positions and increase of non-agricultural positions in the 1977 data. However, the overall 1962 model was still reproduced and father's occupational status was still the main variable affecting current occupational status rather than caste or educational achievement. Although technically classified as "non-agricultural" occupations they nevertheless are intertwined with the agricultural production, processing and sales of agricultural products.

In sum, therefore, stratification in village India is more closely associated with agrarian relations of production rather with industrial relations of production. This is similar to other rural areas of Third World nations as well.

As indicated earlier, status scales for rural agrarian populations are likely to be different from standard scales developed in industrialized developed nations (see e.g. Haller et al., 1972). Hansen and Converse (1976) further caution against any indiscriminate application of universal scales of occupational prestige to sub-populations of Third World nation states.

Status in rural India is derived primarily from one's place in the agrarian means of production. This is indicated in studies of occupational prestige of rural India (see Sharda, 1979). At the apex of the hierarchy are farmers who are owner cultivators, followed by tenant cultivators, village professionals, white collar workers, village artisans and service workers. Villagers rate professionals below farmers. The occupation of a tenant farmer, who owns very little or even no land of his own and is typically illiterate or barely literate, is rated just below the owner cultivator but above all other white collar occupations. Furthermore, a Government of India enquiry revealed that unskilled manual workers prefer agricultural labor rather than non-agricultural labor jobs even when the income returns are significantly lower for agricultural labor (Government of India, 1969). In essence, agricultural occupations of all ranks are preferred over non-agricultural pursuits for a majority of villagers belonging to most castes (Table 3). Most village occupations, especially agricultural ones, require apprenticeship within the family. This may be one of the major reasons for the strong influence of father's occupation on son's current occupation in village India (Table 2).

In addition, there is also an increase in the effect of land ownership on current occupation. Whereas the net effect of family land ownership on current occupation was negative and non-significant ($\beta = -0.025$, $b = -0.247$) in 1962, its effect in 1977, on the other hand, was positive and significant ($\beta = 0.112$, $b = 0.181$) (see Tables 2 and 4). This substantial increase in the effect of land ownership clearly indicates the increasing advantage for sons of landowners in gaining better jobs whether or not they remained illiterate. It seems that sons of small holders whose land may have become uneconomic due to green revolution technology may find it advantageous to lease their land and open small business operations, taxis, grocery or other shops even if they are not well educated. Similarly, all new entrepreneurs, whether owners or managers of workshops or transporters, are people with land. In essence, while tied to their land they are extending their influence to non-agricultural operations as well. Therefore, we see the emerging influence of family landownership on the village India status.

Agriculture and Education

It is important to note that in these Village India data education had no effect on occupational status, either on first job or current occupation in 1962 (Table 2). This is a unique finding. I attribute this effect to low levels of education and the predominance of subsistence level agriculture. Agricultural occupations do not require formal education when they employ only traditional practices. Many farmers distrusted educated laborers and feared that they were not suitable for agricultural jobs. In fact, anecdotal evidence suggest that educated rural youth rarely engage in agricultural pursuits. Often they prefer to be unemployed or migrate to the cities rather than to work on the farms. Many agricultural families preferred their sons to receive apprenticeship within the family rather than to be educated outside the family. There was some change after the green revolution when owner cultivators started realizing that there was a need for education. Education then began to help farmers to relate to the new technology and scientific information about seeds, soils and fertilizers. Large owner-cultivators now producing a surplus realized that they had to deal with modern financial (e.g. rural banks), commercial (grain markets) and bureaucratic (community development agents) establishments. The value of education among landed farmers began to increase. However, the average education was still too low in 1977 (2.28 years). The zero effects of education on current occupation of the respondents in both years (1962 and 1977) were, therefore, in part due to the predominance of farmers in the sample.

Education and Status Allocation

The Free and Compulsory Primary Education Act was enacted in 1961, after independence, to end illiteracy. This Act provided that all children between the ages of 6 and 14 should be enrolled in primary (elementary) schools. Massive efforts were made to open primary schools to make primary education accessible to all villagers. However, the Act had yet to produce its desired effects in the status allocation process by the time these studies were conducted. As we have seen, the average education for the respondents in 1962 was less than one year and in 1977 it was 2.28 years (Appendix B). Educational attainment in both 1962 and 1977 could be predicted only from caste rank (among the variables tested), controlling for the effects of father's occupation and family land ownership. Clearly, this could be explained by caste values about education. Brahmins and other high castes typically inculcate values of education to their children. Lower castes (including former-untouchable), despite the reservation of seats and provision of government scholarships for higher education, often lack values that aspire toward educational attainment (Pimpley, 1976, 1983).

It could be argued that as the level of educational attainment increases and the economy differentiates in village India, the influence of caste on occupational attainment may increase! This has not been the case so far. The village economy did not create enough higher status jobs that require high school education. Thus the only realistic option for those who obtain high school or post-high school education is to leave their respective villages for jobs in urban areas. Educated out-migrants are also disproportionately from high castes. This may be the reason why there were no direct effects of caste on respondent's current occupation in both years (see Table 2).

Links between educational achievement and occupational attainment in village India are still not forged despite the changes in occupational structure due to green revolution (see Tables 2 and 4, and Appendix B). However, some improvements in educational achievement had occurred (Table 6). However, these improvements are rather small. Only 15.6% more respondents had primary or post-primary education compared to the respondents in the 1962 sample. Formal educational standards are a requirement for government positions (e.g. clerical and teaching) and for some professionals, a very small proportion of the village India labor force, but not for the great majority. I have argued that farm people perceive few, if any, benefits of education for their farm operations. (Similar findings were

Table 6. Distribution of Village Indian Males, Ages 20–64, According to Educational Standards, 1962 And 1977.

Sr. No.	Standard	1962	1977	Percent Change
1	Illiterate	82.5 (628)	67.3 (555)	−15.2
2	Literate, no standards	5.0 (38)	4.6 (38)	−0.4
3	Primary	5.8 (38)	7.7 (64)	1.9
4	Middle	4.1 (31)	14.0 (115)	9.9
5	Jr. High	2.4 (18)	4.8 (40)	2.4
6	College	0.2 (2)	1.6 (14)	1.4

Note: 0 – *N*.

also reported by Haller (1957, 1958) for rural Wisconsin residents as of 1948). Neither is the value of education recognized for new skilled manual occupations by employers. Employers do not require education for manual jobs. Illiterate and educated labor are treated alike. They are given on the job training by the owner/managers and more experienced workers. Village India in 1977 was still an agrarian economy. New occupations were incorporated into the village economy but without making education a requisite for those jobs. Patterns of status allocation, therefore, remained largely unaffected by changes in the occupational hierarchy due to the green revolution and by improvements in average education of workers.

CONCLUSIONS

I suggest that the low permeability typically reported in village India in the past has prevailed because it has been affected by the agrarian mode of production, but not much, if at all, by caste. All over the world there is a tendency for farmers to come from farming families. This is true for both those who own land and for those who are landless agricultural laborers. India is no exception to this rule. Children of farmers learn how to farm from their elders. In the past, children of farmers did not acquire as much education as those who came from non-agricultural families despite their high social status in village India. This may change in the future.

Overall, the supply of education remained low both in 1962 and in 1977. The overall effect of education on occupational attainment, therefore, remained practically zero even after the green revolution and inspite of the fact that there was an increase in non-agricultural occupations. Most higher status positions were monopolized by the landed gentry who perceived benefits from new non-agricultural operations. However, the overwhelming majority of new "modern" occupations, especially the manual occupations, were also tied to agrarian economy and hence governed by the status allocation regime of the agrarian mode of production.

In sum, this paper investigated the claims of the two hypotheses. The frequently cited *caste* hypothesis posits that the effects of caste on respondent's first job and current occupation are overwhelmingly positive and that makes the Indian system of stratification a non-permeable. The alternative non-caste *socioeconomic ascription* hypothesis posits that father's occupation (i.e. socioeconomic origins) strongly affects son's first job and current occupation and these effects (both direct and indirect) are greater than the effects of caste.

It is to be noted that there are in fact two forms of the socioeconomic hypothesis. The one the data support is a rather typical agrarian hypothesis: the greater the amount of land and wealth a family possesses, the higher the occupational level the family's son will come to occupy (Lenski, 1966, pp. 226–230). The second one is a rather typical modernization hypothesis applied to agrarian societies: in an industrializing agrarian economy, the greater the amount of landed wealth a family possesses, the higher the level of education a family's sons will obtain, and as a result, the higher the level in the occupational structure they will come to occupy. This form of the hypothesis must be rejected: education was not involved in the status allocation process in either year. In short, neither caste nor such modernization that had occurred by 1977 had noteworthy effects on allocation to occupational status positions. The only substantive determinants were the landed wealth and occupational status origins.

ACKNOWLEDGMENTS

I wish to thank several agencies and persons for help in this project; the Census of India for the 1962 data; the American Institute of Indian Studies for the Award of a Senior Research Fellowship during 1976–1977; the University of Utah for continued support; the University of Wisconsin's Department of Rural Sociology for its support and encouragement where I was an Honorary Fellow when I was writing this article; Joseph W. Elder, Archibald O. Haller, Warren Hagstrom and George A. Miller for providing comments on earlier drafts.

REFERENCES

Anthropological Survey of India (1991). Study shows 1051 backward classes. In: A. A. Engineer (Ed.), *Mandal Commission Controversy*. Delhi: Anajta Publications.

Bailey, F. G. (1957). *Caste and economic frontier*. Manchester: Manchester University Press.

Banerjee, B. (1970). Agriculture as a caste profession. *Man in India*, *50*(September), 240–257.

Barber, B. (1968). Social mobility in Hindu India. In: J. Silverberg (Ed.), *Social Mobility in the Caste System in India: An Interdisciplinary Symposium* (pp. 18–35). The Hague: Mouton.

Beidelman, T. O. (1959). *A comparative analysis of the Jajmani system*. New York: J. J. Augustine.

Bills, D. B., & Haller, A. O. (1984). Socioeconomic development and social stratification: Reassessing the Brazilian case. *The Journal of Developing Areas*, *19*(October), 59–70.

Bills, D. B., Haller, A. O., Kelley, J., Olson, B. B., & Pastore. J. (1985). Class, class origins, regional socioeconomic development and the status attainment of Brazillian men. In: R. V. Robinson (Ed.), *Research in Social Stratification and Mobility* (Vol. 4, pp. 89–127). Greenwich, CT: JAI Press.

Blau, P. M., & Duncan, O. D. (1967). *The American occupational structure*. New York: Wiley.

Burman, B. K. R. (1970). *Social mobility movements among scheduled castes and scheduled tribes in India*. Delhi: Government of India Press (Ministry of Home Affairs).

Census of India (1961). *A guide to the 1961 census publications program*. Delhi: Government of India Press (Ministry of Home Affairs).

de-Barry, W. T. (Ed.) (1958). *Sources of Indian tradition*. New York: Columbia University Press.

Dumont, L. (1970). *Homo hierarchicus: The caste system and its implications*. Chicago: University of Chicago Press.

Duncan, O. D. (1971). Path analysis: Sociological examples. In: H. M. Blalock (Ed.), *Causal Models in the Social Sciences* (pp. 115–138). Chicago: Aldine-Atherton.

Elder, J. W. (1970). Rajpur: Change in the Jajmani system of an Uttar Pradesh village. In: K. Ishwaran (Ed.), *Change and Continuity in India* (pp. 105–127). New York: Columbia University Press.

Elder, J. W. (1996). India's caste system. *Education About Asia*, *1*(2), 20–22.

Epstein, T. S. (1978). Re-surveys and re-studies of rural societies. In: B. Dasgupta (Ed.), *Village Studies in the Third World* (pp. 127–131). Delhi: Hindustan Publication.

Finney, J. M. (1972). Indirect effects in path analysis. *Sociological Methods and Research*, *1*(November), 175–186.

Frankel, F. (1971). *India's green revolution: Economic gains and political costs*. Princeton, NJ: Princeton University Press.

Frankel, F. (1978). *India's political economy, 1947–1977: The gradual revolution*. Princeton, NJ: Princeton University Press.

Freed, S. A. (1963). An objective method for determining the collective caste hierarchy in an Indian village. *American Anthropologist*, *65*, 879–891.

Gartrell, J. (1977). Status, inequality and innovation: The green revolution in Andhra Pradesh, India. *American Sociological Review*, *42*(2), 318–337.

Gartrell, J. (1981). Inequality with in rural communities of India. *American Sociological Review*, *46*(6), 768–782.

Ghurey, G. S. (1961). *Caste, class and occupation*. Bombay: Popular Book Depot.

Gough, E. K. (1960). Caste in a Tanjore village. In: E. R. Leach (Ed.), *Aspects of Caste in South India, Cyclone and North-West Pakistan*. Cambridge: Cambridge University Press.

Gough, E. K. (1966). The social structure of a Tanjore village. In: M. N. Srinivas (Ed.), *India's Villages*. Bombay: Asia Publishing House.

Gould, H. A. (1964). A Jajmani system of North India: Its structure, magnitude and meaning. *Ethnology*, *3*(1), 12–41.

Government of India (1969). *Income of rural labor households*. The National Sample Survey No. 134. Eighteenth Round (Feb. 1963–Jan. 1964). New Delhi: Cabinet Secretariat.

Gupta, A. R. (1984). *Caste hierarchy and social change: A study of myth and reality*. New Delhi: Jyotsna Prakashan.

Haller, A. O. (1957). The influence of planning to enter farming on plans to attend college. *Rural Sociology*, *22*(June), 137–141.

Haller, A. O. (1958). Research problems on the occupational achievement levels of farm-reared people. *Rural Sociology*, *23*(December), 355–362.

Haller, A. O. (1982). Reflections on the social psychology of status attainment. In: R. M. Hauser, D. Mechanic, A. O. Haller & T. S. Hauser (Eds), *Social Structure and Behavior: Essays in Honor of William H. Sewell* (pp. 3–28). New York: Academic Press.

Haller, A. O., Holsinger, D. B., & Saraiva, H. U. (1972). Variations in occupational prestige hierarchies: Brazilian data. *American Journal of Sociology*, *77*(March), 947–956.

Haller, A. O., & Portes, A. (1973). Status attainment process. *Sociology of Education*, *46*, 51–91.

Haller, A. O., & Saraiva, H. U. (1991). Ascription and status transmission in Brazil. In: J. G. Scoville (Ed.), *Status Influences in the Third World Labor Markets*. New York: Walter de Gruyter.

Hansen, D. O., & Converse, J. W. (1976). Cultural milieu and isolation as sources of intrasocietal variation in occupational prestige hierarchies: Recent Brazilian data. *Rural Sociology, 41*(3), 371–381.

Hansen, D. O., & Haller, A. O. (1973). Status attainment of Costa Rica males: A cross-cultural test of a model. *Rural Sociology, 38*(3), 269–286.

Harper, E. B. (1959). Two systems of economic exchange in village India. *American Anthropologist, 61*, 760–778.

Holsinger, D. B. (1975). Education and the occupational attainment process in Brazil. *Comparative Education Review, 19* (June).

Hutton, J. H. (1969). *Caste in India: Its nature, function and origins*. Bombay: Oxford University Press.

Ibbetson, D. (1916). *Punjab castes*. Lahore, India: Government Printing Press.

Kerchkoff, A. C. (1976). The status attainment process: Socialization or allocation? *Social Forces, 55*(2), 369–381.

Kolenda, P. M. (1963). Toward a model of the Hindu Jajmani system. *Human Organization, 12*(XXII), 11–31.

Lenski, G. (1966). *Power and privilege: A theory of social stratification*. New York: McGraw-Hill.

Linton, R. (1936). *The study of man*. New York: Appleton Century.

Maine, H. (1881). *Village communities in the east and the west* (4th ed.). London: John Murry.

Marriott, M. (1959). Interactional and attributional theories of caste ranking. *Man in India, XXXI*(2), 92–107.

Marriott, M. (1961). *Village India*. Bombay: Asia Publishing House.

Marriott, M., & Inden, R. (1974). An enthnosociology of South Asian caste system. University of Chicago Manuscript.

Mayer, A. C. (1968). Caste: The Indian caste system. In: D. L. Sills (Ed.), *International Encyclopedia of the Social Sciences* (Vol. 2, pp. 339–344). The Macmillan Company and the Free Press.

Metcalf, C. (1832). Minute. *Report from Select Committee*, Evidence III, Revenue, APP. No. 84.

Milner, M., Jr. (1994). *Status and sacredness: A general theory of status relations and an analysis of Indian culture*. New York: Oxford University Press.

Nesfield, J. C. (1885). *Brief view of the Caste system of the North-Western Provinces and Oudh, together with an examination of the names and figures shown in the census report*. Allahabad: North-Western Provinces and the Oudh Press.

Pastore, J., Ceotto, E. E., Haller, A. O., Quirinio, T. R., & Carter, T. M. (1975). Occupational wage differentials among university educated technical personnel in a developing economy. *Journal of Vocational Behavior, 7*(1), 113–126.

Pimpley, P. (1976). Social characteristics of scheduled caste students in Punjab. *Indian Journal of Social Work, 37*(1).

Pimpley, P. (1983). The problem of non-attendance of school children (6–14 years) of the scheduled castes in Haryana. Unpublished Report to the Indian Council of Social Science Research. Department of Sociology, Panjab University, Chandigarh.

Richards, J. A., & Hansen, D. O. (1982). Industrialization and changes in the normative bases of student aspirations: Evidence from southern Brazil. *International Journal of Contemporary Sociology, 19*(3–4), 109–123.

Rowe, W. L. (1963). Changing rural class structure and the Jajmani system. *Human Organization, XXII*(1), 41–44.

Sewell, W. H., Haller, A. O., & Portes, A. (1969). The educational and early occupational attainment process. *American Sociological Review, 34*(February), 82–92.

Sharda, B. D. (1970, October 16–18). Jajmani system in a larger setting: Contemporary patterns in Uttar Pradesh, India. Paper read at 19th Annual Meetings of the Mid-west Conference on Asian Affairs held at Indiana University, Bloomington, Indiana.

Sharda, B. D. (1977). *Status attainment in rural India*. Delhi: Ajanta Publications.

Sharda, B. D. (1979). Occupational prestige in rural India. *Rural Sociology*, *44*(4), 705–718.

Sharda, B. D. (1981). Convergence and divergence in stratification processes: Comparison between the rural sectors of India and the United States. *Rural Sociology*, *46*(1), 20–41.

Sharda, B. D. (1986). The effects of 'green revolution' on the occupational attainment of rural India: Evidence from North Indian Villages, 1962–1977. *International Journal of Contemporary Sociology*, *2*(1–2), 1–20.

Silverberg, J. (Ed.) (1968). *Social mobility in the caste system*. The Hague: Mouton.

Sinder, L. (1958). *The inability of urban and rural castes in Northern India under the impact of Mohammedanism*. Unpublished Ph.D. Dissertation, Columbia University (cited in Barber, 1968).

Singh, V. P. (1976). *Caste, class and democracy*. Cambridge: Schenkman Publishers.

Sorokin, P. (1927). *Social mobility*. New York: Harper and Row.

Srinivas, M. N. (1966). *India's villages*. Bombay: Asia Publishing House.

Srinivas, M. N. (1968). A note on sanskritization and westernization. In: R. Bendix & S. Lipset (Eds), *Class, Status and Power* (pp. 552–560). New York: Free Press.

Stein, B. (1968). Social mobility and medieval South Indian sects. In: J. Silverberg (Ed.), *Social Mobility in the Caste System* (pp.78–94). The Hague: Mouton.

Stevenson, H. C. N. (1954). Status evaluation in the Hindu Caste system. *Journal of the Royal Anthropological Institute*, *LXXXI*(1–2), 45–65.

Visaria, L., & Visaria, P. (1995). India's population in transition. *Population Bulletin*, *50*(3), 1–49.

Weber, M. (1958). *Religion of India: The sociology of Hinduism and Buddhism* (trans. H. H. Gerth & D. Martindate). Glencoe, IL: Free Press.

Wiser, W. H. (1936). *The Hindu Jajmani system*. Lucknow: Lucknow Publishing House.

APPENDIX A: RURAL INDIA'S OCCUPATIONAL PRESTIGE SCALE (RIOPS)

0–2	Professional, Technical Proprieters and Administrative	
	Doctor	86
	Engineer	80
	Priest	47
	Proprietor/Businessman	75
	Teacher	78
3	Clerical and Related Services	
	Clerk	62
	Postman	48
	Peon/Daftri	60
4	Sales	
	Photographer	(72)
	Shopkeeper/Grocer	72
	Salesman, Travelling	60
	Sales Assistant/Private service	47
	Vendor	(27)
5	Service	
	Bus/Taxi Driver	59
	Tailor	52
	Carpenter	51
	Weaver	50
	Potter	45
	Watchman	49
	Water Carrier	41
	Barber	29
	Cobbler	28
	Hod Carrier	27
	Ditch Digger	21
	Day Laborer	19
	Sweeper/Janitor	16

6	Farmer, Fishermen, Hunters and Loggers	
	Plantation Owner	100
	Farmer, Owner	90
	Farmer, Tenant	85
	Farmer, Dairy	80
	Logger	48
	Gardner	45
	Hunter and Fisherman	26
	Agricultural laborer/Siri	50
7–9	Skilled and Unskilled Workers	
	Electrician	71
	Millhand	59
	Toolmaker	(58)
	Welder	(58)
	Glass Blower	(58)
	Goldsmith	58
	Blacksmith	58
	Factory Worker	56
	Carpenter	51
	Bicycle repair	51
	Weaver	50
	Rope Maker/Leaf Platter	41
	Plumber	35
	Mason	33
X	Miscellaneous	
	Pensioner, military	64
	Thief/Cattle thief	(16)

Note: Scale Scores are from Bam Dev Sharda, "Occupational Prestige in Rural India," *Rural Sociology*, *44*(4), pp. 705–718, 1979. Scores in ( )s are estimated.

APPENDIX B: COEFFICIENTS OF CORRELATION MEANS AND STANDARD DEVIATIONS OF SELECTED STATUS VARIABLES FOR VILLAGE INDIAN MALE HEADS OF HOUSEHOLDS AGED 20-64 IN 1962 ($N = 761$) AND 1977 ($N = 823$), BY RIOPS SCORES (Appendix A)

	CASTE	FLAND	FOCC	EDU	FJOB	OCC	MEAN	STD. DEV.
CASTE								
1962		0.298	0.426	0.193	0.303	0.310	4.80	2.18
1977		0.259	0.240	0.286	0.325	0.284	5.10	2.10
FLAND								
1962			0.349	−0.33	0.371	0.297	14.53	26.61
1977			0.212	0.65	0.254	0.294	7.60	16.80
FOCC								
1962				−0.010	0.635	0.684	75.40	22.20
1977				0.031	0.495	0.439	63.26	31.33
EDU								
1962					−0.042	−0.050	0.93	2.49
1977					−0.071	−0.048	2.28	3.89
FJOB								
1962						0.734	71.86	25.20
1977						0.681	66.33	27.62
OCC								
1962							73.66	24.86
1977							65.92	27.18

Note: CASTE = Ritual status of Jati Cluster; FLAND = Family land ownership; FOIC = Father's occupational status (RIOPS); EDU = Respondent's educational attainment (# of years); FJOB = Respondent's first job (RIOPS); OCC = Respondent's current occupation (RIOPS).

APPENDIX C: CASTE RANKING

Jati Cluster Rank	Varna	Cluster Name	Jatis (in the Sample)	Ideal Attributes
A. Twice born				
I	Brahman	Brahman	Brahman	Domestic priests, astrologers, men of learning, "true caste" of the varna model, practice vegetarianism, do not share *hooka* (smoking water pipe) with others, chastity, and restrain from widow remarriage. Done *janeoo* at early age.
II	Kshatrya	Rajput	Rajput, Kshatrya	Claim purity of martial blood, practice chastity and restraint from widow remarriage, proud of taking meat and drinks. Done *janeoo* at marriage.
III	Vaishya	Mercantile Vaishya Jatis	Aggarwal, Bania, Kahtri Arora, Sood, and Mahajan	Claim twice born status. Hindus wear *janeeo* at marriage, avoid widow remarriage, and practice vegetarianism.
B. Non-twice born				
IV		(i) Jats; (ii) others Vaishyas: Gujjar and Kambojh	Jats, Jat-sikh	Primarily agriculturalists eat together. The inter-jati hierarchy is based on (i) the practice or lack of practice of widow remarriage, and (ii) the practice or lack of practice of accepting bride price, (iii) practice of *kareva* (remarriage without ceremonies). Jats and Jat-sikhs consider themselves Kshatriyas/Rajputs and far superior to other agricultural jatis. None wear *janeoo*.
V	Shudra	Sat Shudra	Sunar, Tarkhan (Khati, Barhai,Dhaman), Lohar, and Chhimba (Darzee)	Do not practice *kareva*, share water and smoke pipe with each other.
VI		Asat Shudra Jatis (never untouchable)	Nai, Kumhar (Ghumar, Prajapat), Jhiwar, Julaha, and Teli	Twice born jatis do not accept food but accept water from them, nor do they share smoke with them.
VII		Asat (Impure) former Untouchable jatis	Koeri (Kori), Jogi, Dhanuk, Mirasi, Chamar (Ramdasia, Harijan, Arya, Mahajan, Rai/Rai-Sikh), Chuhra (Mazhbi sikh, Balmiki)	Accept food from all other castes and from each except from "peripheral castes" and tribes. They also except offerings from all shrines of Hindus, Sikhs, or Muslims. Within group rankings vary whether a jati (i) begs alms and (ii) removes night soil.
VIII		Segmented jatis (former untouchables)	Lanadia, Bawaria, Majhili	There are no references to these jatis in Ibbetson. They seem to be groups separated from other low jatis as "sects" or locality groups to form independent jatis.

Jati Cluster Rank	Varna	Cluster Name	Jatis (in the Sample)	Ideal Attributes
IX	Formerly "tribe"	Former "criminal tribes"	Bazighar, Gandhila, Sansi	They were wandering tribes, but now settled in villages. Under the British raj, they were classified as "criminal tribes" and associated with prostitution and crimes. They are known to hunt wild animals.

Note: There were a small number (15) of Muslim and Christian respondents who did not state their jati/caste. They were excluded from this classification.

THE FUTURE OF GENDER IN MEXICO AND THE UNITED STATES: ECONOMIC TRANSFORMATION AND CHANGING DEFINITIONS

Patricia Fernández-Kelly

INTRODUCTION

In his celebrated book, *The Meanings of Macho* (1992), anthropologist Matthew C. Gutmann explores the changing character of masculinity in Mexico City. Buffeted by economic crisis in the late 1980s, middle-class men and women increasingly had to pool personal incomes to support families. Unable to sustain long-held patriarchal expectations, men made virtue out of necessity. "Women are becoming independent," explained one of Guttman's informants, "because men are giving [them] freedom to work. Now women and men have to help each other out. That's why they both have opinions" (p. 161).

Thousands of miles away, in California, Chicago and New York, gender relations were undergoing equally momentous transformations. *Downsizing* and the consequent displacement of millions of men from stable employment in the 1980s, dealt a severe blow to notions of masculinity. In *Families in the Faultline* (1994, p. 78), sociologist Lillian Rubin asked, "Why do men talk about wanting to 'wear the pants' and complain about 'ball-busting feminists?' " Because, she claims, those words reflect a widespread resentment about changes caused by

The Shape of Social Inequality: Stratification and Ethnicity in Comparative Perspective
Research in Social Stratification and Mobility, Volume 22, 255–280
© 2005 Published by Elsevier Ltd.
ISSN: 0276-5624/doi:10.1016/S0276-5624(04)22009-1

forces over which working men have little control – their own diminished income, insecurity about the future, the full-time employment of wives, and the new demands women are making on them.

My purpose in this paper is to condense the main findings of research conducted over the last two decades, about the effects of economic change on gender definitions. By using examples from Mexico and the United States, I illustrate the mechanics of larger developments affecting core and peripheral countries. My argument is built around several assumptions. I hold as a premise that gender is a preeminent vector of social organization – a relational process resulting in the unequal distribution of power and other resources on the basis of sexual distinctions. In that respect, gender is akin to race in that it is socially built upon physical differences. To conceptualize gender as a process is to challenge perspectives that view sexual roles and inequalities as reflections of biological constraints. Furthermore, a focus on gender should illuminate the experience of both women *and* men by explaining the character of the relationships between women of varying racial, ethnic and class backgrounds and between those women and their male counterparts (Fernández-Kelly, 1995, p. 144).

Without a dynamic understanding of gender, it is impossible to elucidate the paradoxical realities that marked the end of the 20th Century and ushered in the new millennium. In the United States, the economic upheavals of the 1970s and 1980s brought about an epidemic of plant closings, decreases in manufacturing, and an expansion of services and advanced technology. Unionization rates plummeted as real wages remained stagnant. Companies streamlined and reconfigured, leaving in their wake large numbers of dislocated workers. Pessimistic forecasts of rising social turbulence soon followed. Yet they didn't materialize. That was largely because the massive incorporation of women into the labor force softened the effects of rapid economic change. Instead of revolt, the end of the century witnessed unrivaled prosperity at the aggregate level but also declining standards of living for some segments of American society, especially those formed by rural populations and the urban poor.

Parallel tendencies ensued in Mexico, beset during the same period by a succession of economic setbacks. Miscalculations about the magnitude of oil reserves, followed by a ballooning national debt, eventually led to the devaluation of the peso, high rates of inflation and negative growth rates. Together with stiff monetarist policies imposed by international lending organizations, those changes disrupted the fragile achievements of the previous thirty years. Fledgling middle classes saw their living standards plunge and despair grew among the poor. As in the United States, however, economic debacle in Mexico was not succeeded by violent eruptions.[1] Gender played a part in maintaining the relative calm. To bolster shaky standards of living, women rushed into the formal labor force or supplemented men's earnings through their involvement in the informal sector. In other words, on both sides of the U.S.-Mexican Border, recent economic transitions brought about shifts in the employment opportunities for men and women and this, in turn, altered preexisting conventions regarding the proper role of the sexes.

I expound these ideas in four parts. As a first step to understand subsequent changes, I consider the relationship between industrial expansion and evolving gender identities in the United States at the turn of the 20th Century. During that period, expectations grew that women should become dependent wives and mothers, and men the sole providers of families. This was part of larger social and political trends that brought about the attenuation of class conflict under the aegis of an emergent welfare state. Equivalent trends were taking place in Mexico where the Revolution of 1910 opened up paths for the implementation of enlightened social policies. Later in the century, with the shift toward an information-based economy, other factors began to reshape gender relations in both countries. I consider those new determinants and their effects in the third section, with special attention to the growing atomization of the labor force in terms of sex.

An adequate explanation of gender transformations requires more than a discussion of general currents. Therefore, in the fourth section I discuss five cases drawn from empirical research conducted in Mexico and the United States. My goal is to illustrate the multifarious links between material circumstance and the changing meaning of manhood and womanhood. In the conclusion I summarize the argument and offer a glimpse into the future of gender.

INDUSTRIALIZATION AND THE RISE OF WELFARE LEGISLATION

Trends in the United States

The ascent of industrial capitalism in the latter part of the 19th Century – with its momentous application of new technologies, its rapid expansion of production and markets, and its bloody struggles over terms of employment – culminated in a historical pact between investors and workers through the mediation of the state. In exchange for the compliance of a predominantly male labor force, capitalists accepted government regulations regarding higher wages, better working conditions, stronger unions, and larger benefit packages. Yet the rationalization of industry, and its corresponding social arrangements, was fraught with paradoxes. Here, I briefly consider the plurality of motives and alliances that led to a contested agreement about the role of men and women in the U.S. Then I examine parallel developments in Mexico to create a comparative framework.

Factories and mills rapidly multiplied in the United States during the second half of the 19th Century. In the absence of a welfare state, industry incorporated workers of all sorts, including women and children. Abuse was rampant, leading social observers to press for government intervention. A rich historical literature underscores the relationship between early industrialization and the design of protective legislation meant to ease class tensions while increasing workers' acquiescence. Purposively or not, those laws helped to circumscribe gender roles in the early 1900s. The record left by social reformers of the time gives evidence of a vibrant debate surrounding the proper role of men and women in the home and in the world of paid employment. Florence Kelley, for example, became a topmost leader of the Progressive Movement by fighting for improvements in the treatment of women and children. An indefatigable activist, she pioneered the use of scientific data to sway the U.S. Supreme Court in favor of limits on hours of work for women. With her friend, Louis Brandeis, she influenced the 1908 case of *Mueller versus Oregon* that established women's protected status because of the alleged greater value of their maternal functions by comparison to property rights. Kelley also developed strategies like consumer boycotts of garments produced in sweatshops, and lobbied for legal requirements that employers document worker's ages as a step to end the exploitation of children. Her trajectory illustrates a new relationship between American Civil Society and the State, marked by the prominent role of educated women in the promotion of welfare laws (Sklar, 1995).

As the century advanced, industry in general and heavy industry in particular grew at an accelerated pace. Table 1 shows that, by 1910, jobs in manufacturing represented more than a third of those available in the economy at large. A decade

Table 1. Employment by Industry (in Thousands Except Percentages).

Year	Total	Manufacturing (%)	FIRE[a] (%)	Services (%)
1994	123,060	20,157 (16.4)	8,141 (6.6)	42,986 (34.9)
1990	118,793	21,346 (18.0)	8,051 (6.8)	39,267 (33.0)
1980	99,303	21,942 (22.0)	5,993 (6.0)	28,752 (29.0)
1970	78,678	20,746 (26.3)	3,945 (5.0)	20,385 (25.9)
1960	54,234	16,796 (31.0)	2,669 (4.9)	7,423 (13.7)
1950	45,222	15,241 (33.7)	1,919 (4.2)	5,382 (11.9)
1940	32,376	10,985 (33.9)	1,502 (4.6)	3,681 (11.4)
1930	29,424	9,562 (32.5)	1,475 (5.0)	3,376 (11.5)
1920	27,434	10,702 (39.0)	902 (3.3)	3,100 (11.3)
1910	21,697	7,828 (36.0)	483 (2.2)	2,410 (11.1)

Source: U.S. Bureau of the Census: Statistical Abstract (1996).
[a] Financial, Insurance, Real Estate.

later that figure had grown to 39%, a proportion without equal in the subsequent years.

It was in that context that protective legislation emerged as a two-edged sword. On the one hand, it represented a salutary response to the savage effects of unrestricted labor markets and employers' abuses, but it also created barriers for women's autonomous negotiating capacity. Many social reformers saw labor legislation as a means to limit the misuse and moral defilement of vulnerable workers. At the same time, the new legislation made women, especially mothers, more costly to hire thereby creating new incentives for the employment of men. Progressive reformers thus aided female segregation in the workplace with its underlying presumption that woman's primary ambit is in the home (Lehrer, 1987).

The struggle for the "family wage" as a masculine entitlement was another aspect of the unfolding events. Samuel Gompers, the first president of the American Federation of Labor (AFL), captured rising feelings on that subject when stating, "It is wrong to permit any of the female sex of our country to be forced to work, as we believe that men should be provided with a fair wage in order to keep his female relatives from going to work" (Quoted in Leckie, 1996, p. 12). Male organizations, like the AFL, but also leaders and participants in women's groups like the American Association for Labor legislation (AALL) and the Women's Trade Union League (WTUL) backed the idea that working men should earn enough to support women and children. For that reason, they provided steady pressure for restricting women's involvement in paid work, heightening their dependence on men (Kessler-Harris, 1988, p. 8). Florence Kelley herself saw the family wage as a means to reinforce an order in which husbands would support "the wives throughout life and the children at least until the fourteenth birthday" (Skocpol, 1992, p. 408). Other social activists, like Emile Hutchinson, saw the family wage as a device to safeguard feminine morality. Fears of female licentiousness as a result of contact with men in the workplace permeate the narratives of the time (Smith-Rosenberg, 1984).

There were still other dimensions in the debate over protective legislation and the family wage. Working men, and union leaders and organizers often supported the new laws as a way to make women less competitive and thus improve their own bargaining capacity vis-à-vis capital. According to Skocpol (1992), however, the desire to undercut female competition was not the only, or main, reason behind organized labor's support of protective legislation – instead, wage and hours laws expressed broad aspirations concerning the living standards of American workers as a whole. Unable to dispense with the services of local labor forces, capitalists made concessions and complied with the new legislation. The events of the age highlight the pivotal role of gender in the articulation of class hierarchies but the opposite is also true: given the dialectical character of the process, the struggle for the family wage shows the importance of class distinctions in the workings of gender definitions.

The casting of men as "breadwinners" and women as "housewives" was riddled with tensions caused by the divergent interests of the groups involved. For social reformers and dominant classes, the family wage entailed the possibility of moralizing footloose men by charging them with the support of families, and women by removing them from the perils of paid employment. With the transformation of men into sole "providers," industrialists secured a disciplined labor force but had to comply with costly legislation aimed at improving the working and living conditions of wage earners. This new order made manhood coterminous with submission to the coercive demands of industry but it also enhanced the purchasing power of ordinary Americans (Ehrenreich, 1984; Hartmann, 1987). Finally, working-class women were not passive agents in this process. A retreat into the home and reliance on men's earnings exacerbated their subordination but it also reduced the strains derived from their earlier attempts to combine paid and domestic labor.

The gender self-definitions thus forged prevailed, for almost a century but the period surrounding the Second World War first showed how fluid those identities were. When working men became soldiers, they left behind vacuums in industry and services that were rapidly occupied by women. In addition, almost 300,000 women served in the Army and Navy performing such non-combatant jobs as secretaries, typists, and nurses. Government campaigns stretched the limits of gender definitions by portraying women's employment as a patriotic duty. Capturing the sense of the age was *Rosie the Riveter*, a character promoted by the media to encourage the idea of factory work as an extension of feminine skills (Honey, 1985). Shown in posters as a muscular but winsome operative, Rosie became a new model of womanhood. Evans and Loeb sang of her, "All the day long, whether rain or shine, she's a part of the assembly line. She's making history, working for victory . . . That little girl will do more than a male will do, working overtime on the riveting machine . . ." (New York: Paramount Music Corporation, 1942).

But Rosie the Riveter did not arrive to stay long, at least not immediately. The end of the war brought about new efforts to push women back into the home. Hollywood movies of the 1950s and 1960s are filled with the tales of women who, having experienced the passing allure of career and financial independence, discover true happiness in the voluntary surrender to marriage and family. Still, the heightened participation of women in paid employment during World War II had irreversible effects that became all the more apparent as the structure of the economy changed in the two subsequent decades. Later in this article, I describe that evolution. First I turn my attention to Mexico.

The Mexican Counterpart

Throughout the first half of the 20th Century, parallel but not identical changes were taking place south of the border. There, the legacy of colonial domination and subsequent processes of distorted industrialization delimited employment opportunities for both men and women. As with other Latin American countries, Mexico's landscape has been marked by the presence of a few large cities hovering, like misshapen giants, over a vast but underutilized countryside. Unabated rural-urban migration gives evidence of stagnant opportunities in the rural sector throughout the 1900s. The introduction of machinery to accelerate and expand agricultural production left large numbers of workers without means of survival. In cities, the incapacity of industry to incorporate the available supply of migrant labor, as well as government's inability to enforce protective legislation, led to the expansion of unregulated economic activity, the so-called "informal economy" (Portes, 1989).

The imbalance between the urban and rural sectors had a powerful effect on individual choices. Although many men were ejected from the countryside as a result of the mechanization of agricultural production, it was mostly women who had to leave their hometowns in search of survival. Contrary to a widespread impression, it has been women, not men, who have constituted the majority of rural-urban migrants in Mexico and other parts of Latin America. Diminished opportunities in the rural sector and growing demand for domestic workers in large cities partly explain that trend. Alone and often without protection, young women from small towns and villages in Latin America faced multiple dangers. The life paths described by Chaney and Bunster (1888) for Lima, Peru, are also typical of Mexico City. Young servants were defenseless against the sexual advances of men in the homes where they worked or in the streets they traversed in their free time. When they became pregnant, they were routinely dismissed. Many became peddlers or market vendors, occupations that allowed them to eke out a living while simultaneously looking after their children.

The realities surrounding paid domestic work in cities like Mexico have always been harsh. Perhaps for that reason they have provided a steady well of inspiration for popular culture, including soap operas or *telenovelas*. One of the most famous was *Simplemente María*, a series that galvanized the attention of viewers throughout Latin America for more than a decade in the 1970s and 1980s. It told the story of a beautiful girl from the Peruvian countryside loved by her employers' son. Although the young man plans to marry María, social convention and tortuous intrigue stands in the way of the couple's happiness. Fired from her job and expecting a child, María vows to defy all odds. Slowly but determinedly she uses her sewing skills to become an internationally famous couturier. When the time of her revenge is at hand, she forgives her tormentors. In her virtue and success she thus fulfills the dreams of hundreds of thousands of women throughout the hemisphere. *Simplemente María* was a staggering success because it gave tangible voice to the yearnings of the popular classes in Latin America, especially women.

Patriarchal ideologies, affirming male supremacy and women's subordination, have been commonplace in Mexico since Pre-Colonial times but often difficult to sustain for various reasons. As revealed by *Simplemente María*, among the most vulnerable sectors, dire need has always pushed women, as well as children, into formal and informal employment. At the top end of the class hierarchy women of means – who were able to delegate domestic responsibilities on servants – could secure paid or unpaid employment in prestigious occupations like government and education. In Mexico it has been mostly among the tottering middle classes that the ideal of men as sole providers and women as housewives has been realized. In that country the confinement of women to the domestic sphere has been as much the product of patriarchal ideology as the effect of limited employment opportunities for both men *and* women.

A narrow focus on Mexican patriarchy would make it difficult to understand conspicuous developments in the early 1900s. The first popular revolution of the 20th century – occurred in Mexico in 1910 – gave birth to a populist state that set into law progressive ideas concerning labor relations and women's employment.[2] Subsequent legislation made generous provisions in public health, social security, minimum wages and severance payments for workers of both sexes. Furthermore, Mexican Law earmarked special allotments for women, especially mothers, including access to subsidized childcare centers – *guarderías infantiles* – generous maternity leaves with full pay, and lactation periods during working hours. Mexican Labor Law is among the most enlightened and forward looking in the world. Unfortunately, it was never fully enforced, partly because of limited government resources and also because of the pressures brought forth by employers unwilling to comply with costly legal requirements.

Starting in the 1940s and gaining momentum a decade later, import-substitution industrialization (ISI) opened up new possibilities for men and women's employment. A nationalist rhetoric grew out of government efforts to invigorate industry and decrease Mexico's dependence on external economic forces (Evans, 1996). The goal of import substitution was to replace expensive imports with domestic products, especially those in heavy industry. ISI had positive effects in Mexico and other parts of Latin America. It was responsible for periods of high expansion in manufacturing. As Table 2 shows, the share of industrial output as part of total gross domestic product in the four largest countries grew rapidly between 1950 and 1967.

Table 2. Latin America: Share of Industrial Product in the Total Gross Domestic Product (in Percentage).

	1950	1960	1967
Total	18.7	21.7	23.1
Argentina	29.4	32.2	34.1
Chile	21.2	23.7	25.8
Mexico	19.9	23.3	25.6
Brazil	15.1	21.4	21.6

Source: Industrial Development in Latin America, *Economic Bulletin for Latin America*, Vol. XIV, No. 2, 1969.

Rapid industrial development in Mexico required a trained labor force that would earn wages large enough to support families and expand aggregate demand. Thus, the idea of the family wage that had informed the aspirations of the American working class at the turn of the century gained strength in Mexico as part of government attempts to achieve economic independence. As with other large economic projects, ISI was associated with definitions about the proper role of men and women in the organization of production. Moreover, it was as part of the efforts to modernize industry that the Mexican government gave new impetus to policies in education, health, housing, and transportation. Opportunities grew in the most advanced sectors of the economy for both men and women. A new middle-class with its eyes turned to ways of life favored in the United States began to appear in the urban landscape.

In the next section I explore the forces that reshaped the fortunes of workers on both sides of the border in the latter part of the 20th Century.

GENDER IN THE ERA OF ECONOMIC INTERNATIONALIZATION

Trends in the United States

In the late 1960s, the economic and political order that had produced the world's most affluent proletariat in the United States began to crack. Computer technology and cheap, rapid transportation freed employers from spatial constraints and reduced their dependence on local work forces. With increased frequency, companies began to relocate industrial operations to areas of the world where wages were low and workers docile. Able to roam the planet in search of optimal conditions, employers had few incentives to continue paying a family wage to the common man. During the 1970s and 1980s an epidemic of plant closings caused massive layoffs in places like New York, Pittsburgh, Baltimore and Detroit. The sputtering smokestacks that had dotted the old industrial landscape gradually went still.

Estimates of the time give an idea of the profound character of industrial restructuring. After conducting the first influential study on the subject, economists Bennett Harrison and Barry Bluestone concluded that, "somewhere between 32 and 38 *million* jobs were lost during the 1970s as the direct result of private disinvestment in American business" (Bluestone & Harrison, 1982, p. 9). Large manufacturing firms eliminated more than 900,000 jobs a year beginning in the mid-1970s, simply in the course of closing domestic branch plants. Harris (1984) calculated a total loss of 3.5 to 4 million jobs between 1978 and 1982 – one out of every four positions in large manufacturing facilities. Others noted with alarm the rising tendency of companies to renege on past commitments to workers under the guise of "efficiency" and "flexibility" (Gordon, 1996; Harrison, 1999). Writings by Michael Piore and Charles Sabel (1990) took a more optimistic view, emphasizing the opportunities that the new economy was creating for individual entrepreneurs. Despite varying positions, there was consensus about the irrevocable nature of economic change in the United States (Harrison & Bluestone, 1988).

Figure 1 shows the dramatic rearrangement of the U.S. economy since mid-century. In 1960 one third of all jobs were found in manufacturing and services represented only a small fraction of employment (13.7%). By 1999 the percentages were almost precisely the reverse. Only 16.4% of jobs were in manufacturing and more than a third were in services (34.9).

Deindustrialization, as an effect of globalization, was more than an economic strategy to lower production costs – it was also a massive political shift that altered

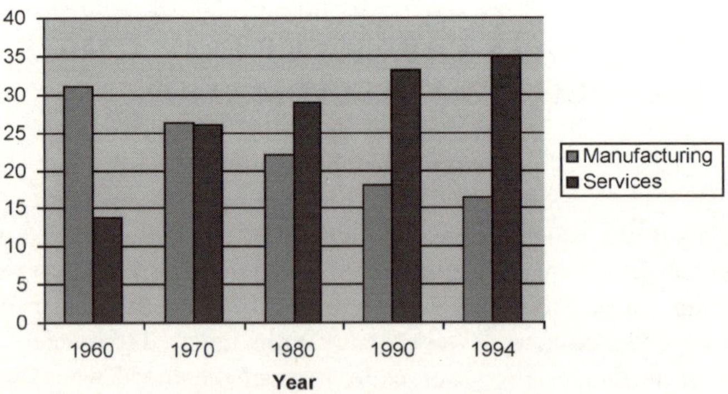

Fig. 1. Employment in Manufacturing & Services, 1960–1994. *Source:* U.S. Bureau of the Census: Statistical Abstract.

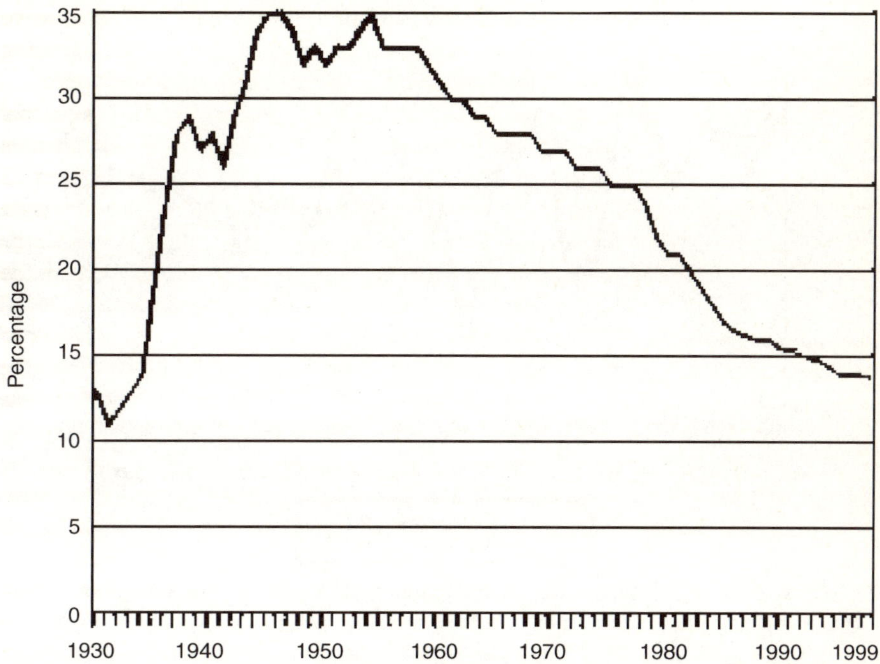

Fig. 2. Union Density 1930–1999. *Source:* Bureau of the Census, *Historical Statistics of the United States, Colonial Times to 1970*; Bureau of Labor Statistics, *Handbook of Labor Statistics* Bulletin 2070, December 1980; and Bureau of Labor Statistics, *Employment and Earnings*, January, various years, 1983–2000. Prepared by the AFL-CIO.

the tenuous balance of power between employers and workers to the advantage of the former. Dropping unionization rates give evidence of that. Figure 2 synthesizes information about trends in union membership between 1930 and 1999. Almost 33% of American workers belonged in unions in 1970. That figure fell to 18% in 1980 and to an abysmal 13% at the end of the 20th Century. In other words, the period that saw the acceleration of global investments and the transfer of productive activities from the United States to less developed countries, also witnessed a massive depletion of labor organizations and a decline in their bargaining capacity vis-à-vis employers.

Minimal or negative increases in hourly wages give yet another indication of workers' waning fortunes throughout the period of transition. Figure 3 shows that, especially during the 1980s, real hourly wages dropped as much as 6% in the United States. It was only after 1995 that they rebounded, partly as a result of vigorous activity in the financial and speculative sectors.

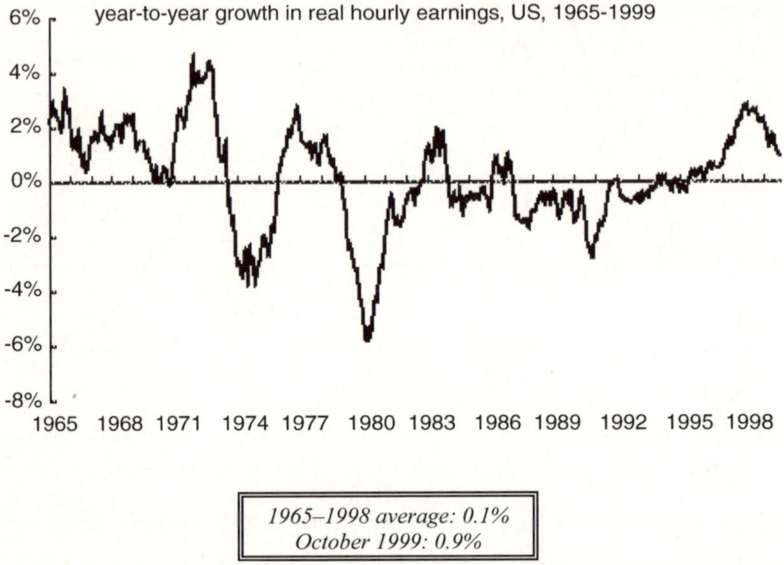

Fig. 3. Year-to-Year Growth in Real Hourly Earnings, U.S., 1965–1999. *Source:* Historical Statistics of the United States, National Bureau of Economic Research; U.S. Bureau of Labor Statistics.

The transition from basic industry to services and information had numerous subsidiary effects. Diminishing opportunities for native-born workers in blue-collar employment paralleled increases in automation and greater reliance on domestic and international subcontracting as means to disperse the economic and political risks of production. In competitive industries like garment, but also in advanced sectors like electronics, subcontracting chains connected large companies with small firms and even individuals doing piecework at home (Ward, 1988; Fernández-Kelly & Sassen, 1994). Ironically, the rising demand for personal services and customized products on the part of new professional classes stimulated the employment of immigrants. As Sassen (2000) has noted most incisively, economic innovations led to the reconfiguration of urban landscapes. Old centers of industry, like New York, rebounded as "global cities" where fast-growing world trade is coordinated and where professionals coexist with low-skilled immigrants and displaced native-born workers.

A major consequence of rapid economic change was the growth in the number of two-earner households among both professional *and* working-class populations. In the aftermath of the feminist mobilization that started in the 1970s, new generations saw women's advances in education and employment as birthrights not privileges.

The sexual revolution of the 1960s lifted the stigma of premarital sex and divorce, expanding women's options. Extensions to Civil Rights legislation made sexual discrimination a matter of legal concern. Innovations in contraceptive methods and legalized abortion further increased women's capacity to compete on an equal footing with men in the labor market. By the end of the 1970s, normative images of executives in "power suits," clutching briefcases, and marching confidently into the workplace had replaced those of mothers in aprons. The new professional woman burst into the scene as a culmination of yearnings for emancipation but, as we will soon see, she had troubles all her own.

The situation was somewhat different for working-class women whose entrance into the labor force was not determined exclusively by a desire for self-fulfillment. As the capacity of men to earn a family wage declined, those women entered the labor force primarily to enhance family earnings (Spalter-Roth et al., 1990). Female labor force participation increased from 20% in 1900 to 55% in 1988, with much of the growth among mothers in families with annual earnings below $20,000. By 1988, 67% of mothers who were single parents, 65% of mothers in dual-parent families, and 53% of mothers of children under three years of age were in the labor force (Hayshed, 1997). Those proportions continued to increase during the 1990s. Figure 4 shows that, by the end of the century, 60% of adult women were working outside the home with that figure representing an unprecedented 46% of the total labor force (Smith & Bachu, 1998). Furthermore, those figures do not include women working in the informal economy and, therefore, underestimate the actual number of women working for pay.

Economic change eroded the material foundations that had held together the notion of males as family providers and women as subordinate wives and mothers. The effects were felt in every aspect of culture. During the last two decades of the 20th Century the popular media, television in particular, obsessively reviewed emerging definitions with alternate glee and horror. Phil Donahue, the man who created the modern-day TV talk show, donned skirts more than once while discussing the new sensibilities surrounding gender. His influential programs contributed to create a new climate of tolerance for sexual minorities, including homosexuals and trans-sexual men and women. Just as Donahue was exploding the myths of unchangeable masculinity and femininity, innovative marketing campaigns presented images of men sensitively holding babies and women in army fatigues forcefully clutching rifles. Unisex fashion and haircuts further expressed a new yearning for gender equality. Controversies about women in the military began to demolish the last bastion of male exclusivity.

Several paradoxes marked the massive entrance of American women into the labor force. Despite their growing importance as income earners, they continued to assume the lion's share of domestic obligations, especially with respect to

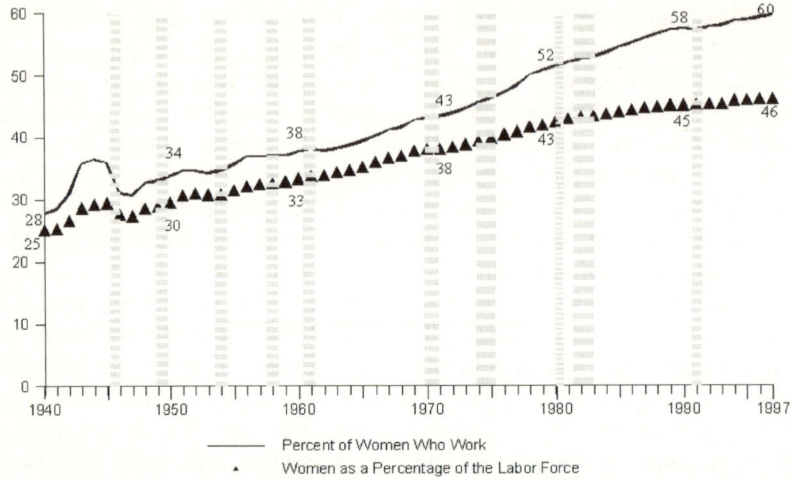

Fig. 4. Women's Labor Force Participation and Business Cycles: 1940–1997. *Note:* Labor force participation includes those who work full or part-time, or are unemployed. Recession years are indicated by the horizontal lines. *Sources:* (a) 1940–1947 rates, U.S. Bureau of the Census, 1960, Series D 13–25; 14 years old and over; (b) 1948–1997 rates, U.S. Bureau of Labor Statistics, website extract, 1998; 16 years old and over; (c) Business cycles, 1940–1996, Statistical Abstract of the United States, Table 895, 1998.

childcare. Juggling the demands of home, motherhood, and paid employment remains a defining influence for women of all kinds. This has had a powerful impact on the national ethos, giving rise to a culture of anxiety centered on family life and parental responsibilities. Especially since the late 1980 a string of legal suits surrounding the physical and sexual abuse of children by service providers in day care facilities and private homes exposed intense ambivalence among and about working mothers. A rumble of discontent throughout the land covertly or explicitly blames working mothers for problems ranging from teenage pregnancy to rising youth crime. Ironically the shift towards a global economy has not eliminated the old contradictions surrounding earlier patriarchal arrangements.

What has changed, however, is the expectation that domestic and reproductive work should be women's only responsibilities. People of both sexes now expect everyone to be at least potentially able to support him or herself and make substantial contributions to the household. The new mores reflect, to some extent, value systems that grew in the aftermath of the Women's Movement but they are also the effect of deep economic transformation that has resulted in the atomization of the labor force in terms of sex.

The Mexican Counterpart

The radical transitions provoked by globalization were also felt in Mexico where import-substitution industrialization had gained currency at mid-century. From the outset, liberal economists had denounced ISI because of its reliance on cumbersome protectionist measures. Although it is true that ISI did not meet all its objectives, it did not fail entirely. Instead, the attempts to expand national industry were cut short by new international pressures that provoked what sociologist Anibal Quijano (1976) called a new "opening to the exterior." The hope for self-sufficiency was replaced by a growing interest in export-oriented manufacturing. Mexico's *Maquiladora* Program, in full bloom by the 1970s, was the main exemplar of this trend. It consisted of government incentives to facilitate foreign investments in the production of exportable goods, mainly garments and electronics products. Assembly plants, known as *maquiladoras*, were allowed by government to operate along the U.S.-Mexico border as directly owned subsidiaries or subcontractors of foreign corporations, most of them located in the United States. Many of the jobs eliminated north of the border as a result of deindustrialization ended up transformed in Mexican *maquiladoras*.

The program soon became the fastest growing sector of the Mexican economy and the second largest source of foreign exchange (Cravey, 1999). In subsequent years *maquiladoras* grew into the largest experiment in export led industrialization and an early blueprint for the North American Free Trade Agreement. The shift from import-substitution to export-led industrialization entailed a re-constitution of the labor force in terms of gender. For over thirty years, *maquiladoras* have hired an overwhelming majority of women – about 85% of their total labor force. This carries momentous implications because, both in the United States and Mexico, hiring of women has been associated with declining terms of employment for working people in general. Employers pay women comparatively low wages and expect them to leave their jobs when getting married or pregnant. Women, in turn, tend to see themselves mainly as mothers and wives, not workers. Because they tend to occupy low positions in the labor market, they have had little power to organize and bring about improvements in working conditions.

In the next section I further explain how economic changes on both sides of the U.S.-Mexico border have affected specific sectors of workers.

AN ARRAY OF ILLUSTRATIVE CASES

The large transformations that simultaneously resulted in widespread plant closures in the United States and a shift towards export-oriented industrialization in Mexico had major consequences for working men and women in various classes and segments of production. Race and ethnicity also played a part, defining the position of various groups in the reconfigured panorama. Here I focus on five cases – three from the United States and two from Mexico – in an effort to illustrate details and outcomes in this new age.

Zoe's Dilemma

In 1992, shortly after the election of William Jefferson Clinton to the presidency, of the United States, the name of Zoe Baird galvanized national attention. A successful lawyer married to a Yale professor, she was the epitome of professional womanhood and Clinton's choice for the post of Attorney General. Her rise to prominence was as rapid as her downfall. In congressional hearings that recreated a trial by fire, Baird was forced to confess that she and her husband had engaged the services of a Peruvian couple, of dubious residency status, to care for their infant son. Adversaries of the Clinton administration rapidly portrayed Zoe Baird as a lawbreaker unfit for public service and succeeded in removing her from the political scene. More interesting, however, were other aspects of the case that received negligible attention at the time. Zoe Baird, her husband and newborn child represent a new class on the ascent, formed by two-earner households of means and education. The circumstances that led to her undoing exemplify the plights that professional women now confront.

The same forces that caused declines in manufacturing over the last three decades led to an unprecedented demand for specialized and professional workers. Baird was one of several million who benefited from those trends and her type of employment was defining of the moment. Throughout the 1980s and 1990s, jobs grew most rapidly in the FIRE sector whose apt acronym designates Finance, Insurance, and Real Estate. Figure 5 shows the rapid growth of the FIRE sector since the mid 20th century. Between 1960 and 1994 its labor force almost quadrupled. By 1994 jobs in finance, insurance and real estate represented 7% of total employment.

Although comparatively small on the aggregate, FIRE encompasses some of the most lucrative and demanding jobs in the nation. As the economy internationalized, cities like New York, and Los Angeles became the locus for the administration of international markets and their multiple derivations. Even smaller places like Hartford, Massachusetts, became the seat of corporate clusters. *Aetna*, the insurance giant where Zoe Baird accepted a lucrative position shortly before her advent to public notoriety, was located in that city.

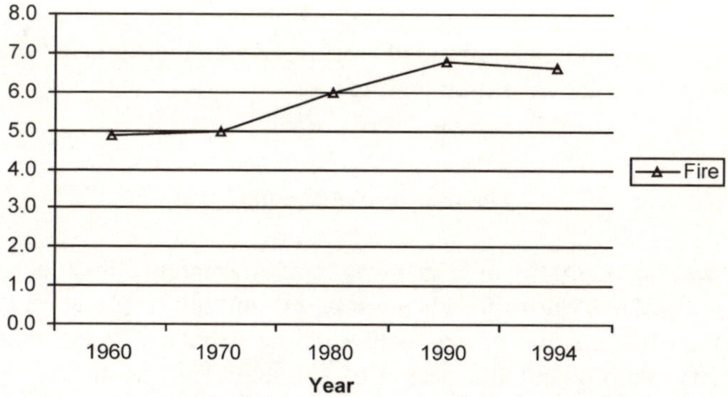

Fig. 5. Employment in Finance, Insurance & Real Estate 1960–1994. *Source:* US Bureau of the Census: Statistical Abstract.

Professional salaries grew as fast as the demand for what Robert Reich (1990) called "Symbolic workers." Lavish earnings were only matched by the exigencies of long working weeks. The growth in the number of households formed by high-powered professionals fueled the demand for unskilled and semiskilled workers in numerous niches of production. From designer's clothing and custom-made furniture to chefs with an international flair, specialized caterers, dog walkers, personal trainers, nannies, and *au pairs*, all were occupations that met the needs, and relied on the elevated purchasing power, of the new technocratic class. More importantly, the dilemmas created by maternity and childcare induced a new demand for "live-in" service providers to fill the spaces and functions left empty by the employment of professional women (Hondagneu-Sotelo, 2001).

Zoe Baird's circumstances were emblematic of the period. Well connected and charismatic, she had been a diligent student and a valuable employee who spent most of her time in the office. In the early 1990s her salary surpassed $150,000 and her duties multiplied. Her new job with *Aetna* required an hour commute by car from her New Haven home. Responsible and forward-looking, she had long postponed motherhood. Her first son had arrived after careful planning. Unable to find native-born workers willing to provide, live-in services, she had depended on immigrants. The scandal that followed exposed some of the difficulties surrounding the new professional woman. Condemned and vilified, Baird faced a series of unflattering characterizations. Many saw her as the incarnation of all that is wrong with the women's movement: in the pursuit of material success, she had supposedly abdicated maternal responsibilities. Her comeuppance was met with satisfaction by those who saw her as a representative of a new ruling cadre whose prerogatives depended on the exploitation of vulnerable immigrants. Yet Baird's case was of a more simple character. For all the glitter of her life, she confronted the same dilemmas that even the most humble of working women must face: how to attend to the needs of children while maintaining jobs.

The Immigrant Machine

Baird's ordeal also shows the growing interdependence between the new professional classes – whose fortunes are based on high levels of education and employment in information-based sectors of the economy – and recent waves of immigrants. Throughout the period of economic reconstitution sketched in the previous sections, immigration to the United States grew rapidly reaching unprecedented levels by the end of the century. Figure 6 presents the contour of legal immigration to the U.S. since 1901. By the year 2000, approximately 10% of the American population was foreign-born reaching levels close to those that prevailed early in the 20th Century. Official figures, however, do not take into consideration a substantial number of immigrants, mostly from Mexico and other parts of Latin America, who arrive and remain in the country illegally.

There is nothing new about the persistent arrival of foreigners into a country whose very identity is coterminous with immigration. Nevertheless, since the 1970s several dimensions of the phenomenon have changed. Earlier immigrants, mostly from Europe arrived into American cities that provided a bounty of industrial jobs. Historically, large farms and agricultural firms in the Southwest absorbed Mexicans. Many children of immigrants joined unions and moved into jobs of higher status. By the 1980s the old paths for economic and social mobility were

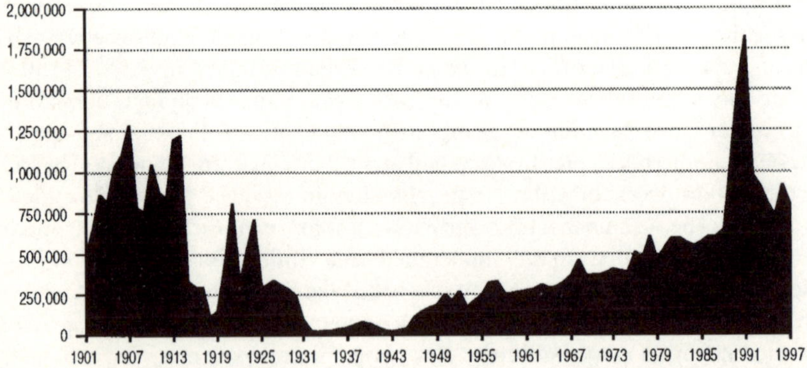

Fig. 6. Legal Immigration: Fiscal Years 1901–1997. *Source:* U.S. Department of Justice: Immigration and Naturalization Service Annual Report (1998).

not available to new immigrants who, therefore, often ended providing a myriad of services, or working informally, in global cities.

In New York, Los Angeles and Miami, immigrants became preferred providers of labor in small firms specializing in the production of goods ranging from clothing to electronics products. Many of the firms that hired them had opened factories in overseas locations. Moving assembly jobs to less developed countries paralleled the growing employment of immigrants at home. The maintenance of plants in strategic U.S. points presented additional advantages derived from the proximity to opportunity markets. By combining "outsourcing" with reduced production in the United States, employers improved their competitive stance. Subcontracting arrangements proliferated as producers sought to reduce costs. The growth of informal work during the 1980s also created vigorous demand for foreign-born workers.

In global cities and sectors where professionals clustered in response to the new demands for high skilled labor, immigrants became the logical candidates for menial positions, especially those linked to reproductive labor and the provision of personal services. The number of immigrant *au pairs*, nannies and live-in maids in American cities increased over the last twenty years after a long period in which paid domestic service had all but disappeared (Hondagneu-Sotelo, 2001).

The new exigencies altered somewhat the character and composition of immigrant flows. Women are now more likely than in the past to migrate alone leaving behind children in the care of relatives. In many American cities, and even in suburbia immigrants, men as well as women, constitute the foundation of economic activity. Although their presence is often undetected, they form a strategic sector whose modest wages and vulnerable status presents multiple advantages to employers.

Under any circumstances, migration is a jarring experience that forces individuals to make major adjustments. Gender relations among immigrants are particularly susceptible of impact. A new literature on this subject suggests that migration can be a powerful vehicle for women to acquire added leverage vis-à-vis men in their homes. Because immigrants must pool scarce resources in order to survive, women's contributions acquire greater value than in the countries of origin. Immigrant women have been notable in their capacity to reproduce cultural practices in areas of destination, thus enabling the adaptation of their families into churches and community organizations (Hondagneu-Sotelo, 1994).

In addition, the growing need for paid domestic workers – and other like occupations – enables women to acquire an independent income, small as it may be. That, in turn, has allowed them to refashion their own identity. Ironically, the fragile position of immigrants as a whole can expand the negotiating capacity of women and force men to modify patriarchal expectations. It is therefore not surprising that when interviewed, men who are more likely to yearn for their hometowns and countries of origin where, they imagine, manly prerogatives remain unquestioned. Women, on the other hand, soon see the benefits of added independence in the United States. The outlook is not entirely sanguine, however, because the tensions brought about by changing roles can also exacerbate conflict. In some communities, domestic violence, alcoholism and other maladaptive symptoms are the result.

The Disappearance of Manhood

The adjustments immigrants must make to adapt in areas of destination are fraught with tension. Yet there are groups for whom the drama of gender has acquired even larger proportions. African Americans living in impoverished neighborhoods have long faced obstacles in the labor market. A history marked by residential segregation and racial exclusion has limited their options in the United States to a larger extent than any other group. Changes in the global economy exacerbated even further the conditions surrounding this nation-within-a-nation. The decline of manufacturing that resulted from the transition to an "information-based economy" broadened the gulf between those able to benefit from the new opportunities and those left behind. Clustered in inner cities with collapsing infrastructures, negligible investment, and appalling school conditions, new generations of black Americans are more likely to be permanently unemployed than their ancestors. They increasingly constitute a "non-working" class whose very existence challenges every previous hope for assimilation. Figure 7 provides a comparative look at unemployment rates by race between 1980 and 1998. Consistently, and despite upturns in the nation's economic performance, black unemployment nearly doubles that of whites. Unaccounted for in these calculations are thousands of African Americans who have fallen entirely out of the labor force.

William Julius Wilson (1996) first called attention to the unique effects of the "disappearance of work." One of them has been the virtual collapse of notions of masculinity dependent on paid employment. Instead of accepting demeaning jobs, whose forebears held without sizeable benefits, impoverished African-American men often seek meaning and power outside of the limits of legality. Not able to support families, or hold rank vis-à-vis women, they often redefine the meaning of success by rejecting marriage and emphasizing independence and sexual prowess. Women's yearnings for security and intimacy are viewed as potential entrapments. Hip-hop and rap music often gives voice to gender resentment. The portrayal of women as "hos" and "bitches" is but a veiled expression of men's loss of masculine status. In those circumstances, women too have had to rethink the meaning of

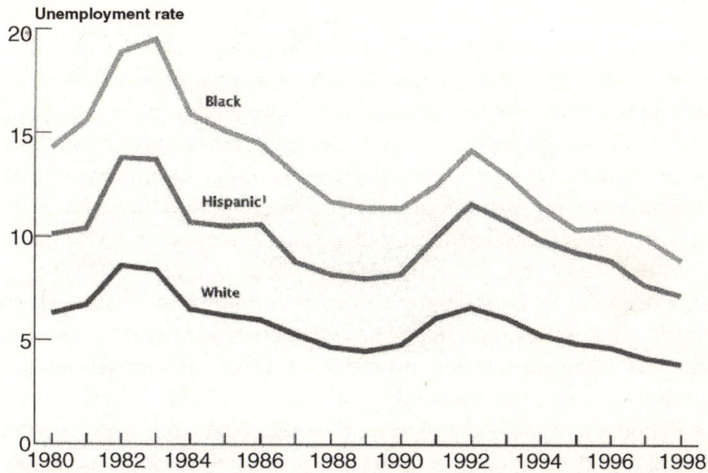

Fig. 7. Unemployment Rate, by Race and Hispanic Origin: 1980–1998. *Note:* [1]Persons of Hispanic origin may be of any race. *Source:* Chart prepared by U.S. Census Bureau. For data, see Table 651.

womanhood by emphasizing personal autonomy and rejecting romantic illusions of marriage and family. Singular for its level of atomization, the experience of urban blacks underscores the role of gender as a pivotal force in the organization of social groups, even those most vulnerable.

Women of the Maquiladoras

Twenty years ago, when Mexico's *maquiladora* program was still new, researchers and public officials often saw it as a temporary solution for rising levels of unemployment along the northern border. The abrupt termination of the *Bracero* program, which had enabled Mexican men to enter the United States as guest workers, heightened joblessness and the possibility of popular turmoil (Fernández-Kelly, 1983). As described earlier in this chapter, Mexico's government reacted by creating incentives to foreign investment in export manufacturing. *Maquiladoras* multiplied rapidly but, against the expectations of many, those plants did not create jobs for displaced men; instead, they targeted young, single women as preferred providers of labor.

Although the story is by now familiar, it is well to remember that the preference for women as providers of labor was part and parcel of a larger strategy on the part of employers to retain competitiveness in a global setting. At the local level, however, women's employment brought about a number of effects worth considering. It increased the capacity of many to make contributions to fragile households mostly formed by parents and siblings. Single mothers were also represented among *maquiladora* workers. Some evidence suggests that towards the end of the 1970s, another type of household was gaining prominence – that formed by single women living together and pooling income to defray shared expenses. Most of those women were recent migrants from rural towns and villages in nearby Mexican states.

Maquiladora work is unlike other forms of employment – even other types of factory work – for its level of intensity and requirement to engage in repetitive operations over extended periods of time. Low wages and reduced opportunities for promotion increase the probability of rapid turnover. Women employed in Mexico's export-processing plants tend to rotate frequently from one employer to the next as a way to assuage tedium and maximize personal advantage. This, however, limits their capacity to benefit from government regulations that reward long-term employment. In other words, export-processing industrialization bears little resemblance to earlier forms of manufacturing that grew during the nineteenth and early twentieth centuries in Europe and the United States. Those early stages eventually led to improvements in standards of living. After several decades of existence, there is little evidence that *maquiladora* work will achieve similar objectives. That is because present conditions make it possible for companies to employ workers in less developed countries without having to consider them as potential consumers of their products. The disconnection between markets and production is having a profound impact upon living standards. Televisions assembled in the Mexican border find their way into the homes of middle-class people in advanced industrial countries. There are few incentives to expand the buying capacity of workers in less developed areas.

Starting in the early eighties, *maquiladoras* faced a series of labor shortages in cities like Tijuana and Ciudad Juarez. A common explanation was that the growth in employment had exhausted the labor supply. Yet there were other factors at work. The devaluation of the Mexican peso and high rates of inflation reduced the appeal of factory work for many women in need to maintain already low standards of living. In search of higher wages they shifted to other forms of employment in U.S. border cities, mainly as domestics. Paradoxically, the spaces left empty by women in the *maquiladoras* was occupied by a growing number of young men who had been ejected from small rural communities by policies of austerity imposed by the Mexican government in response to the support of international development organizations. Although men have never constituted the predominant labor force in export-processing plants, their increasing numbers in them point to the deteriorating conditions of employment for large numbers of Mexican men.

The New Meanings of Macho

In his influential account about the changing conditions surrounding men in Mexico City, Matthew Gutmann (1996) notes the extent to which gender relations have been altered in the last two decades. Economic crisis and neo-liberal economic policies narrowed the options of most urban families. Unable to survive without women's financial contributions, men were forced to make new adjustments, especially with regard to childcare. Surprisingly, economic crisis has not led to an epidemic of fractured families and households. The opposite has often occurred. Women in need of holding jobs have often incorporated younger relatives, cousins in particular, to care for children and do housework in exchange for room and board and the possibility of continuing their education. Thus, an expansion in the number of household members has often been an adaptation to economic exigencies (Chant, 2002).

In these circumstances, men too have had to assume responsibilities that were formerly an exclusive female domain. New values associated to the merits of paternity and father's care seem to be emerging in a country known for its patriarchal pride.

CONCLUSION

In this article I have summarized findings about the relationship between economic change and gender identities over the last century. By comparing events taking place in the United States and Mexico I have tried to gain insight into larger trends occurring in advanced and less developed countries. My analysis shows that gender is not a secondary process but a central aspect in the articulation of class hierarchies. On both sides of the U.S.-Mexico border, the purposive definition of realms of activity for men and women were decisive for organizing production and shaping the interactions between capital and labor.

In the United States, women's employment outside the home became the contested terrain around which welfare legislation was passed in the early 1900s. Heated debates of the time revealed two interdependent dimensions. One was the intent to protect a growing proletariat from the abusive practices of employers. Another was to circumscribe the roles of men and women as part of the effort to enhance working-class standards of living. The casting of men as sole family providers and women as specialized mothers and wives was, therefore, not only an expression of continuity with respect to value systems forged earlier on but also a selective organizational strategy suited to changing economic conditions.

Social reformers of the Progressive Movement were especially active in bringing about modern approaches towards the treatment of women and children. By restricting hours of work and kinds of employment appropriate to those two populations, government curtailed the supply of labor available to employers and enhanced the negotiating capacity of a predominantly male labor force. Manly feelings of solidarity between workers and employers further enhanced the capacity of industry to expand production. Thus, protective legislation for women and the family wage, as a male entitlement, helped to form a working class whose prosperity was without precedent. It also charged men with the sole support of families and made women entirely dependent on men's earnings.

Despite the socially created boundaries between male and female employment, women's participation in the labor force increased during the Second World War and ebbed immediately afterwards only to rise again in the 1960s. In their efforts to retain competitiveness, manufacturing firms threatened by foreign competition increasingly tapped new pools of labor, especially those formed by women, in less developed countries. Mexico's *maquiladoras* became a classic example of that process. At the same time, in advanced economies, the shift from manufacturing to services and advanced technology led to the proliferation of new jobs bearing characteristics long associated with female employment. The streamlining of corporations, the decline of unionization, *outsourcing*, and the growth of contingent work increased the probability of women's employment but also the tendency for men to work in feminized occupations. As with the early years of industrial expansion in the 1900s, changing gender definitions were a pivotal aspect of the reorganization of production in the latter part of the 20th Century.

The examples provided in the previous section point to the several facets of contemporary gender relations. The most general trend consists of greater atomization of the labor force in terms of sex. The disappearance of the family wage entails a new expectation that all workers, regardless of ascribed characteristics or domestic involvement, will assume responsibility for the maintenance of at least one person: him or herself. As more women join the world of employment, they face the promise of added autonomy and economic self-reliance. At the same time, the deteriorating conditions of work in several economic sectors raise concerns about the full meaning of gender atomization. As suggested by the case of Mexico's *maquiladoras* and dramatically exposed by the situation of inner-city blacks, the transformation of gender relations is fraught with dangers.

Perhaps most importantly, the new economic arrangements that followed globalization did not bring about solutions for the intractable tensions between the demands of paid work and home, especially the care of children. To bring about a solution to this most endurable contradiction is a major challenge in the new century.

NOTES

1. An exception was the Zapatista Movement that began in 1994 in the southernmost state of Chiapas but even that most dramatic manifestation of public discontent was sedate by comparison to insurrections of the past.

2. The Mexican Revolution preceded the Russian uprising by seven years. The two explosions shared several features in common, including a growing frustration among the popular classes over the concentration of land in the hands of a small oligarchy.

REFERENCES

Bluestone, B., & Harrison, B. (1982). *The deindustrialization of America: Plant closings, community abandonment, and the dismantling of basic industry*. New York: Basic Books.

Chaney, E., & Bunster, X. (1888). *Sellers and servants*. New York: Bergin & Garvey.

Cravey, A. J. (1999). *Women and work in Mexico's Maquiladoras*. New York: Rwman & Littlefield.

Ehrenreich, B. (1984). *The hearts of men: American dreams and the flight from commitment*. Garden City, NY: Doubleday.

Evans, P. (1996). *Embedded autonomy: States and industrial transformation*. Princeton, NJ: Princeton University Press.

Fernández-Kelly, M. P. (1983). *For we are sold, my people and I: Women and industry in Mexico's frontier*. Albany, NY: State University of New York Press.

Fernández-Kelly, M. P. (1995). Labor force recomposition and industrial restructuring in electronics: Implications for free trade. *Hofsra Labor Law Journal*, *10*(2), 623–717.

Fernández-Kelly, P., & Sassen, S. (1994). Recasting women in the global economy: Internationalization and changing definitions of gender. In: C. E. Bose & E. A. Belen (Eds), *Women in the Latin American Development Process*. Philadelphia: Temple University Press.

Gordon, D. M. (1996). *Fat and mean: The corporate squeeze of working Americans and the myth of managerial "downsizing."* New York: Free Press.

Harris, C. S. (1984). The magnitude of job loss from plant closings and the generation of replacement jobs: Some recent evidence. *Annals of the American Academy of Political and Social Science*, *475* (September), 15, 19.

Harrison, B., & Bluestone, B. (1988). *The great u-turn: Corporate restructuring and the polarizing of America*. New York: Basic Books.

Hartmann, H. I. (1987). Changes in women's ecoomic and family roles in post-World War II United States. In: L. Beneria & C. R. Stimpson (Eds), *Women, Households, and the Economy* (pp. 33–64). New Brunswick, NJ: Rutgers University Press.

Hayshed, H. V. (1997). Developments in women's labor force participation. *Monthly Labor Review Online*, *120*(9).

Hondagneu-Sotelo, P. (1994). *Gendered transitions: Mexican experiences of immigration*. Berkeley, CA: University of California Press.

Hondagneu-Sotelo, P. (2001). *Domestic: Immigrant workers cleaning and caring in the shadows of affluence*. Berkeley: University of California Press.

Honey, M. (1985). *Creating Rosie the Riveter: Class, gender, and propaganda during World War II*. Boston: University of Massachusetts Press.

Kessler-Harris, A. (1988). *A woman's wage: Historical meanings ad social consequences*. Lexington, KY: University Press of Kentucky.

Lehrer, S. (1987). *Origins of protective labor legislation for women, 1905–1925*. Albany, NY: State University of New York Press.

Piore, M. J., & Sabel, C. F. (1990). *The second industrial divide*. New York: Basic Books.

Portes, A. (1989). Latin American urbanization during the years of the crisis. *Latin American Research Review, XXIV*(3), 7–44.

Sassen, S. (2000). *Cities in a World economy*. Thousand Oaks, CA: Pine Forge Press.

Skocpol, T. (1992). *Protecting soldiers and mothers: The political origins of social policy in the United States*. Cambridge, MA: Harvard University Press.

Smith, K. E., & Bachu, A. (1998, October). Women's labor force attachment patterns and maternity leave: A review of the literature. Paper presented at the Annual Meeting of the Southern Demographic Association. Annapolis, MD.

Smith-Rosenberg, C. (1984). *Disorderly conduct: Visions of gender in victorian America*. New York: Alfred A. Knopf.

Spalter-Roth, R., Hartmann, H. I., & Andrews, L. M. (1990). Mothers, children, and low-wage work: The ability to earn a family wage. Revised version of a paper presented at the 85th meeting of the American Sociological Association. Washington, DC.

Ward, K. (1988). Women and the global economy. In: B. Gutek, A. Stromberg & L. Larwood (Eds), *Women and Work Annual Review* (Vol. 3, pp. 17–48). Urbana, IL: University of Illinois Press.

Wilson, W. J. (1996). *When work disappears: The World of the new urban poor*. New York: Alfred A. Knopf.

DO ETHNIC ENCLAVES BENEFIT OR HARM LINGUISTICALLY ISOLATED EMPLOYEES?

M. D. R. Evans

ABSTRACT

Sociologists often take for granted that segregation and social closure automatically entail disadvantage, so the hypothesis that partially separated ethnic sub-economies, or "ethnic enclaves," might thrive and benefit their workers and employers was a radical departure from past thinking (Wilson & Martin, 1982; Wilson & Portes, 1980). Since then, controversy has raged over why entrepreneurs set up businesses in immigrant enclaves, and over the consequences enclave-based business has for employers and employees. The bulk of the evidence suggests that most immigrant entrepreneurs are "pulled" by the opportunities presented by ethnic resources to open enclave businesses, although some are also "pushed" by mainstream employers' discrimination. The consequences of enclave employment are less clearly established. This paper seeks to clarify the issue by using multi-level analysis of unit-record Census data to assess the impact of availability of co-ethnic employment on job quality for immigrant employees from all the wide variety of non-Anglophone countries represented in Australia. I control for the effects of human capital characteristics by using a detailed specification of location and quantity of education and work force experience. The results

The Shape of Social Inequality: Stratification and Ethnicity in Comparative Perspective
Research in Social Stratification and Mobility, Volume 22, 281–318

© 2005 Published by Elsevier Ltd.

ISSN: 0276-5624/doi:10.1016/S0276-5624(04)22010-8

show a strong interaction between availability of co-ethnic employment and individual English-language skill such that immigrants not fluent in English get substantially better jobs if they belong to a group containing a large proportion of entrepreneurs. The effect of availability of ethnic employment is much weaker among immigrants with middling levels of English fluency, and the quality of job opportunities of immigrants fully fluent in English are unrelated to the availability of co-ethnic employment. These results are fully consistent with the "communications costs" hypothesis, but inconsistent with either the co-ethnic predation hypothesis or the discrimination hypothesis.

INTRODUCTION

An ideal-type rational market sorts workers according to their productivity. Workers eagerly seek the best jobs they can get. Employers seek good workers and wish to pay as little as possible for them. Other employers with the same goals are competing in the same market. Their bidding against one another sets the "going rate" and the typical quality of worker for different jobs.

Evolutionary theories see the "free-market societies" of the twentieth-century West as approximating this model: Problems of mutual trust had been sufficiently resolved by the late 19th century that employers need no longer rely on the bonds implied by ascriptive and particularistic ties, but instead could focus solely on acquiring workers whose skills and abilities would enhance their business's profits (Blau, 1956; Eisenstadt, 1964; Parsons & Smelser, 1956). Economists argue that competition would then strongly tend to weed out discriminatory firms, and that firms hiring on the basis of skill would flourish (Becker, 1971). But other theorists raised a major challenge with segmented market theories initially proposed in the late 1960s and early 1970s.

Segmented market theories generally posit a fractured labor market in which some core (or "primary") segments offer good jobs, pleasant working conditions, and high wages, in contrast to other segments where job ladders are truncated, employment is insecure, and pay is low (Bonacich, 1979; Bowles & Gintis, 1975; Burawoy, 1985; Edwards, 1975). The core segments are inaccessible from the peripheral (or "secondary" segments). Why would anyone accept a job in the secondary labor market? One possibility is simply that there are fewer jobs in the primary sector than there are job seekers: workers can be thought of as queued up according to productivity-related characteristics, and as being chosen from the queue until all the primary-market jobs are gone. The remaining workers must accept jobs in the secondary market. This possibility raises no particular problems for evolutionary theory. But if workers are allocated to the primary or secondary labour markets on the basis of ascriptive characteristics or (the absence of) particularistic ties, that undermines evolutionary theory's basic hypothesis of the triumph of universalism and merit. The "split labour market hypothesis" proposed the possibility that ethnicity is the crucial determinant of access to the core sector or relegation to the periphery (Bonacich, 1979) and related work explicitly posited an unbreachable barrier between the two sectors (Piore, 1979).

But are labour markets fragmented on the basis of ethnicity necessarily disadvantageous to outgroup members? Are they necessarily at odds with evolutionary theory? It is possible that particular ethnic groups may have cultural traditions or established skills that enable them to be particularly successful in certain lines of work, or niches (Hechter, 1978; Light, 1979). And for some ethnic groups, it has been argued, fragmentation has facilitated small business development, which in turn enables "ethnic businesses" to offer conditions of work, efficient employer-employee matches, and a range of jobs comparable to what is available in the broader labour market (Portes & Jensen, 1987, 1989; Wilson & Martin, 1982; Wilson & Portes, 1980).[1] If ethnic markets are fragmented by specialized tastes and preferences, that provides an opportunity for ethnic entrepreneurs to use their existing skills and social networks to serve those markets (e.g. Li & Li, 1999; Masurel et al., 2002). And if mainstream labour-market employers' response to communications costs is to make worse job offers to non-fluent applicants, then co-ethnic employers have a window of opportunity: They can make offers that beat those available in the broader market but are still lower than what equally skilled applicants who are fluent in the dominant language can command in the broader market.[2] The evidence favours these two hypotheses (Evans, 1989), so it seems fair to say that markets subdivided into ethnic niches can provide opportunities for immigrant entrepreneurs, that it can provide them with what Light (1984; see also Min & Bozorgmehr, 2000) dubbed "entrepreneurial resources."[3] But, researchers are beginning to ask, what about the employees (Spencer & Bean, 1999)?

Several strands of prior research suggest that markets which include ethnic niches may benefit employees, provided that the boundaries between the ethnic niche and the mainstream labor market are permeable. For example, Evans (1989) has argued that mainstream job opportunities set a floor on what co-ethnic entrepreneurs can offer: If their job offers are worse than what immigrant workers could get in the broader labor market, then immigrants will leave for the broader market. Thus, the accessibility of employment in the mainstream economy may obviate the kind of co-ethnic exploitation that might occur if strong barriers were to keep immigrants out of the mainstream market (Sanders & Nee, 1987). Conversely, others have argued that competition from the ethnic economy undermines discrimination in the mainstream economy (Evans & Kelley, 1991). Bailey and Waldinger (1991) have argued that enclave employment serves important training and socialization functions for many immigrants, with both employers and employees clearly recognizing their employment relationship as a way-station from which most employees will transit to the mainstream economy. Nee, Sanders and Sernau (1994), too, emphasize what they call "porous ethnic boundaries," with enclave employment being a destination state for some immigrants, and a stepping stone to the mainstream economy for many others. In all these approaches, the availability of alternatives, the permeability of boundaries, the *partial* separation of niche economies from the mainstream form important constraints on employers' behavior. From a slightly different angle, all these "accessibility" and "porosity" arguments suggest the crucial importance of social networks spanning mainstream and ethnic communities. Both survey evidence and qualitative research suggest that immigrants who work in the enclave tend to concentrate their social ties within it more than do their peers who work in the mainstream economy, but closure is far from complete, at least in the limited number of settings that have been examined thus far (Fong & Ooka, 2002; Sanders et al., 2002).

THEORY: COMMUNICATION COSTS

I propose that in the mainstream labor market, workers with weak language skills get worse jobs and lower pay than their fluent peers, because difficulties in communication impose costs on employers from the dominant ethnic group. Immigrants not fluent in the dominant language may also pose inadvertent costs to mainstream employers because, even when words are successfully translated, closely related but unstated expectations may not be fully shared (Goldscheider & Kobrin, 1980; Ladbury, 1984; Werbner, 1984); the only evidence to date is qualitative and it supports this penumbra hypothesis (Dyer & Ross, 2000).[4] Some indirect supporting evidence comes from an in-depth study of mainstream firms indicating that efficiency was lowest in fairly balanced mixed-language groups, higher where one language was almost exclusive (Baker & Wooden, 1992). This is basically a rational-choice explanation of ethnic difference (Hechter, 1986), informed by Stinchcombe's (1990) understanding of the primacy of information-processing in organizations.[5]

In all the industrialized countries investigated thus far, substantial research shows that linguistically isolated workers (those not fluent in the dominant language) get worse jobs and lower incomes than do other workers who are fluent but otherwise have similar human capital characteristics. The "cost" of linguistic isolation to the worker is substantial: Net of education and experience, an immigrant who does not speak the dominant language at all loses around 10–15% in occupational status and, in analyses not controlling for occupational status, earnings compared to a fluent peer (Carliner, 1981; Chiswick & Miller, 1999, 2002; Evans, 1987; Kossoudji, 1988; McManus et al., 1983; Stolzenberg, 1990; Tainer, 1988).

One explanation is that non-fluent workers get worse labor-market outcomes because majority group employers find them more costly. For example, it takes longer to convey instructions when communication is uncertain, and non-fluent workers may make more mistakes because of faulty communication. In addition, mainstream employers may find it difficult to assess the cognitive skills of non-fluent job applicants – a central concern in hiring (Stinchcombe, 1990) – and may, as a result, offer these applicants worse jobs. But a co-ethnic entrepreneur who can communicate easily with them bears no such cost. As a result, the ethnic capitalist can prosper by hiring co-ethnic workers, offering them better jobs than the broader market offers but – at least initially – slightly lower status jobs than majority group workers with equivalent skills hold in the broader market. Workers with language difficulties can thus do better working for a co-ethnic: Immigrant entrepreneurs need to entice employees out of the broader labor market by making job offers at least slightly better than the job offers available in the broader market.[6] The theory does not claim that these will be wonderful offers – they could be very low offers – but only that they will be better than what the broader market offers. Prior exploratory research of particular groups in single settings has shown that many immigrant employers make use of co-ethnics who are not proficient in the dominant language (Kim, 1981; Light & Bonacich, 1988; Model, 1988). Supporting indirect evidence is that immigrants from smaller or more dispersed groups tend to acquire better dominant language skills than their peers from concentrated groups (Chiswick & Miller, 2001). More generally, research into differences among immigrant groups reveals that the larger an ethnic group's linguistically isolated workforce, the more likely are members of that group to set up businesses (Evans, 1989).[7] Moreover, there seems to be some "path-dependence" in that particular ethnic groups' small tendencies towards specializing in particular modes of incorporation seem to intensify over time, at least in the short run (Logan et al., 2000; Raijman & Tienda, 2000).[8]

This theory views labor markets in advanced industrial societies as partially differentiated: Language difficulties impair workers' opportunities in the broader market, but their isolation is not complete and they receive a steady flow of information about opportunities in the broader market (see also Werbner, 2001). Note that the theory posits that ethnic business will have no benefit to fluent employees in the ethnic group. Thus:

Hypothesis. Communications costs. For immigrants whose mother tongue is not that of the host society, the availability of employment with entrepreneurs of the same ethnic group will:

(i) strongly enhance the job opportunities of workers least fluent in the dominant language;
(ii) moderately enhance the job opportunities of workers partially fluent in the dominant language; and
(iii) have no effect at all on the job opportunities of workers fluent in the dominant language.

ALTERNATIVES

My preferred communication-costs theory is inconsistent with a number of other plausible arguments which lead to quite different hypotheses about the effects of ethnic business niches within a broader labor market. Notable among them are widely accepted theses about majority group discrimination against ethnic minorities and claims about exploitation by co-ethnics. Figure 3 gives a graphical representation of the communications-costs hypothesis, and its two leading alternatives.

Alternative 1

Ethnic discrimination hypothesis: Majority-group employers' ethnic prejudices lead them to discriminate against immigrant workers in hiring, job quality, and pay (Gordon, 1964). An alternative (with the same outcome in this case) is that majority-group owners cynically inflame the ethnic prejudices of their majority-group workers in order to procure their acquiescence in discriminating against immigrant workers in hiring, job quality, and pay, thereby reducing labor costs (Bonacich, 1979). If there is ethnic discrimination (of either of these types), then immigrants – fluent and non-fluent alike – will have an incentive to seek shelter with co-ethnic employers. If so, then the availability of enclave employment should benefit immigrant workers regardless of their facility in the dominant language. This hypothesis predicts a substantial positive main effect of the availability of ethnic employment, no main effect of language fluency, and no interaction effect of language fluency and availability of co-ethnic employment.[9] A graphical representation of these predictions is in Fig. 3.

Alternative 2

Co-ethnic predation hypothesis. If the boundaries between the ethnic niche and the broader labor market completely impede the flow of information about job opportunities, then ethnic entrepreneurs will have a strong incentive to exploit their employees, and the discipline of offers from the broader market will not prevent them from doing so (Sanders & Nee, 1987). In this case, the availability of opportunities in the ethnic labor market should have no effect on the opportunities of fluent immigrants (who escape its clutches), and should provide no benefit (indeed, perhaps even harm) to linguistically isolated employees. In this view, linguistic isolation stunts the growth of social networks, thereby reducing the number and diversity of "weak ties" immigrants have. This, according to Granovetter's hypothesis emphasizing the role of social networks in the job search process (1973), should seriously impair the occupational opportunities of linguistically isolated immigrants. This hypothesis predicts a negative or zero main effect of the availability of co-ethnic employment, a substantial positive effect of language skills, and a large negative interaction effect such that linguistically isolated workers are much worse off in groups with many entrepreneurs because the ethnic labor markets in such groups are more nearly self-contained. These predictions are presented graphically in Fig. 3.

PRIOR RESEARCH

Prior research on this topic consists mainly of case studies of particular ethnic groups, rather than comparisons among groups, but within-group contrasts between enclave and mainstream workers are also relevant to the theory. Some are largely consistent with my theory (Evans, 1987; Lewin-Epstein & Semyonov, 1994; Portes & Jensen, 1989), others offer ambiguous results (Nee et al., 1994; Zhou & Logan, 1989). It seems fair to say that results to date have been inconclusive about the impact of co-ethnic employment on occupational success. I should like to propose the tentative conclusion that this array of conflicting results comes about because co-ethnic employment affects the opportunities of differently endowed immigrants differently, enhancing the opportunities of linguistically isolated immigrants but offering no special advantages to immigrants fluent in the dominant language.

Proxying enclave employment by residential location, Edin et al. (2003) find substantial gains to earnings of the least skilled immigrants for living in ethnic enclaves. The analysis adjusts for selectivity in residential location. This is somewhat indirect evidence in terms of my hypothesis, because language skill is not directly measured, but the pattern of results is consistent with the communications-costs hypothesis.

Studying the situation of Arabs in Israel and defining employment sector by workplace location, Lewin-Epstein and Semyonov (1994) find that knowledge of

Hebrew has a nil effect (or a negative effect, depending on the model) on the occupational status of Arab men in the Arab sector, no effect on the occupational status of men in the mainstream economy, and, for Arab women, a negative (or null, depending on the model) effect on occupational status in the Arab sector and a negative effect in the mainstream economy. Controlling for occupational status, they find a residual positive effect of knowledge of Hebrew on annual earnings in the mainstream economy for both men and women, and, for, men, a smaller (about one half the size) effect in the ethnic niche (with a zero effect for women within the ethnic niche). Although one cannot be sure from the analysis as presented, their findings on occupational status (at least for men) are broadly consistent with the notion that those who do not speak Hebrew are better off in the enclave, and those who do speak Hebrew are better off in the mainstream economy.

Nee et al. (1994) study assesses the situation of Asian immigrants in Los Angeles. Their findings unfortunately are sensitive to specification, but at least in some models suggest that immigrants with weak English language skills are more likely than their fluent peers to have a co-ethnic boss. They also found an insignificant effect of having a co-ethnic boss on hourly earnings, but they only tested this for the immigrant population as a whole, rather than interacting it with respondent's language skill.

Studying the situation of Cubans in Miami, Portes and Jensen (1989, Tables 6 and 7) find strong effects of English language skills on earnings when occupational status is not controlled for both enclave (residential definition) and non-enclave workers, but no effect on earnings when occupational status is controlled. That is consistent with the view that language skills are most important at the hiring stage, that they affect the quality of the jobs that entrepreneurs offer to aspiring employees.

Zhou and Logan (1989) study Chinese immigrants in the New York area, and proxy ethnic niche employment in several different ways (one at a time): by residence in New York City vs. elsewhere, by place of work in New York City vs. elsewhere, and by an industrial sector definition. Their analyses reveal unstable and mixed effects of English language skills on earnings, net of occupation. This result is consistent with the view that English language skills are important in the calibre of jobs that immigrants get at the hiring stage, but that subsequent within-organization careers flow on in well-worn pathways from the hiring stage with English language skills have no independent continuing impact. Unfortunately, this interpretation must remain tentative as the authors do not show the results predicting occupation, so there is no way to tell from their results whether English language skill affects job quality.

Studying Australia, Evans (1987) distinguished five broad groups of immigrants according to the size of their ethnic business communities and found that the effect of language skills on occupational status was smaller in the groups where business ownership was common. That evidence is consistent with the communications-costs theory, but does not amount to a systematic test, because the possibility of spurious correlation cannot be dismissed with just five groups.

In all this prior research, differences among immigrant groups have been discussed discursively, but not systematically measured and incorporated into the analyses. This paper seeks to go a step further, by beginning to measure the availability of co-ethnic employment across the full range of immigrant groups in a society, and investigating whether the occupational opportunities of immigrants are affected by the availability of co-ethnic employment.

Correctly to test the hypothesis, one needs to address the determinants of immigrants' socioeconomic attainments using a suitable array of causal variables including both the focal variables of the hypothesis and an exhaustive set of controls.

I focus on occupational status as the dependent variable for two related reasons. First, prior theory and hypotheses point to hiring decisions and hence job quality as the key to the detrimental effect of language fluency on the socioeconomic career – it is thought that mainstream employers' offers of worse jobs to non-fluent employees that harms their prospects and income, rather than that employers pay non-fluent employees less than their fluent peers in the same jobs. Second, existing research (as discussed above substantively in the section on prior research) mostly: (1) finds a detrimental effect of non-fluency when occupational status is the dependent variable; (2) finds a detrimental effect on income when occupational status is omitted from the equation; and (3) finds no effect on income when occupational status is in the equation. These findings strongly suggest that occupation should form the focus of an inquiry into how language skill affects status attainment.

The requisite causal variables of primary interest are an individual-level measure of English language skill and a measure of the prevalence of business ownership in the group, and their interaction. I use the prevalence of business ownership as my indicator of the availability of co-ethnic employment, because the theories I address emphasize the behavior of owners. Alternative possibilities are addressed in the Data, Methods, and Measurement section, below.

The control variables are particularly important in papers using contextual variables, because a reasonably exhaustive array of them substantially reduces the risk of spurious correlations linking the focal variables to occupational status. To control these effects, I follow Evans and Kelley's (1991) model that measures education in years completed and differentiates local from foreign education (and allows curvilinearities in both), includes an indicator for possession of a recognized trade qualification, measures years of local labor force experience and years of foreign labor force experience separately (and includes curvilinearities in both), and measures of rural or urban residence and citizenship.

SETTING

Australia is a good site for testing this hypothesis because her immigrants are diverse and recent. First, some immigrants to Australia closely resemble the dominant Anglo-Celtic majority, but many others differ greatly, and in a variety of ways. Importantly for testing the communications-costs hypothesis, there are immigrants from a variety of non-English speaking countries, which greatly reduces the risk of conflating particular cultural characteristics with English language competence. Third, this ethnic diversity is recent and it remains a politically charged issue (Kelley, 1996). As late as 1947, barely 2% of the population were born outside Australia and the British Isles; the main waves of non-Anglo-Celtic migration began only after World War II and are still coming (Price, 1986). Australians' attitudes towards immigrants are relatively middle-of-the-road in international terms (Evans & Kelley, 1998). In terms of disadvantage, it may be noteworthy that a great deal of qualitative research indicates substantial misunderstandings and conflicts between immigrant workers and labor unions to which they belong (e.g. Griffin & Testi, 1997).

The availability of co-ethnic employment, as indicated by the prevalence of entrepreneurs within one's ethnic group (percentage self-employed with employees), varies widely among immigrant groups. Naturally not all immigrant employers hire co-ethnic employees, but the point for the purposes of this article is that they *could* hire such employees if they found it in their interest to do so. Entrepreneurship is very high among the Chinese (17%), the Greeks (10%), the Italians (10%), and the Lebanese (9%). It is very low among the Maltese (3%), the Sri Lankans (3%), and the then-newly-arrived Vietnamese (0%). For comparison, the figure among native-born Australians is 6%.

Appendix A gives more details.

DATA

Data are from the Public Use Sample of individual records from the 1981 Australian Census (Australian Bureau of Statistics, 1983). The analysis is restricted to employees (working men and women who are neither solo self-employed nor business owners employing others).

This dataset is also excellent for my purposes because it includes an unusually rich array of measures of human capital characteristics, including items that allow one to distinguish foreign education from local education, and to distinguish foreign labour force experience from local experience.[10] Good measurement of education is crucial in this analysis, because prior research shows that immigrants' educational attainment is strongly correlated with their skills in the dominant language in many countries (e.g. Chiswick & Miller, 1992; Jasso & Rosenzweig, 1990), so inadequate measurement of education would risk attributing to language skill effects that properly belong to education.

Having a good set of control variables is important here because it helps to ensure that observed effects of group characteristics are not artefacts of inadequate individual-level measurement. The measurement of the control variables is described in Appendix B.

As a final data-and-sample-definition issue, let me note that I have chosen to include both women and men in the analysis. Statistical purity would, by contrast, incline one to omit women because adequate measurement of their labour force experience is not available in these data, and it is known that women's labor force participation patterns and their determinants differ among immigrant groups in Australia (Evans & Lukic, 1998; Kim, 1998). Nonetheless, labor force experience is less important to occupational status than to other aspects of status attainment such as supervisory responsibility, ownership, or income. Moreover, immigrant men's employment experiences have been studied more widely than immigrant women's, so we have more to learn about immigrant women. As a result, I judge it worthwhile including women.

MEASUREMENT

Ethnic Groups

The Public Use Sample of the 1981 Australian Census distinguishes nearly 100 birthplace countries. This level of detail is vital to this paper because it enables me to compute group characteristics from countries that share a language (so that, for example, Portugal can be grouped with Brazil, rather than Portugal being pre-grouped into Southern Europe and Brazil pre-grouped into South America as is the case in many other datasets, including Public Use Samples of later Australian Censuses). This is necessary to test the "communications costs" hypothesis in which shared native language lowers ethnic entrepreneurs' costs of employing co-ethnic employees who are not fluent in the dominant language [1]. Note that deriving the group-level characteristics from language groups does not guarantee any particular outcome: Prior research has documented instances of highly differentiated ethnic and racial identity within some linguistic groups (e.g. Denton & Massey, 1989), and if these identities are common and take precedence over instrumental concerns with ease of communication in the workplace then the effect of the prevalence of entrepreneurs in one's language group on one's occupational status should be nil.

I focus on immigrants who were born in countries where English is not the main language (Appendix A gives a list of the constituent countries), since the communications costs hypothesis applies only to workers at risk of linguistic isolation.

The Availability of Co-Ethnic Employment

Good estimates of the extent of the linguistic sub-economy are vital to this paper, and fortunately Census data again provide them. I do not have data on whether or not the individual respondents whose data I mainly analyze themselves work in a business where their native tongue is spoken. Instead, I proxy the potential availability of employment within each language group – by the contextual characteristic "percentage of employers" (who own a business and employ others) in the language group's labor force. More specifically, for each language group, I computed the percentage of employers (who own a business and employ others) in the group's labour force. Next, I attached this information to the individual records, so that each respondent's data now includes a new variable measuring the prevalence of business ownership in his or her language group. Note that even middling levels of entrepreneurship may be more consequential than they at first appear if they involve a strong tendency towards co-ethnic hiring: If 11% of the ethnic labour force are entrepreneurs who employ, on average, 1 co-ethnic each, then 22% of the groups' labour force will be working inside the ethnic economy; if they employ an average of 2 co-ethnics each, then 33% of the group's labour force will be working inside the ethnic economy, and so on.

Note also that the model does not *assume* that immigrant entrepreneurs are likely to hire workers from their language group. If they do not hire such workers, then the regression coefficients for the percentage of employers in the language group and for the interaction of that with language fluency will not be statistically significant. If they hire such workers and co-ethnic predation occurs, then the effect of the percentage of employers and its interaction with individual language skill will show that non-fluent workers are worse off in groups with many entrepreneurs – the effects will be jointly significant statistically and the interaction will be significant (but the main effect of percentage entrepreneurs could be either significant or not). If they hire such workers and such workers come to them because of discrimination in the mainstream labor market, then there will be a strong positive effect of the percentage of employers in the group, but no interaction effect of this with individual language skill because according to the discrimination hypothesis enclave employment benefits all immigrants alike by shielding them from discrimination. If employers hire co-ethnic workers and the communications-costs hypothesis holds, as I have argued, then the regression coefficients for the main effect and for its interaction with language skill will show that non-fluent workers are better off when co-ethnic employment is widely available – the effects will be jointly significant statistically and the interaction effect will be significant (the main effects for the percentage of employers could be either significant or not).

Use of an indirect, proxy measure provides a conservative test of my hypothesis. Prior research shows that not all immigrant entrepreneurs are enthusiastic about hiring co-ethnic workers, and so my "availability" estimates are over-estimates to some unknown degree that may vary in an unmeasured way among language groups. This probably mainly introduces random noise into the observed relationships involving this variable, thereby most likely biasing the results against finding significant communications-costs effects (since the random noise will exaggerate the standard errors).

Alternative Measurement Possibilities: Residential Concentration. An alternative proxy measure for the availability of co-ethnic employment would be residential concentration, but I decided against this for three reasons: (1) It is an even more indirect measure than the percentage of entrepreneurs, which increases the likelihood of failing to find an effect that is really present; (2) No available dataset contains both the necessary geographic detail for an individual-level measure of residence in an area of ethnic concentration and the necessary level of detail on country of origin, migration timing, education and occupation; and (3) Empirically, the areas of greatest ethnic concentration in Australia in this period were those where very recently arrived immigrants in need of special assistance, especially refugees, were temporarily settled and they were areas of rather low employment (co-ethnic or otherwise) because, in this period, government benefits encouraged language and job training for these immigrants, rather than immediate employment. Most immigrants moved on from these areas quite quickly (Bureau of Immigration, Multiculturalism, and Population Research, 1996). Clearly, these areas were not what we normally mean by ethnic or linguistic niches or enclaves, because of the importance of government benefits compared to the labor market.

Alternative Measurement Possibilities: Industrial Concentration. A second alternative possibility would be to proxy linguistic niche employment by employment in an industry that also employs many other immigrants who share a mother tongue. I have decided against this strategy mainly because it involves even more indirect measurement than the prevalence of business ownership on which I rely. It would also be an uncomfortable choice in terms of the social history of migration and settlement, because many of the well-known examples of industrial concentration in Australia involved the migrants joining an existing industrial working class as employees of native capitalists – the steel industry of the Sydney region and automobile manufacturing and cannery work in Victoria being examples that spring to mind. In short, some of the best known examples of industrial concentration have involved very little immigrant business ownership, and hence do not correspond to linguistic enclave economies in the usual sense (although they are very interesting in many other respects).

Fluency in the Dominant Language

English language skills vary widely. They are measured by a self-report item asking respondents first if they speak a language other than English at home, respondents who answered "yes" were then asked how well they speak English: "Not at all," "Not well," "Well," "Very well." The Census was generally administered as a postal paper-and-pencil questionnaire, and was available in 35 languages. Staff with specialized language skills were available for assistance. I have coded respondents who reported that they speak only English at home as speaking English "Very well."

Prior research using this variable has found that one can effectively code the categories as equi-distant points on a single continuous measure without loss of information (Evans, 1987; Miller, 1989). Moreover, calibration of this crude Census question against diverse finely differentiated measures in detailed survey work suggests that this measure, although subject to more random measurement error than a multiple-item index, reproduces quite closely the correlations that the better measures have with criterion variables (Chiswick & Miller, 1998).

MODEL

I predict individuals' occupational status, incorporating the variables described above into Evans and Kelley's model of immigrants' status attainment (1991), thus capturing other potentially important individual-level human capital characteristics. (Appendix B provides definitions of all these variables. Descriptive statistics and a correlation matrix are available for further analysis from www.international-survey.org.) Incorporating such a detailed model into this analysis is important because it provides a tough test of the existence of the contextual effect by reducing the chances that the contextual effect might merely reflect unmeasured individual characteristics. Using a model that fairly exhaustively accounts for individual-level influences thus allows one to be more confident about findings of contextual or group-level effects.

Because there are fewer "cases" of language groups than there are of immigrants, I use a multi-level model to ensure that the estimates of the standard errors on the contextual effect and its interaction are not "over optimistic" as they can be in ordinary estimates of such effects from OLS (e.g. DiPrete & Forristal, 1994). In the event, it makes little difference here, as the OLS results are close to the technically more correct multi-level results (OLS results available from the author upon request). Because of the independent invention of multi-level models in many different fields and subfields, nomenclature is non-standard, but the type of model is sometime called a "variance components model.".

The model is fairly long, so it is more readily grasped if we build it up in chunks of closely related components.

Focal Effects

To assess the communications-costs hypothesis, one must assess the main effects of the availability of co-ethnic employment and of individual-level language skill, and their interaction: The potential availability of employment within the language group (GroupEmployers %) is measured, as described above, by the prevalence of entrepreneurship in the group.[11] English language skill (EngSkil) is measured as described above.

The final substantively crucial measure for this analysis is a multiplicative interaction term which allows the effect of respondent's individual English language skill to vary depending on the nature of the group's enclave economy. The extent of the enclave economy (Enclave %) is measured, as described above, by the prevalence of entrepreneurship in the group; English language skill (EngSkil) is measured as described above; and the interaction term is simply the product of the two:

$$\text{Interaction} = (\text{EngSkil})\,(\text{GroupEmployers}\,\%)$$

This focal part of the model is designed to assess the claim that an immigrant's occupational status is a function of their own skill in English, the size of their group's subeconomy, and, especially, the interaction of the two:

$$\begin{aligned}\text{OccupationalStatus} = {} & b_0 + b_1\text{GroupEmployers}\,\% + b_2\text{EngSkil} \\ & + b_3(\text{EngSkil})\,(\text{GroupEmployers}\,\%) + \cdots \end{aligned} \quad (1)$$

Note that this is not meant to be the complete model, and that the full model is estimated simultaneously, not piece by piece. Describing it piece by piece is merely an expository device for focusing on one section at a time so that each part of the model and the text describing it are near one another.

Next, I will describe, block by block, the array of control variables in the model. Note that b_0 in the multilevel model is not a constant; it is discussed further at the end of the modelling section.

Education: Foreign and Domestic

Education is measured in years, estimated from a detailed series of questions (see Appendix B). In Australia, as elsewhere, education has a strong effect on occupational status, with the later years of education – toward the end of secondary school and into university – mattering more than earlier years (e.g. Evans & Kelley, 1991).

Many immigrants migrated as children and completed their education in Australia. But most obtained their schooling in their home countries and came to Australia afterwards. The education they obtained overseas has a lower payoff than Australian education, mainly because of the lower quality of schooling in the home countries that provided most of Australia's non-English speaking immigrants in this period, the lack of country-specific knowledge (compared to their peers who were educated in Australian schools), and lack of contacts and other "social capital" relevant to the Australian labor market (Evans & Kelley, 1991).

The result is a pattern similar to the hypothetical, stylized Fig. 1. (1) Education obtained in the host nation increases occupational status (line AA′). However, the effect is not linear but increases at higher levels of education (on the right). To capture this curvilinearity, the model includes a quadratic term (EducationSquared) in addition to the usual linear term (Education). (2) Education obtained overseas prior to immigration shows a similar pattern but the occupational returns start lower (point B in Fig. 1). This is captured in the model by a binary variable indicating where the education was obtained (EducatedInAustralia). (3) The returns to overseas education may differ both in slope and curviture (BB′ in the figure). These differences are captured in the model by allowing different coefficients for both the linear and the quadratic term, obtained by including two interactions: (Education) times (EducatedInAustralia) and (EducationSquared) times (EducatedInAustralia).

In addition to academic education, we allow for the effects of vocational training leading to a recognized manual qualification – these are held mostly in the skilled crafts and some semi-skilled occupations (Australian Mission, 1969). Australia

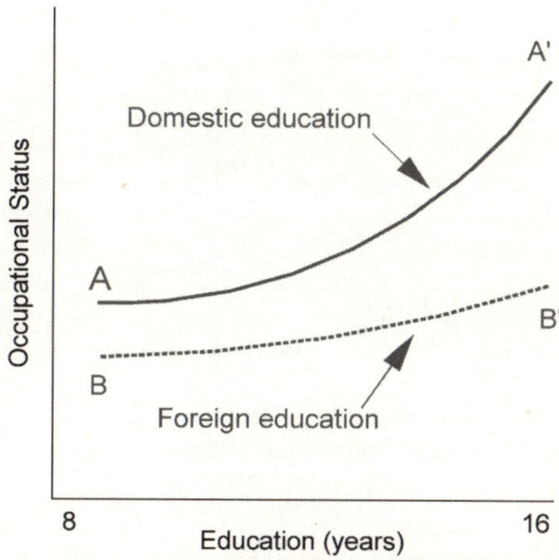

Fig. 1. Model: Education Component.

has elaborate systems of agreement for mutual recognition of qualifications with many other countries. The educational part of the model is then:

$$\begin{aligned}&\text{Occupational status}\\&\quad = \cdots + b_4\text{Education} + b_5\text{EducationSquared}\\&\qquad + b_6\text{EducatedInAustralia} + b_7(\text{Education})\,(\text{EducatedInAustralia})\\&\qquad + b_8(\text{EducationSquared})\,(\text{EducatedInAustralia})\\&\qquad + b_9\text{HasAFormallyRecognizedManualQualification} + \cdots \qquad (2)\end{aligned}$$

Labor Force Experience: Foreign and Domestic

In Australia, as elsewhere, labor force experience can increase occupational status, typically more rapidly early in the career and less rapidly later (shown schematically in Fig. 2, line AA′). For immigrants who came as children and so have experience only in the Australian labor market, I model this by conventional linear (AustralianExperience) and quadratic (AustralianExperienceSquared) terms.

Most immigrants worked for a period in their home countries before migrating, and it is unlikely that this experience overseas would have the same impact on their Australian job as would experience in Australia (line BB′ in Fig. 2). The model caters for this possibility with separate terms for foreign labor force experience (ForeignExperience) and its square (ForeignExperienceSquared).

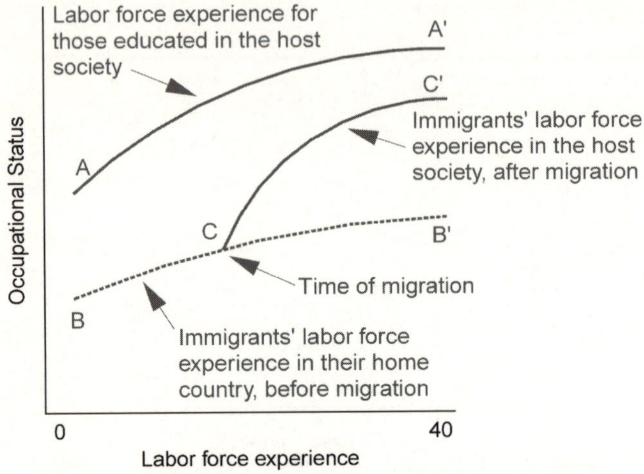

Fig. 2. Model: Labor Force Experience Components.

After coming to Australia, immigrants will start to accumulate experience in the Australian labor market (line CC′ in Fig. 2). However, there is no good reason to assume that the returns to experience for these mid-career immigrants would be identical to returns for childhood immigrants at an equally early stage in their Australian career. For example, for a 30-year-old immigrant the first 3 years in Australia might increase their skills much more rapidly than the first 3 years of labor market experience for an 18-year-old Australian-educated immigrant. To cater for such possibilities, the model allows separate slope and curvature for the Australian labor force experience of migrants who came to Australia as adults: (AustralianExperience) times (EducatedInAustralia) and (AustralianExperienceSquared) times (EducatedInAustralia).

Thus the labor force experience part of the model is:

$$\begin{aligned}\text{OccupationalStatus} &= \cdots + b_{10}\,\text{AustralianExperience} + b_{11}\text{AustralianExperienceSquared}\\ &\quad + b_{12}(\text{AustralianExperience})\,(\text{EducatedInAustralia})\\ &\quad + b_{13}(\text{AustralianExperienceSquared})\,(\text{EducatedInAustralia})\\ &\quad + b_{14}\text{ForeignExperience} + b_{15}\text{ForeignExperienceSquared} + \cdots \qquad (3)\end{aligned}$$

Control Variables

To cater for differences between migrants in residential location, family arrangements, and legal citizenship, I add a number of controls:

$$\text{OccupationalStatus} = \cdots + b_{16}\text{Married} + b_{17}\text{Citizen} + b_{18}\text{RuralResidence} + b_{19}\text{Female} + \text{Error} \quad (4)$$

Final Model

The final model combines all these elements:

$$\text{OccupationalStatus} = b_0 + b_1\text{EngSkil} + b_2\text{GroupEmployers}\,\% + b_3(\text{EngSkil})\,(\text{GroupEmployers}\,\%) + \text{Eq. (2)} + \text{Eq. (3)} + \text{Eq. (4)} \quad (5)$$

In this model b_0 is not a constant. Rather,

$$b_{0ij} = B_0 + u_{0j} + e_{0ij}$$

where j indexes the language groups and i indexes the individuals. The results are shown in Appendix C.

Extensive sensitivity tests show that the effects are robust, with multilevel results being very close to the OLS analogues, except for the standard errors on the focal variables which are, as they should be, rather larger in the multi-level estimation. I used the program ML-Win to estimate the multi-level model.

In the text, I mainly discuss predicted values, a kind of simple simulation of the average occupational status of people who differ in language skill and in the availability of employment in their language group, but who are identical in all other variables included in the model.[12] The result is a three-dimensional surface showing how occupational status varies with language skill and availability of ethnic employment, controlling for human capital characteristics.

RESULTS

Controls: Human Capital and Background

First, note that the model is well-behaved, robust over inclusion or omission of the key interaction term, and when estimated in OLS has an R-squared of about 0.46

which is towards the upper end of the usual range for this topic. Note that the models replicate a known feature of the connection between education and occupational status, namely a strong linear main effect that is kicked up at the higher end (captured by the quadratic term) reflecting the fact that the gains to each year of education are greater at higher levels of education. The results also replicate the standard result that foreign education pays off a bit worse in general than education in the host country, especially at the top. Also following well-documented patterns from other research, labor force experience in the new country yields much greater gains in job quality than does foreign labor force experience (Appendix C).

Thus the model captures important features of the process of immigrants' occupational attainment, which provides confidence that work-related compositional characteristics are as thoroughly controlled as possible.

Focal Results: Language and Employment Availability

Occupational status increases strongly with language skill: immigrant men and women who are fluent in English get better jobs than their linguistically isolated peers (Fig. 3, details in Appendix C).

Importantly, the model replicates the standard results for skills in the host country's language. If we re-estimate the model without the key interaction in order to compare with prior research, fully fluent immigrants get jobs about 13 points out of 100 (on average) better than do immigrants with no English, net of education and experience (results available from the authors upon request). By comparison, a year of education is worth roughly 4 status points. The size of the language skill effect is consistent with most earlier research.

Turning now to the key hypotheses, recall that the communications-costs hypothesis predicts a strong language fluency effect for immigrants whose language groups contain few entrepreneurs, a moderate-sized language effect for those in groups with a middle sized language enclave, and only a small effect for those from language groups abounding in entrepreneurs. By contrast, recall that the discrimination hypothesis predicts that immigrants at all levels of language fluency benefit when there is an ethnic niche economy, and the co-ethnic predation hypothesis predicts that the least fluent immigrants will get worse jobs if they belong to groups with many entrepreneurs than if their group offers few jobs and they must seek them in the mainstream economy. These quite different predictions about how the availability of co-ethnic employment and individual language skill jointly affect job quality are shown graphically in the upper panel of Fig. 3.

The results, net of many other potentially confounding forces, strongly confirm the communications-costs hypothesis (Fig. 3, bottom panel).

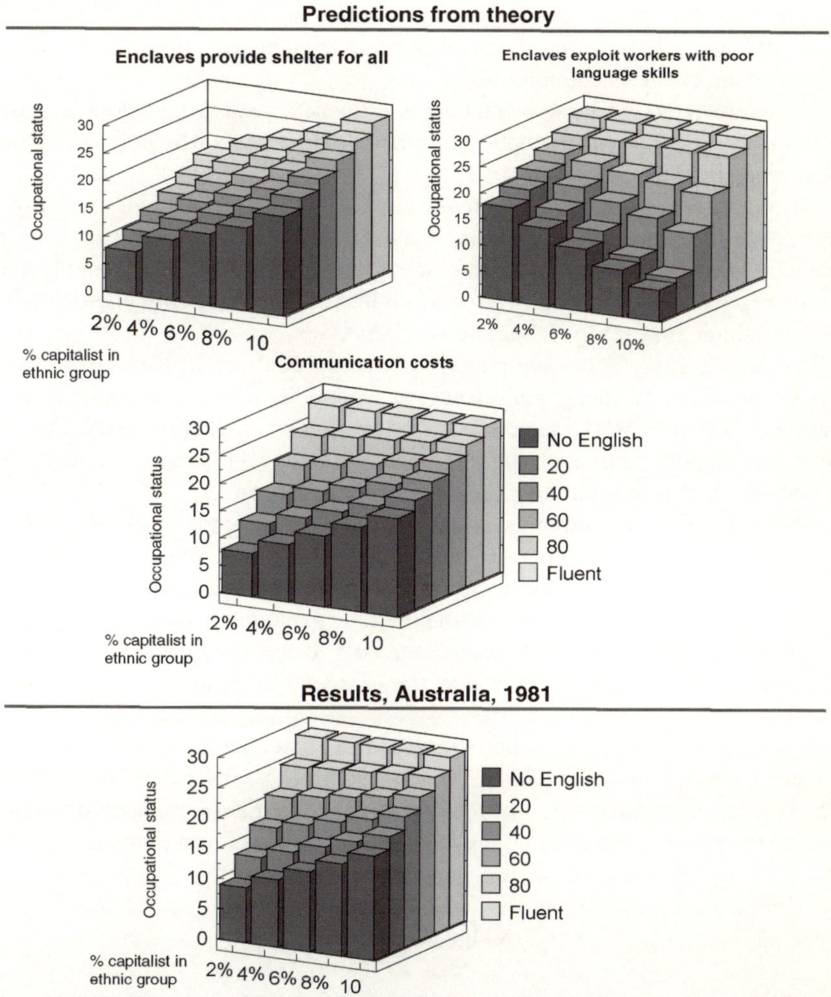

Fig. 3. Communications Costs, Discrimination, or Co-Ethnic Predation? Hypotheses and Multi-Level Model Results Giving the Effect of Availability of Employment in the Language Group on Occupational Status for Immigrants of Differing Degrees of Fluency in English.

As predicted by the communications-costs hypothesis, the "cost" of poor English skills varies by the availability of ethnic employment, being about 20 status points in groups with very low levels of entrepreneurship (Fig. 3, lower panel, back row of bars), but shrinking by half to 10 status points in groups with very high levels of entrepreneurship (Fig. 3, front row of bars). Thus, lack of English skill has a price in all immigrant groups, but the price is much smaller for workers in the groups with thriving business communities.

These predicted values show that the non-fluent workers who belong to groups with many entrepreneurs get better jobs than do their peers who belong to groups with little ethnic business.

Also conforming to the communications-costs hypothesis, the quality of employment of those with middling English skills is less dependent upon the availability of employment in their ethnic group, and the quality of employment for immigrants fluent in English is independent of the availability of employment in their ethnic group (Fig. 3, back row of bars).

The results (Fig. 3, bottom panel) are clearly contrary to the discrimination hypothesis which predicted gains from the availability of enclave employment for immigrants at all levels of language skill (Fig. 3, top panel, upper left). This does not mean that discrimination in the broader market never occurs, but only that it is not sufficiently common to have a major impact on careers.

The results (Fig. 3, bottom panel) are clearly contrary to the co-ethnic predation hypothesis which predicted especially low occupational status for workers in groups with many business owners (Fig. 3, top panel, upper right). Such a finding does not guarantee that no instances of co-ethnic exploitation occur, but in order to observe the results we do, such instances must be heavily outweighed by instances of co-ethnic bosses paying better than the broader market to immigrants with no English.

It should also be mentioned that there are further systematic group characteristics affecting immigrants' occupational status waiting to be discovered in these data. The fact that Ω_u remains significant says that there remain significant differences among the groups even after taking enclave opportunities into account.

DISCUSSION

My results show that the availability of employment in a linguistic enclave leads to substantially better jobs for linguistically isolated immigrant workers who do not speak the dominant language fluently, slightly benefits immigrant workers who are moderately skilled in the dominant language, and has no effect on the occupational opportunities of immigrant workers who are fully fluent in the dominant language. Workers not fluent in English get much worse jobs if they belong to a group that lacks an ethnic niche economy; the non-fluent pay a middle-sized penalty in groups with a middling level of entrepreneurship, and they are penalized only a little if they belong to a group with a flourishing linguistic sub-economy. These results are consistent with my communications-costs hypothesis.

The results are contrary to the co-ethnic predation hypothesis which predicts that impermeable boundaries between an immigrant groups and the host society allow enclave entrepreneurs to exploit linguistically isolated workers even more than majority entrepreneurs would do. If the co-ethnic predation hypothesis were correct, the results would show that linguistically isolated workers are worse off if they belong to a group with a high level of entrepreneurship (in contrast to the actual beneficial effect). The generally beneficial effect does not rule out spectacular, isolated instances of co-ethnic predation, instead the regression result simply means that there are more instances of benefit than harm. It seems reasonable to interpret this finding as reflecting the permeability of boundaries between the ethnic group and the host society.

The results are also contrary to the discrimination hypothesis predicting that the shelter from majority-group employers' discrimination provided by the linguistic niche economy would offer benefits to immigrant workers, fluent and non-fluent alike. If this hypothesis had turned out to be correct, the results would have shown a strong positive impact of availability of enclave employment on occupational status for all immigrants. But, instead, the results reveal a substantial interaction with linguistic isolation: Availability of co-ethnic employment benefits the linguistically isolated, but does not benefit immigrants fluent in the host society's language. This, of course, does not mean that discrimination never occurs, but rather means that it is not the main reason immigrants seek work in the linguistic enclave.

Thus, the evidence supports the communications-costs hypothesis. One interesting implication is that barriers to small business ownership will principally impair the occupational careers of the most vulnerable immigrants who do not speak the dominant language well. If barriers to small business ownership – such as elaborate licensing or accounting requirements – stunt the growth of ethnic sub-economies (Waldinger et al., 1990), then the immigrants who would otherwise work there must instead seek jobs in the broader labor market. My results suggest that this would have no effect on the occupational status of the immigrants fluent in the dominant language, would force moderately fluent immigrants into slightly worse jobs, and would substantially downgrade the occupational status of immigrants who cannot communicate in the dominant language. This might be mitigated for non-fluent upper middle class immigrants in societies which have substantial immigrant-oriented social programs (Lewin-Epstein & Semyonov, 1994), but it seems reasonable to expect that this would not override the main outcome: a larger effect of skills in the dominant language, specifically a downgrading of the labor market opportunities of the least skilled.

My results suggest that the existence of an ethnically fragmented economy does not necessarily pose a challenge to market theory and evolutionary theory: It is rational for immigrants to set up shop when they have a market with specialized tastes and preferences to serve and a linguistically isolated labor force to draw upon (Evans, 1989). The results presented in this paper suggest that working in a linguistic sub-economy, at least for a short while upon arrival in a new country, can be rational for immigrants who arrive without skills in the dominant language.

These results contribute to a growing body of evidence suggesting that the existence of alternatives and the permeability of boundaries may be crucial in preventing career damage from ethnic discrimination. Thus, for example, alternatives were not available in one of the clearest instances of ethnically-based wage discrimination in the advanced societies – against Mediterranean immigrants in the early days of the "guest worker" program in Germany – in this case, immigrants were effectively prohibited from starting businesses. In such instances, workers cannot "vote with their feet" (or at least cannot do so and stay in their new society), and so such societies can come to correspond more closely to a split labor market. By contrast, most other advanced societies (and Germany after its early experiment) have allowed immigrants to start businesses, thereby – by accident or design – undermining discrimination by providing alternatives.

This reinforces the results from a variety of other research emphasizing the importance of boundary-spanning social networks and of permeable boundaries in developing immigrants socioeconomic opportunities (e.g. Bailey & Waldinger, 1991; Nee et al., 1994; Werbner, 2001). It may be useful to blend these concepts in a notion of "accessibility of alternatives." And it will be interesting to see whether it also holds in other countries that partial social closure (or partial separation) is beneficial to immigrants with little skill in the host language.

In terms of policy, my result that linguistically isolated workers climb higher on the occupational ladder if they belong to an immigrant group with a thriving small business community might seem to suggest shifting intake criteria. In particular it might seem to suggest that would-be immigrants who cannot speak the usual language of the society should be excluded unless their group has a strong niche economy. But reflection will show that this would be ill-advised for two reasons: (1) At least in the contemporary West, policies explicitly differentiated on the basis of ascriptive characteristics (such as ethnicity) tend to be illegitimate in the eyes of the public; (2) Prior research shows that ethnic community size is an important source of immigrant entrepreneurship in linguistically distinct communities (Evans, 1989), which would seem to suggest that among immigrants who do not speak the dominant language, those from big ethnic groups stand the best chance of occupational success, and so should get higher priority in the entrance queue. But it is worth noting that being big has disadvantages as well as advantages. In particular, if minority ethnic groups get large enough to elicit competitive feelings from the majority group, prejudice tends to be exacerbated (Blalock, 1967; Kunovich & Hodson, 2002), ethnic conflict to erupt (Olzak, 1989, 1992; but see Belanger & Pinard, 1991) and, possibly, wage discrimination in the mainstream economy to increase (Catanzarite, 2002; Frisbie & Neidert, 1977; Stolzenberg & D'Amico, 1977; Tienda & Lii, 1987). And through the mechanisms of residential segregation and assortative mating, large groups are more likely to be self-perpetuating (Fong & Ooka, 2002; Jasso & Rosenzweig, 1990; Stevens, 1992). Further, at high levels of residential concentration, entrepreneurship opportunities may actually decline (Fischer & Massey, 2000).

The model that I have used in this paper controls very thoroughly for individual-human capital characteristics, and finds important effects of one group characteristic – the availability of co-ethnic employment – on occupational attainment among non-fluent immigrants. But, it is also noteworthy that the results also show that there are further systematic differences among immigrant groups that need to be discovered in future research. A promising approach might be to elaborate this model by devising for all immigrant groups measures of additional concepts that case studies have highlighted. For example, vertical integration has been emphasized in the success of the Cuban enclave in Miami (Portes, 1989; Wilson & Martin, 1982; Wilson & Portes, 1980), but no-one has yet attempted to document the extent of vertical integration of firms for the whole range of immigrant groups, let alone to explore its impact on capitalists' and workers' jobs, security, and incomes. It would also seem reasonable to explore the consequences of certain aspects of the migration pattern, e.g. its continuity, sex composition, etc., may affect small business opportunities. And features of the local labour market, too, may affect both general opportunities for small business and particular opportunities for immigrant entrepreneurs (Light & Rosenstein, 1995), with a kind of (possibly complicated) path-dependence shaping opportunities over time (Gulati & Gargiulo, 1999; Logan et al., 2000; Model, 1997; Raijman & Tienda, 2000). These possibilities seem likely to expand the array of influences on occupational attainment, but none of them runs against the communications-costs hypothesis and hence none seems likely to alter the interaction between availability of employment in the linguistic enclave and individual language skill found here.

NOTES

1. Comparative studies of the conditions of work prevailing in mainstream and ethnic firms are rare, but one study in Australia comparing industrial relations practices finds that there was a good deal of variation within the mainstream and ethnic groups but no systematic differences between them (Callus & Knox, 1993).

2. This is not to say that individual human capital characteristics are irrelevant to ethnic entrepreneurship, indeed, research shows favourable human capital characteristics raise the chances of business ownership among immigrants (Evans, 1989; Li, 2001). The argument is that aside from individual characteristics, contextual factors exert a separate influence.

3. These could also include facilitate transnational entrepreneurship (Fernandez & Kim, 1998; Portes et al., 2002) through shared language and existing social networks making communication and search costs lower than for potential competitors in the mainstream economy.

4. Some indirect supporting evidence can be drawn from the finding that there is some ethnic clustering of residence, for at least some ethnic groups even after their employment has moved on from the enclave into the mainstream economy, and in cases where housing discrimination does not appear to be the cause (Logan et al., 2002). This suggests that at least some immigrants positively value interacting with co-ethnics.

5. In this paper, I address only structural factors. An entire research tradition not addressed here concerns the effects of enduring cultural traits and of particularism in the form of ethnic solidarity (e.g. Min & Bozorgmehr, 2000; Kim, 1999). I believe that including these features in a fully elaborated model would add to the explained variance, but not substantially alter the parameters estimated in this paper, because it seems reasonable to assume that those cultural characteristics are uncorrelated with the structural characteristics of interest here. Moreover, in the assessment of ethnic entrepreneurship, to date structural factors have somewhat been neglected in favor of cultural features (Rath & Kloosterman, 2000).

6. Sanders et al. (2002) point out that co-ethnic employment frees linguistically isolated immigrants from the dependence on social networks that is often necessary for a non-fluent worker to acquire a job in the mainstream labor market.

7. This is a net effect, apart from individual labour market characteristics; those matter too, separately (Evans, 1989; Li, 2001).

8. It is possible that this path dependence also leads entrepreneurially successful groups to persist as employers even when their own group's supply of working class labour has dried up or moved up the occupational ladder. That issue is beyond the scope of the paper, but some of these across-ethnic group "immigrant" enclaves have very interesting dynamics (e.g. Light et al., 1999).

9. Although their substantive import is vastly different, for the analyses undertaken for this paper, the discrimination hypothesis yields predictions identical to those provided by the "synergy" hypothesis that an abundance of entrepreneurs encourages the kind of successful vertical integration that distinguishes the Cuban enclave in Miami, and thereby provides benefits to all the immigrants, regardless of their skills in the dominant language working in the ethnic enclave. But I have labelled this hypothesis the "discrimination" hypothesis because the actual connection of the prevalence of entrepreneurs to vertical integration is only a matter of conjecture at this point. Hence an empirical result contrary to the prediction has a clear an interpretable meaning for the discrimination hypothesis, but for the synergy hypothesis a result contrary to prediction might mean that the synergy hypothesis of benefits for all is not true, but it is just as likely to mean that prevalence of entrepreneurs is not an important source of synergy. In the face of this ambiguity, it seems more prudent to focus on the discrimination hypothesis, reserving the question of synergy for data with more direct measures.

10. This cannot be done so accurately in subsequent Australian Censuses, because, although the requisite data is collected at the necessary level of detail, the Australian Bureau of Statistics groups it into coarse categories in the publicly released data files.

11. There are a number of multi-lingual countries, for which the coding is necessarily somewhat arbitrary. In these cases, I have coded the country to the language spoken by most immigrants from that country (for example, Switzerland is coded to the German group, because most Swiss immigrants in Australia are German-speaking). The expert advice of Charles Price and James Jupp was very helpful in developing this coding, and I have also relied upon *The Australian People: An Encyclopedia of the Nation, Its People, and Their Origins* (Jupp, 1988).

12. To obtain predicted values, I first set all the control variables to their means, set the variables of interest to a beginning point (say, individual language skill = 0 and % capitalist in language group = 2), multiplied by their regression coefficients, and added up. I then repeated that task for each combination of values of the variables of interest.

REFERENCES

Australian Bureau of Statistics (1983a, b, c). Making Sense of Census '81 (Order No. 2140.0); Census '81: Education Qualifications (No. 2149.0); Census '81: Occupation (No. 2148.0). Canberra: Australian Government Publishing Service. Available from: Information Services, ABS, PO Box 10, Belconnen, ACT 2616, Australia.

Australian Mission to Study Methods of Training Skilled Workers in Europe (1969). *The training of skilled workers in Europe: Report of the Australian tripartite commission*. Canberra: Australian Government Publishing Service.

Bailey, T., & Waldinger, R. (1991). Primary, secondary, and enclave labor markets. *American Sociological Review*, *56*, 432–445.

Baker, M., & Wooden, M. (1992). *Immigrant workers in the communication industry*. Canberra: Australian Government Publishing Service.

Becker, G. S. (1971 [1957]). *The economics of discrimination* (2nd ed.). Chicago: University of Chicago Press.

Belanger, S., & Pinard, M. (1991). Ethnic movements and the competition model. *American Sociological Review*, *56*, 446–457.

Blalock, H. S. (1967). *Toward a theory of minority group relations*. New York: Wiley.

Blau, P. (1956). *Bureaucracy in modern society*. New York: Random House.

Bonacich, E. (1979). The past, present, and future of split labor market theory. In: C. B. Marrett & C. Leggon (Eds), *Research in Race and Ethnic Relations*. Greenwich, CT: JAI Press.

Bowles, S., & Gintis, H. (1975). *Schooling in capitalist America*. New York: Basic Books.

Burawoy, M. (1985). *The politics of production*. London: New Left Books.

Bureau of Immigration, Multiculturalism and Population Research (1996). *Migrants and religion*. Canberra: Australian Government Publishing Service.

Callus, R., & Knox, M. (1993). *The industrial relations of immigrant employment*. Canberra: Australian Government Printing Service.

Carliner, G. (1981). Wage differences by language group and the market for language skills in Canada. *Journal of Human Resources*, *16*, 384–399.

Catanzarite, L. (2002). Dynamics of segregation and earnings in brown-collar occupations. *Work and Occupations*, *29*(3), 300–345.

Chiswick, B. R., & Miller, P. W. (1992). Language in the labor market. In: B. R. Chiskwick & M. Koster (Eds), *Immigration, Language, and Ethnic Issues*. Washington, DC: American Enterprise Institute.

Chiswick, B. R., & Miller, P. W. (1998). Language acquisition. *International Migration Review*, *32*(4), 877–900.

Chiswick, B. R., & Miller, P. W. (1999). Language skills and earnings among legalized aliens. *Journal of Population Economics*, *12*(1), 63–89.

Chiswick, B. R., & Miller, P. W. (2001). A model of destination-language acquisition: Application to male immigrants in Canada. *Demography*, *38*(3), 391–409.

Chiswick, B. R., & Miller, P. W. (2002). Immigrant earnings: Language skills, linguistic concentrations and the business cycle. *Journal of Population Economics*, *15*(1), 31–57.

Denton, N. A., & Massey, D. S. (1989). Racial identity among Carribean Hispanics. *American Sociological Review*, *54*, 790–808.

DiPrete, T. A., & Forristal, J. D. (1994). Multi-level models: Methods and substance. *Annual Review of Sociology*, *20*, 331–357.

Dyer, L. M., & Ross, C. A. (2000). Ethnic enterprises and their clientele. *Journal of Small Business Management*, *38*(2), 48–66.

Edin, P.-A., Fredriksson, P., & Aslund, O. (2003). Ethnic enclaves and the economic success of immigrants – Evidence from a natural experiment. *Quarterly Journal of Economics* (forthcoming).

Edwards, R. C. (1975). The social relations of production in the firm and labor market structure. In: R. C. Edwards, M. Reich & D. M. Gordon (Eds), *Labor Market Segmentation* (pp. 3–26). Lexington, MA: D.C. Heath.

Eisenstadt, S. N. (1964). Social change, differentiation, and evolution. *American Sociological Review*, *29*, 375–386.

Evans, M. D. R. (1987). Language skill, language usage, and opportunity: Immigrants and enclave economies in Australia. *Sociology*, *21*, 253–274.

Evans, M. D. R. (1989). Immigrant entrepreneurship: Effects of ethnic market size and isolated labor pool. *American Sociological Review*, *54*, 950–962.

Evans, M. D. R., & Kelley, J. (1991). Prejudice and discrimination in the labor market? Socioeconomic attainments of immigrants in Australia. *American Journal of Sociology*, *97*(3), 721–759.

Evans, M. D. R., & Kelley, J. (1998). Australian attitudes to immigrants: A 24-nation comparison. *Australian Social Monitor*, *1*(1), 11–14.

Evans, M. D. R., & Lukic, T. (1998). Immigrant women's labor force participation in Australia: The impact of resources and family-level cultural practices on immigrant women's workforce participation. *Gender Issues*, *16*(4), 52–83.

Fernandez, M., & Kim, K. C. (1998). Self-employment rates of Asian immigrant groups: An analysis of intragroup and intergroup differences. *International Migration Review*, *32*(3), 654–681.

Fischer, M. J., & Massey, D. S. (2000). Residential segregation and ethnic enterprise in U.S. metropolitan areas. *Social Problems*, *47*(3), 408–424.

Fong, E., & Ooka, E. (2002). The social consequences of participating in the ethnic economy. *International Migration Review*, *36*(1), 125–146.

Frisbie, W. P., & Neidert, L. (1977). Inequality and the relative size of minority populations. *American Journal of Sociology*, *82*, 1007–1030.

Goldscheider, C., & Kobrin, F. E. (1980). Ethnic continuity and the process of self-employment. *Ethnicity*, *7*, 256–278.

Gordon, M. (1964). *Assimilation in American life*. NY: Oxford University Press.

Granovetter, M. (1973). The strength of weak ties. *American Journal of Sociology*, *78*, 1360–1380.

Griffin, G., & Testi, J. (1997). Immigrant workers and enterprise bargaining in Australia. *Journal of Intercultural Studies*, *18*(2), 109–125.

Gulati, R., & Gargiulo, M. (1999). Where do interorganizational networks come from? *American Journal of Sociology*, *104*(5), 1439–1493.

Hechter, M. (1978). Group formation and the cultural division of labor. *American Journal of Sociology*, *84*, 293–318.

Hechter, M. (1986). Rational choice theory and the study of race and ethnic relations. In: J. Rex & D. Mason (Eds), *Theories of Race and Ethnic Relations* (pp. 264–280). New York, NY: Cambridge University Press.

Jasso, G., & Rosenzweig, M. (1990). *The new chosen people: Immigrants in the United States*. New York: Russell Sage.

Jones, F. L., & McDonnell, P. (1977). Measurement of occupational status in comparative analysis. *Sociological Methods and Research*, *5*, 437–459.

Jupp, J. (Ed.) (1988). *The Australian people: An encyclopedia of the nation, its people, and their origins*. Sydney: Angus & Robertson.

Kelley, J. (1996). Ethnic sympathies and politics in Australia, 1995. *Wwa: Worldwide Attitudes* 1996-01-15, 1–8.

Kelley, J., & McAllister, I. (1984). Immigrants, socio-economic attainment, and politics in Australia. *British Journal of Sociology*, *35*, 387–405.

Kim, I. (1981). *New urban immigrants: The Korean community in New York*. Princeton, NJ: Princeton University Press.

Kim, D. Y. (1998). Ethnic enclaves, middleman minorities, ethnic niches, self-employment: A call for conceptual clarification. Presented at the Eastern Sociological Society Annual Meeting, Philadelphia, March.

Kim, D. Y. (1999). Beyond co-ethnic solidarity: Mexican and Ecuadorean employment in Korean-Owned businesses in New York City. *Ethnic and Racial Studies*, *22*(3), 581–605.

Kossoudji, S. (1988). English language ability and the labor market opportunities of Hispanic and East Asian immigrant men. *Journal of Labor Economics*, *6*, 205–228.

Kunovich, R. M., & Hodson, R. (2002). Ethnic diversity, segregation, and inequality: A structural model of ethnic prejudice in Bosnia and Croatia. *Sociological Quarterly*, *43*(2), 185–212.

Ladbury (1984). Choice, chance, or no alternative: Turkish Cypriots in business in London. In: R. Ward & R. Jenkins (Eds), *Ethnic Communities in Business: Strategies for Economic Survival* (pp. 105–124). Cambridge: Cambridge University Press.

Lewin-Epstein, N., & Semyonov, M. (1994). Sheltered labour markets, public sector employment, and socio-economic returns to education of Arabs in Israel. *American Journal of Sociology*, *100*, 622–651.

Li, P. S. (2001). Immigrants' propensity to self-employment: Evidence from Canada. *International Migration Review*, *35*(4), 1106–1128.

Li, P. S., & Li, Y. (1999). The consumer market of the enclave economy: A study of advertisements in a Chinese daily newspaper in Toronto. *Canadian Ethnic Studies*, *31*(2), 43–60.

Light, I. H. (1979). Disadvantaged minorities in self-employment. *International Journal of Comparative Sociology*, *20*, 31–45.

Light, I. H. (1984). Immigrant and ethnic enterprise in North America. *Ethnic and Racial Studies*, *7*, 195–216.

Light, I. H., Bernard, R. B., & Kim, R. (1999). Immigrant incorporation in the garment industry of Los Angeles. *International Migration Review*, *33*(1), 5–25.

Light, I. H., & Bonacich, E. (1988). *Immigrant entrepreneurs: Koreans in Los Angeles, 1965–1982*. Berkeley, CA: University of California Press.

Light, I. H., & Rosenstein, C. (1995). Expanding the interaction theory of entrepreneurship. In: A. Portes (Ed.), *The Economic Sociology of Immigration* (pp. 166–212). New York: Russell Sage.

Logan, J. R., Alba, R. D., Dill, M., & Zhou, M. (2000). Ethnic segmentation in the American metropolis: Increasing divergence in economic incorporation, 1980–1990. *International Migration Review*, *34*(1), 98–132.

Logan, J. R., Alba, R. D., & Zhang, W. (2002). Immigrant enclaves and ethnic communities in New York and Los Angeles. *American Sociological Review*, *67*(2), 299–322.

Masurel, E., Nijkmamp, P., Tastan, M., & Vindigni, G. (2002). Motivations and performance conditions for ethnic entrepreneurship. *Growth and Change*, *33*(2), 238–260.

McManus, W., Gould, W., & Welch, F. (1983). Earnings of Hispanic men: The role of English language proficiency. *Journal of Labor Economics*, *1*, 101–130.

Miller (1989). Unemployment patterns in the youth labour market. Australian Economic Papers 1986: 222–235.

Min, P. G., & Bozorgmehr, M. (2000). Immigrant entrepreneurship and business patterns: A comparison of Koreans and Iranians in Los Angeles. *International Migration Review*, *34*(3), 707–738.

Model, S. (1988). Mode of entry and ethnic composition of firms. *Sociological Forum*, 110–126.

Model, S. (1997). An occupational tale of two cities: Minorities in London and New York. *Demography*, *34*(4), 539–550.

Nee, V., Sanders, J. M., & Sernau, S. (1994). Job transitions in an immigrant metropolis: Ethnic boundaries and the mixed economy. *American Sociological Review*, *59*(6), 849–872.

Olzak, S. (1989). Labor unrest, immigration, and ethnic conflict: Urban America 1880–1915. *American Journal of Sociology*, *94*, 1303–1339.

Olzak, S. (1992). *The dynamics of ethnic competition and conflict*. Stanford, CA: Stanford University Press.

Parsons, T., & Smelser, N. J. (1956). *Economy and society*. Glencoe, IL: Free Press.

Piore, M. (1979). *Birds of passage: Migrant labor and industrial societies*. Cambridge, England: Cambridge University Press.

Portes, A. (1989). *Immigrant America*. Berkeley, CA: University of California Press.

Portes, A., Guarnizo, L. E., & Haller, W. J. (2002). Transnational entrepreneurs: An alternative form of immigrant economic adaptation. *American Sociological Review*, *67*(2), 278–298.

Portes, A., & Jensen, L. (1987). What's an ethnic enclave? *American Sociological Review*, *52*, 768–771.

Portes, A., & Jensen, L. (1989). The enclave and the entrants. *American Sociological Review*, *54*, 929–949.

Price (1986). Refugees and mass migration: Australia. *International Migration Review*, *20*(Spring), 81–86.

Raijman, R., & Tienda, M. (2000). Immigrants' pathways to business ownership: A comparative ethnic perspective. *International Migration Review*, *34*(3), 682–706.

Rath, J., & Kloosterman, R. (2000). Outsiders' business: A critical review of research on immigrant entrepreneurship. *International Migration Review*, *34*(3), 657–681.

Sanders, J., & Nee, V. (1987). Limits of ethnic solidarity in the enclave economy. *American Sociological Review*, *52*, 745–773.

Sanders, J., Nee, V., & Sernau, S. (2002). Asian immigrants' reliance on social ties in a multiethnic labor market. *Social Forces*, *81*(1), 281–314.

Spencer, D., & Bean, F. D. (1999). Self-employment concentration and earnings among Mexican immigrants in the U.S. *Social Forces*, *77*(3), 1121–1147.

Stevens, G. (1992). The social and demographic context of language use in the United States. *American Sociological Review*, *57*, 171–185.

Stinchcombe, A. (1990). *Information and organizations*. Berkeley: University of California Press.

Stolzenberg, R. M. (1990). Ethnicity, geography, and occupational achievement of Hispanic men in the United States. *American Sociological Review*, *55*, 143–154.

Stolzenberg, R. M., & D'Amico, R. (1977). City differences and nondifferences in the effect of race and sex on occupational distribution. *American Sociological Review*, *42*, 937–950.

Tainer, E. (1988). English language proficiency and the determination of earnings among foreign born men. *Journal of Human Resources*, *23*, 108–122.

Tienda, M., & Lii, D. T. (1987). Minority concentration and earnings inequality: Blacks, Hispanics, and Asians compared. *American Journal of Sociology*, *93*, 141–165.

Waldinger, R., Aldrich, H. E., & Ward, R. (1990). *Ethnic entrepreneurs: Immigrant business in industrial societies*. Beverly Hills: Sage.

Werbner, P. (1984). Business on trust: Pakistani entrepreneurship in the Manchester garment trade. In: R. Ward & R. Jenkins (Eds), *Ethnic Communities in Business: Strategies for Economic Survival* (pp. 189–210). Cambridge: Cambridge University Press.

Werbner, P. (2001). Metaphors of spatiality and networks in the plural city: A critique of the ethnic enclave economy debate. *Sociology*, *35*(3), 671–693.

Wilson, K., & Martin, W. A. (1982). Ethnic enclaves: A comparison of the Cuban and Black economies in Miami. *American Journal of Sociology*, *88*, 135–160.

Wilson, K., & Portes, A. (1980). Immigrant enclaves: An analysis of the labor market experiences of Cubans in Miami. *American Journal of Sociology*, *86*, 295–319.

Zhou, M., & Logan, J. (1989). Returns on human capital in ethnic enclaves: New York City's Chinatown. *American Sociological Review*, *54*, 809–820.

APPENDIX A: AVAILABILITY OF EMPLOYMENT WITHIN LANGUAGE GROUPS

Nearly all immigrants to Australia from Africa in this period were from English-speaking, British-origin families, and so are omitted here. The Singaporese immigrants are heavily of Chinese origin, and so are included in the Chinese language group (rather than as Malay speakers). The Taiwanese immigrants are heavily ethnic Chinese (rather than ethnic Taiwanese) and so are coded into the Chinese language group. The Cypriot immigrants are heavily of Greek origin and so are coded into the Greek language group. The Swiss immigrants are coded into the German language group (because most of them have German as a first language). Many of the Israeli immigrants (at the time) were the children of Polish immigrants and connect with the Polish community in Australia, so they are coded into the Polish language group. The Australian Census reports data from Yugoslavia as a whole, rather than its constituent republics for this date, but most of the immigrants were either Croats or Serbs whose languages are mutually intelligible (whatever their political differences), and so the fact that they are combined causes no problem from the point of view of the communication costs hypothesis. Deciding what to do about the Slav languages is difficult (although not seriously consequential for this analysis since there are only a few Slav immigrants from any country other than Poland which is an important source country and from then-Yugoslavia which is also an important source country), because native speakers typically insist on large differences, but experts are divided on the issue of mutual intelligibilty. The compromise I have adopted is to make estimates for Poland and Yugoslavia separately, but to make a combined estimate for the other Slav countries. I have coded Finland and Estonia into a single language group on the grounds that their languages are sufficiently mutually intelligible to work as a "trade language," but I recognize that this could be argued either way (it isn't really important to the analysis because there are only a handful of immigrants from either place). The Egyptian immigrants are also a difficult case, as many of the older ones are of Greek mother tongue, but not so among the more recent ones. Unfortunately, the proportions remain a matter of speculation. I have opted to estimate the Egyptian rate separately, but one could certainly argue the case other ways. Again, the numbers involved are very small, so the choice is not very consequential for the analysis.

It is thus important to note that there is abundant measurement error in this variable. The principal effect of that error should be to inflate the standard error on the parameter estimate of the effect of this variable on occupational status, that is, to make it less likely that the model will find a statistically significant impact of accessibility of employment within the language group on individual occupational status. Thus, to the extent that there is an impact on the analysis, it is to bias the results against the communications cost hypothesis.

The percent entrepreneurs for each language group are shown country by country in Table A.1.

Table A.1. Percent Entrepreneur (Conducting Own Business with Employees) in Australia by Birthplace, Immigrants from Non-English Speaking Countries, 1981.[a]

Country	% Entrepreneur
Albania	5.25
Argentina	5.63
Austria	6.90
Bangladesh	2.67
Belgium	8.13
Bolivia	5.63
Brazil	5.62
Bulgaria	5.25
Burma	6.10
Cambodia	6.10
Chile	5.63
China	17.35
Colombia	5.63
Cook Islands	6.38
Cyprus	10.02
Czechoslovakia	5.25
Denmark	8.13
Ecuador	5.63
Egypt	3.31
Estonia	6.68
Fiji	6.38
Finland	6.68
France	8.13
Germany	6.90
Greece	10.02
Gulf States	3.31
Hong Kong	17.35
Hungary	6.10
India	7.14
Indonesia	5.26
Iran	2.67
Iraq	3.31
Israel	7.55
Italy	9.77
Japan	6.10

Table A.1. (*Continued*)

Country	% Entrepreneur
Kiribati & Tuvalu	6.38
Korea	6.10
Laos	0.00
Latvia	5.25
Lebanon	9.91
Lithuania	5.25
Malaysia	5.26
Malta	2.88
Mexico	5.63
Nauru	6.38
Netherlands	7.37
New Caledonia	6.38
Norway	8.13
Pakistan	2.67
Papua New Guinea	6.38
Paraguay	5.63
Peru	5.63
Philippines	5.26
Poland	7.55
Portugal	5.62
Romania	5.25
Singapore	17.35
Solomon Islands	6.38
Spain	6.52
Sri Lanka	7.14
Sweden	8.13
Switzerland	6.90
Syria	3.31
Taiwan	17.35
Thailand	0.00
Timor	6.38
Tonga	6.38
Turkey	3.31
Ukraine	5.25
Uruguay	5.63
Vanuatu	6.38
Venezuela	5.63
Vietnam	0.00
West Indies & Caribbean	5.63
Western Samoa	6.38
Yugoslavia	3.66

[a] Public Use Sample of the 1981 Census.

APPENDIX B: VARIABLES AND SCORING

Dependent Variable

Occupational status. Socioeconomic status of respondent's occupation. The Australian Bureau of Statistics detailed occupational classification is recoded into the ANU-2 occupational status scale. In practice, it is similar to the widely used Duncan SEI scale for the United States, and compares well with other occupational status scales in cross-national analyses (Jones & McDonnell, 1977). I use a linear transformation of the scale which gives a more intuitive metric, ranging approximately from zero to 100.

Focal Variables

The two focal causal variables are: (1) the aggregate or contextual variable percentage self-employed with employees; and (2) the interaction of this with individual English language skill.

Controls: Education

Education. Years of primary, secondary, and tertiary education. Years of primary and secondary education are computed from information on the age at which respondent left primary or secondary school (available in the Census) and usual age to start school in the respondent's home country. Starting age is 5.5 for Australia, but varies widely among the countries which have sent immigrants to Australia, ranging from 5.5 for the Federal Republic of Germany and the U.K. up to 7.5 for many of the Scandinavian countries and Yugoslavia. In developing estimates of starting age, we have drawn heavily on the work of the Australian Mission to Study Methods of Training Skilled Workers in Europe (1969) and, for Eastern Europe, on the expert advice of Krzystof Zagorski. We then develop estimates of years of tertiary education, based on detailed information about the highest degree, diploma, or certificate. Apprenticeships in Australia involve some formal training, so we have estimated the increment they contribute to educational attainment. In converting the information about tertiary education into estimates of years, we have relied on the expert advice of Don Anderson. Finally, we add our estimates of years of school to years of tertiary education to arrive at an estimate of years of education.

Educated in Australia. Scored one for men educated in Australia. Immigrants who arrived in Australia before completing their education are scored one. Immigrants who arrived in Australia after completing their education are scored zero. Age left school is asked directly, and provides the necessary information for respondents who had no tertiary education. For respondents who had tertiary education, the years of tertiary education (see education, variable 3) are added to age left school to estimate age at end of education. Age at arrival is estimated as current age minus years since arrival in Australia (asked in a direct question, available in single years). The (very few) immigrants who completed their education and arrived in Australia in the same year are scored as having been educated abroad.

Education squared. Years of education (variable 3) minus 10, quantity squared. Subtracting a number near the mean before squaring reduces computational inaccuracies due to rounding error in estimation of the regression coefficients, but otherwise leads to predictions mathematically equivalent to those obtained using a conventional squared term (e.g. Kelley & McAllister, 1984).

Australian education: Interaction. Education (variable 3) times educated in Australia (variable 4).

Australian education squared: Interaction. Education squared (variable 5) times educated in Australia (variable 4).

Trade qualifications. Scored one for respondents who completed a recognized apprenticeship program or otherwise obtained a formal "trade certificate," and scored zero for all others. Trade qualifications are required in order to obtain employment in many skilled blue collar occupations, and in some semi-skilled blue collar and low service occupations, as well.

Controls: Labor Force Experience

Australian labor force experience. Years in the Australian labor force is estimated as age minus age completed education for immigrants who were educated in Australia (variable 4). For those who were educated abroad, this is estimated as current age minus age at arrival in Australia.

Australian labor force experience squared. Australian labor force experience (variable 9) minus 20, quantity squared. Subtracting 20 before squaring reduces computational inaccuracies (see variable 5).

Foreign labor force experience. Years in overseas labor force. This is scored zero for immigrants educated in Australia (see variable 4). For immigrants who were educated abroad, this is scored as the difference of age at arrival in Australia and age at the completion of education.

Foreign labor force experience squared. Years in overseas labor force (variable 11) minus 10 quantity squared. We subtract 10 before squaring to reduce computational inaccuracies (see variable 5).

Australian labour force experience for immigrants who began their career in Australia Interaction. Australian labour force experience (variable 9) times educated in Australia (variable 4).

Australian labour force experience squared for immigrants who began their career in Australia Interaction. Australian labour force experience squared (variable 10) times educated in Australia (variable 4).

Other Controls

English fluency. Self-rated competence, coded as an equal-interval continuous variable ranging from a low of 0 for "speaks no English," 33 for speaks English "not well," 67 for speaks English "well," and 100 for "speaks only English" and "speaks English very well."

Married Scored one for currently married, zero for others.

Citizen Scored 1 if Australian citizen, otherwise zero.

Rural is a dummy variable coded 1 for rural residence and 0 for urban.

Female is a dummy variable coded 1 for women and 0 for men.

APPENDIX C: PARAMETER ESTIMATES

Multilevel model predicting occupational status from the availability of enclave employment, individual language skill, their interaction, and controls.

	Coefficient	Standard Error
Focal variables		
1. Availability of employment in language group (% employers in group)	1.0630	0.2774
2. English language fluency (0–100)	0.1896	0.0228

	Coefficient	Standard Error
3. Interaction: Availability of employment by individual language fluency (v1 × v2)	−0.0103	0.0029
Control variables		
4. Education (years)	3.9708	0.0958
5. Educated in Australia (0 or 1)	3.5140	4.7350
6. Education squared ($[v4–10^2]$)	0.4071	0.0142
7. Australian education (v4 × v5)	−0.1285	0.3720
8. Australian education squared (v6 × v4)	0.1136	0.0571
9. Trade qualifications (0 or 1)	−0.3066	0.6568
10. Australian labor force experience (years)	0.1361	0.0356
11. Australian labor force experience squared ($[v10–20]^2$)	−0.0088	0.0030
12. Foreign labor force experience (years)	0.0058	0.0394
13. Foreign labor force experience squared ($[v12–10]^2$)	0.0007	0.0032
14. Australian labor force experience for immigrants who began their career in Australia (v10 × v5)	−0.0797	−0.0953
15. Australian labor force experience squared for immigrants who began their career in Australia (v11 × v5)	−0.0110	0.0061
16. Married (0 or 1)	2.6343	0.5655
17. Citizen (0 or 1)	0.1232	0.5434
18. Rural (0 or 1)	1.2641	1.1003
19. Female (0 or 1)	−1.1543	0.4732
20. B_{0ij}, where $B_{0ij} = -34.5270\ (2.5571) + u_{0j} + e_{0ij}$ and $[u_{0j}] \sim N(0, \Omega_U)$: $\Omega_U = [6.9705\ (2.4940)]$ and $[e_{0ij}] \sim N(0, \Omega_E)$: $\Omega_E = [291.0905\ (5.1651)]$		
Fit (log likelihood, iterative generalized least squares)	54303.41	
N of Cases	6375	

Data source: Australian Census, 1981 Public Use Sample.

ECONOMIC CHANGE AND THE LEGITIMATION OF INEQUALITY: THE TRANSITION FROM SOCIALISM TO THE FREE MARKET IN CENTRAL-EAST EUROPE

Jonathan Kelley and Krzysztof Zagorski

ABSTRACT

This article takes advantage of a unique historical opportunity, the transformation of Central-East Europe with the collapse of Communism, to address a fundamental question in the social justice-equity-legitimation research tradition: how strong is the link between a nation's economy and its citizens' normative judgments concerning income inequality? We argue: (1) that the transition from a socialist economy to a free market economy should increase normative support for income inequality; (2) that to the extent that people perceive differences in pay actually to be large, they will believe more inequality to be morally legitimate; and (3) that normative support for income inequality will be higher among better educated people and among those in higher status jobs. We find that normative support for inequality increased dramatically. In Communist times the Polish and Hungarian publics favored less inequality than citizens of Western nations thought right; but within

The Shape of Social Inequality: Stratification and Ethnicity in Comparative Perspective
Research in Social Stratification and Mobility, Volume 22, 319–364
© 2005 Published by Elsevier Ltd.
ISSN: 0276-5624/doi:10.1016/S0276-5624(04)22011-X

a decade after the fall of Communism they favored much more inequality than Westerners think right. These normative changes did not arise from socioeconomic or demographic change in population structure but in large part from perceived changes in actual income inequality. Our data are from the World Inequality Study, which pools data from the International Social Survey Programme and other projects; there are 18 representative national samples in six Central-East Europe nations (N = 23,260) and, for comparison, 32 in Western nations (N = 39,956).

Income inequality is a central feature of modern society, a central focus of research in social stratification and labor economics, a key source of political conflict in many nations, and the topic of much philosophical analysis and prescriptive argument (e.g. Aristotle, 322BC; Blau & Duncan, 1967; Franklin et al., 1992; Rawls, 1971; Sen, 1973). Recently a flourishing tradition of empirical research on the origins and development of people's norms about the distribution of income has developed under the rubrics of "social justice," "equity," or the "legitimation of inequality" (Alwin, 1987; Berger et al., 1972; Gijsberts, 1999; Jasso, 1980; Kelley & Evans, 1993; Kluegel & Smith, 1986; Kluegel et al., 1995; Moore, 1992; Zagorski, 1994). This literature shows that in all countries studied so far – poor as well as rich, socialist as well as capitalist – there is near consensus among the general public about how much ordinary workers should earn, and consensus that elite occupations should be paid more than ordinary workers, but widespread disagreement about how much more and why (Haller, 1990; Kelley & Evans, 1993; Svallfors, 1993).

This article takes advantage of a unique historical opportunity, the transformation of Central-East Europe[1] with the collapse of Communism, to address a fundamental question in the social justice-equity-legitimation line of research: how strong is the link between the nation's economy and its citizens' normative judgments concerning income inequality? In Western nations the birth of a market-oriented economy occurred generations ago, far beyond the reach of modern survey research, and moreover stretched over a period of generations. In Central-East Europe it is happening over a brief span of years, in clear view of our eyes and our surveys. This provides an unprecedented scientific opportunity to use systematic survey data to study the links between the economy and individual norms.

This article also addresses a political dilemma faced by Central-East European nations and many other democracies in the developing world: it is by no means clear that the early stages of economic growth, during which inequality inevitably grows (Kelley & Klein, 1982, pp. 184–190; North & Thomas, 1973), can easily coexist with democracy. Nonetheless, both theoretical considerations (Hirschman, 1981; Offe, 1991) and empirical evidence (Zagorski, 1994) suggest that such coexistence is not only indispensable for political and economic change but also possible. However, if the public finds the new inequality morally objectionable, a populist attack on it becomes a potent political appeal that could easily bring into power governments that hinder political transformation and economic growth, to the long-run disadvantage of all. Indeed, Britain during the first industrial revolution was not fully democratic, nor were most continental European nations around the turn of the century when they were first industrializing, nor are most contemporary Asian "tiger" economies. It is the beginning of the process that seems most fragile, but once underway, there seem to be reciprocal reinforcing relations between political and economic freedom on one hand, and inegalitarian attitudes on the other.

This article shows how the shift from an objectively egalitarian command economy under Communism toward a free-market economy in Central-East Europe dramatically changed the public's norms about income inequality. The data show that the result was rapidly growing acceptance of inequality, taking public opinion far from the egalitarian norms of the past. But these changes were no swifter than the rapid growth in actual inequality. So, our analysis shows that the potential conflict between economic development and democracy still exists, but is now no greater that it was in the past despite the dramatic growth in actual inequality.

Data are from the World Inequality Study, a project pooling data from the International Social Survey Programme, the International Survey of Economic Attitudes, and other projects (Kelley et al., 2003). There are 18 surveys, all representative national samples, in six Central-East European nations with 23,260 cases in all. For comparison, we also analyze 32 representative national samples of Western nations, with 39,956 cases.

THEORY

The Setting

In recent years in both Central-East European and Western nations there has been a marked shift toward more free-market economies: (1) After the fall of Communism in 1989–1990, more market-oriented economies have emerged throughout East, Central-East and Central Europe (Clauge & Rausser, 1992). These changes have been most dramatic in Poland, where early "shock treatment" shifted the economy rapidly in a market direction (Balcerowicz, 1994; Bartholdy & Flemming, 1993;

Lipton & Sachs, 1990). Hungary and the Czech Republic are almost as advanced in their economic transformation, although the changes were more gradual there (Adam, 1993; Koves, 1992; Thomas, 1992). The subsequent return to power in Poland and Hungary of elected coalitions dominated by reformed ex-Communists has slowed the rate of change but not stopped it; (2) Economic rationalists (and their political allies under various labels) have led the way to substantial economic reform in Australia, the USA, and many other Western countries in the past decades (Capling & Galligan, 1992; King & Lloyd, 1993; Pusey, 1991; Yergin & Stanislaw, 1998).

By creating new opportunities and by undermining older government policies that had both favored blue-collar workers and imposed many constraints on would-be entrepreneurs, these market-oriented changes increased income inequality in Central-East European nations.[2] For the general logic by which inequality grows, examples from other times and places, and the influence of initial conditions, see Gerber and Hout (1998), Kelley and Klein (1982, pp. 184–190), Nee and Matthews (1996), or North and Thomas (1973).[3] In particular, the incomes of high-status jobs requiring university education rose (Beskid et al., 1995; Danziger & Gottschalk, 1994; Headey et al., 1995; Murphy & Welch, 1994).[4] How, then, do ordinary people evaluate the resulting inequality of income?

Self-Interest and the Moral Evaluation of Income Inequality

That people's economic views are shaped by their self interest, their "pocketbook," is a familiar assumption, common to Marx, classical economics, and sociological functionalism (e.g. Davis & Moore, 1945). Stretching the time horizons forward, expectations of personal benefits to come in the future also provide a motive for accepting the market and inequality, even for people who have not so far benefited from it (the "tunnel model": Hirschman, 1981; Offe, 1991; Zagorski, 1994).

Implications of self-interest considerations for the legitimacy of inequality are not entirely certain, since it was somewhat unclear at the time which groups would benefit, and which would lose, from the emergence of a market economy in formerly Communist nations. But it seems likely that people were experiencing and perceiving generally better prospects to the well educated rather than the poorly educated, to those in higher status jobs rather lower status jobs, to supervisors and the self-employed rather than ordinary employees, to those already prosperous rather than the poor, and to the middle class rather than the working class. If so, these groups can be expected to take a more benign view of income inequality, hoping themselves to benefit in the long run.

"Intellectual" Considerations Relevant to Acceptance of the Free Market

Simple self-interest may not be the only, or even the main motivation. For example, much evidence suggests that ordinary people shape their political decisions more by their perception of the general good of the nation as a whole than by simple self-interest (e.g. Eulau & Lewis-Beck, 1985; Lewis-Beck, 1988). Thus people who believe that, for the population at large, the free market is legitimate, efficient, or reasonable will hold a more sympathetic view of it and its consequences, including inequality. There are several reasons for this:

- The assumption that market reform will in the long run be beneficial to most people, bringing Central-East Europe closer to the visibly superior standard of living in the West, is a strong reason for accepting it for the public good, regardless of one's personal prospects (Frentzel-Zagorska, 1993; Lewis-Beck, 1988; Mason, 1995; Zagorski, 1994).
- Intellectual attraction to the merits of a free market has the same consequence. The intellectual ascendancy of neo-classical free market economic reasoning (represented, for example, by Schultz's (1980) Nobel Lecture; Yergin & Stanislaw, 1998), has led to a near consensus among the elite in many nations favoring only a limited role for government in the economy (e.g. Frentzel-Zagorska & Zagorski, 1993; Putnam et al., 1993, pp. 28–38), although ordinary citizens in Central-East Europe do not share this view (Sikora & Kelley, 1999).

Consequences of Accepting the Free Market

Accepting something new also implies some acceptance of its consequences. For example, if you decide to build yourself a new house, that implies also accepting some intrinsically attractive consequences (e.g. having more space), accepting some consequences of uncertain intrinsic worth (e.g. living in a new neighbourhood), and accepting some intrinsically undesirable consequences (e.g. having to pay a new mortgage). Similarly, accepting the free market provides strong grounds for also accepting its varied consequences. These include competition; minimal government regulation; relatively free trade; the rule of law; willingness to let employment in uncompetitive industries decline and to let weak firms expire; provision of health and welfare benefits by government or by insurance rather than entirely by the firm (so job losses do not imply destitution); and many others. We argue that income inequality is one of the free market's inevitable consequences: it is both a pre-requisite for the free market – providing motivation for workers to invest in training and to work hard – and a consequence of the free market – arising out of differences in workers' resources, effort, talent, and luck. As a result, those who accept the free market will tend also to find inequality legitimate on the pragmatic ground that it is inevitably part and parcel of the attractive free-market package.[5]

Moral Authority of the Market Ideal

There are also moral reasons that can lead to the same conclusion. The four styles of moral reasoning commonly used in Western societies include the authoritative mode invoking the moral sanction of some legitimate authority (Bellah, 1974; Potter, 1972; Tipton, 1982).[6] Historically, the most familiar example of the authoritative mode is a church pronouncing on moral issues. But in modern societies legitimate authority is, in addition, sometimes national (for example, appeals to the American way of life as a justification for free speech), and sometimes political (for example, party loyalties shaping voter's attitudes on political issues, e.g. Nie et al., 1979) and also, we suggest, sometimes economic. Specifically, appeal to the legitimacy of the free market can be used to morally justify its diverse consequences, including inequality (Yergin & Stanislaw, 1998). Appeal to theological individualism can have the same effect (Davis & Robinson, 1999).

Rewards to Productivity

Following Aristotle in the Nicomachean Ethics, we suggest that many people will accept the general principle that rewards ought to be proportional to productivity: That people whose skill, effort or ability enable them to produce more ought to be rewarded in proportion; and that equal pay for unequal contributions is unjust. If Aristotle was correct about his time, this norm dates back to the origins of Western civilization. In a world of small, independent producers – like most of the Western world from Aristotle's time through the nineteenth century – the principle is a natural one, involving little more than abjuring theft and eschewing economic discrimination. For example, if you work twice as hard as I do, or twice as skillfully, and so make twice as many sandals as I, you will have twice as many to sell at the end of the day, and so twice the income I have. Twice as much, that is, unless buyers discriminate against you by offering a premium for my sandals – thus wasting their own money, since discrimination in a competitive market is costly to those who do it (Becker, 1971; Ehrenberg & Smith, 1982, pp. 401–412) – or unless governments impose tax, license or regulatory policies that achieve the same effect indirectly.

This view is close to the "marginal productivity theory of distribution" or "neo-classical distribution theory" systematized by nineteenth century liberal economists (e.g. Adam Smith, 1776[1937]; for a summary of some difficulties see, for example, Frank, 1985, Chap. 6; Thurow, 1975, Chap. 2). Sociological functionalists make very similar arguments (Davis & Moore, 1945), with similar uncertainties (e.g. Tumin, 1953), and similar empirical consequences (Stinchcombe, 1963). Some philosophical arguments lead to similar conclusions (e.g. Nozick, 1974). The hypothesis of widespread public acceptance of productivity norms is strongly supported by decades of research in experimental social psychology showing that rewarding "inputs" is one of the important ways to achieve justice or fairness in social exchange (e.g. Berger et al., 1972; Walster et al., 1978).

This theory implies that changes in productivity will cause changes in people's views about legitimate earnings. Thus if a change in circumstance increases an occupation's impact on productivity, then its legitimate earnings will increase correspondingly (Stinchcombe, 1963). So if the emergence of a free market, full of opportunity and risk, in place of the rigidities of a command economy increases the payoff of good management and good government, then the earnings thought legitimate for managers and government officials will grow correspondingly. This argument assumes: (1) that these increases reflect greater gains in the productivity of high-status workers than in the productivity of workers in low status occupations, as neo-classical economic theory implies; (2) that the general public correctly perceives these increases (as we demonstrate below); and (3) that the public attributes these changes to growth of productivity or believe that they increase the common good.[7] Alternative explanations – for example, political privilege, bureaucratic favoritism, corruption, or crime – may be part of the story part of the time, but are implausible as general principles.[8] Insofar as these assumptions hold, the earnings regarded as legitimate for high status occupations should rise correspondingly.

Implications. Most of these essentially "intellectual" considerations are more likely to be known to, and understood by, the educational elite than by ordinary citizens, and more by the prosperous than the poor. They are also more likely to be understood by people working in high status, cognitively complex occupations that afford a wide overview of economic change, rather than by people in routine, narrowly focused manual jobs. That implies a link between education and acceptance of inequality, and between occupational status and acceptance of inequality. But it does not imply any particular link with supervision, business ownership, government employment, or subjective social class. In contrast, arguments based on self-interest imply a link between views about inequality and supervisory position, business ownership, government employment, and subjective class, as well as a link with education, income, and status.

Summary of Hypotheses

Thus we have argued that:[9]

Hypothesis 1. The transition from a socialist economy to a free market economy will increase normative support for income inequality.

Hypothesis 2. To the extent that people perceive differences in pay actually to be large (and attribute them to productivity, or believe they increase the common good), they will believe inequality to be morally legitimate.

Hypothesis 3. In the transition from a socialist economy to a free market economy: (a) normative support for income inequality will be higher among better educated people rather than the poorly educated, among the prosperous more than among the poor, and among those in higher status jobs rather lower status jobs (for both intellectual and self-interested reasons); while (b) normative support for income inequality will be higher among supervisors and the self-employed rather than ordinary employees, and among the middle class rather than the working class (for self-interested reasons).

While "existentialist" theory assumes that the drive for consistency between perceptions of petrified reality and its legitimation results in petrified attitudes that are difficult to change even when the perceptions begin to change, an alternative hypothesis is that perceptions of fast and radical changes would create painfully acute cognitive dissonance, if the norms did not also change in tandem. In contrast to these rigid formulations, another argument is that people seek "optimum arousal" stemming from reducing cognitive dissonance to a moderate level (Berlyne, 1960; Frentzel, 1965) rather than seeking total dissonance reduction (Festinger, 1964). In this view, if the system as a whole is felt to be legitimate, the "normal gap" between perceived levels of inequality and norms concerning them may hold steady or even increase during periods of change. Thus, we also argue that:

Hypothesis 4. The perception of rapidly growing inequality leads to the legitimation of more inequality than was accepted in the past. The gap between perceived and accepted inequality may even grow. As a consequence, given system legitimacy, perceptions of inequality determine its legitimation to a great extent, though this determination is far from perfect or complete.

Rejected Alternative Theories

There are several plausible alternative theories which are inconsistent with our arguments. We will suggest that all of them should be rejected.

- *Egalitarianism.* Radically egalitarian views reject anything – not just productivity – as a legitimate basis for inequality. Examples are the strong egalitarianism of early Christianity, some economists and moral philosophers (e.g. Rawls, 1971; Sen, 1973, pp. 77–106), many revolutions, and most utopian communities. Some have argued that egalitarian norms are widespread in modern societies, especially socialist ones (Bell, 1972, p. 40; Jasso, 1980). This directly contradicts our Aristotelian hypothesis.
- *Enlightenment.* A persuasive argument can be made that the general tenor of intellectual and cultural change in the 19th and 20th centuries – the *zeitgeist* of the time – is liberal and egalitarian (e.g. Chirot, 1986; Robinson & Bell, 1978). Starting with the conservative, religious, highly stratified, often aristocratic societies of the 18th century, over the course of the 19th and 20th centuries scientific progress, secularization, economic growth, the spread of democracy, the expansion of the welfare state, and related changes have undermined tradition, religion, privilege, and economic inequality. A natural implication is that people's norms about inequality are, over time, becoming more egalitarian. This is in contrast to our Aristotelian prediction that changes over time are becoming less egalitarian.
- *Existential Theories.* "Existential" arguments posit that whatever is factually the case comes in time to be accepted normatively – that habit, familiarity, and comparison with the perceived rewards of similar others confer legitimacy (Berger et al., 1972, p. 139; Heider, 1958, p. 235; Gijsberts, 1999, pp. 51–80; Homans, 1974, p. 250). During Communism's 40 year reign, income differences were much smaller than in the West and the white collar jobs held by the "intelligentsia" were downgraded (Domanski & Zagorski, 1991; Kraus & Hodge, 1987). The dominant elite glorified manual labor, especially in heavy industry. Thus if values come from habit and experience, Central-East Europeans would hold much more egalitarian views than Westerners. While this might change after the fall of Communism – just three or four years before our surveys – a lifetime of experience and propaganda would, on existential arguments, fade only slowly. So existential arguments imply that differences in earnings will continue to be illegitimate in the formerly socialist societies of Central-East Europe, changing only gradually toward the greater acceptance of inequality typical of market societies. This conflicts with our prediction that rapid economic change produces rapid changes in norms.

DATA

Our data are from the World Inequality Study, a project pooling data from the International Social Survey Programme, the International Survey of Economic

Attitudes, and other projects into a single harmonized file suitable for cross-cultural and over-time analyses (Kelley et al., 2003).[10]

The International Social Survey Programme (ISSP)

Most of the data are from the 1987–1988, 1992–1993, and 1999–2000 "Social Inequality" modules of the International Social Survey Programme.[11] These surveys mostly began with interviews with a stratified random sample followed by a leave- behind self-completion questionnaire with the ISSP items; several were conducted entirely by mail and some entirely by interview. Australia's survey was a simple random sample but the other surveys involved various forms of clustering. Completion rates averaged around 60%, counting losses at the interview and the drop-off stages (for details on the sampling techniques and response rates for each country, see www.issp.org). These rates compare favorably with recent experiences in many industrial nations (e.g. the highly regarded 1989 International Crime Victim Survey averaged 41% over 14 nations [van Dijk et al., 1990]). These data have been widely used in international comparisons (e.g. Kelley & Evans, 1995).

As this paper focuses on changes over time, we restrict analysis to nations with data in two or more time periods. (1) The ISSP participants[12] in Central-East Europe include: Lilia Dimova (1999) who conducts the Agency for Social Analyses' annual survey of Bulgarian social trends; Ludmila Khakhulina and Tatjana Zaslavskaya (1999) who design the Center for Public Opinion and Market Research's annual survey of Russian social trend; Brina Malnar and Nikos Tos (1999) who direct an annual survey of social trends in Slovenia; Peter Mateju and Michal Illner (1999) of the Annual Social Survey of the Institute of Sociology, Academy of Sciences of the Czech Republic; Peter Robert (1999) and who worked with Tamas Kolosi (Kolosi and Robert, 1989) Annual Social Survey of the Social Research Informatics Center TARKI, Hungary. (2) ISSP participants in the West include Jos Becker and Masja Nas (1999) of the Annual Opinion Survey of the Sociaal en Cultureel Planbureau, the Netherlands; Bogdan Cichomski and Pawel Morawski (1999) of the Polish General Social Survey; James A. Davis, Tom W. Smith and Mike Hout (1999) of the General Social Survey, USA; Alan Frizzell and Heather Pyman (1999) Carleton University Annual Survey, Canada; Philip Gendall (1999) of the Department of Marketing, Massey University, New Zealand; Max Haller and Franz Hoellinger (1999) who conduct the Biennial Survey of the Institut fuer Soziologie der Universitaet Graz, Austria. Janet Harkness, Peter Ph. Mohler and Michael Braun (1999) of the All Bus Survey, Germany; Roger Jowell, Sharon Witherspoon and Lindsay Brook (1999) of the British Social Attitudes Survey, Britain; Kelley and Evans (1999) of the International Social Science Survey, Australia; Mahar Mangahas, Mercedes Abad, Linda Luz Guerrero, Felipe

Miranda, Steven Rood and Ricardo Abad (1999) of the Annual Social Weather Stations Survey of Social Attitudes in the Philippines; Knut Kalgraff Skjak, Bjørn Henrichsen, Knud Knudsen and Vigdis Kvalheim (1999) of the Norwegian Social Science Data Services' Annual Opinion Survey; and Stefan Svallfors and Jonas Edlund (1999) who conduct an annual survey of Changing Swedish Attitudes and Values. (3) The Zentralarchiv fuer Empirische Sozialforschung at the University of Koeln (1994) and the Spanish data archive (Diez-Medrano, 2002) painstakingly cleaned the data; their files were, with extensive modifications and refinements, incorporated into the World Inequality Study (Kelley, Evans & Sikora, 2003) sketched in (Evans and Kelley, 2002).

The International Survey of Economic Attitudes and Other Surveys

This paper also uses data from the International Survey of Economic Attitudes (ISEA), a collaborative international project begun in 1991 (Kelley et al., 1998), which has conducted surveys in Australia, Bulgaria, Finland, Hungary, the Netherlands, and Poland.[13] The ISEA survey methodology is similar to that of the ISSP, in most cases done by the same survey organization. Several other surveys, not part of the ISEA or ISSP are also used, as detailed below.

Poland

Our most extensive Central-East European data are from Poland, including one survey from the Communist era. Six Polish data sets are used: (1) The first is from the 1987 Social Structure Survey conducted on a national stratified random sample by a team of researchers from the Institute of Sociology, the University of Warsaw and the Institute of Philosophy and Sociology, Polish Academy of Sciences (Slomczynski et al., 1989). There are 807 cases. The questions analyzed here were asked only of those currently employed; however analysis of other Polish (and Hungarian) surveys shows that the employed do not differ appreciably from the rest of the population on the issues at hand; (2) The second Polish survey was conducted by the survey unit of the Polish Academy of Sciences as a post-election panel in the 1991 election survey organized by the Academy's Institute of Political Studies (Gebethner & Raciborski, 1992; Kelley et al., 1993). The first wave of the panel was a nationally representative, stratified random sample conducted just before the parliamentary elections in 1991. The completion rate was 85% and the sample is representative of the population in age, sex, education, and rural vs urban residence. Demographic and background variables are from this wave. Attitudinal data are from the second wave conducted in December 1991 as a panel on the first. The completion rate was over 90% and the sample is representative of the population in age, sex, education, and rural vs urban residence. There are 1,519 cases; (3) The third and fifth Polish surveys were from the 1992 and 1999 rounds of the ISSP (Cichomski & Morawski, 1999); (4) The fourth and sixth Polish surveys were conducted in 1994 and (as a panel based on it) in 1997 as a part of International Survey of Economic Attitudes by the Centre for Social Opinion Research (CBOS), Warsaw, a highly regarded quasi-government agency. Completion rates were over 90% in the first survey and 78% in the second, with 2,127 cases and 1,669 cases respectively.

Hungary

The three Hungarian surveys, including one in the Communist era, were collected by Tarsadalomkutatasi Informatikai Egyesules (TARKI), Hungary's ISSP member and leading academic survey center. Their surveys were based on stratified random samples drawn using the official "personal number system" identifying each resident: (1) The first and third Hungarian surveys were conducted as part of the 1987 and 1999 ISSP surveys (Robert, 1999). There are 2,606 cases; (2) The second Hungarian survey constituted a part of the TARKI 1992 Social Mobility Panel (TARKI, 1993). Face-to-face interviews were conducted in May and June 1992 by trained interviewers; the completion rate was 82%. The background and demographic data used in the analysis are from this wave of the survey. Attitudinal data are from the second wave, a panel on the first conducted in October 1992 by face-to-face interviews with respondents still contactable at the original addresses; the completion rate was 86%. Both the original and panel samples are representative of the population in age, sex, and place of residence (TARKI, 1993). There are 1,250 cases.

Western Nations

(1) The eight Australian surveys were collected in by the International Social Science Survey, Australia's leading academic survey and the Australian ISSP member (Kelley & Evans, 1999). Three surveys included an ISSP module and the rest included the ISEA. All were based on simple random samples of Australian citizens drawn from the compulsory electoral roll using a slight modification of Dillman's Total Response Method (1993) with up to four follow-up mailings, two with fresh copies of the questionnaire, over a six to nine month period. Several surveys included a panel component. Comparison of mail and face-to-face surveys using the same questionnaire suggests that mail produces identical or sometimes superior results (Bean, 1991; Visser et al., 1996). Completion rates were 60–65%, which compares favorably with recent experience in the USA (Dillman, 1993, p. 234) and many industrial nations (e.g. van Dijk et al., 1990). There are 17,079 cases in all. The surveys are representative of the population in sex, age, education, occupation, labor force status, and other variables that can be compared with the census (Bean, 1991, 1995). (2) There are three surveys of the Netherlands, one

the 1987 ISSP (Becker & Nas, 1987) and the second by the ISEA group largely replicating the 1992 ISSP (Gijsberts & Ganzeboom, 1996). The third, part of the ISEA, was in 1998 (Nieuwbeerta et al., 1998). There are 1,638, 993 and 790 cases respectively. All are random samples and representative of the population in age, sex, education and occupation.

MEASUREMENT

Legitimate Earnings

The legitimate earnings questions have been extensively tested and shown to have good measurement properties in a dozen diverse nations (Kelley & Evans, 1993, pp. 88–93; see also Sarapata, 1963; Verba & Orren, 1985, Chap. 8). They are from the International Social Survey Programme's 1992 "Inequality-II" module, in turn a refinement of its 1987 "Inequality-I" module. The wording:

> Next, what do you think people in these jobs ought to be paid – how much do you think they should earn each year before taxes, regardless of what they actually get . . .

	Please write in how much they ought to earn each year
a. First, about how much do you think a skilled worker in a factory ought to earn?	$ ___________ dollars
b. A doctor in general practice?	$ ___________ dollars
etc . . .	

Further occupations followed, covering the full range from the lowest to the very highest: (1) *Blue collar workers*: "Unskilled worker in a factory" and "skilled worker in a factory." We use these occupations as the baseline to which other occupations are compared;[14] (2) The *economic elite*: "the owner-manager of a large factory," and "the chairman of a large nation-wide corporation;" (3) *Professionals*: a "lawyer" and a "doctor in general practice;" (4) *Elite government* officials: "A cabinet minister in the {national} government" and "a judge in the {nation's highest appellate court}."[15]

Answers to these questions were in local currency units. We express these as a ratio to each respondent's views about the proper income for two low status occupations (averaged): unskilled workers and skilled factory workers. For example, suppose a respondent thinks unskilled workers should earn \$20,000 and skilled workers \$30,000, for an average of $(\$20,000 + \$30,000)/2 = \$25,000$. If the same respondent thinks that a lawyer ought to earn \$50,000, we treat that as $\$50,000/\$25,000 = 2$, i.e. twice as much as for low status jobs.

Several points should be noted about this definition: (1) The use of a ratio is usual in this context (Arts et al., 1995; Kelley & Evans, 1993). It abstracts away from currency units (e.g. zlotys or dollars) and allows cross- national comparability; (2) A ratio also abstracts away from absolute levels of pay (which vary substantially between richer and poorer nations), to focuses directly on the *relative* income hierarchy. For example, if a Australian thinks that professionals should earn $50,000, which is roughly twice the average unskilled wage in Australia, we take that to be the same as a Pole saying professionals should earn 20,000 zlotys which is about twice the average Polish unskilled wage, even though the $50,000 buys much more than the 20,000 zlotys; (3) We make no adjustment for taxes. Tax incidence studies suggest that in most countries the actual incidence of all taxes combined is approximately a flat percentage of income. If so, adjustment for taxes would not affect the ratios we use and our results would be unchanged.

For these figures, we use a denominator specific to each respondent – the respondent's own views about unskilled and skilled workers.[16] We do this with some hesitation since ratios (or difference scores, as they are in our log formulation) can be problematic. However because of the rapid social change, vast inflation, and currency changes in Central-East Europe during this period, the public's knowledge of actual income levels in local currency units is uncertain. Some seem to have thought in terms of price levels that prevailed six months or a year before the interview, while others made larger or smaller adjustments for inflation. We eliminate these sources of error by taking the ratio to the respondent's own perceptions, since their time-frame and inflationary perceptions appear in both numerator and denominator, and so cancel out. In our judgment, the advantages of this approach outweigh the disadvantage of using ratio scores. Specifically, for each respondent, i, we calculate:

$$\text{legitimate income of ordinary workers}_i = \frac{\text{income unskilled workers ought to earn}_i + \text{income skilled workers ought to earn}_i}{2} \quad (1)$$

We then divide respondent *i*'s answers on the legitimate earnings of other occupations by this figure and take the natural log of the result. For example, for a lawyer:

$$\text{legitimate income of lawyer}_i = \ln \frac{\text{income a lawyer ought to earn}_i}{\text{legitimate income of ordinary workers}_i} \quad (2)$$

Analyzing the logarithm implicitly assumes that people think mainly in percentage terms, treating, for example, a 10% raise in a lawyer's income as similar to a 10% raise in a secretary's, even though the absolute amount is quite different. This approach is strongly enjoined by theory, past research on these questions, and studies of income inequality (Arts et al., 1995; Jasso, 1980; Kelley & Evans, 1993).

A variety of plausible alternative specifications lead to the same conclusions. Specifically, a lawyer's income could be measured simply in local currency units (although metric coefficients are then not comparable across countries), or their log (comparable slopes, but not intercepts), or converted into U.S. dollars at parity purchasing power. Or it could be measured relative to the average income of unskilled workers in that country, or what the respondent believes unskilled workers actually earn, or alternatively by the log of either of those. All lead to the same substantive conclusions (as in previous research using similar items: Kelley & Evans, 1993, Appendix); complete results are available on request.

Attitude Structure

The incomes people believe to be legitimate for various elite occupations are highly correlated both in Central-East Europe and in the West (Table 1). Previous research found similarly high correlations among a diverse range of elite occupations (Kelley & Evans, 1993, pp. 89–93). Analysis earlier Polish and Australian surveys with a more extensive list of occupations confirms the generality of these patterns. In particular further distinctions between government and private sector employment – for example, skilled worker in a government factory versus skilled worker in a private factory, or director of a government owned bank versus director of a private bank – mattered little to respondents.

Factor analysis clearly shows a single factor both in Central-East Europe and in the West (Table 1, last column). Furthermore, all six items have very similar correlations with a range of criterion variables, as they should on the classic psychometric measurement model for a single homogenous factor. Note, however, that the pattern of correlations in Central-East Europe differs from that in the West, particularly with respect to historical period, education, and age. Also in Central-East Europe, views about medical doctors are less closely tied than other occupations to the underlying factor, a departure from Western patterns that has long been noted.[17]

A scale averaging all six items has excellent reliability, with alphas around 0.90 in both Central-East Europe an in the West. Specifically, the scale is:

$$\begin{aligned}
&\text{legitimate income of elite occupations}_i = \\
&\quad \text{mean (legitimate income of chairman}_i\text{, legitimate income of factory owner}_i\text{,} \\
&\quad \text{legitimate income of lawyer}_i\text{, legitimate income of doctor}_i\text{,} \\
&\quad \text{legitimate income of judge}_i\text{, legitimate income of cabinet minister}_i\text{)} \qquad (3)
\end{aligned}$$

where the legitimate income of lawyers, etc, are as defined in Eq. (2).

Table 1. Legitimate Earnings of Various Occupations: Correlations, Means, Standard Deviaitons and Principal Axis Factor Loadings in Six Central-Eastern European Nations (23,260 Cases) and 10 Western Nations (39,956), 1987–2001.[a]

	Correlations						Factor Loading
	Chair	Factory	Lawyer	Doctor	Judge	Cabinet	
A: Central-East Europe							
Chair, large corporation	1.00						0.84
Factory owner	0.73	1.00					0.78
Lawyer	0.64	0.59	1.00				0.80
Doctor	0.55	0.48	0.64	1.00			0.67
Judge, highest court	0.68	0.67	0.69	0.55	1.00		0.86
Cabinet minister	0.64	0.58	0.58	0.50	0.70	1.00	0.76
Criterion variables							
Time	0.27	0.13	0.24	0.14	0.22	0.20	–
Male	0.08	0.09	0.06	0.05	0.06	0.06	–
Age	0.02	−0.01	0.01	0.05	0.02	0.04	–
Education	0.21	0.20	0.14	0.16	0.18	0.17	–
Family income	0.16	0.20	0.21	0.17	0.17	0.13	–
Mean (geometric)[b]	4.22	5.97	2.99	2.27	4.38	4.12	–
Standard deviation	0.77	0.91	0.66	0.51	0.70	0.66	–
B: West							
Chair, large corporation	1.00						0.76
Factory owner	0.60	1.00					0.75
Lawyer	0.59	0.60	1.00				0.82
Doctor	0.58	0.56	0.69	1.00			0.76
Judge, highest court	0.62	0.60	0.68	0.58	1.00		0.81
Cabinet minister	0.59	0.58	0.58	0.54	0.65	1.00	0.75
Criterion variables							
Time	−0.03	0.14	0.14	−0.07	0.04	−0.09	–
Male	0.12	0.10	0.01	0.01	0.06	0.05	–
Age	0.17	0.13	0.14	0.16	0.17	0.16	–
Education	0.05	−0.06	−0.09	−0.05	−0.06	−0.05	–
Family income	0.17	0.15	0.13	0.14	0.13	0.16	–
Mean (geometric)	3.83	3.33	2.75	2.86	3.57	2.91	–
Standard deviation	0.74	0.79	0.54	0.53	0.58	0.62	–

[a] Source: World Inequality Study, incorporating data from the International Social Survey Programme, the International Survey of Economic Attitudes, and other sources. The number of cases varies depending on missing data and because not every occupation was included in all surveys.

[b] Example: Central-East Europeans on average think that the chairman of a large corporation should earn 4.22 times as much as a factory worker (column 1). The legitimate earnings of a chairman is measured in a logarithmic metric, with a raw mean of 1.44; the geometric mean is exp(1.44) = 4.22.

Measurement: Class and Background Variables

We measure class and stratification position broadly, combining ownership of the means of production and authority in the workplace (the heart of Marx's and Dahrendorf's conceptions of class and their modern descendants, e.g. Wright, 1985), with education, occupational status, and income (the heart of the "SES" tradition: Blau & Duncan, 1967). Combined additively, they give a powerful, flexible model of class well suited to comparative research with both conceptual and empirical advantages over typological approaches (Kelley, 1990; Kelley & Evans, 1995). Details are in the measurement appendix.

Measurement: Historical Period

We measure historical period by the date each survey was conducted. The earliest surveys were in 1987, still in the Communist era in Central-East Europe, and the latest in 2001. The largest number of surveys are in 1987/1988, 1992/1993, and 1999/2000. There are Communist era data for Poland and Hungary (as well as many Western nations). By 1992/1993 – still only a few years after the fall of Communism in 1989 – there are data for six Central-East European nations (see Table 2).

Measurement: Other Variables

We control for *age, sex, subjective social class*, and *labor force participation* (measurement details are in the appendix). Measurement of *perceived earnings* of various occupations is described in the text below.

METHOD

Potential Bias Due To Missing Data

Our key questions about legitimate earnings are difficult, requiring a dollar or other currency unit figure as the answer. This requires more knowledge and thought than traditional survey questions, so there is more missing data than usual, averaging 10–15%, compared to around 10% for family income and under 5% for most other questions. In designing the questionnaire, we chose these questions because they give richer data than the alternatives and allow more persuasive comparisons among countries, but the amount of missing data is a worry. However, a detailed analysis shows that non-response is predominantly random, as also found in earlier analyses of these data (Kelley & Evans, 1993, pp. 118–120), so no substantial difficulty arises (details available on request).

Table 2. Legitimate Earnings of Various Occupations: Geometric Means for Central-East European and Western Nations, 1987–2001.[a]

	Scale: All Items Pooled[b]	Chairman, National Corporation	Factory Owner	Lawyer	Doctor	Judge, Highest Court	Cabinet Minister	Cases
Eastern Europe								
All Eastern Europe pooled								
Communist era	2.56	2.69	–	–	2.03	–	3.25	3,063
1990–1995	3.45	4.10	5.53	2.62	2.19	3.88	3.92	10,846
1996–2001	4.19	5.12	6.50	3.45	2.46	5.02	4.72	9,351
Russia[c]								
1990–1995	3.64	6.14	6.05	2.11	2.08	4.00	4.38	1,761
1996–2001	4.66	7.90	6.81	3.56	2.27	6.93	6.92	1,400
Poland								
Communist era	2.51	2.68	–	–	1.94	–	3.15	713
1990–1995	3.35	3.85	5.46	2.68	2.09	3.72	3.52	4,868
1996–2001	4.77	5.60	8.55	3.96	2.51	5.89	5.47	2,460
Czech Republic[c]								
1990–1995	2.82	2.86	4.90	2.02	1.75	3.28	3.55	1,066
1996–2001	4.41	5.31	7.48	3.49	2.38	5.69	4.62	1,701
Hungary								
Communist era	2.57	2.70	–	–	2.05	–	3.28	2,350
1990–1995	5.30	6.32	7.20	4.37	3.55	5.87	6.63	1,154
1996–2001	6.40	8.51	10.18	5.62	3.85	7.03	6.81	1,054
Bulgaria[c]								
1990–1995	2.94	2.88	4.17	2.51	2.09	3.42	3.50	1,012
1996–2001	2.57	2.59	3.16	2.28	2.01	2.79	2.97	1,792
Slovenia[c]								
1990–1995	3.17	3.73	–	–	2.31	–	3.79	985
1996–2001	3.70	4.59	5.55	2.91	2.52	3.96	3.48	944
Western nations								
Communist era	3.31	4.06	2.35	2.25	3.09	3.39	3.23	11,307
1990–1995	3.07	3.62	3.38	2.78	2.73	3.52	2.70	15,802
1996–2001	3.33	3.90	3.81	2.97	2.85	3.64	2.91	12,847

[a] Source: World Inequality Study, incorporating data from the International Social Survey Programme, the International Survey of Economic Attitudes, and other sources. The number of cases varies depending on missing data; the numbers shown are for the overall scale. Example: Central-East Europeans in the Communist era on average thought that high status occupations should earn 2.56 times as much as a factory worker (row 1, column 1). Legitimate earnings are measured in a logarithmic metric, with a raw mean of 0.94; the geometric mean is $\exp(0.94) = 2.56$.

[b] Legitimate earnings are measured by an additive scale averaging answers about the legitimate earnings of the six elite occupations, each expressed as (the logarithm of) a ratio to the legitimate earnings of skilled and unskilled factory workers. If not all questions were answered, the mean is of those that were answered.

[c] No Communist era data available.

Missing data is treated by the pair-wise present method, which is generally preferable to the usual alternatives (Hertel, 1976; Joreskog & Sorbom, 1988, pp. 1:12–1:17; Little, 1992, pp. 1229–1231).

Model

The model, estimated by OLS is:

$$\begin{aligned}&\text{legitimate income of elite occupations}_i = \\ &\quad a + b_1\text{Time} + b_2\text{Male} + b_3\text{Age} + b_4\text{Education} + b_5\text{FamilyIncome} \\ &\qquad + b_6\text{SubjectiveClass} + b_7\text{OccupationalStatus} + b_8\text{Supervisor} \\ &\qquad + b_9\text{PettyBourgeoisie} + b_{10}\text{Entrepreneur} \\ &\qquad + b_{11}\text{GovernmentEmployee} + e \qquad (4)\end{aligned}$$

To cater for possible interactions, we estimate the model separately for Eastern and Western Europe, and (in other analyses) separately for each Central-East European nation. Some models replace the scale for elite occupations Eq. (3) with each occupation separately. Models estimated for the whole population including those not in the labor force (for whom occupation-related variables are not defined) replace the labor force variables (7–11 in Eq. (4)) with a single indicator of labor force participation.

A more general estimate of changes over time allows for non-linearities by adding a quadratic, time squared, to the model:

$$\text{legitimate income of elite occupations}_i = (\text{Eq.4}) + \text{TimeSquared}_i + e \qquad (5)$$

This model is reported in Fig. 1, as are analogous results for time changes in perceived inequality estimated from the analogue to Eq. (5). In practice, time changes in legitimate inequality in Central-East Europe are linear, so our main model remains Eq. (4). However changes in perceived inequality in the East, as well as all changes in the West, have a small but statistically significant curvilinearity, as shown in Fig. 1.

Finally, to estimate the impact of changes in perceived income inequality, we add a term measuring respondents' perception of actual income inequality.[18] For example, for lawyers we estimate:

$$\text{legitimate income of lawyers}_i = (\text{Eq.4}) + \text{PerceivedEarnings}_i + e \qquad (6)$$

The "perceived earnings" term is somewhat different (in ways described later) than the corresponding terms in the equations treating the legitimate income of business or government occupations.

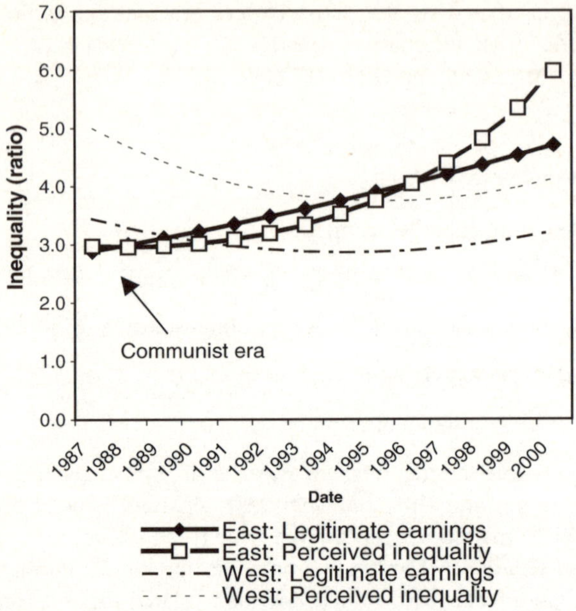

Fig. 1. Legitimate Earnings of Elite Occupations in Central-Eastern Europe and in the West and Perceived Earnings of Elite Occupations. *Note:* Adjusted for differences in background and social structure. Predicted values from Eq. (5), estimated by OLS.

DESCRIPTION

Baseline: Inequality at the End of the Communist Era

Towards the end of the Communist era in the late 1980s, norms about legitimate earnings were quite egalitarian in Central-East Europe, at least judging from the two countries for which data exist, Poland and Hungary (Table 2). They believed that high status occupations like "chairman of a large national company" or "cabinet minister in the national government" should earn around 2.5 times as much as ordinary workers. In contrast, the public in Western nations held less egalitarian norms, thinking the elite should earn 3 or 4 times as much as ordinary workers (see also Kelley & Evans, 1993, pp. 97–100). These differences are in part due to differences in social structure – Central-East Europeans had, on average, less education and lower status jobs than Western Europeans – but even after adjusting for that, Central-East Europeans had more egalitarian values, save perhaps for government officials.[19]

Changes in Central-East Europe After the Fall of Communism

With the shift toward a market economy after the fall of Communism in 1989, normative support for income inequality increased sharply (Table 2).[20] We have the fullest data for Poland and Hungary, so let us begin there.

Poland

By 1991 Poles believed that those in high status occupations deserved to earn around 3 times as much as ordinary workers, up from 2.5 times as much just a few years before. Thus in the brief period between the fall of Communism at the end of the 1980s and our survey in 1991, Poles' norms shifted from one of the most egalitarian known in the literature to a level close to the inegalitarian norms of the West.

As the shift toward a market economy grew apace during the Polish "shock treatment" of the early 1990s (Balcerowicz, 1994), norms about inequality continued to change in concert. By late 1994, Poles had come to believe that those in high status occupations deserved to earn around 3.5 times as much as ordinary workers, rising close to 3.7 times as much by 1997 and fully 7 times as much by 1999, far more than Westerners think proper.

Most dramatically, by 1999 Poles had come to feel that the "owner/manager of a large factory" should earn 14 times as much as an ordinary worker. This is a vast sum, almost four times what they thought right less than a decade before and twice what Westerners think is right (Table 2). This – and the similar if less dramatic change in the pay thought right for corporate chairmen – may come about because factories are key positions in classical free market capitalism, and the hoped-for engine of economic growth in post-Communist economies. Their performance is crucial during the chaotic and uncertain birth of a new economic system, rich with opportunities for future prosperity but equally replete with the treacherous shoals leading to disaster. In such circumstances, good management is highly productive and amply rewarded by the market.

There were similar changes for other elite occupations. But medical doctors, who Poles continue to think should be modestly paid, are a partial exception.

Hungary

The same patterns appear in Hungary (Table 2). By 1992, the egalitarian norms of the past had been replaced by support for inequality close to the higher levels acceptable in the West. This change took Hungary from one of the most egalitarian nations known – one clearly below the Western range – to a position well within the Western range. And by 1999 they accepted much more inequality than Westerners think proper.

Among the most dramatic norms in Hungary, as in Poland, concern the "owner/managers of large factories." There were no private factories in Hungary in 1987 so the factory owner question was not asked then. But in 1987 Hungarians thought it right for cabinet ministers in the national government, many of whom had responsibility for dozens of factories, to earn only 2 or 3 times what ordinary workers earned. By 1992 Hungarians already thought factory owners ought to earn 7 times as much as ordinary workers and by 1999 no less than 10 times as much. This is a huge sum, far beyond anything the Hungarians thought proper in Communist times and over twice as much as Westerners think proper for their factory owners.

Russia

In the 1990s, changes in Russia, the largest Central-East European nation, appear to be broadly similar to those in Poland (Table 2). We have no Communist era data for the USSR, but assuming Russian opinion was similar to Communist era Polish opinion is probably a reasonable guess. In any case, by the early 1990s, Russians thought that elite occupations should earn, on average, about 3.6 times as much as ordinary workers, rising sharply to 4.7 times as much by the end of the century.

Czech Republic, Bulgaria and Slovenia

Changes in the smaller Central-East European nations show a more mixed pattern (Table 2). There are no Communist era data for any of them, so there is again considerable uncertainty.

In the early 1990s, opinion in the Czech Republic was (still?) quite equalitarian, Czechs thinking elite occupations should earn just 2.8 times as much as ordinary workers – little different from Polish opinion in the Communist era. But by the end of the century, this increased sharply to 4.4 times as much, just a little less than Poles or Russians then thought proper.[21]

Bulgaria is very different. In the early 1990s, they thought elite occupations should earn 2.9 times as much as ordinary workers, noticeably less than Poles or Russians then thought proper. But by the end of the century, opinion had shifted slightly against inequality – in the opposite direction to changes in the rest of Central-East Europe – with Bulgarians thinking the elite should get just 2.6 times as much as ordinary workers.

Finally, in Slovenia changes in the 1990s appear to be small and mixed. There is acceptance of much higher pay for corporation chairmen, acceptance of a little more for doctors, but a decline in the pay thought right for cabinet ministers.

Parallel Changes Following Economic Reform in the West?

The general shift in economic policy in Britain, Australia, and many other Western nations in the late 1980s and 1990s was away from a highly regulated "social-market" type of economy toward a less regulated free market economy. In many ways this parallels the more dramatic changes in Central-East Europe. The data suggest the possibility of a slight change toward accepting more inequality in Australia,[22] Norway (Knudsen, 2001) – a country almost as equalitarian as Communist-era Central-East Europe – and some other European nations (Gijsberts, 1999, pp. 51–80). But other nations show different patterns. Overall, there is perhaps a slight decline in support for inequality from the late 1980s to early 1990s, followed by a slight rise in support for it toward the end of the century (Table 2).

None of these results makes any adjustment for structural changes following the end of the Communist era. It is to these that we now turn.

ANALYSIS

The end of Communism led to a variety of structural changes in the labor market, more in some nations than in others. Most notable was the emergence of private entrepreneurs, the growth of the petty bourgeoisie, and the decline of employment in government owned-industry. It might be that these structural changes alone explain the growing acceptance of inequality, without any deeper sea-change in Central-East European values.

In addition, long run trends toward higher educational levels and an aging population continued unabated in both East and West. There were changes in the distribution of income as well. Any of these could confound the comparison between the Communist era and later times. These complications need to be taken into account. That is done in Table 3, which estimates the models of Eq. (3) and Eq. (4) by ordinary least squares regression.

Adjustment for Structural Changes

In the event, ongoing changes in education, age composition, and family income do not account for changes since the fall of Communism (Table 3, column 1). After taking them into account, very large time changes remain. Indeed, time changes are the single most important influence on views about the legitimate earnings of elite occupations, with $\beta = +0.28$. This is in sharp contrast with the West, where time changes are minor and in the opposite direction, with $\beta = -0.03$ (column 12).

Nor do changes in the labor market account for changes in views about legitimate earnings in the post-Communist era (Table 3, column 2). On the contrary, time

Table 3. Legitimate Earnings of High Status Occupations in Six Central-East European (23,260 Cases) and Ten Western Nations (39,956 Cases) with Data From at Least Two Time Periods, 1987–2001.[a]

	Central-East Europe			Central-East European Nations[b]						Western Nations		
	Beta (1)	Beta (2)	b (3)	Russia b (4)	Poland b (5)	Czech R. b (6)	Hungary b (7)	Bulgaria b (8)	Slovenia b (9)	b (10)	Beta (11)	Beta (12)
Social change												
Time (Decades since 1989)	0.28	0.27	0.37	0.44	0.54	0.58	0.87	−0.23	ns	−0.05	−0.04	−0.03
Background and status												
Male	0.06	0.08	0.10	0.15	0.10	0.08	0.12	ns	0.09	0.08	0.08	0.07
Age (decades)	0.11	0.10	0.05	ns	0.08	0.05	0.08	ns	0.08	0.06	0.14	0.17
Education (years)	0.19	0.14	0.03	0.03	0.04	0.02	0.04	0.02	0.03	ns	−0.02	ns
Family income (ratio)	0.17	0.16	0.06	0.07	0.02	0.03	0.06	0.04	0.03	0.07	0.20	0.23
Subjective class	ns	ns	ns	ns	ns	ns	ns	0.16	ns	−0.13	−0.05	−0.04
In labor force (0 or 1)	0.03	–	–	–	–	–	–	–	–	–	–	−0.07
Social class[c]												
Occupational status (0 to 1)	–	0.08	0.18	0.20	0.20	0.16	0.14	ns	ns	0.08	0.04	–
Supervise (0 or 1)	–	ns	ns	ns	ns	ns	ns	0.10	ns	ns	ns	–
Petty bourgeoisie (0 or 1)	–	ns	ns	ns	−0.08	ns	ns	ns	ns	0.12	0.06	–
Entrepreneur (0 or 1)	–	ns	ns	ns	ns	ns	0.71	ns	ns	ns	ns	–
Government worker (0 or 1)	–	ns	ns	ns	ns	ns	0.19	ns	ns	−0.10	−0.09	–
Constant	–	–	0.39	0.56	0.23	0.27	0.06	0.77	0.47	0.83	–	–
R^2	0.16	0.18	0.18	0.12	0.27	0.31	0.49	0.11	0.15	0.09	0.09	0.09
Scale reliability, alpha[d]	0.905	0.905	–	0.900	0.906	0.895	0.901	0.887	0.857	–	0.899	0.901
Cases	23,260	14,574	14,574	2,031	5,023	1,771	2,831	1,692	1,226	25,102	25,102	39,956
Population, million	–	–	–	148	39	10	10	8	2	–	–	–

Note: ns – not significantly different from zero at $p < 0.01$, two-tailed.

[a] *Source:* World Inequality Study, incorporating data from the International Social Survey Programme, the International Survey of Economic Attitudes, and other sources. The Western nations are Australia, Canada, West Germany, the Netherlands, New Zeland, Norway, the Philippines, Sweden, Great Britain, and the USA. Columns 1 and 12 are from Eq. (3) and columns 2–11 from Eq. (4).

[b] Listed in order of population size.

[c] For those in the labor force only.

[d] Legitimate earnings are measured by an additive scale averaging answers about the legitimate earnings of six elite occupations (chairman of a large national corporation; owner-manager of a large factory; lawyer; doctor in general practice; judge in the nations's highest court; and cabinet minister in the national government), each expressed as (the logarithm of) a ratio to the legitimate earnings of skilled and unskilled factory workers. If not all questions were answered, the mean is of those that were answered. Some early surveys asked only three occupations (chairman, doctor, and cabinet minister). Reliabilities are standardized item alphas.

changes remain large, and are still by far the most important influence, with $\beta = +0.27$. In concrete terms, every decade since the fall of Communism in 1989 has produced an increase in the legitimate earnings of elite occupations of around 47% (column 3; $\exp(0.37) = 1.47 = 47\%$ increase by 1999). This is a dramatic change.

The changes in Central-East Europe seem to have occurred at about the same rate throughout the period since the fall of Communism (Fig. 1).[23] In particular, there is no clear evidence for a disproportionate response to the sudden and unexpected fall of Communism, nor the "shock therapy" that some Central-East European nations underwent in the years immediately following. If anything, it may even be that changes were most rapid toward the end of the century, about 10 years after the fall of Communism. In Poland, the country for which we have the longest series of surveys, this appears to be the case ($t = 19.8, p < 0.001$).[24] But for Hungary, with the next best data, exactly the opposite pattern prevails ($t = -15.8$, $p < 0.001$). Thus no firm conclusion is warranted.

In Western nations, in contrast to Central-East Europe, there is no substantial change in the legitimate earnings of elite occupations over the last decade of the century (Table 3, column 10 and Fig. 1). If anything, there may have been a slight *decline* from the end of the 1980s to the middle 1990s, followed by an equally small increase through the end of the century (the curvilinearity is significant: $t = 16.4, p < 0.001$).[25]

Differences in Central-East Europe

These patterns are clear in the larger Central-East European nations but not in all of the smaller ones. In Russia, with a population of around 150 million, the legitimate earnings of elite occupations rose by 55% in the decade following the end of Communism (Table 3, column 4; $\exp(0.44) = 1.55 = 55\%$). In Poland, with a population near 40 million – and more extensive marketization of the economy – change was even more rapid: 72% (= exp(0.54)). The same was true in the Czech Republic (79%) and even more dramatically in Hungary (139%). These latter two are both smaller nations, with populations around 10 million, with relatively extensively marketized economies.

However, in small (2 million), generally Westernized Slovenia, there was no statistically significant change, although their norms were not especially egalitarian at the beginning. And in Bulgaria, with a population of 8 million and little marketization, the legitimate earnings of elite occupations actually *declined* 21% between 1992 and 1999.[26] It is not clear why these two nations depart from the general pattern. One possibility is that the citizens of smaller nations are more likely to take as a reference group the norms and behavior other nations rather than responding to the internal developments in their own economy.

Overall, it seems likely that the general pattern of growing acceptance of inequality applies to the majority of the population of the formerly Communist Central-East European nations, although not to every nation, particularly not all the smaller ones.

Changes in Views about Specific Occupations

The same general pattern holds for all six occupations available in our data (Table 4, panel 1). Changes over time are largest for views about the legitimate pay of the chairman of a large national corporation ($\exp(0.46) = 58\%$ increase) and around 35% for other occupations. Somewhat surprisingly, the growth in legitimate earnings for cabinet ministers in the national government is just as high as for other elite occupations, despite that fact that the actual power of cabinet ministers has declined since the Communist era, as the centralized and authoritarian "dictatorship of the proletariat" faded unlamented into history.

Doctors are an exception to the general pattern: the legitimate pay of a "doctor in general practice" increased by only 14% in Central-East Europe since the end of the Communist era. As we noted before, doctors have long been somewhat of a special case in Central-East Europe. But this is not true of all professional occupations: the pay thought legitimate for lawyers increased by a substantial 43%.

Social Structure and Legitimate Earnings

Education

The most important socioeconomic influence on norms in Central-East Europe is education: the well educated have long been more hostile to Communism and more sympathetic to market reforms than the less educated (Frentzel-Zagorska & Zagorski, 1993; Zaborowski, 1995). They are also substantially more willing to endorse high pay for elite occupations of all types, $\beta = 0.19$ overall (Table 3, column 1) or $\beta = 0.14$ even after adjusting for their better occupational outcomes (column 2). For example, a university educated Central-East European would, on average, favor paying elite occupations 23% more than someone with the same background and occupation who left school at age 16.[27] The effect is larger in Poland and Hungary (about 32%); about the same in Russia and Slovenia; and less in the Czech Republic and Bulgaria (about 15%; columns 4–9). By contrast, well

Table 4. Legitimate Earnings of Various Occupations in Eastern Europe, 1987–2001. 6 Nations with Data from at Least Two Time Periods; Resnondents in the Labor Force Only.[a]

	Business Occupations		Professional Occupations		Government Occupations	
	Chairman, National Corporation b (1)	Factory Owner b (2)	Lawyer b (3)	Doctor b (4)	Judge, Highest Court b (5)	Cabinet Minister b (6)
Panel 1: Basic model						
Time (Decades since 1989)	0.46	0.24	0.36	0.13	0.34	0.30
Male	0.12	0.15	0.07	0.06	0.09	0.09
Age (decades)	0.06	0.06	0.04	0.05	0.06	0.06
Education (years)	0.03	0.03	0.01	0.02	0.03	0.03
Family income (ratio)	0.07	0.09	0.08	0.05	0.06	0.05
Subjective class	ns	ns	ns	−0.10	ns	ns
Occupational status (0 to 1)	0.29	0.17	0.08	0.11	0.12	0.16
Supervise (0 or 1)	ns	0.08	ns	ns	ns	ns
Petty bourgeoisie (0 or 1)	ns	ns	ns	ns	ns	ns
Entrepreneur (0 or 1)	ns	ns	ns	ns	ns	ns
Government worker (0 or 1)	ns	−0.08	−0.06	−0.04	ns	0.06
Constant	0.35	0.79	0.45	0.19	0.58	0.51
R^2	0.16	0.09	0.13	0.10	0.11	0.10
Panel 2: Controlling for perceptions of the actual amount of inequality[b]						
Time (Decades since 1989)	0.30	ns	0.06	−0.06	0.07	0.09
Perceptions	0.43	0.56	0.44	0.28	0.54	0.41
Other variables[c]	–	–	–	–	–	–
Cases	13,747	10,705	11,031	14,320	10,801	13,441

Note: ns – not significantly different from zero at $p < 0.01$, two-tailed.

[a]Russia, Poland, Czech Republic, Hungary, Bulgaria and Slovenia. Number of cases varies depending on missing data and because not every occupation was included in all surveys. Source: World Inequality Study, incorporating data from the International Social Survey Programme, the International Survey of Economic Attitudes, and other sources.

[b]Measured by the perceived earnings of other occupations. To avoid part-whole artifacts, for business occupations this is the perceived earnings of professional and government occupations; for professional occupations, it is the perceived earnings of business and government occupations; and for government occupations, the perceived earnings of business and professional occupations.

[c]Controlled but not shown: male, age, education, family income, subjective class, occupational status, supervise, petty bourgeoisie, entrepreneur, and government worker.

and poorly educated Westerners have much the same views on inequality (Table 3, columns 11 and 12).

The fact that educational differences persist in Central-East Europe even after adjusting for the better jobs education brings, and that there are no corresponding educational differences in the West,[28] both suggest that the education effect is not self-interest – although the well educated do stand to gain more than the poorly educated from marketization – but something else. One plausible candidate is the greater knowledge and understanding that well-educated Central-East Europeans have of economics, and the intellectual predominance of market economy ideas in the public discourse of Central-East Europe.

Demography and Stratification Position

Demographic influences on legitimate earnings are modest in magnitude both in general (Table 3) and for each specific occupation (Table 4). This is consistent with previous findings (Gijsberts, 1999; Kelley & Evans, 1993):

- Men favor somewhat higher earnings for high status occupations than do women, by roughly 10%. The difference is largest in Russia and Hungary, but evident everywhere, including in the West. The only exception is Bulgaria. Men are especially generous to business occupations (Table 4, columns 1 and 2), but less so to professional occupations (columns 3 and 4).
- Older respondents are noticeably more supportive of inequality in both Eastern nations ($\beta = 0.11$) and, especially, in Western nations ($\beta = 0.17$). But the effect varies in size from nation to nation, disappearing entirely in Russia and Bulgaria. It is about the same size for all six occupations. This is a life-cycle effect, with people becoming more supportive of inequality as they age.[29]
- Family income has a large effect, with the more prosperous in both East ($\beta = 0.17$) and West ($\beta = 0.23$) favoring higher pay for elite occupations. The effect is largest in Russia and Hungary, but is evident in all Central-East European nations. It appears to be a bit stronger for business occupations than for government occupations, with professional occupations somewhere in between.
- Subjective social class hardly matters in Central-East Europe. The exceptions are Bulgaria (where the upper classes favor higher pay for the elite) and doctors (for whom the lower classes favor higher pay). In the West, those subjectively identifying with the upper classes actually favor less pay for the elite than equally well-educated, high status and prosperous people who identify with the lower classes.
- There is little difference between those in the labor force and others. In the East, they are fractionally more supportive of high pay for elite occupations, but in the West slightly less supportive.
- Those in higher status occupations favor higher pay for elite occupations, both in the East and the West. The difference modest: a professional, themselves at the top of the occupational hierarchy would, on average, favor higher pay those in elite jobs. The difference is larger in Russia and Poland, 22%, but absent in Bulgaria and Slovenia. It is largest for business occupations, especially chairman (34%); middling for government occupations; and – surprisingly – smallest for professional occupations (8–10%).

Class Position

Other class differences are modest:

- Supervisors support no higher pay for elite occupations than anyone else, save in Bulgaria. But they would pay factory owners a modest 8% more than others think proper.
- The petty bourgeoisie – the solo self-employed – are still rare in most of Central-East Europe. But so far as we can tell, they do not have distinctive views about legitimate earnings save in Poland, where they would pay elite occupations 8% less than others think right. In the West, in contrast, the more numerous and long established petty bourgeoisie seem to have adopted more pro-business values and would pay the elite 13% more.
- Entrepreneurs – private business owners with employees – are also still exceedingly rare in Central-East Europe. Their views do not yet seem to be very distinctive, save perhaps in Hungary where they would pay the elite far more than others think proper.
- Government workers, still numerous in Central-East Europe, are not very distinctive. Only in Hungary do they differ from workers in private firms, preferring to pay the elite 21% more, surprisingly. Throughout the East, they would pay lawyers, doctors and factory owners a little less than others think right. In the West, in contrast, government workers would pay the elite 10 or 11% less than private employees think right.

The fact that all these differences are small – especially compared to the influence of education and occupational status – suggests that norms about legitimate earnings are only in small part a matter of self-interest (Hypothesis 3a) rather than "intellectual" considerations (Hypothesis 3b).

Perceptions of the Actual Level of Inequality

We also measured perceptions of how much occupations are thought *actually* to earn:

> We would like to know what you think people in these jobs *actually* earn . . .
>
> ≫ Please say how much you think they *usually* earn each year, before taxes.
>
> ≫ Many people are not exactly sure about this, but your best guess will be close enough.
>
> a. First, *about* how much do you think a skilled worker in a factory earns? $ ________ dollars
>
> etc . . .

A series of other occupations followed, with wording parallel to that for the legitimate occupational earnings questions. Following the methods used in the analysis of legitimate earnings, we express each respondent's answers to these questions as (the natural log of) the ratio his or her perceptions of elite earnings to his or her perceptions of the actual earnings of ordinary workers (similar to Eqs (1) and (2)).[30]

The growth of income inequality in Central-East Europe is clearly perceived by the public (Fig. 1). The perceived earnings of elite occupations roughly doubled over the decade after the fall of Communism, from around 3 times the income of ordinary workers to 6 times that. The growth was more rapid toward the end of the 1990s than it was in the first few years after the fall of Communism ($t = 15.23$, $p < 0.001$).[31]

Changes in the West followed a very different pattern (Fig. 1). At the end of the 1980s, the Western public perceived the elite in their countries to earn about 5 times as much as ordinary workers – far more than Easterners thought their elite earned. But then inequality in the West was perceived to have declined for the next few years, up to 1995, with the elite's income dropping to less than 4 times ordinary workers'. Then it stabilized or perhaps rose slowly again through the end of the century.

The Gap Between Perceived and Legitimate Earnings

At the end of the Communist era, amount of inequality the Central-East European public thought existed in their societies was about what they thought was morally proper: they felt that the elite ought to earn, and did actually earn, about 3 times as much as ordinary workers (Fig. 1). Then over the next half a dozen years, their feelings about how much the elite ought to earn rose steadily while the elite's actual pay lagged a bit behind. Only in 1996 did norms and reality come once again into agreement. After that the elite's actual income – at least, as perceived by the Central-East European public – grew much more rapidly. By the and of the century, the public thought the elite actually earned about 6 times as much as ordinary workers but felt that they ought to earn only 4 or 5 times as much.

One consequence of these parallel changes is that in many post-Communist societies, there has been little change in public opinion on broad questions about "whether there is too much inequality in our society" or whether the government should have "reducing inequality" as a goal for public policy (e.g. Zaborowski, 1994, 1995).[32] But by the end of the century, the society to which the questions refers is in fact very unequal, much more so than in Communist times, so the meaning of the answers is quite different. There is nothing inconsistent in this: people can perfectly well hold that inequality ought to be higher now than it was in Communist days (for example, that the elite's earnings should increase from 2 times ordinary workers' earnings to 4 times), but simultaneously hold both that it was about right in Communist times (when it ought to be 2, and actually was 2) and right ten years later (when it ought to be 4 and actually was 4).

In politics, questions of income inequality sometimes concern specific occupations (e.g. cabinet ministers earn too much), sometimes broader groups of occupations (e.g. the elite has too much money) and sometimes constitute a broad global issue about the amount of "inequality in the society as a whole" (corresponding to point 4 in Fig. 4). The links between specific "micro" norms on earnings and the society-wide outcome are complex (Jasso, 1994), as yet poorly understood, although politically important in many nations. We reserve our analysis of them for a future paper. In this first paper, we concentrate on norms and perceptions about the earnings of specific occupations and groups of occupations, important issues in themselves and an essential first step in understanding the role of income inequality in the politics of post-Communist societies.

The gap between perceptions and norms in the West shows a quite different pattern (Fig. 1). At the end of the 1980s, the Western public thought the elite actually earned about 5 times as much as ordinary workers, but that it ought to earn only 3.5 times as much. Over the next few years, the public thought the elite's income actually declined, from 5 to less than 4; but at the same time the public's norms about how much the elite ought to earn also declined, from 3.5 to less than 3. So the gap between reality and public norms did not change greatly. Later, toward the end of the century, the public perceived the elite's income as growing, but also felt that some growth was legitimate. So the gap stayed much the same.

Do Actual Changes in Inequality Explain Normative Changes?

If we assume that the public believes differences in earnings largely reflect productivity – as they do according to classical economic theories about competitive markets – Aristotelian norms then imply a strong link between *perceptions* of occupational earnings and normative *acceptance* of earnings differentials (Hypothesis 2). Thus when people perceive changes in actual income of different occupations, they should endorse corresponding changes in the occupation's legitimate earnings. To see whether this is so, we expand our basic model (Eq. (5)) to include a measure of perceived earnings (Eq. (6)).

Technical Complications

However, the perceived earnings term in Eq. (6) raises some difficult technical issues. For an occupation such as doctor (and other elite occupations) the difficulty is that there is correlated error between estimates of a doctor's legitimate income and perceptions of their actual income. If, for example, one respondent is thinking of a highly trained, high-tech doctor in a university teaching hospital while another respondent is thinking of a modest, elderly family doctor in a small rural village, there will be a strong, artifactual correlation between perceived and legitimate income simply because of this heterogeneity in the kinds of doctors the two respondents are thinking of. This will bias upward the estimates of the effect of perceived income on legitimate income. Our estimates suggest that this bias is large, perhaps as much as doubling the effect (details available on request).

We therefore omit the perceived income of doctors from the version of Eq. (6) predicting the legitimate earnings of doctors. We also omit the perceived income of lawyers, a closely related professional occupation, and use only the perceived incomes of business occupations (chairman, factory owner) and government occupations (judge, cabinet minister).[33] In effect, we use these as instruments in estimating the perceived income of doctors. Similarly, for business occupations we estimate perceived inequality using only professional and government occupations and for government occupations, we use only business and professionals.

Consequences of Changes in Perceived Inequality

The evidence that perceptions of occupational earnings shape normative acceptance of earnings differentials is strong (Table 4, panel 2). Indeed, their effect is stronger than any other influence in our model. These results imply that if marketization increases an elite job's pay by $1000, then that job's legitimate pay will rise by roughly $500. This rise is largest for factory owner and judge, around $700, and smallest for doctors, around $300.

These results are consistent with other evidence from a number of Central-East European nations using different measurement and methods (Alwin et al., 1995; Arts et al., 1995). They are also consistent with Hypothesis 2.

Changes in perceived inequality probably explain most, but not all, of the increase in legitimate inequality in Central-East Europe since the fall of Communism. However, the results vary considerably from occupation to occupation, and the technical complications are serious, so no unequivocal conclusion is warranted.[34]

- For corporation chairman, the impact of time drops from 0.46 (Table 4, panel 1, row 1) to 0.30 (panel 2, row 1). This suggests that about a third of its effect is due to changes in perceived inequality.[35]
- For lawyer, judge, and cabinet minister the impact of time drops even more sharply, suggesting that 70 or 80% of time's effect is due to changes in perceived inequality. And for factory owner, all of the effect seems to be due to changes in perceived inequality.

- For doctors the small time effect, 13%, is more than accounted for by changes in perceived inequality. Central-East Europeans seem to think that doctors' pay should fall about 6% further behind the pay of other elite occupations.

These results are consistent with Hypotheses 2.

Rejected Alternative Theories

Our results are inconsistent with the predictions of a number of other theories and therefore argue against these theories.

- *Egalitarianism*. The strict egalitarian rejection of any inequality whatsoever is clearly not shared by ordinary people in Central-East Europe. They did not hold completely egalitarian views even in the past – despite the ideological egalitarianism of Communism, its sustained propaganda for equality, and very low levels of actual inequality in Communist society – even though they were more egalitarian than most Westerners. Even less do they hold such views in the present.
- *Enlightenment*. The general tenor of change in Central-East Europe since the fall of Communism is certainly not toward the liberal and egalitarian ideals of the enlightenment. Whether this is one symptom the beginning of a long term reversal of the trend in economic and welfare areas, or is only a temporary reversal in the general liberal trend, itself to be reversed in a decade to two, is not clear from our data.
- *Existential Theories*. Our results are not consistent with the existential argument that whatever is factually the case for a long time comes to be accepted normatively and remains accepted for even a longer time. That argument implies that the egalitarian legacy of 40 years of Communism would change only gradually. Yet in fact there was no gradual, long term decline in egalitarian views, but rather a sudden, dramatic shift.

CONCLUSION

Our data suggest that the transition from a Communist command economy led the public abruptly to change its view about inequality, at least in the larger Central-East European nations and most, but not all, of the smaller nations. So far as we can judge from the Polish and Hungarian data, the Central-East European public held strongly egalitarian norms up to the last days of Communism. But within two or three years of its fall, amidst the first tentative steps toward a market economy, they seem to have shifted far toward the much less egalitarian norms found in the West. And as free markets developed further, ideals continued to change. Just a decade later, at the end of the 20th century, Central-East Europeans accept substantially more income inequality than most Westerners think right.

Much more speculatively, our argument leads to a prediction about future trends in attitudes toward inequality in Central-East Europe. Our argument assumes that Central-East Europeans are fundamentally similar to Westerners, so that differences in their norms about inequality are just a reflection of their different circumstances. We assume that the present objectively high level of inequality reflects the unusual opportunities, and unusual risks, that accompany the disintegration of the command economy and the emergence of a new, untried, but potentially much more productive market economy. These opportunities and risks mean that the differences between good and bad economic leadership have huge consequences and so imply that the public with think it right to reward them highly. But after this formative period, eventually the market will develop and mature, leaving few unusual opportunities and few unusual risks, eventually converging on the usual Western pattern. Productivity differences will then be little different than in Western economies, and so attitudes about income inequality will, on Aristotelian arguments, gradually become similar to Western patterns. This implies that norms in Central-East Europe will eventually converge on the usual Western pattern. But they will converge from above, not below.

Political Implications

As a market economy gradually sprang up after the fall of Communism, acceptance of income inequality in Poland and Hungary grew rapidly, taking public opinion far from the egalitarian norms of the past. But the actual amount of inequality also seems to have grown rapidly – indeed the public mostly think it grew even more rapidly. So there has been relatively little change in public opinion on broad questions about "whether there is too much inequality in our society" or whether the government should have "reducing inequality" as a goal for public policy.

This has important political implications. In the past, populist anti-inegalitarian political appeals were popular, but not overwhelmingly popular. If public attitudes toward inequality had remained unchanged to the objectively much more inegalitarian present, then the discrepancy between what the public wants and what the reality is would have grown vastly, and the populist appeal might well have become irresistible. That attitudes have shifted so quickly means that there is now much more scope for market-oriented reform than would otherwise have been the case.[36] Thus even in the early stages of economic development when objective inequalities often grow rapidly and are perceived as such, democracy and inequality can coexist. However, the growing gap between perceived and accepted inequalities – even if the latter grow too – may stimulate some dissatisfaction. This may have contributed to electoral victories of ex-Communist parties in Central-East Europe in the last decade.

NOTES

1. The older nomenclature was "Eastern Europe" but usage is now varied and sometimes conflictual, with both normative and substantive issues involved. We wish to take no views here on these matters, and so adopt the neutral, if ponderous, "Central-East" usage.

2. There was, of course, already inequality in state socialist societies before marketization, some based on political and bureaucratic advantages of a sort that would be undermined by the changes accompanying marketization (e.g. Zhou & Suhomlinova, 2001). That reduces inequality, *ceteris paribus*. But, net of that there was rising earnings inequality in the early 1990s (e.g. Gerber & Hout, 1998).

3. There were similar but much less marked changes from liberalizing policies in the West (Harrison & Bluestone, 1990; Johnson et al., 1995; Smeeding et al., 1993).

4. In the absence of institutional change, the early stages of capitalist economic development probably do not in themselves increase inequality (Kelley & Haller, 2001; Lindert, 2000, the references given there).

5. This acceptance may, however, be limited to a relatively short "extraordinary period" (Balcerowicz, 1994) during which people are willing to sacrifice their short-term interests in favor of long term, possibly altruistic goals (as, for example, fighting Communism and building a new democratic order).

6. The other three are the deductive mode, deriving morality from general principles held to be universally valid; the expressive mode, judging actions as morally right or wrong according to one's immediate emotive reaction; and the consequentialist mode, assessing rights and wrongs by their results.

7. Our data demonstrate sharp changes in the public's perceptions of the earnings of high status jobs. We have no direct evidence that they attribute this to changes in productivity, although that is consistent with the general tenor of public attitudes toward economic transformation and the market economy (e.g., Frentzel-Zagorska & Zagorski, 1993; Zagorski, 1994) and with direct evidence in our Polish, Bulgarian, Finnish and Australian surveys that the public regards private companies as more economically efficient than state-owned ones.

8. Government privilege and bureaucratic favoritism of course remain, although less in Poland and Hungary than in many other post-Communist nations. The decline in the government's influence and the growth of the private sector reduce the bureaucracy's influence compared to the command economy of the past.

9. For related arguments and persuasive data, see Gijsberts (1999, pp. 51–80).

10. This project was supported by a grant from the Australian Research Committee's *Research Infrastructure Equipment and Facilities Scheme* (RIEF) to the Melbourne Institute of Applied Economic and Social Research, University of Melbourne (Dawkins et al., 2000).

11. The Drafting Committee for all three of these modules was chaired by M. D. R. Evans and one of us (Kelley).

12. References are given only to the most recent survey, usually 1999. Details are in the references.

13. Full citations are given only for the latest survey. The Finish survey, available only for one time period, was not used in this analysis.

14. Earlier surveys included "farm laborer," which is a useful addition, but it is not available in the 1999 round of surveys. In the interests of comparability over time, we therefore omit it.

15. The phrases in brackets varied to reflect local nomenclature. For example, in the USA judge was "judge in the Supreme Court" (the highest U.S. court) while in Australia it was "judge in the High Court" (Australia's highest court).

16. We use this rather than a constant that is the same for all respondents – for example, the society-wide mean income of unskilled workers used in previous analyses of these data by Kelley and Evans (1993).

17. In Poland and Hungary in the Communist era, and probably throughout Central-East Europe, the earnings thought proper for doctors were less than in Western nations. This is a long standing difference. Doctors, professors and similar professional occupations not involved in the production of physical goods were treated as a pure cost to the economy in the Communist's system of national accounts (like welfare transfers), not counted as a valuable service, much less as investment; and their actual pay was abysmal. Routine while collar jobs were also less valued than in capitalist societies and skilled workers more highly valued (Kraus & Hodge, 1987).

18. Our model assumes that perceptions influence norms, rather than the other way around. This follows theory and the usual models (e.g. Homans, 1974; Kluegel et al., 1995). However the opposite causal order could be argued (Headey, 1991). The dramatic change in perceptions of inequality following the fall of Communism described later in this paper, and found in other studies on many other aspects of inequality (e.g. Zaborowski, 1995), combined with the only modest shift in norms in the same period, is more consistent with our assumption than with the opposite.

19. OLS estimates from a pooled model using Eq. (4) with the addition of an East European dummy variable gives $t = 19.9$ for chairman; $t = 28.6$ for doctor and $t = 5.3$ for cabinet minister, all significant at $p < 0.001$. However OLS underestimates the standard error (Eastern Europe is a country-level rather than individual level variable) and so overestimates the t-values.

20. For a different view see Listhaug and Aalberg (1999).

21. See also Rehakova (1997).

22. For other analyses of attitudes to inequality in Australia, see Austen (1999); Borland (1999); Evans and Kelley (2002); Headey (1991); and Kelley and Evans (1993).

23. Based on Eq. (5), which allows for curvilinear effects by including a time quadratic.

24. Based on Eq. (5) estimated for Poland alone, using six surveys with 8,041 cases. The corresponding estimate for Hungary is based on three surveys.

25. Based on Eq. (5) estimated for Western nations only, with 32 surveys and 25,102 cases.

26. There is a lively debate about just how much of a transition to a market economy and how much of a change in living standards the end of Communism brought to Bulgaria, in part because there are continuing debates about the degree to which GNP and other

living-standards measures were inflated towards the end of Communism. If so, then the actual or anticipated gains in living standards associated with marketization that are legitimating in equality in the other countries might be absent there – not that the causal process is different, but that the level of marketization is so low it has not generated any legitimation.

27. Viz a difference of $16 - 9 = 7$ years of education, times the effect of education: $\exp(7 \times 0.03) = 1.23 = 23\%$ more (Table 3, column 3).

28. Indeed, the Western evidence suggests that the well educated are if anything *less* favorable to inequality than poorly educated Westerners in comparable jobs (Table 3, column 11).

29. When the age difference was first discovered in data for a single point at time, it seemed likely to be reflecting a secular trend toward more equalitarian attitudes (Kelley & Evans, 1993; Kluegel et al., 1995). Our multi-time period data rule out that important possibility.

30. How accurate these perceptions are, especially in the unsettled economies of Central-Eastern Europe, is debatable. Our impression is that they are, at least in aggregate, reasonably accurate. In particular, they do not vary much according to respondents' own social characteristics, thus behaving more like facts than values. But whether or not these questions fully reflect reality, they are still real in their consequences.

31. Estimated from a model analogous to Eq. (5), based on 14,538 cases.

32. Our results are based on standard questions about the earnings of specific occupations which are widely in the social justice-equity-legitimation literature (e.g. Kelley & Evans, 1993; Kluegel et al., 1995; Zentralarchiv, 1989, 1994). They do not directly ask about inequality in the society as a whole but instead build up a picture of the whole as the sum of many concrete, specific micro level parts. A different approach to inequality, common in political contexts, is to ask broad global questions about the amount of "inequality in the society as a whole."

33. Measured by an additive scale analogous to Eq. (3).

34. Sensitivity tests with alternate measurement of the perceptions variables are consistent in showing that perceptions have a very strong effect on norms. However, the size of the remaining time effect is sensitive to measurement decisions.

35. Viz $(0.46-0.30)/0.46 = 36\%$.

36. Moreover, a good case can be made that attitudes to inequality shape attitudes to many other political policies that can serve as a means of reducing inequality, for example views on unemployment policy or government ownership of industry (Luo, 1998; Sikora, 2000).

ACKNOWLEDGMENTS

We thank Clive Bean, Peter Dawkins, M. D. R. Evans and Janina Frentzel-Zagorska for their comments. Parts of the paper draw on Kelley and Zagorski (2002).

REFERENCES

Adam, J. (1993). Transformation to a market economy in the former Czechoslovakia. *Europe-Asia Studies*, *45*(4), 627–646.

Alwin, D. F. (1987). Distributive justice and satisfaction with material well-being. *American Sociological Review*, *52*, 83–95.

Alwin, D. F., Gornev, G., & Khakhulina, L. (1995). Comparative referential structures, system legitimacy, and justice sentiments: An international comparison. In: J. R. Kluegel, D. S. Mason & B. Wegener (Eds), *Social Justice and Political Change: Public Opinion in Capitalist and Post-Communist States* (pp. 109–130). New York: Aldine de Gruyter.

Aristotle (1985 [322 BC]). *Nicomachean ethics* (trans. T. Irwin). Indianapolis: Hackett Publishing.

Arts, W., Hermkens, P., & van Wijck, P. (1995). Justice evaluation and income distribution in East and West. In: J. R. Kluegel, D. S. Mason & B. Wegener (Eds), *Social Justice and Political Change: Public Opinion in Capitalist and Post-Communist States* (pp. 131–150). New York: Aldine de Gruyter.

Austen, S. (1999). Norms of inequality. *Journal of Economic Issues*, *33*(2), 435–442.

Balcerowicz, L. (1994). Economic transition in central and Eastern Europe: Comparisons and lessons. *Australian Economic Review*, *1*, 47–59.

Bartholdy, K., & Flemming, J. (1993). Statistical review: Economic developments and prospects in Eastern Europe and the former Soviet Union. *Economics of Transition*, *1*(3), 366–381.

Bean, C. S. (1991). Comparison of national social science survey data with the 1986 census. *National Social Science Survey Report*, *2*(6), 12–19 (ISSN 1031-4067).

Bean, C. S. (1995). Update: Comparison of national social science survey data with the census. Working Paper, International Survey Center, Research School of Social Sciences, the Australian National University.

Becker, G. S. (1971 [1957]). *The economics of discrimination* (2nd ed.). Chicago: University of Chicago Press.

Becker, J., & Nas, M. (1987). *ISSP, The Netherlands*. Codebook and Machine-readable Data File. The Hague: Sociaal en Cultureel Planbureau.

Bell, D. (1972). On meritocracy and equality. *The Public Interest* (Fall), 40.

Bellah, R. N. (1974). New religious consciousness and the crisis in modernity. In: C. Y. Glock & R. N. Bellah (Eds), *The New Religious Consciousness* (pp. 333–352). Berkeley: University of California Press.

Berger, J., Zeldich, M., Anderson, B., & Cohen, B. (1972). Structural aspects of distributive justice: A status value formulation. In: J. Berger, M. Zelditch & B. Anderson (Eds), *Sociological Theories in Progress* (Vol. 2, pp. 119–146). Boston: Houghton Mifflin.

Berlyne, D. E. (1960). *Conflict, arousal and curiosity*. New York: McGraw-Hill.

Beskid L., Milic-Czerniak, R., & Sufin, Z. (1995). *Polacy a nowa rzeczywistosc ekonomiczna: Procesy przystosowania sie w mikroskali*. Warsaw: IFiS PAN Publishers.

Blau, P., & Duncan, O. D. (1967). *The American occupational structure*. New York: Free Press.

Borland, J. (1999). Earnings inequality in Australia: Changes, causes and consequences. *Economic Record*, *75*, 177–202.

Capling A., & Galligan, B. (1992). *Beyond the protective state: The political economy of Australia's manufacturing industry policy*. Cambridge: Cambridge University Press.

Chirot, D. (1986). *Social change in the modern era*. New York: Harcourt Brace Jovanovich.

Cichomski, B., & Morawski, P. (1999). *ISSP, Poland*. Codebook and Machine-Readable Data File. University of Warsaw: ISS (Institute for Social Studies).

Clauge, C., & Rausser, G. C. (Eds) (1992). *The emergence of market economies in Eastern Europe*. Cambridge, MA: Blackwell.

Danziger, S., & Gottschalk, P. (Eds) (1994). *Uneven tides: Rising inequality in America*. New York: Russell Sage.

Davis, J. A., Smith, T. W., & Hout, M. (1999). *ISSP, United States*. Codebook and Machine-Readable Data File. Chicago: National Opinion Research Center.

Davis, K., & Moore, W. E. (1945). Some principles of stratification. *American Sociological Review, 10*, 242–249.

Davis, N., & Robinson, R. V. (1999). Their brothers' keepers? Orthodox religionists, modernists, and economic justice in Europe. *American Journal of Sociology, 104*, 1631–1665.

Dawkins, P. et al. (2000). *International economic and social unit-record database*. Australian Research Committee, Research Infrastructure Equipment and Facilities Scheme grant to the Melbourne Institute of Applied Economic and Social Research, University of Melbourne.

Diez-Medrano, J. (2002). *ISSP 1999: Inequality-III*. Codebook and Machine-Readable Data File. JD Systems, Madrid, Spain.

Dillman, D. A. (1993). The design and administration of mail surveys. *Annual Review of Sociology, 17*, 225–249.

Dimova, L. (1999). *ISSP, Bulgaria*. Codebook and Machine-Readable Data File. Sofia: Agency for Social Analyses.

Domanski, H., & Zagorski, K. (1991). Mechanizmy dystrybucji zarobkow w roznych systemach: Australia, Polska, Stany Zjednoczone, Wegry. *Studia Socjologiczne, 3–4*, 122–123.

Ehrenberg, R. G., & Smith, R. S. (1982). *Modern labor economics*. Glenview, IL: Scott, Foresman.

Eulau, H., & Lewis-Beck, M. S. (1985). *Economic conditions and electoral outcomes: The United States and Western Europe*. New York: Agathon Press.

Evans, M. D. R., & Kelley, J. (2002). *Australian economy and society 2001: Education, work, and welfare*. Sydney: Federation Press.

Evans, M. D. R., Kelley, J., & Kolosi, T. (1992). Images of class: Public perceptions in Hungary and Australia. *American Sociological Review, 57*, 461–482.

Festinger, L. (1964). *Conflict, decision and dissonance*. Stanford: Stanford University Press.

Frank, R. H. (1985). *Choosing the right pond: Human behavior and the quest for status*. Oxford: Oxford University Press.

Franklin, M., Mackie, T., & Valen, H. (Eds) (1992). *Electoral change: Responses to evolving social and attitudinal structures in western countries*. Cambridge: Cambridge University Press.

Frentzel, J. (1965). Cognitive consistency and self-concept. *The Polish Sociological Bulletin, 1*, 71–86.

Frentzel-Zagorska, J. (1993). The road to a democratic political system in post-Communist Eastern Europe. In: J. Frentzel-Zagorska (Ed.), *From a One-Party State to Democracy: Transition in Eastern Europe* (pp. 165–193). Amsterdam: Rodopi.

Frentzel-Zagorska, J., & Zagorski, K. (1993). Polish public opinion on privatisation and state interventionism. *Europe-Asia Studies, 45*(4), 705–728.

Frizzell, A., & Pyman, H. (1999). *ISSP, Canada*. Codebook and Machine-Readable Data File. Ottawa: Carleton University Survey Center.

Gebethner, S., & Raciborski, J. (1992). Wybory '91 a Polska Scena Polityczna. (Elections '91 and the Polish Political Scene). Warszawa: Wydawnictwo Fundacji Polska w Europie.

Gendall, P. (1999). *ISSP, New Zealand*. Codebook and Machine-Readable Data File. Palmerston North: Department of Marketing, Massey University.

Gerber, T., & Hout, M. (1998). More shock than therapy: Market transition, employment and income in Russia, 1991–1995. *American Journal of Sociology, 104*, 1–50.

Gijsberts, M. (1999). *The legitimation of inequality in state-socialist and market societies, 1987–1996*. Ph.D. Thesis, University of Utrecht, Thela Thesis.

Gijsberts, M., & Ganzeboom, H. (1996). Replication in the Netherlands of the international social survey programme's: Ideology of inequality/social inequality, Round 2. Netherlands: Interuniversity Consortium for Sociological Research (ICS): Occasional Papers and Document Series.

Haller, M. (1990). Societal types and attitudes towards inequality. Paper presented to the World of Congress of Sociology, Madrid.

Haller, M., & Hoellinger, F. (1999). *ISSP, Austria*. Codebook and Machine-Readable Data file. Graz: Institut fuer Soziologie der Universitaet Graz.

Harkness, J., Mohler, P. P., & Braun, M. (1999). *ISSP, Germany*. Codebook and machine-readable data file. Mannheim: Zentrum für Umfragen, Methoden und Analysen.

Harrison, B., & Bluestone, B. (1990). Wage polarisation in the U.S. and the 'flexibility' debate. *Cambridge Journal of Economics*, *14*(3), 351–373.

Headey, B. (1991). Distributive justice and occupational income. *British Journal of Sociology*, *42*, 581–596.

Headey, B., Andorka, R., & Krause, P. (1995). Political legitimacy vs. economic imperatives in system transformation: Hungary and East Germany 1990–1993. *Social Indicators Research*, *36*(3), 247–273.

Heider, F. (1958). *The psychology of interpersonal relations*. New York: Wiley.

Hertel, B. R. (1976). Minimizing error variance by introducing missing data routines in survey analysis. *Sociological Methods and Research*, *4*(4), 169–184.

Hirschman, A. O. (1981). *Essays in trespassing: Economics in politics and beyond*. Cambridge: Cambridge University Press.

Homans, G. (1974). *Social behavior: Its elementary forms* (Rev. ed.). New York: Harcourt Brace & Jovanovich.

International Labor Office (1969). *International standard classification of occupations*. Geneva: International Labor Office.

Jasso, G. (1980). A new theory of distributive justice. *American Sociological Review*, *45*, 3–32.

Jasso, G. (1994). What is just? *Contemporary Sociology*, *23*, 707–709.

Johnson, D., Manning, I., & Hellwig, O. (1995). *Trends in the distribution of cash and non-cash benefits*. Australian Government Printing Service.

Joreskog, K. G., & Sorbom, D. (1988). LISREL 6. Chicago: Scientific Software, Inc.

Jowell, R., Witherspoon, S., & Brook, L. (1999). *ISSP, Britain*. Codebook and Machine-Readable Data File. London: Social and Community Planning Research.

Kelley, J. (1990). The failure of a paradigm: Log-linear models of social mobility and Kelley replies to Muller. In: J. Clarke, S. Modgil & C. Modgil (Eds), *John Goldthorpe: Consensus and Controversy* (pp. 319–346, 349–357). London: Falmer Press.

Kelley, J. (1992). Social mobility and politics in the anglo-American democracies. In: F. C. Turner (Ed.), *Social Mobility and Political Attitudes: Comparative Perspectives* (pp. 21–49). New Brunswick, NJ: Transaction.

Kelley, J., & Evans, M. D. R. (1993). The legitimation of inequality: Attitudes towards inequality in nine nations. *American Journal of Sociology*, *99*, 75–125.

Kelley, J., & Evans, M. D. R. (1995). Class and class conflict in six western democracies. *American Sociological Review*, *60*, 157–178.

Kelley, J., & Evans, M. D. R. (1999). *ISSP, Australia*. Codebook and Machine-Readable Data File. Melbourne: Melbourne Institute of Applied Economic and Social Research, University of Melbourne.

Kelley, J., Evans, M. D. R., & Sikora, J. (2003). *World inequality study*. Codebook and Machine Readable Data File, 1987–2001. Melbourne Institute of Applied Economic and Social Research, University of Melbourne.

Kelley, J., Evans, M. D. R., Zagorski, K., Kolosi, T., & Wnuk-Lipinski, E. (1993). *International survey of economic attitudes: Poland, December 1991*. Canberra: International Survey Center, Institute of Advanced Studies, the Australian National University.

Kelley, J., & Haller, A. O. (2001). Working class wages during early industrialization: Brazilian evidence. *Research in Social Stratification and Mobility, 18*, 119–161.

Kelley, J., & Klein, H. S. (1982). *Revolution and the rebirth of inequality: A theory applied to the national revolution in Bolivia*. Berkeley and Los Angeles: University of California Press.

Kelley, J., & Zagorski, K. (2002). Changing attitudes toward income inequality in East and West. In: M. D. R. Evans & J. Kelley (Eds), *Australian Economy and Society 2001: Education, Work, and Welfare* (Chap. 18). Sydney: Federation Press.

Kelley, J., Zagorski, K., & Evans, M. D. R. (1998). *The measurement of economic attitudes: Concepts and scales in the international survey of economic attitudes*. University of Melbourne: Melbourne Institute of Applied Economic and Social Research.

Khakhulina, L., & Zaslavskaya, T. (1999). *ISSP, Russia*. Codebook and Machine-Readable Data File. Moscow: Center for Public Opinion and Market Research.

King, S., & Lloyd, P. (Eds) (1993). *Economic rationalism: Dead end or way forward*? Sydney: Allen & Unwin.

Kluegel, J. R., Mason, D. S., & Wegener, B. (Eds) (1995). *Social justice and change*. New York: Aldine de Gruyter.

Kluegel, J. R., & Smith, E. R. (1986). *Beliefs about inequality: American's views of what is and what ought to be*. New York: Aldine de Gruyter.

Knudsen, K. (2001). How large should inequality be? Attitudes on legitimate earnings in Norway: 1992–1999. *Tidsskrift For Samfunnsforskning, 42*(4), 507–536.

Kolosi, T., & Robert, P. (1989). *Hungary: ISSP 1988*. Budapest: Tarsadalomkutatasi Informatikai Tarsulas.

Koves, A. (1992). *Central and East European economies in transition. The international dimension*. Boulder: Westview Press.

Kraus, V., & Hodge, R. W. (1987). Economy, polity, and occupational prestige. *Research in Social Stratification and Mobility, 6*, 113–139.

Lewis-Beck M. S. (1988). *Economics and elections: The major western democracies*. Ann Arbor: University of Michigan Press.

Lindert, P. H. (2000). Trends in income inequality during industrialization. Paper presented to the Twelfth World Congress of the International Economic History Association.

Lipton, D., & Sachs, J. (1990). Creating a market economy in Eastern Europe: The case of Poland. *Brookings Papers on Economic Activity, 1*, 75–147.

Listhaug, O., & Aalberg, T. (1999). Comparative public opinion on distributive justice – Study of equality ideals and attitudes toward current policies. *International Journal of Comparative Sociology, 40*(1), 117–140.

Little, R. J. A. (1992). Regression with missing X's: A review. *Journal of the American Statistical Association, 87*, 1227–1237.

Luo, X. W. (1998). Solving unemployment? A comparative study of Great Britain and the United States. *International Journal of Public Opinion Research, 10*(2), 121–144.

Malnar, B., & Tos, N. (1999). *ISSP, Slovenia*. Ljubljana: Ljubljana University.

Mangahas, M., Abad, M., Guerrero, L. L., Miranda, F., Rood, S., & Abad, R. (1999). *ISSP, The Philippines*. Codebook and Machine-Readable Data File. Quezon City: Social Weather Stations.

Mason, D. S. (1995). Justice, socialism, and participation in the post-Communist states. In: J. R. Kluegel, D. S. Mason & B. Wegener (Eds), *Social Justice and Political Change: Public Opinion in Capitalist and Post-Communist States* (pp. 49–80). New York: Aldine de Gruyter.

Mateju, P., & Illner, M. (1999). *ISSP, Czech Republic*. Codebook and Machine-Readable Data File. Prague: Institute of Sociology, Academy of Sciences of the Czech Republic.

Moore, D. (1992). Entitlement and justice evaluations: Who should get more, and why. *Social Psychology Quarterly*, *54*, 208–223.

Murphy, K. M., & Welch, F. (1994). Industrial change and the rising importance of skill. In: S. Danziger & P. Gottschalk (Eds), *Uneven Tides: Rising Inequality in America* (pp. 101–132). New York: Russell Sage.

Nee, V., & Matthews, R. (1996). Market transition and societal transformation in reforming state socialism. *Annual Review of Sociology*, *22*, 401–435.

Nie, N. H., Verba, S., & Petrocik, J. R. (1979). *The changing American voter* (enlarged ed.). Cambridge: Harvard University Press.

Nieuwbeerta, P., Gijsberts, M. I. L., & Ganzeboom, H. B. G. (1998). *Social and economic attitudes in the Netherlands 1998*. Netherlands: Interuniversity Consortium for Sociological Research (ICS): Occasional Papers and Document Series.

North, D. C., & Thomas, R. P. (1973). *The rise of the western World: A new economic history*. Cambridge: Cambridge University Press.

Nozick, R. (1974). *Anarchy, state, and utopia*. New York: Basic Books.

Offe, C. (1991). Capitalism by democratic design? Democratic theory facing the triple transition in East Central Europe. *Social Research*, *58*(4), 865–892.

Potter, R. B., Jr. (1972). The logic of moral argument. In: P. Deats (Ed.), *Towards a Discipline of Social Ethics, Boston* (pp. 93–114). Boston University Press.

Pusey, M. (1991). *Economic rationalism in Canberra*. New York: Cambridge University Press.

Putnam, R. D., Leonardi, R., & Nanetti, R. (1993). *Making democracy work: Civic traditions in modern Italy*. Princeton: Princeton University Press.

Rawls, J. (1971). *A theory of justice*. Cambridge, MA: Harvard University Press.

Rehakova, B. (1997). Income and justice: Tolerance of the Czech society to income inequalities in 1992 and 1995s. *Sociologicky Casopis*, *33*(1), 69–86.

Robert, P. (1999). *ISSP, Hungary*. Codebook and Machine-Readable Data File. Budapest: TARKI.

Robinson, R. V., & Bell, W. (1978). Equality, success, and social justice in England and the United States. *American Sociological Review*, *43*, 125–144.

Robinson, R. V., & Kelley, J. (1979). Class as conceived by Marx and Dahrendorf: Effects on income inequality and politics in the United States and Great Britain. *American Sociological Review*, *44*, 38–57.

Sarapata, A. (1963). Iustum pretium. *Polish Sociological Bulletin*, *1*, 41–56.

Schultz, T. W. (1980). Nobel lecture: The economics of being poor. *Journal of Political Economy*, *88*, 639–651.

Sen, A. (1973). *On economic inequality*. Oxford: Clarendon.

Sikora, J. (2000). *Attitudes to government ownership: Australia, Finland, Bulgaria and Poland in the mid-1990s*. Ph.D. Dissertation, Institute of Advanced Studies, the Australian National University.

Sikora, J., & Kelley, J. (1999 [published 2002]). Attitudes to private and public ownership in East and West: Bulgaria, Poland, Australia and Finland, 1994/1997. *The Soviet and Post-Soviet Review, 26*(1), 13–42.

Skjak, K. K., Henrichsen, B., Knudsen, K., & Kvalheim, V. (1999). *ISSP, Norway*. Codebook and machine-readable data file. Bergen: NSD (Norwegian Social Science Data Services).

Slomczynski, K., Bailecki, I., Domanski, H., Janicka, K., Mach, B. W., Sawinski, Z., Sikorska, J., & Zaborowski, W. (1989). *Struktura Spoleczna: Schemat Teoretyczny i Warsztat Badawczy* Warsaw: IFIS PAN.

Smeeding, T. M., Saunders, P., Coder, J., Jenkins, S., Fritzell, J., Hagenaars, A. J. M., Hauser, R., & Wolfson, M. (1993). Poverty, inequality, and family living standards impacts across seven nations: The effect of noncash subsidies for health, education and housing. *Review of Income and Wealth, 39*(3), 229–256.

Smith, A. (1776 [1937]). *The wealth of nations*. New York: Modern Library.

Stinchcombe, A. (1963). Some empirical consequences of the Davis-Moore theory of stratification. *American Sociological Review, 28*, 805–808.

Svallfors, S. (1993). Dimensions of inequality: A comparison of attitudes in Sweden and Britain. *European Sociological Review, 9*, 267–287.

Svallfors, S., & Edlund, J. (1999). *ISSP, Sweden*. Codebook and Machine-Readable Data File. Umea: Department of Sociology, University of Umea.

TARKI [Tarsadalomkutatasi Informatikai Egyesules] (1993). TARKI Social Mobility Survey, 1992: Documents (P. Robert, principal investigator) Budapest: TARKI (H1132 Budapest, Victor Hugo U. 18–22, Hungary).

Thomas, S. (1992). The political economy of privatisation: Poland, Hungary and Czechoslovakia. In: C. Clauge & G. C. Rausser (Eds), *The Emergence of Market Economies in Eastern Europe*. Cambridge, MA: Blackwell.

Thurow, L. C. (1975). *Generating inequality: Mechanisms of distribution in the U.S. economy*. New York: Basic.

Tipton, S. M. (1982). *Getting saved from the sixties: Moral meaning in conversion and cultural change*. Berkeley: University of California Press.

Treiman, D. J. (1977). *Occupational prestige in comparative perspective*. New York: Academic Press.

Tumin, M. M. (1953). Some principles of stratification: A critical analysis. *American Sociological Review, 18*, 387–393.

van Dijk, J. J. M., Mayhew, P., & Killias, M. (1990). *Experiences of crime across the world: Key findings from the 1989 international crime survey*. Deventer and Boston: Kluwer Law & Taxation.

Verba, S., & Orren, G. R. (1985). *Equality in America: The view from the top*. Cambridge, MA: Harvard University Press.

Visser, P. S., Krosnick, J. A., Marquette, J., & Curtin, M. (1996). Mail surveys for forecasting: An evaluation of the Columbus Dispatch poll. *Public Opinion Quarterly*, 60.

Walster, E. G., Walster, W., & Berscheid, E. (1978). *Equity: Theory and research*. Boston: Allyn & Bacon.

Wright, E. O. (1985). *Classes*. London: Verso.

Yergin, D., & Stanislaw, J. (1998). *The commanding heights: The battle between government and the marketplace that is remaking the modern world*. New York: Simon & Schuster.

Zaborowski, W. (1994). Beliefs about inequality: Changing income hierarchy in Poland. In: M. Alestalo, E. Allardt, A. Rychard & W. Wesolowski (Eds), *The Transformations of Europe: Social Conditions and Consequence* (pp. 207–218). Warsaw: IFiS Publishers.

Zaborowski, W. (1995). *Orientacje Egalitarne w Spoleczenstwie Polskim w Latach 1988–1993*. Warsaw: IFiS Publishers.

Zagorski, K. (1994). Hope factor, inequality, and legitimacy of systemic transformations: The case of Poland. *Communist and Post-Communist Studies*, *27*(4), 357–379.

Zentralarchiv fuer Empirische Sozialforschung (1989). *International social survey programme: ISSP 1987, Social inequality* (Codebook ZA-No.1680, 2nd ed.). Koeln: Zentralarchiv fur Empirische Sozialforschung der Universitat zu Koeln. (Bachemer Strasse 40, D-5000 Koeln 41, Germany).

Zentralarchiv fuer Empirische Sozialforschung (1994). *International social survey programme: ISSP 1992, Social inequality-II*. Koeln: Zentralarchiv fur Empirische Sozialforschung der Universitat zu Koeln. (Bachemer Strasse 40, D-5000 Koeln 41, Germany).

Zhou, X., & Suhomlinova, O. (2001). Redistribution under state socialism: A USSR and PRC comparison. *Research in Social Stratification and Mobility*, *18*, 163–204.

Zhou, X., & Suhomlinova, O. (2001). Redistribution under state socialism: A USSR and PRC comparison. *Research in Social Stratification and Mobility*, *18*, 163–204.

APPENDIX: MEASUREMENT

The Class-Status-Power Model

Objective class is measured by Kelley's extension of the Blau-Duncan model to include ownership and authority (Kelley, 1992, pp. 23–34; Kelley & Evans, 1995; Robinson & Kelley, 1979). Details:

Ownership and Control Aspects of Class:

Petty Bourgeoisie are defined as self-employed without employees; they are scored 1 and all others zero.

Entrepreneurs (capitalists in Marx's class scheme) are defined as self-employed with employees. Most, of course, run very small businesses.

Supervisory authority is scored 1 for those who supervise others and zero for everyone else.

Government employees are coded 1 and others 0.

SES Aspects of Class

Education is years of education. There are many arguments over how best to measure education, perhaps especially in the Eastern European context. Years of education has the great advantage of being a single information-packed measure which should only be set aside in favour of multiple categorical indicators if there is empirical evidence that years of education is not performing well – the traditional Occam's Razor criterion that the simpler is to be preferred to the complex unless the simpler can be demonstrated not to work. In our context, if years of education were not an appropriate measure in Central-Eastern Europe, then that should show up empirically as weaker correlations between education and dependent variables in Central-East Europe than in the West. But actually, the correlations are *larger* in Central-East Europe than in the West (Table 1). We therefore conclude that years of education is a suitable measure of education for this analysis. It is possible that expanded measurement of education including such variables as educational track and academic performance would add to the variance explained, but that possibility cannot be pursued here as they are not in these databases.

Occupation refers to present occupation for those currently employed, or to past occupation for those not now employed. Preliminary analysis showed that including a "no occupation" dummy variable in the analysis made little difference to the substantive results and so it was, for simplicity, omitted.

In most surveys, occupations were initially coded into the 4 digit International Standard Classification of Occupations (International Labor Office, 1968 or 1988) with a few local extensions. In some surveys, a standard 3 digit (or better) census code was used. We then recoded occupations into the 14 categories of Treiman's (1977, pp. 203–208) International Standard Classification of Occupations and thence into Kelley's (1990, pp. 344–346) Worldwide Status Scores, which are conceptually similar to Duncan's SEI scores.

Family Income is measured in local currency, expressed as a ratio of the average income of full-time blue collar workers (for comparability between nations).

These various dimensions are not sufficiently correlated to justify combining them into a single indicator, as categorical schemes implicitly assume (Kelley, 1992, pp. 23–34; Kelley & Evans, 1995). Moreover different dimensions of class are influential in different zones of social life, so combining them into one coarse categorical indicator would lose important information, and would prevent one from discovering which aspect matter more in the legitimation of inequality. Accordingly, we prefer to measure class as a set of variables rather than shoe-horning them into an ill-fitting categorical schema.

Measurement of Other Variables

Male is scored 1 for men, 0 for women.

Age is measured in years.

Subjective class is a 10 category self-placement, with one end labelled "top" and the other "bottom" (e.g. Kelley & Evans, 1995). The word "class" is deliberately because of its party political overtones in many European nations (Evans et al., 1992).

RACE, SOCIOECONOMIC DEVELOPMENT AND THE EDUCATIONAL STRATIFICATION PROCESS IN BRAZIL

Danielle Cireno Fernandes

ABSTRACT

This study investigates the determinants of educational stratification in Brazil. It draws on theories of educational and racial inequality to examine the impact of economic development on educational stratification and the role of race in that process. Using a nation-wide probability sample (PNAD-1988), I find no evidence of any overall trend toward the equalization of educational opportunities over the past decades, but rather a mixed pattern of increasing and decreasing effects. Further, socioeconomic transformations brought about by the process of industrialization have not lessened the effect of race as one of the main determinants of educational stratification in Brazil. There is strong evidence that it may even have increased.

1. INTRODUCTION

This study examines the determinants of educational attainment in Brazil. I ask two specific research questions: (1) What impact does economic development

The Shape of Social Inequality: Stratification and Ethnicity in Comparative Perspective
Research in Social Stratification and Mobility, Volume 22, 365–422
© 2005 Published by Elsevier Ltd.
ISSN: 0276-5624/doi:10.1016/S0276-5624(04)22012-1

have on educational stratification; (2) What is the role of race in this process?

Brazil is a late-comer in educational development and expansion, even compared to other Latin American countries (Castro, 1989). Portuguese and Brazilian culture have seen schooling as a valuable resource for social standing and political power, sometimes even more important than economic resources (Carvalho, 1980), since the early stage of colonization (Ribeiro, 1984). Education has also seen as the basis for the homogeneity and unity of a powerful political elite which characterizes the political and social structure peculiar to Brazil (Carvalho, 1980).

The educational level of the Brazilian population is relatively low in comparison with other Developing Countries. During the last century, Brazil's educational system expanded, but produced results not even close to an acceptable level (6 years of schooling in average).[1] The level of schooling, together with vast educational inequality, drastically handicaps economic growth and is understood as one of the main causes of income inequality (Langoni, 1973). Nonetheless, the educational system has expanded considerably in the last 40 years, allowing rates of growth higher than those observed in other Latin American countries. However, it has been rather an unusual expansion. In recent decades, post-secondary education has received much more investment from the government than primary or even secondary levels. During this time, Brazil has established a system of public universities that is today recognized as one of the best among Developing Countries. Yet, education is unevenly distributed within the country at all levels (Barcelos, 1992). This inequality is even deeper among groups with socioeconomic and racially distinct heritage (Hasenbalg & Valle Silva, 1990; Souza & Valle Silva, 1994). Thus, a public university system has emerged at the expense of lower levels, allowing an educational system with vast inequality as its main characteristic.

Despite low and uneven educational attainment, the accelerated rhythm of industrialization and urbanization of the last decades has dramatically altered the shape of the Brazilian social structure (Pastore, 1982; Pastore & Haller, 1993). However, cumulative evidence has shown that increasing levels of industrialization and modernization do not eliminate the effects of race and skin color as a criterion of social selection and generation of social inequalities. In fact, recent research has shown that in spite of these social transformations, the non-white population is still exposed to systematic social disadvantages in such indicators as infant mortality, life expectancy, opportunity of upward social mobility, participation in the formal labor market, income distribution, and, more important for our analysis, educational attainment. Barcelos (1992, p. 55) concluded that the Brazilian educational system is facing a strong crisis. "The crisis is serious and has color," he wrote.

A single year of education in Brazil is an important achievement, given the low educational level of the population. Each additional year is well rewarded by the market (Haller & Saraiva, 1992), even in rural areas where occupational opportunities requiring formal qualifications are scarce (Neves, 1997). In fact, Brazil has one of the world's highest rates of economic return to education, reaching an increment of 12%, and perhaps as high as 16%, of earnings per year, depending upon the region in which one lives. A completed level (elementary, secondary or college) provides even greater advantage (Haller & Saraiva, 1992).

This paper is intended to establish the determinants of educational stratification in Brazil. It is divided into seven sections, including this introduction. In the next section I discuss theoretical perspectives on economic development and their relation with educational expansion, and provide an overview of theoretical explanations of the process of racial inequality in Brazilian society. Section three describes the hypotheses, and section four describes the data, sample, variables, and methodological issues. In section five and six I present the statistical analysis. Section seven discusses the theoretical implications of the study.

2. THE THEORIES

2.1. Theoretical Perspectives on Education and Socioeconomic Development

The impact of socioeconomic development on educational expansion has been the focus of several studies. For the Modernization (or Industrialization) view, education is the main vehicle to distribute social gains to individuals brought by socioeconomic development: an achieved (no longer ascribed) process of status distribution in the social mobility process. This view is derived from Parsonian Sociology (Parsons, 1970; Treiman, 1970). Standard empirical sources include Blau and Duncan (1967), Holsinger (1975), Hauser and Featherman (1976), Featherman and Hauser (1978), and Kuo and Hauser (1995), among others.

In contrast, in the Social Reproduction view, educational expansion is the main channel through which capitalist development perpetuates class antagonism by selecting and training individuals to perform occupational roles that merely reflect their families' social position. Thus education is seen as an instrument of social domination (Bowles & Gintis, 1976; Edwards, 1979). Instead of increasing "universalism" in the status attainment process, educational expansion is the path "in which 'ascriptive' forces find ways of expressing themselves as 'achievement' "

(Halsey, 1977, p. 184). Education can also be understood as a way in which Cultural Capital is transmitted and acts as a powerful vehicle of social reproduction (Bourdieu, 1973; Collins, 1979).

I turn next to the application of these theoretical perspectives for educational expansion and its relation with economic development. I pay special attention to the role of racial inequality.

2.1.1. The Meritocratic Hypothesis

Modernization theorists suggest that educational systems expand in response to the functional requirements of industrial society and that education plays an increasingly important role in the process of status attainment (Lenski, 1966; Treiman, 1970). As educational requirements rise with economic development, educational qualifications become more important for occupational placement. Also, with modernization and the expansion of the educational system, educational selection tends to become more meritocratic and less ascriptive. Hence, inequality of educational opportunity, as measured by its dependence on socioeconomic and socio-cultural characteristics, should decrease across all educational levels over time. This line of thought, also called the "Functional Paradigm,"[2] sees schooling as representing a rational and efficient way of sorting and selecting talented people, where the most able and motivated attain the highest positions. Schools teach the kind of cognitive skills and norms that are essential for the performance of adult roles (Debreen, 1968). The learning of cognitive skills is necessary for the fulfillment of economic positions in a society increasingly dependent on knowledge fundamental for economic growth (Treiman, 1970).

The meritocratic hypothesis states that equality of opportunity in the long run would be recognized by: (a) an increase in the association between educational and occupational status; (b) a decrease in the association between parents' social status and the social status of their children; and (c) a decrease in the association between parents' social status and their children's educational achievement. This hypothesis does not predict less inequality in status attainment, but rather a rationality to the process of status attainment (Goldthorp, 1996; Hurn, 1993). According to Blau and Duncan (1967), the path to this process is to be seen as the current status of an individual is more and more determined by higher educational attainment and experience in the labor market, usually measured by the first job, rather than being inherited from one's parents. Of course, equality of opportunity in schooling plays a key role. In this view, social selection through education is related to the trend towards increasing universalism in the socially selective process.

My focus here is on this last proposition of the Meritocratic Hypothesis, which implies that as societies become more and more developed or industrialized, the educational attainment of children of socially privileged and less socially privileged parents should become increasingly similar (Holsinger, 1975; Treiman, 1970). In fact, some research indicates a stable pattern of more educational equality, at least for the graded school levels.[3] According to Featherman and Hauser (1978, p. 249 Hauser):

> [. . .] Trends towards *equality of educational opportunity* are manifestly obvious within the graded school years, as rising levels of socioeconomic background have enabled an increasing proportion of nearly every stratum and social category to complete the full extent of publicly financed education beyond the compulsory minimum.

Kuo and Hauser (1995), however, point out that this apparent equality may not pertain to all ethnic groups:

> Effects of social background on schooling have declined among Blacks and Whites, but the pattern of change appears different in the two groups. Among Blacks, the data are consistent with a sharp, global decline in the effect of all background characteristics between cohorts born from 1937 to 1946 and all earlier cohorts, but among Whites there appears to have been a gradual decline in the effects of just three specific background characteristics: farm background, southern origin and family structure[4] (Kuo & Hauser, 1995, p. 156).

Despite the world expansion of education (Mayer et al., 1993), equalization of educational opportunities has long been challenged in sociological theory and research, indeed ever since it was proposed (Boudon, 1974; Bourdieu, 1973; Bourdieu & Passeron, 1977; Bowles & Gintis, 1972, 1976; Collins, 1971; Halsey, 1977; Jencks et al., 1972; Thurow, 1975). Halsey (1977), for example, in analyzing educational expansion in Britain, showed that what had been observed is an increase in the dependency of educational attainment on parental educational and occupational status, despite an increase in the effect of education on occupational status. Sewel and Shah (1967) demonstrated that socioeconomic background exerts an influence independent of measured intelligence on both college entrance and college graduation. Heyns (1974) doubted the tendency towards universalism in social selection proposed by Blau and Duncan (1967) as education expands, because even if educational attainment (in terms of the last year of formal schooling completed) becomes more universal, stratification within schools will be inherent within the school system. She presented evidence of an association between track assignment and children's social background in the American educational system. Gamoran and Mare (1989), after a careful analysis, confirm a class bias in track assignments in American schools. Kerckhoff (1993) found such an association for Britain. For Brazil, Haller and Saraiva (1991) found that social origins are powerful status allocation mechanisms and that their effect increases with economic development, instead of decreasing as suggested by Lenski (1966) and Treiman (1970). According to the latter, the effect of status origins on offsprings' occupational and educational attainment should decrease with economic development and the effects of individuals' education on his or her status should rise. What Haller and Saraiva (1991) found in Brazil was an increase in both processes, and they called the anomaly the *educational monopoly hypothesis*.

> This hypothesis holds that as development proceeds, the linkage between education and other status variables becomes progressively closer. This increasing linkage between qualification and rewards is evident to large proportion of the population, obviously including the families of the higher status levels. More than other families, these families have the resources to put their offspring through many long years of formal education and to pay to get them into the best schools. So the rich tend to monopolize educational opportunities [. . .]. The higher the level of development, the greater the role of education as a mediator of influence on one's status origins on one's status (1991, pp. 83–84).

A recent view argues that even though school produce skills that are important in production and as consequence increase earnings, skill enhancement explains only part of this contribution to individual earnings. "Schooling also raises earnings by its effects on individual's norms and preferences, making the prospective worker more attractive to the employer by attenuating problems of work incentives and labor discipline" (Bowles & Gintis, 2000).

Those who raised doubts about the causal relationship between economic development and the equalization of educational opportunities also find support in the general theory of social stratification, where social structures are generally understood as stable, and "stratification regimes have in-built sustaining properties, as well as powerful defenders" (Erikson & Jonsson, 1996, p. 67). Sorokin (1927) claimed that social mobility is little influenced by societal differences and transformations, an assertion that has been supported by mobility studies (Eriksson & Goldthorp, 1992; Featherman et al., 1975) and also in comparative studies about trends on educational inequality (Shavit & Blossfeld, 1993) and studies which argue for constancy in the structure of occupational hierarchy across societies (Treiman, 1968).

2.1.2. Social Reproduction Theories of Education

Social Reproduction theorists see educational expansion as a process that excludes low social classes or low status-groups from desirable occupational positions. Selection and allocation in the labor market based on educational credentials are used to maintain the privilege of dominant social groups (Bowles & Gintis, 1976; Collins, 1971, 1979). Educational attainment, then, is part of a lager process of legitimization of class structure.[5]

This view, like the Modernization perspective, focuses on the causal relationship of the socialization role of education and its selective function. However, the social consequences are quite different to those predicted by Modernization proponents. The expansion of the educational system at lower levels will be supported by the dominant groups, making it available for children of ethnic minorities and working class origins. On the other hand, if the dominant groups want to maintain their social privileges in the society, they must retain their advantage of access to higher educational credentials. Hence, although the Modernization and Social Reproduction approaches agree that the educational distribution, whether the result of functional imperatives of industrialization (Bowles & Gintis, 1976; Parsons, 1970; Treiman, 1970) or an outcome of competition among status groups (Collins, 1971, 1979), leads to greater equality of educational opportunities at the lower levels of the educational system, they disagree about the trends in inequality at higher levels of education. While Modernization theorists predict a decreasing trend on educational inequality at all levels, Social Reproduction theorists predict a constant trend or even an increasing importance of social origins determining higher levels of the educational hierarchy as economic development moves forward.

Proponents of Cultural Capital theory (Bourdieu, 1973; Bourdieu & Passeron, 1977) state that children from families with a low level of parental education are likely to lack cultural resources such as dominant social values, attitudes, and language skills that help them to acquire higher educational achievement. In this perspective, cultural capital is the main mechanism for social reproduction in modern societies. With the democratization of modern societies, as a consequence of economic development, the demands for equality of educational opportunity and meritocratic selection increase, and high-status families lose their capacity to directly guarantee high social position for their children. Cultural capital consists of goods transmitted by pedagogic actions within families. Cultural capital is related to all cultural investments of the family outside the regular educational system. Parental educational and economic resources are good indicators of family cultural capital, though one does not necessarily predict the other (Katsillis & Rubinson, 1990). Thus, in this view, social origins, especially parental educational level, would *not* lose their importance in determining educational stratification as a consequence of economic development.

2.1.3. Challenges from New Trends

Empirical analyses have shown that the relationship between family background and educational expansion as societies face economic development do not point to a decreasing pattern. This relationship does not change at the same rate for all educational levels or for all measures of social origins, neither does it for all racial groups at the same time (Mare & Winship, 1988; Olneck, 1979). Hauser and Featherman (1976) found that the variability of schooling had decreased and that there had been a noticeable increase in the length of schooling for U.S. men born in the first half of the century. However, the effect of only some social origins had declined (parental education and family size persisted as significant predictors of educational attainment). Featherman and Hauser (1978) also found a decline across cohorts in the effect of social background on educational attainment, but this did not hold for all educational levels. The decrease was observable on measures of quantity of graded school attained, but not on post-secondary levels. Kuo and Hauser (1995) found that, even though the effects of measured and unmeasured family background characteristics on the educational attainment of Blacks and Whites had declined, this pattern affected racial groups differently. Among Blacks all measures of background effects have declined, among Whites the trend replicates the earlier findings of Hauser and Featherman (1976).[6]

Mare (1980, 1981, 1993)[7] introduced a new perspective that helps clarify the causal relationship between social origins and educational attainment. He shows that models which estimate inequality of educational attainment by using the latest year of formal schooling completed as the dependent variable[8] do not make a distinction between those individuals who completed a given transition level and those who did not. Such models fail to distinguish two processes: the expansion of the educational system and the selection and allocation of students into the educational system. Mare proposes a model to measure changes in inequality of educational opportunity with parameters that are not affected by the degree of educational expansion or contraction. This model views educational attainment as a sequence of transitions (from primary to secondary level, from secondary to post-secondary level etc.). At each level, the individual can either make the transition or discontinue. He formulates the model as a set of logit regressions, and estimates the effects of exogenous variables on the log odds of making that transition. Only those individuals who made the earlier transition will be part of the sample that will be used to estimate the log odds of making the next transition.

For the United States, Mare found a pattern of a decreasing effect of background on educational attainment from the lowest to the highest transition within the same cohort, but a relatively constant pattern across cohorts within each transition, meaning that social barriers are stronger on the earlier steps of educational life and weaker at higher levels. Factors other than family background, like ability or motivation, not measured in his analysis, determine educational attainment for the highest transitions. These effects remain relatively constant across cohorts, which implies that economic development does not have a predetermined effect on patterns of educational inequality. People who made the highest transitions are more homogeneous in terms of ability and motivation (always unmeasured), which reduces the effect of observed socioeconomic origins variables. This hypothesis also implies that, with the increasing upgrading of educational attainment due to expansion of education, growing proportions of successive cohorts reach higher levels of schooling, so their selectivity declines and the homogeneity of unmeasured factors will be lower than it had been for earlier cohorts. Thus, the effects of socioeconomic origins on one's educational transitions would increase across cohorts within each transition "...This is the same mechanism, albeit in reverse, that partly accounts for declines in fathers' schooling effects *within* cohorts" (Mare, 1993, p. 371).

Mare and Winship (1988) show that the odds of making a given transition vary by race, but less is known about how race and ethnicity interact with educational expansion and allocation. An exception is Shavit and Kraus (1990). Using Mare's model, they found a declining effect of ethnicity across cohorts in Israel. They conclude that "the decline in the ethnic difference in entry into secondary education was not accompanied by an equalization of opportunities at the higher educational levels" (p. 138). It was rather a consequence of the expansion of the vocational secondary education which enabled growing proportions of the Sephardim (an ethnic group with lower socioeconomic origins) to get into vocational education. Thus, students of subordinate-group origins were diverted from attaining higher education by various means, from the expansion of nonacademic educational alternatives to raising the admission standards and tuition fees of universities. This is a finding already suggested by Shavit (1984) for Israel, and also by Karabel (1972) in analyzing the role of ethnic groups on American educational expansion.

Although this hypothesis has been tested in Less Developed Countries or nations that have suffered great socioeconomic change in the last half century, most of these countries have passed through a strong process of educational expansion, and, to a certain degree, a decrease in income inequality, as is the case for Taiwan, Italy, Poland, Hungary and Israel (Shavit & Blossfeld, 1993). Remaining unknown is how socioeconomic background affects educational attainment through the individual life course in Brazil, a society that has faced an increase in income inequality and an unusual educational expansion in the last decades.[9]

2.2. Theoretical Perspectives on Race and Educational Inequality

2.2.1. Racial Inequality in Brazil

Research on racial inequality in Brazil has re-emerged in the last fifteen years or so.[10] Race relations in Brazil had been seen in a context of "racial democracy," where all races are assumed to live together with no conflict or social inequality directly associated to *race relations*, but rather to *class relations*. Notions of "the pacific people" and "the cordial man" are deeply embedded in Brazilian society, and there is said to be a tendency for conciliation and compromise rather than conflict. An example of this is, until recently, the complete lack of racial/ethnic social movements that would claim for equality of social opportunities.

However, research has demonstrated that after more than 100 years since the abolition of the black slavery system:

(1) The Non-White[11] population is still over-represented at the bottom of the social hierarchy. People with the same level of schooling, but of different racial heritages do not have the same chances of occupational opportunity (Andrews, 1992; Hasenbalg, 1979; Hasenbalg & Valle Silva, 1988, 1991).
(2) Controlling by family background (parents' income, parents' schooling, etc.), the Non-White population shows an educational attainment significantly inferior to that of the White population. A larger proportion of Non-Whites study in schools with shorter classroom hours (Rosemberg, 1986).

The image of racial and ethnic harmony is a part of the ideological concept of Brazilian (human) nature which serves to absorb tensions, and it is also a way to anticipate and control some areas of social conflicts (Hasenbalg, 1979; Hasenbalg & Valle Silva, 1988). This ideology also helped to cover the fact that the Non-White population has suffered (Beozzo, 1983; Fotaine, 1985; Huntington, 1982/1983; Lovell, 1991; Reeve, 1977; Rosemberg, 1987; Valle Silva, 1992).[12]

2.2.2. The Declining Significance of Race: Marxist and Neoliberal Approaches

The myth of "racial democracy" seems to have worked very effectively against the Non-White population (Hasenbalg & Valle Silva, 1988). Some, however, hold that racial differences in Brazil are nothing more than the heritage of past social relations of slavery (Fernandes, 1969, 1972a; Ianni, 1966). While Hoetink (1973) denies a causal relationship between the slavery social relation and the racial order which developed after abolition, under this Marxist perspective there is a *direct* causal relationship between slavery and the social relations that developed thereafter. In other words, current differences between Whites and Non-Whites may be explained by background disadvantages inherited from the past. However, with the development of a modern capitalist economy, Brazil would approach a real racial democracy. Today's racial order, which underlies its social stratification, is an "archaic" inheritance of slavery system, and will disappear as capitalist development moves forward. Thus, Fernandes (1969) does not see racial prejudice and discrimination as a result of the current social status of the Non-White population. In his words:

> The persistence of prejudice and discrimination constitutes a cultural backwardness phenomenon [. . .] it has nothing to do with competition or rivalry between Blacks and Whites

> nor even with the possible or real aggravation of racial tensions. It is the expression of the mechanics that in fact perpetuate the past into the present (p. 295). (My translation.)

This perspective, like the Industrialism Thesis, has an optimistic view of the future of racial relations. However, inequality of social opportunity would not decrease with development, as the Industrialism Thesis predicts, because racial relations would be gradually translated into class relations. Thus capitalist development would stratify, even though race would lose its power as a determinant of social stratification. For Fernandes (1964, pp. 292–299).

> It seems impossible to know how Brazilian racial relations will develop in the distant future. It seems likely that the dominant tendencies will bring the establishment of an authentic racial democracy. (Our translation.)

Following about the same logic but using a weberian approach, Wilson (1978) sees a decline in the significance of race as a determinant of one's position in the economic arena. For him, economic class has become more important than race in determining job placement and occupational mobility. Education plays a key role in this process. In this sense, the traditional racial struggle for power and privilege has shifted away from the economic sector, and is now concentrated in the socio-political order. Increasing inequality within Blacks' class structure is a demonstration that the life chances of the Black group have become increasingly a consequence of class affiliation with capitalist development, and not racial heritage.

Like the Marxist and Weberian views, Neoclassical economics holds that racial discrimination will disappear when free markets are developed. One of the most prominent analyses of the importance of race discrimination as a social barrier is Becker (1957). According to Becker, racism is essentially a problem of tastes and attitudes. White employers are understood to have "tastes" for discrimination. They are willing to lose income in order to be associated with other Whites rather than with Blacks. Since not only White employers, but also White workers prefer not to be associated with Black workers, they require monetary compensation for such an association, which maintains the wage of the Black worker below that of the White worker. However, marginal productivity analysis shows that White employers hire fewer Black workers than efficiency would require, since Black workers are less expensive. For Becker, long-run market forces will lead to the end of racism against Blacks because less discriminatory employers will be able to operate at lower costs by hiring equivalent Black workers at lower wages, thus driving more discriminatory employers out of business. Under this view, capitalist development is incompatible with racism or any other kind of discriminatory practices that would jeopardize the "logic" of capitalist development in a market economy. In other words, capitalist economic development de-stratifies society.

2.2.3. Uneven Development: the Persistence of Race

The durability of race divisions casts doubt on much social theory. Parsonian sociology, Orthodox Marxism, Neoclassical and Modernization Theory offer distinct pictures of the contradiction in the industrialization process, but they agree that modernization alone acts as an equalizing process. Under these views, industrialization acts as an overriding socioeconomic and political force that replaces traditional, religious, familial, ethnic and political attributes that would otherwise shape societies' social stratification systems.

An opposing view holds that capitalist development itself brings neither social equalization of the racial structure, nor an equalization of social opportunities. A forceful version of the theory represented in Reich (1994) concludes that racism continues to serve the needs of the capitalist system. For him, although an individual employer might gain by refusing to discriminate and hiring Blacks below the Whites' wage rate, this does not mean that the capitalist class as a whole would profit if racism were eliminated and labor were more efficiently allocated without regard to skin color. The logic of his argument is that racism divides the working class and weakens the strength of workers when bargaining with employers for better wages. For Reich,

> . . . the economic consequences of racism are not only lower incomes for blacks, but also higher incomes for the capitalist class coupled with lower incomes for white workers. Although the capitalist class may not have conspired consciously to create racism, and although capitalists may not be its principal perpetuators, nevertheless racism does support the continued well-being of the [American] capitalist system (p. 470).

Another radical interpretation of the persistence of social inequality in the face of capitalist development is given by the "Dependency" or "Underdevelopment" literature.[13] This theory suggests a pattern of generated and functionally uneven development that perpetuates, rather than eliminates, social inequalities. Capitalist development creates inequalities, unevenness, and underdevelopment. "Backward" areas or groups are not anomalies; they are part of the same process. Economic development and underdevelopment are the opposite sides of the same coin. They are the products of a single, but dialectically contradictory, economic structure and process of capitalist accumulation.[14] This process generates *core* and *periphery* sides, which could be areas within a single society or societies representing areas of a larger economic system or even groups of people interacting in the same labor market, which are two[15] dissimilar elements within the same overall framework.

Greenberg (1980) argues that these patterns of core and periphery relations and uneven development do not by themselves constitute a theory of race or ethnic relations. They give rise, however, to two important propositions. First, this theory implies that economic growth creates or even exacerbates social divisions and inequalities. Second, *archaic* social relations, like those based on race and ethnicity, are potentially compatible with, and perhaps functional to, capitalist development. What is most important in this theoretical explanation is not only a disparity between core and periphery, but the relationship between them: the development and growth of the core is dependent upon the stagnation and marginality of the periphery. Persistence or even an increase in the traditional and archaic bases of social inequality is a functional necessity of capitalist development.

According to Hasenbalg (1979), it would be useless to try to determine the degree to which the creation or even perpetuation of racist institutions and practices is a necessary condition for capitalist development. On the other hand, there is no reason to believe that industrialization and capitalist development eliminate race as a criterion underlying social relations in the productive system.

To continue, the perpetuation of racial prejudice and discrimination towards the Non-White population must be interpreted as a function of the symbolic and material interests of the dominant White group. Thus it is a set of social mechanisms destined to reproduce the social structure that explains this racially symbolic function in multi-racial societies during and after slavery systems. In Hasenbalg's (1979, p. 77) words:

> Class' societies give a new function to racial discrimination and prejudice: racism practices whether legally institutionalized or not, tend to disqualify the Non-White groups from competition for the highly desired social positions that are a result of the capitalist development and of the class structure differentiation. (My translation.)

2.2.4. Race and the Meaning of Education in Brazil

Hasenbalg and Valle Silva (1990) analyzed factors that determine the educational attainment of students who attended private and public schools and concluded that there is a significant amount of racial disparity in educational attainment opportunities that cannot be accounted for by socioeconomic background. This difference could hardly be explained by the process of discrimination, which occurs within schools, but it might be explained by factors which operate among families. Besides all the economic difficulties faced by the Black population, who, in Brazil, are concentrated in the low income classes, educational attainment could have a different significance or meaning for different racial groups. According to Pinto (1987, p. 8), other problems related to the racial condition itself handicap educational opportunities of young Blacks. For her,

> Black parents are aware of school and professional difficulties which their children may face, so they may discourage any high ambition in them. Discouragement may come about not only because of an evaluation of the objective conditions, but also because of their own assimilation of the negative representation [imposed by society's racial discrimination], leaving to a feeling of inferiority which would limit parents' aspirations for their children. (My translation.)

Other students of race relations in Brazil see the significance of education for Black families as having two different faces, which are not always explained by socioeconomic dimensions. For Azevedo (1953) and Pereira (1967), Black families are aware of the value of education, so much so that sometimes everybody in the family will make sacrifices to have at least one go further in school. On the other hand, the awareness of the difficulties that Blacks have in reaching the same socioeconomic position as Whites from the same class could lead to a contrary posture towards education. Bergmann (1978) describes parents who deliberately limit their sons' and daughters' social ambitions, independently of their socioeconomic background. Bastide (1952) refers to Blacks' sense of reality, which lead them to assume a negative attitude towards education because of their perceptions that, with a diploma, life could be harder, and, as a consequence, Blacks are less educated and *choose* less qualified occupations. Under this view, the pure fact of being Black or Mulatto decreases one's chances for educational opportunity, because of a deep prejudice and discrimination regarding skin color. Having a better socioeconomic background or being born in a more economically developed society has no effect on this situation.

One way to test these arguments is to analyze the process of educational stratification. Inequality of educational attainment in Brazil has been shaped by race. This is true not only for access to education (Barcelos, 1992; Dias, 1979; Fundação Carlos Chagas, 1986; Hasenbalg, 1987; Rosemberg, 1991), but also for racial discrimination within the school system (Figueira, 1988; Gonsalves, 1985; Instituto de Recursos Humanos João Pinheiro, 1988). Hence, the goal of the present research is to analyze educational stratification in Brazil by examining how race relations have changed in response to industrialization and development, and how this has been reflected in the distribution of educational attainment.

3. HYPOTHESES

There are three hypotheses relevant to my research questions: (1) economic development *de-stratifies* society; (2) economic development *stratifies society*; and (3) economic development *does not have a predetermined effect* on the society's stratification patterns.

Hypothesis 1. Economic Development *De-Stratifies Society*.

Modernization Theory suggests that *economic development de-stratifies society*. Access to education will become less and less selective as economic development moves forward. Based on this view, we will test the hypothesis that:

(1) The effect of ascriptive variables as determinants of educational attainment has decreased over the course of the century. Not only the effects of socioeconomically ascriptive variables, but individual attributes such as race and sex should decrease under this hypothesis.

Hypothesis 2. Economic Development *Stratifies Society.*

The Declining Significance of Race view suggests that economic development *stratifies society, but race gives way to socioeconomic background as the main determinant of social stratification.* Families who are more able to invest in educational qualifications for their offspring could obtain better social status for them, independent of racial origins. Based on this view, we will test the hypothesis that:

(2) Race will have a decreasing effect on educational attainment across cohorts. At the same time, socioeconomic background should have an increasing effect on educational attainment across cohorts for all racial groups.

The Uneven Development Thesis *suggests that economic development does not eliminate race as the bases for social stratification*, it in fact *can exacerbate the race effect*. Based on this view, we will test the hypothesis that:

(3) There is an increase in the effect of race on educational attainment across cohorts which is independent of socioeconomic background. Under this view, *archaic* ascriptive determinants of social segmentation, as race, are requirements for capitalist development.

Hypothesis 3. Economic Development *Does Not Have a Predetermined Effect* on Society's Stratification Patterns.

The Differential Selection Process proposed by Mare (1980, 1981) suggests that *economic development does not change patterns of social stratification.* Educational stratification patterns actually do not change as a consequence of the social process generated by economic development. They may appear to do so, but it is simply a consequence of the upgrading of socioeconomic background of those who are in the educational structure. Based on this view, we will test the hypotheses that:

(4) There will be a decrease in the effect of social origins, including race, on school transitions within each cohort. The higher the educational level individual reaches, the less it will be explained by ascriptive measures, and this pattern should be stable across cohorts.

4. METHODOLOGICAL CONSIDERATIONS

4.1. Data

The data come from the Brazilian National Household Sample Survey (PNAD). The PNAD-1988 was especially designed for analyses of social stratification, mobility, education, and the labor market. Its sample has about 290,000 observations for the country as a whole. The data derive from *stratified, multistage cluster samples of households*. This means that the samples were selected using a multistage procedure. In this case, three selection stages were used, based on geographical units: primary units (*municípios*); secondary units (*setores censitários*); tertiary units: households. However, given that our focus on completed educational attainment in Brazil, we restrict our sample to individuals who were 25 years old and over at the time of data collection. In fact, 98% of those selected using this filter declared they were not studying at the time they were surveyed. Because of specific social stratification information, such as parents' educational attainment and father's occupation were asked only of the head of households and their spouses, an extra filter is used. The final sample consists of about 109,000 individuals, or 83.2% of those 25 years old or older in 1988 who were heading a household or were their spouses.

PNAD does not include individuals from the rural areas of the Brazilian Amazon Frontier, due to the enormous difficulties of accessing the interior of this region. Nevertheless, the Brazilian Amazon, even though representing more than half of the country's territory, holds only about 7% of the Brazilian population, and its level of urbanization is quite high – about 75%. The information is from the 1991 census, only three years after our data were gathered. This may represent another problem of sample selectivity bias, but it is not great enough to invalidate the findings. It is important that "Rural" and "Urban" mean something different in Brazil than in the United States. The word "rural" really stands for farm residents or farm workers living in tiny settlements.

A final consideration is that it may be risky to use statistical results from *stratified, multistage cluster samples* as they were *simple random samples* – i.e. that conclusions from *stratified, multistage cluster samples* cannot be interpreted in the same way as *simple random samples* (Hasenbalg & Valle Silva, 1991; Mare, 1980). This is because statistical estimations from *stratified, multistage cluster samples* in general understate standard errors. A remedy for this has been proposed by Goldberg and Cain (1982), and widely applied by many researchers (see Gamoran, 1987). Specifically, I use a t ratio greater than 3.00 in my statistical analyses.

The high quality of the PNAD is well-known. Much work has been done based on PNAD (produced by the "Instituto Brasileiro de Geografia e Estatística"{the Brazilian Census Bureau}), and their quality has been well-verified (Haller & Saraiva, 1992; Hasenbalg & Valle Silva, 1988; Pastore, 1982; Ramos, 1993; Telles, 1992a, b, 1993).

4.2. Models' Specifications and Variables

I examine educational stratification using a cohort analysis. Eleven cohorts are used, each covering five years.[16] Besides a basic descriptive analysis to account for the process of educational expansion in the country and its main peculiarities, my analysis will use both OLS Regression and Logistic Regression.

In the OLS regression, the dependent variable is the educational attainment[17] of individuals. Education is treated as a continuous variable ranging from 0 to 17 years. Race, gender, and socioeconomic background are the independent variables. The same model is applied for each cohort.

OLS regression Model

Model 1

Years of Education $= \alpha + \beta_1$ Mother's Education $+ \beta_2$ Father's Education $+ \beta_3$ Father's Occupation $+ \beta_4$ Father's Occupation (dummy)$+ \beta_5$ Mulatto (dummy) $+ \beta_6$ Black (dummy) $+ \beta_7$ Asian (dummy) $+ \beta_8$ Gender (dummy) $+ \beta_9$ Urban Origin (dummy) $+ \varepsilon$.

Variables

Education

I measure education as the number of years of formal education successfully completed. Successfully completed years of education, in the case of Brazil, is different from the number of years one has attended school. In the Brazilian educational system (this is valid for every state in the country), if a student does not achieve a pre-determined standard he/she will fail. As a consequence, there are in Brazil, for example, children who have been attending school for five or six years but who have successfully completed only two or three years of education. They are recorded here as having two or three years of schooling.

Mother's and Father's Education

Parental education is measured as the number of years of formal education successfully completed. Data about the *exact* number of years of parents' education are not available. The original data on parents' schooling for the 1988 sample were coded as: I – no schooling at all; II – literate; III – incomplete lower elementary school;[18] IV – complete lower elementary school; V – incomplete upper elementary school; VI – complete upper elementary school; VII – complete high-school; VIII – complete college education. I followed a strategy introduced by Bills and Haller (1984) to use the following numbers to represent years of schooling: 0 (no schooling); 1 (literate); 2 (incomplete low elementary); 4 (complete low elementary); 6 (incomplete upper elementary); 8 (complete upper elementary); 11 (complete high-school); 16 (complete college). This scheme incorporates a little unreliability, but less than the original coding would.

Father's Occupational Status

This information was gathered from the following question: "What occupation did your father have when you started to work?" Occupational status is measured using an index of socioeconomic status for occupational status developed by Valle Silva (1985). This index, initially based upon the 1970 census (Valle Silva, 1974), was later updated from the 1980 census as a way to correct the high level of heterogeneity of some occupational titles (Valle Silva, 1985). Valle Silva's scale is not a prestige scale, but a socioeconomic index that combines occupation, education and income.[19] The procedure used to construct this scale has been quite conventional in the sociological literature.[20]

Father's Occupational Status will also be coded as a *dummy* variable, 1 for those who answered the question about father occupation, zero for those who did not. This variable will be used to control for any possible selection bias that father's occupational status may cause, the variable shows a high rate of missing values, around 50.0%.

Race

The Race measure is based on four categories: Whites, Mulattos, Blacks and Asians. They are represented as a set of dummy variables, using Whites as the reference group. In some models, race was categorized into a *single* dummy variable. In this case the variable *Race* has the value of 1 for either Whites or Asian racial categories and the value zero for Mulattos or Blacks. Data generation about race has produced a strong debate in Brazil. Many have questioned the possibility of having reliable data on race, as race is never clearly defined in Brazilian society. Some even question whether it is necessary to gather this kind of information, since "Brazilian society is a racial democracy and race does not account for social inequality." Therefore a note on this is necessary.

As elsewhere, the measurement of race in Brazil is controversial. Indicators of race distinction are not constant across societies, or even in the same society. It changes as the society seeks its own identity. According to Araujo (1987), there is an interaction between information production and social identity construction; thus, the debate about a "racially unambiguous classification" remains open. Nonetheless, Brazil has collected information about race off and on since the census of 1872.[21] The debate about the right way to represent the significance of race in Brazilian society has always been heated. Since, among the Brazilian population, *skin color* is the most frequent category used to distinguish races, much criticism was directed to the impossibility of using only a few classifications such as the four used by the Brazilian Institute of Geography and Statistics (IBGE) – Whites, Blacks, Mulattos, and Asians.[22] In 1976 IBGE included in the PNAD two items about race. One was a open-ended question intended to elicit the terminology used by the people who were interviewed, i.e. a self classification, the way they defined their own race, or skin color. The other, a closed question, was also a self-classification, the exact four classification groups used by IBGE described above. Despite the huge number of terms the open question allowed, 95% of the answers fell into just seven categories. These seven terms were collapsed into the four classification groups. IBGE has retained this four-classification system. Until the 1990s when it was increased to five classification categories: Whites, Mulattos, Blacks, Asians and Indians.

Gender

This is coded 0 for women, and 1 for men.

Urban Origin

Educational opportunities vary for those born and educated in urban vs. rural areas. This variable is based on a question about the area where the respondent had lived until age 15. The values are: 0 for rural area, and 1 for urban area.

Logistic Regression Model

My Logistic Regression approach assesses the odds of making education grade transitions (Mare, 1980, 1981, 1993; Shavit & Blossfeld, 1993). This model measures the probability of making a transition to the next level of school given that the individual had successfully completed the previous level. For each transition the probability of passing through that transition is estimated using only persons who were successful in the earlier transitions. The independent variables are the same as were used on the OLS model.

$\mathrm{Ln}[P\ (\mathrm{Transition}=1)/1\text{-}P(\mathrm{Transition}=1)] = \alpha + \beta_1$ Mother's Education + β_2 Father's Education + β_3 Father's Occupation + β_4 Father's Occupation (dummy) + β_5 Race (dummy) + β_6 Gender (dummy) + β_7 Urban origin (dummy) + ε.

Educational Transition

The dependent variable in this model is level of education successfully completed, but it measures school transitions, so it is a categorical and not a continuous variable.

(I) From zero to one completed year, estimating any access to formal schooling.
(II) From one to four completed years, estimating the chances of finishing the lower elementary level.
(III) From four to eight completed years, estimating the chances of finishing the upper elementary level.
(IV) From eight to eleven completed years, estimating the chances of finishing high school.
(V) From eleven to any higher completed years, estimating the chances of getting any post secondary education.

Table 1. Descriptive Statistics for Each Variable (Brazil – 1988).

Variable	No. of Observation	Lowest	Highest	Mean	Std. Dev.
Education	109.684	0.00	17.00	4.697	4.462
Mother's education	109.684	0.00	16.00	1.655	2.572
Father's education	109.684	0.00	16.00	2.020	2.974
Father's occupation	109.684	1.81	88.75	8.142	6.493
Father's occupation (dummy)	109.684	0.00	1.00	0.4951	0.500
Black (dummy)	109.656	0.00	1.00	0.054	0.226
Mulatto (dummy)	109.656	0.00	1.00	0.394	0.489
Asian(dummy)	109.656	0.00	1.00	0.005	0.069
Gender	109.684	0.00	1.00	0.482	0.500
Urban origin	98.869	0.00	1.00	0.470	0.500
Socioeconomic origin	109.684	–0.902	8.681	0.00	1.00
Race (dummy)	109.656	0.00	1.00	0.552	0.497

Note: Education: successfully completed years of education. Mother's education: successfully completed years of education. Father's education: successfully completed years of education. Father's occupation: index of occupational status. Father's occupation (dummy): 1 = answered father's occupation; 0 = did not answer. Black: race Black = 1; else = 0. Mulatto: race Mulatto = 1; else = 0. Asian: race Asian = 1; else = 0. Gender: women = 0; men = 1. Urban origin: urban = 1; rural = 0. Socioeconomic origin: factor weighted index of mother's and father's education and father's occupational status. Race: White and Asian = 1; Black and Mulatto = 0.
Source: PNAD-1988.

4.3. Variables' Descriptive Statistics

Tables 1 and 2 describe the variables and the zero-order correlations among them.

Table 2. Zero-Order Correlation Matrix of all Variables (Brazil – 1988).

Variables	Education	Mother's Education	Father's Education	Father's Occupation	Father's Occupation (Dummy)	
Part 1						
Education	1.000					
Mother's education	0.558	1.000				
Father's education	0.538	0.713	1.000			
Father's occupation	0.333	0.389	0.484	1.000		
Father's occupation (dummy)	0.101	0.048	0.031	0.000	1.000	
Black	−0.093	−0.072	−0.068	−0.032	−0.014	
Mulatto	−0.206	−0.164	−0.167	−0.086	0.011	
Asian	0.060	0.049	0.058	0.004	0.011	
Gender	0.025	0.005	0.004	−0.022	0.395	
Urban origin	0.514	0.361	0.353	0.253	−0.079	
Race	0.245	0.194	0.195	0.099	−0.055	
Variables	Black	Mulatto	Asian	Gender	Urban Origin	Race
Part 2						
Education						
Mother's education						
Father's education						
Father's occupation						
Father's occupation (dummy)						
Black	1.000					
Mulatto	−0.193	1.000				
Asian	−0.017	−0.056	1.000			
Gender	0.007	0.005	0.007	1.000		
Urban origin	−0.005	−0.086	0.003	−0.043	1.000	
Race	−0.265	−0.895	0.063	−0.008	0.087	1.000

Note: Education: successfully completed years of education. Mother's education: successfully completed years of education. Father's education: successfully completed years of education. Father's occupation: index of occupational status. Father's occupation (dummy): 1 = answered father's occupation; 0 = did not answer. Black: race Black = 1; else = 0. Mulatto: race Mulatto = 1; else = 0. Asian: race Asian = 1; else = 0. Gender: women = 0; men = 1. Urban origin: urban = 1; rural = 0. Race: White and Asian = 1; Black and Mulatto = 0.

Source: PNAD-1988.

5. THE DETERMINANTS OF EDUCATIONAL STRATIFICATION IN BRAZIL: OLS ANALYSIS

Most important for my analysis are the patterns of transformations that the statistical models represent as economic development moves forward. The oldest cohort was born between 1882 and 1913 and the youngest between 1959 and 1963, a period of 81 years, almost a century. Due to the sheer amount of data, I summarize the regression coefficients in figures. This provides a clearer picture of the transformations at stake.

5.1. Testing Hypothesis One – The Meritocratic Hypothesis

Hypothesis number one states that the effect of socioeconomic background variables on educational attainment has decreased over the course of last century as a consequence of economic development. In order to assess this hypothesis, I observe how the coefficients that estimate such variables behave over time. Table 3 summarizes the OLS regression results for the eleven cohorts. Figures 1 through 9 summarize the changes in the coefficients. Analyzing those figures suggests that:

(a) Most coefficients are significant and show a *t ratio* bigger than 3.00, except for the coefficient of the Asian variable at earlier cohorts, and for coefficients of the gender variable for later cohorts. All coefficients show a clear pattern of change, as is shown from Table 3 and Figs 1 through 8.
(b) The effects of parental education and father's occupation lose their explanatory power over educational attainment as time goes by (Figs 1–3). This is especially true for those cohorts which experienced educational attainment after the World War II, when the process of industrialization was being established.
(c) Gender shows a clear decrease in its effect, meaning that women face fewer social barriers to educational attainment as education expands.
(d) However, not all of these changes point towards a complete acceptance of Hypothesis 1. Race and urban origin show the opposite pattern. Having an urban origin emerges as a very strong advantage in relation to the chances of educational opportunity. In fact, urban origin appears to be the highest coefficient in all regressions, and its effects on educational inequality appear to increase as economic development moves forward.
(e) Race effects show an interesting pattern. Being an Asian increases one's chances of educational attainment relative to Whites. Being Black or Mulatto,

Table 3. Unstandardized and Standardized OLS Regression Coefficients of the Determinants of Educational Attainment in Brazil, by Cohorts.

Independent Variables	Cohort 1 (1882–1913)	Cohort 2 (1914–1918)	Cohort 3 (1919–1923)	Cohort 4 (1924–1928)	Cohort 5 (1929–1933)	Cohort 6 (1934–1938)
Part 1						
Mother's education	0.480*	0.388*	0.453*	0.442*	0.531*	0.411*
	[0.284]	[0.233]	[0.256]	[0.247]	[0.287]	[0.219]
	(0.035)	(0.032)	(0.029)	(0.025)	(0.022)	(0.021)
Father's education	0.286*	0.325*	0.295*	0.332*	0.246*	0.322*
	[0.226]	[0.270]	[0.231]	[0.246]	[0.173]	[0.214]
	(0.027)	(0.024)	(0.021)	(0.019)	(0.018)	(0.018)
Father's occupation	0.102*	0.0384	0.0787*	0.06198*	0.0556*	0.0481*
	[0.077]	[0.045]	[0.086]	[0.085]	[0.073]	[0.064]
	(0.020)	(0.011)	(0.011)	(0.008)	(0.007)	(0.007)
Father occupation (dummy)	0.484*	0.124	0.225	0.141	0.132	0.517*
	[0.054]	[0.015]	[0.029]	[0.019]	[0.017]	[0.066]
	(0.140)	(0.122)	(0.097)	(0.084)	(0.076)	(0.070)
Mulatto	−0.724*	−0.818*	−0.930*	−0.799*	−0.941*	−1.042*
	[−0.112]	[−0.120]	[−1.29]	[−0.106]	[−0.120]	[−0.130]
	(0.099)	(0.099)	(0.084)	(0.076)	(0.071)	(0.067)
Black	−0.762*	−0.829*	−1.218*	−1.144*	−1.213*	−1.248*
	[−0.063]	[−0.059]	[−0.089]	[−0.076]	[−0.075]	[−0.078]
	(0.181)	(0.199)	(0.158)	(0.150)	(0.142)	(0.131)
Asian	−1.161	−0.501	0.797	0.579	0.255	1.224
	[−0.010]	[−0.005]	[0.007]	[0.008]	[0.004]	[0.019]
	(1.654)	(1.387)	(1.237)	(0.734)	(0.566)	(0.515)
Gender	0.472*	0.409*	0.568*	0.560*	0.545*	0.351*
	[0.075]	[0.062]	[0.082]	[0.077]	[0.071]	[0.045]
	(0.094)	(0.097)	(0.084)	(0.078)	(0.073)	(0.069)

Table 3. *(Continued)*

Independent Variables	Cohort 1 (1882–1913)	Cohort 2 (1914–1918)	Cohort 3 (1919–1923)	Cohort 4 (1924–1928)	Cohort 5 (1929–1933)	Cohort 6 (1934–1938)
Urban origin	1.777*	2.064*	2.143*	2.182*	2.412*	2.459*
	[0.256]	[0.292]	[0.293]	[0.287]	[0.305]	[0.307]
	(0.109)	(0.107)	(0.090)	(0.080)	(0.075)	(0.070)
Intercept	−0.353	0.432*	0.216	0.580*	0.916*	1.150*
	(0.185)	(0.121)	(0.122)	(0.088)	(0.083)	(0.079)
R^2	0.448	0.478	0.491	0.478	0.416	0.432
Adjusted R^2	0.446	0.477	0.490	0.477	0.416	0.432
N	2638	2758	4074	5617	7303	8956
Independent Variables	**Cohort 7 (1939–1943)**	**Cohort 8 (1944–1948)**	**Cohort 9 (1949–1953)**	**Cohort 10 (1954–1958)**	**Cohort 11 (1959–1963)**	**Brazil**
Part 2						
Mother's education	0.483*	0.470*	0.446*	0.403*	0.336*	0.454*
	[0.259]	[0.261]	[0.260]	[0.257]	[0.246]	[0.263]
	(0.020)	(0.017)	(0.015)	(0.013)	(0.013)	(0.006)
Father's education	0.300*	0.301*	0.277*	0.278*	0.257*	0.281*
	[0.195]	[0.196]	[0.183]	[0.196]	[0.202]	[0.188]
	(0.017)	(0.15)	(0.014)	(0.013)	(0.012)	(0.005)
Father's occupation	0.0481*	0.0433*	0.0309*	0.0195*	0.0201*	0.0330*
	[0.074]	[0.065]	[0.048]	[0.033]	[0.036]	[0.049]
	(0.005)	(0.005)	(0.005)	(0.004)	(0.004)	(0.002)
Father occupation (dummy)	0.514*	0.763*	0.765*	0.745*	0.532*	0.920*
	[0.060]	[0.083]	[0.081]	[0.081]	[0.062]	[0.103]
	(0.068)	(0.065)	(0.061)	(0.060)	(0.060)	(0.022)

Mulatto	−1.155*	−1.105*	−1.235*	−1.193*	−1.136*	−1.012*
	[−0.131]	[−0.120]	[−0.130]	[−0.128]	[−0.132]	[−0.111]
	(0.066)	(0.062)	(0.058)	(0.056)	(0.056)	(0.022)
Black	−1.554*	−1.402*	−1.837*	−1.638*	−1.639*	−1.486*
	[−0.085]	[−0.067]	[−0.087]	[−0.079]	[−0.082]	[−0.076]
	(0.135)	(0.138)	(0.127)	(0.124)	(0.129)	(0.046)
Asian	0.432	1.790*	3.177*	2.474*	3.388*	1.970*
	[0.007]	[−0.025]	[0.047]	[0.030]	[0.039]	[0.026]
	(0.439)	(0.457)	(0.403)	(0.488)	(0.543)	(0.176)
Gender	0.402*	0.0504	0.0255	−0.183*	−0.249*	−0.050*
	[0.047]	[0.006]	[0.003]	[−0.020]	[−0.029]	[−0.006]
	(0.067)	(0.064)	(0.060)	(0.059)	(0.060)	(0.022)
Urban origin	2.597*	2.866*	3.194*	3.074*	2.763*	3.017*
	[0.299]	[0.315]	[0.342]	[0.334]	[0.318]	[0.338]
	(0.068)	(0.065)	(0.060)	(0.058)	(0.058)	(0.022)
Intercept	1.396*	1.723*	2.275*	2.924*	3.444*	1.673*
	(0.074)	(0.071)	(0.067)	(0.064)	(0.065)	(0.024)
R^2	0.468	0.477	0.485	0.471	0.448	0.483
Adjusted R^2	0.467	0.477	0.484	0.470	0.448	0.483
N	10423	12457	14810	15722	14084	98842

Note: Standard error in parenthesis. Education: successfully completed years of education. Mother's education: successfully completed years of education. Father's education: successfully completed years of education. Father's occupation: scale of occupational status. Lowest = 1, highest = 100. Father's occupation: dummy: answered = 1; did not answered = 0. Gender: Women = 0; Men = 1. Black: Black = 1; else = 0. Mulatto: Mulatto = 1; else = 0. Asian: Asian = 1; else = 0. Urban origin: Rural = 0; Urban = 1.

Source: PNAD-1988.

*$|t| > 3.00$.

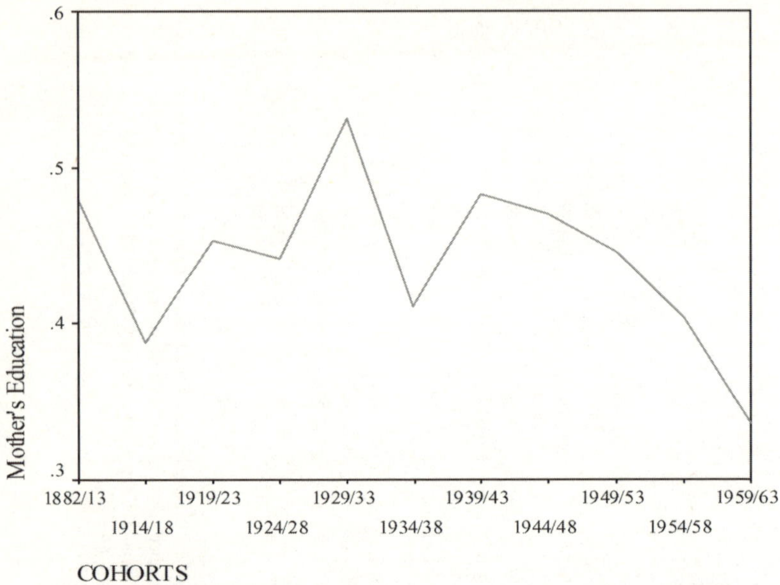

Fig. 1. Unstandardized Coefficient of Mother's Education by Cohorts.

however, handicaps one's chance of educational attainment, more so as economic development moves forward, as compared to being White.[23]

(f) Both Blacks and Mulattos face increasing social barriers in access to education as economic development moves forward, but this affects Blacks more than Mulattos.

The evidence thus far suggests that Hypothesis 1 *is not fully supported*. The equalizing effect of the industrialization process predicted by the Modernization theory is jeopardized, as important ascriptive measures increase their effect on educational attainment as economic development moves forward. The effects of urban origin, as well as the effect of race, are measured net of parental education, father's occupation, and sex. This means that being born Black or Mulatto constrains one's chances of educational attainment independently of other measured variables: parents' educational level, father's occupational status, etc. Socioeconomic transformations brought by the industrialization process even worsen this situation. Thus as development proceeds, the effects of one's educational and occupational status origins and those for women appear to have decreased. But the educational barriers to Blacks and Mulattos and those raised in rural areas increased. (Lipton, 1977, who holds that the specific type of development faced by Less Developed Countries has led to an urban bias has noted

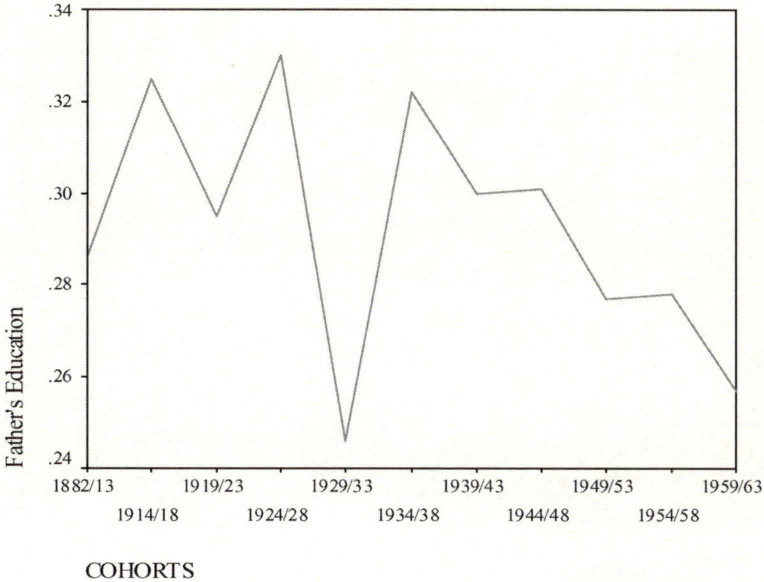

Fig. 2. Unstandardized Coefficient of Father's Education by Cohorts.

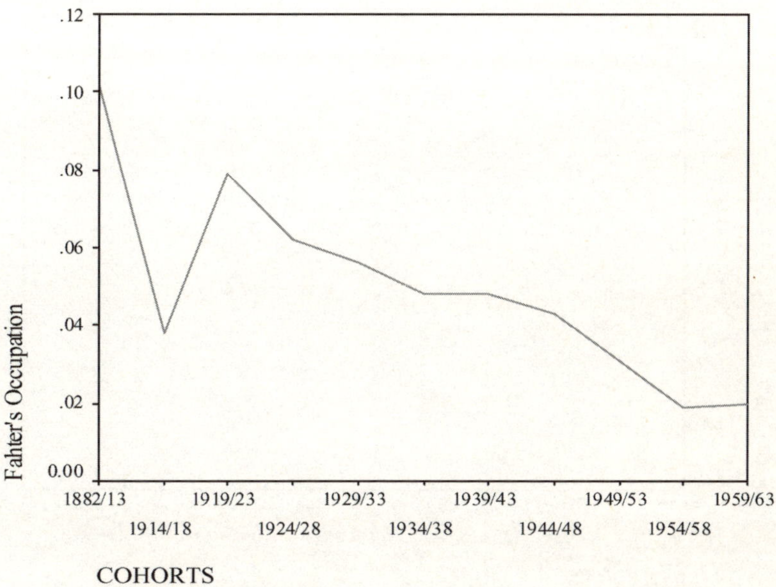

Fig. 3. Unstandardized Coefficient of Father's Occupation by Cohorts.

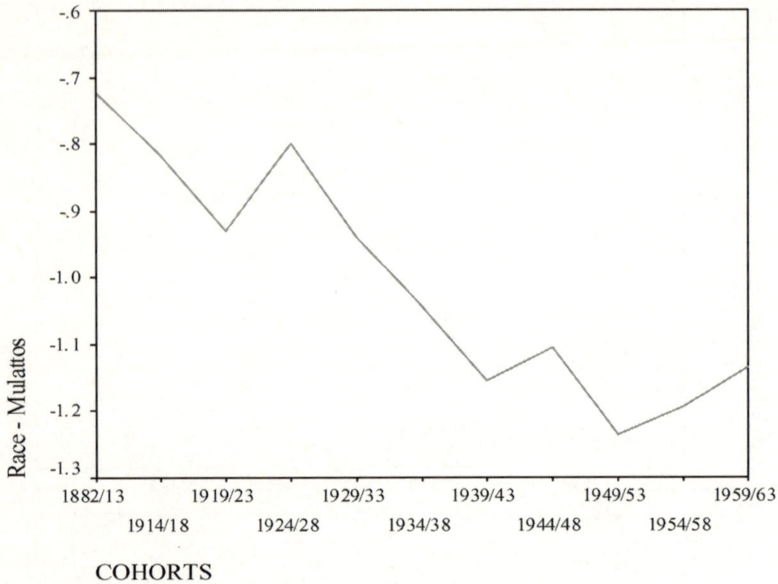

Fig. 4. Unstandardized Coefficients of Race – Mulattos by Cohorts.

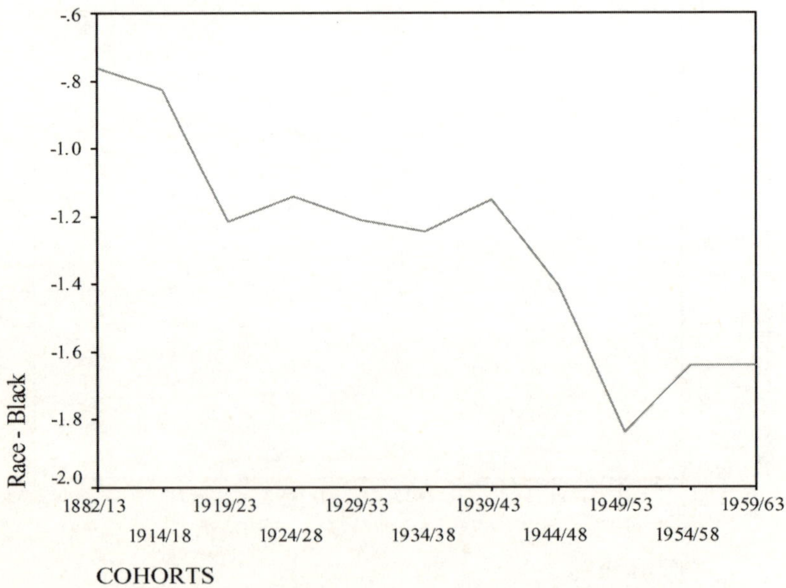

Fig. 5. Unstandardized Coefficients of Race – Blacks by Cohorts.

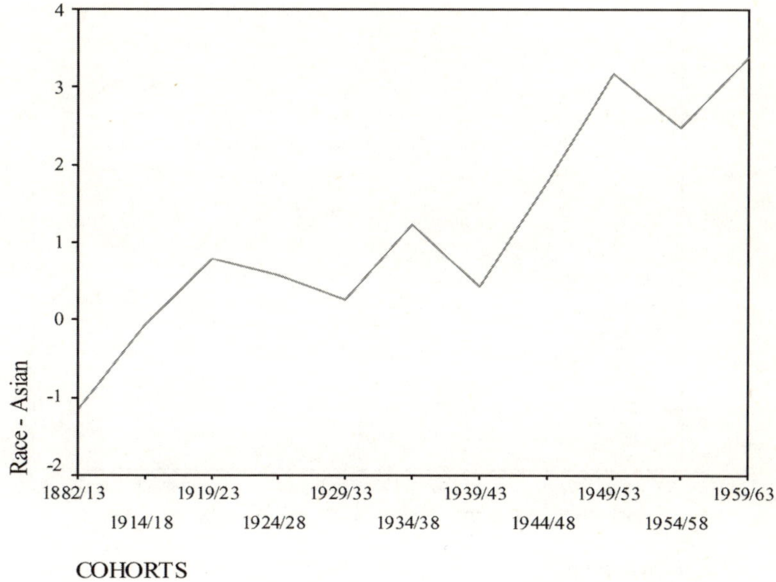

Fig. 6. Unstandardized Coefficients of Race – Asians by Cohorts.

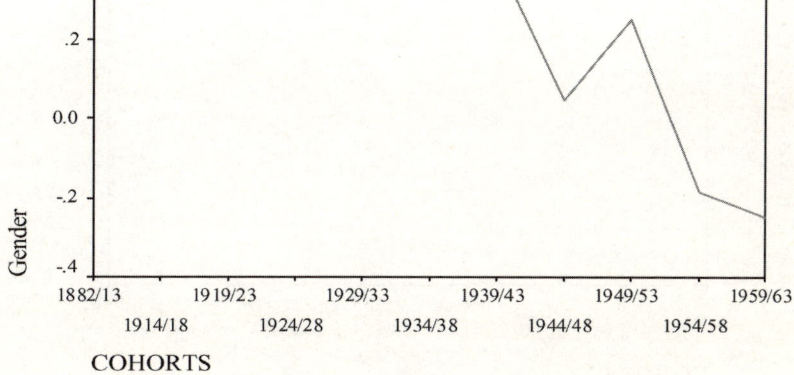

Fig. 7. Unstandardized Coefficients Gender by Cohorts.

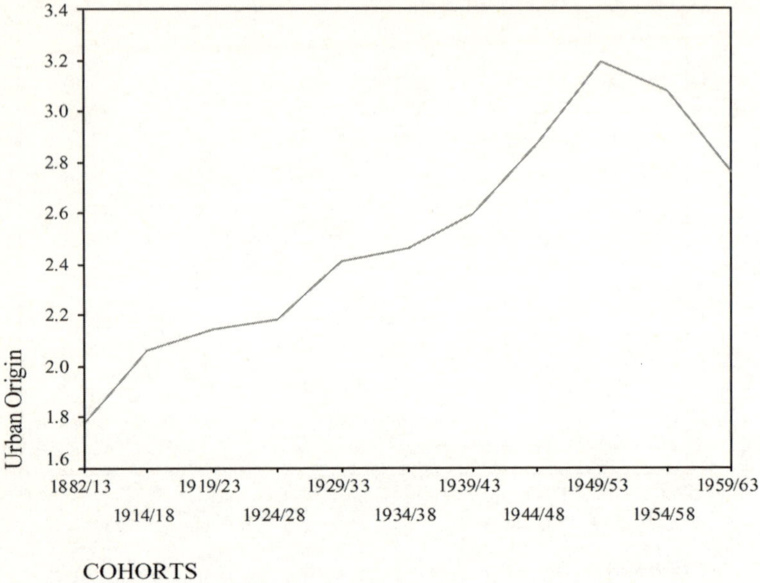

Fig. 8. Unstandardized Coefficients Urban Origin by Cohorts.

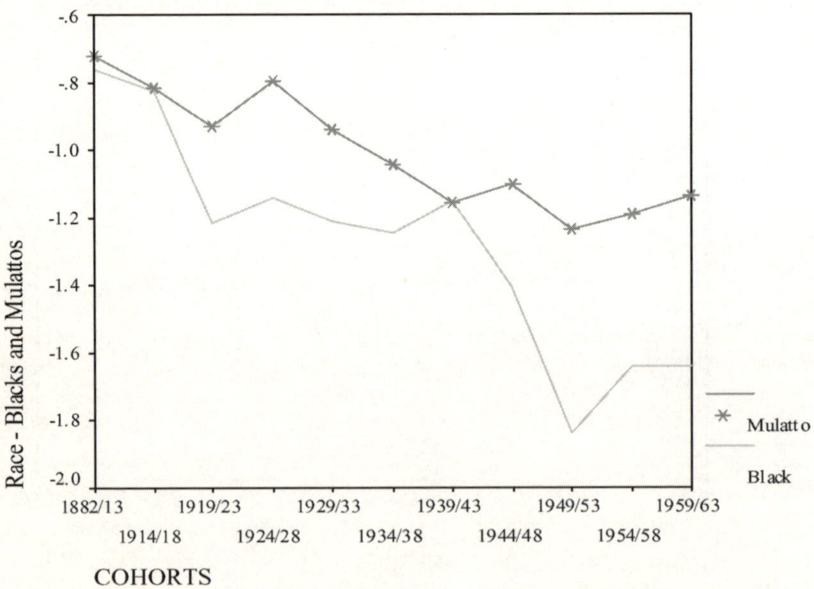

Fig. 9. Unstandardized Coefficients of Race – Blacks and Mulattos by Cohorts.

the increasing effect of urban origins as a consequence of economic development. Few schools are available to those who live outside the cities.)

5.2. Testing Hypotheses Two and Three – The Declining Significance of Race and Uneven Development

Table 3 and Figs 1 through 9 summarize the results of the OLS regression pertinent to Hypotheses 2 and 3. After analyzing those figures, I come to the following conclusions:

(a) Race does *not* have a decreasing effect as economic development moves forward, as the Declining Significance of Race Hypothesis predicted. In fact, it has a significant increasing effect for all racial groups. However, the pattern of effects is different for each group. The impact the social and economic transformations brought by the industrialization process helps Asians' educational attainment more than any other racial group. On the other hand, these same transformations handicap Blacks' and Mulattos' educational opportunities, independent of their socioeconomic background. This pattern is even stronger for Blacks than for Mulattos. It is important to note that the rate of change of these coefficients was highest when the process of industrialization was already established, from the early 1940s on, than before.

(b) Most socioeconomic background variables *do not* show a pattern of increasing importance as explaining educational attainment as the Declining Significance of Race Hypothesis predicts, except for urban origin. Those who lived in rural areas until they were 15 years old face stronger barriers in relation to access to education, than those who lived in urban areas, independent of racial origin and other socioeconomic ascriptive determinants, as we noted earlier when assessing the first hypothesis.

Based on this evidence, I conclude that the Declining Significance of Race Hypothesis *is not supported*. On the other hand, the Uneven Development Hypothesis *is supported*. The shape of the transformations of the coefficients points toward the opposite of what the former hypothesis predicts and in favor of what the latter hypothesis predicts, except for urban origin. For the Uneven Development View's explanation, the pattern of increasing importance of race in explaining educational inequality as economic development moves forward is based on the following logic. First, economic growth creates or even exacerbates social divisions and inequalities. Second, *archaic* social relations, like the ones based on race and ethnicity, are potentially compatible with, and perhaps functional to, capitalist development. Under this perspective, persistence or even an increase in the traditional and archaic basis of social inequality is a functional necessity of capitalist development, as the dominant social groups keep the best social opportunities (Greenberg, 1980).

The idea is that more economic development would bring social and economic competition between the individuals as a way to get the best opportunities. The intensity of this competition creates incentives for Whites to raise racial barriers as a way to eliminate Blacks and Mulattos from the few good opportunities in the occupational hierarchy (Degler, 1971; Hasenbalg, 1979). For Hasenbalg (1979), it would be useless to try to determine *the degree to which* the creation or even perpetuation of racist institutions and practices is a necessary condition for capitalist development. On the other hand, there is no reason to believe that industrialization and capitalist development will eliminate race as a criterion, which underlies social relations in the productive system as the proponents of the Parsonian sociology, Orthodox Marxism, Neoclassical, and Modernization Theories suggest. In fact the durability of race divisions casts doubts on these assumptions. According to Hasenbalg (1979), the perpetuation of social prejudice and discrimination towards the Non-White population may be best understood as a consequence of the symbolic and material interests of the White dominant group. This is represented in a set of social mechanisms, as, for example, increasing barriers toward access to education, which tend to reproduce this aspect of the social structure.

6. PERSISTENT BARRIERS: RACE AND EDUCATIONAL INEQUALITY IN BRAZIL

6.1. Introduction

To this point, patterns of educational stratification in Brazil do not point toward an equalization of educational opportunities. In fact, we saw mixed results in predicting social barriers to educational attainment as economic development moves forward: parents' education, father's occupational status, and gender seem to be fading; race and rural-urban origin seem to be intensifying their effects. Moreover, the effect of being Black or Mulatto increases with development; they become increasingly different from each other, at the expense of Blacks (Fig. 9). These findings point toward a pattern of increasing inequality in educational opportunities for people from different racial groups. Whites and Asians not only have a smoother path toward educational attainment, but chances are also more equalized within this group than within the Non-White group.

I continue to analyze the determinants of educational stratification using a methodology different from the OLS strategy. This procedure will allow us to see the determinants of educational inequality, as separated from the process of educational expansion. In this way, we could see whether patterns of educational inequality are stable or are associated with economic development, independent of the degree of educational expansion. This is a unique procedure, introduced by Mare (1980), and used by several researchers applying data from different countries (Shavit & Blossfeld, 1993).

6.2. Testing Hypothesis Four – Challenges from New Trends

Hypothesis 4 states that there will be a decrease in the effect of social origin on school transitions within each cohort, and a stable pattern of these effects in each transition across cohorts. The higher the educational level an individual reaches, the less it will be explained by status origins, including those that are ascriptive measures, and this pattern is constant over time.

For each level I estimated twelve logit regressions, one for each cohort and one for the country as a whole. The transitions used as the dependent variable for each set of logit regressions are listed below. Table 4 shows the inter-cohort changes in educational attainment and other descriptive measures. Tables 5 through 9 and Figs 10–17 summarize the results from the Logistic regressions. The results suggest that:

(a) Most coefficients are highly significant except for the gender variable in most of the analysis and for some of the coefficients of the oldest cohorts estimating the highest transitions. Most social origin measures show a decreasing pattern from the lowest to the highest school transitions. Parents' schooling, father's occupational status, and even urban origin all show a smooth declining pattern, though at different rates. One year of mother's education raises the odds of making the first transition by almost 60%. This rate decreases to 8% at the last transition, which represents the chances of getting any kind of post secondary education, given that one has finished high school. Living in urban areas until 15 years of age increases by 320% the chances of getting the first year of education. By contrast, this increase is just 35% in the chances of making the last transition that is getting any kind of post secondary education. Father's education and occupational status follow the same pattern.

(b) Not all social origins measures follow a decreasing trend. Gender is not statistically significant in most transitions. Race shows a different pattern. Looking at the first three transitions, we could identify the same pattern

Table 4. Inter-Cohort Change in Educational Attainment and Other Descriptive Measures.

Race	Cohorts	Mean	Standard Deviation	*N*
Part 1				
Whites	1882–1913	2.7650	3.7863	1664
	1914–1918	3.1164	3.9066	1673
	1919–1923	3.3737	3.9333	2575
	1924–1928	3.7578	4.0667	3485
	1929–1933	4.1678	4.2334	4386
	1934–1938	4.4858	4.2639	5358
	1939–1943	5.1600	4.1600	6277
	1944–1948	5.7909	4.7965	7557
	1949–1953	6.6191	4.8066	8967
	1954–1958	7.1385	4.6190	9519
	1959–1963	7.2620	4.2244	8528
	All cohorts	5.6560	4.6789	59989
Mulattos	1882–1913	0.8638	2.0162	1065
	1914–1918	1.1703	2.1937	1149
	1919–1923	1.3625	2.3808	1589
	1924–1928	1.9021	2.7916	2264
	1929–1933	2.2053	3.0800	3045
	1934–1938	2.4859	3.2170	3813
	1939–1943	2.8839	3.5504	4481
	1944–1948	3.4723	3.8625	5544
	1949–1953	4.2250	4.1414	6611
	1954–1958	4.8625	4.1278	7111
	1959–1963	5.2113	3.9209	6552
	All cohorts	3.5572	3.8849	43224
Part 2				
Blacks	1882–1913	0.7164	1.9116	201
	1914–1918	1.1236	2.0712	178
	1919–1923	0.9585	1.7073	289
	1924–1928	1.4624	2.4109	372
	1929–1933	1.7877	2.7135	471
	1934–1938	2.2896	2.9417	618
	1939–1943	2.4422	2.9243	658
	1944–1948	3.1551	3.3555	688
	1949–1953	3.6474	3.6347	848
	1954–1958	4.3826	3.7608	868
	1959–1963	4.7061	3.5045	722
	All cohorts	2.9584	3.3923	5912

Table 4. *(Continued)*

Race	Cohorts	Mean	Standard Deviation	*N*
Asians	1882–1913	3.4000	4.7889	10
	1914–1918	6.0000	5.1769	11
	1919–1923	6.2727	4.5897	22
	1924–1928	6.0000	4.5166	31
	1929–1933	5.6458	4.1024	48
	1934–1938	6.5962	4.1409	52
	1939–1943	7.8471	4.6072	85
	1944–1948	8.9324	4.6533	74
	1949–1953	10.7556	4.5870	90
	1954–1958	10.7842	4.7211	65
	1959–1963	11.2791	4.1537	43
	All cohorts	8.5198	4.9508	531

Source: PNAD-1988.

present in the effect of the other social origins measures described above: that is a sharp decline in the effect of race from one transition to the next. For example, being White or Asian increases one's chances of getting one year of education by a little more than 100% in relation to Non-Whites.

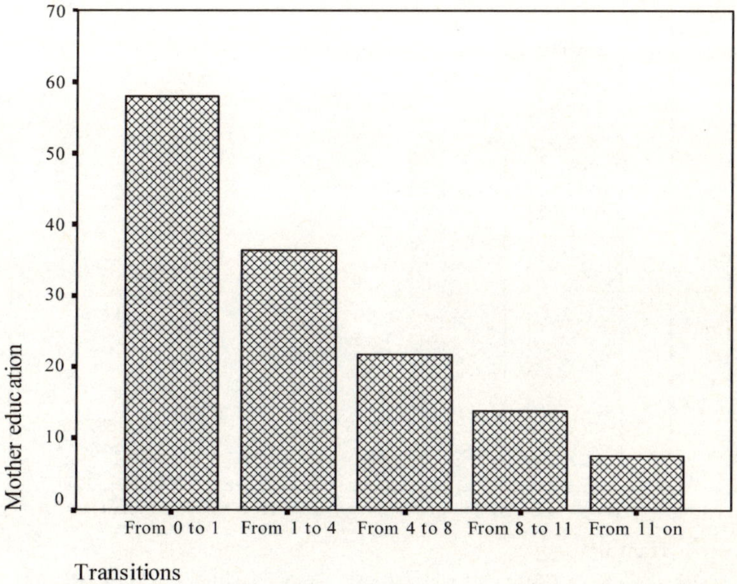

Fig. 10. The Effects of Mother's Education on the Odds of Making School Transitions.

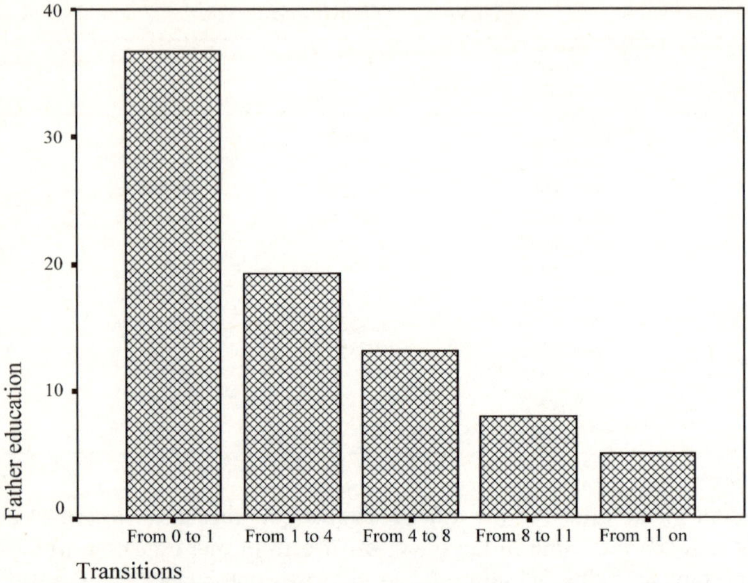

Fig. 11. The Effects of Father's Education on the Odds of Making School Transitions.

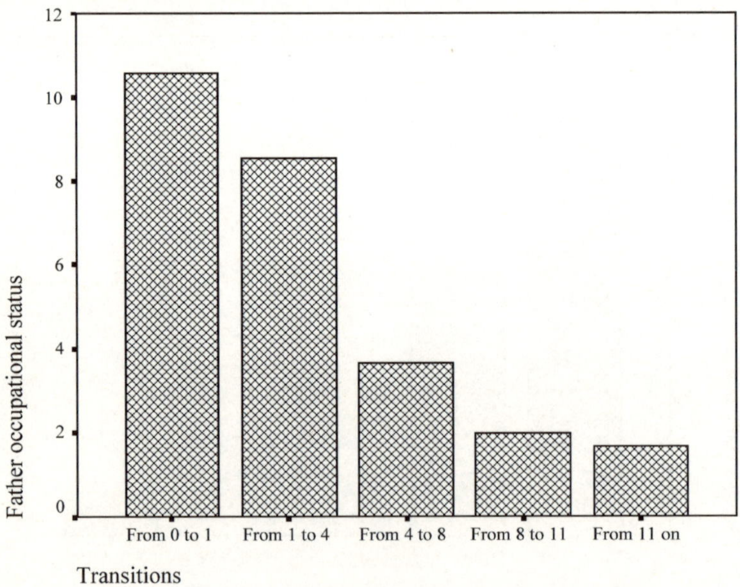

Fig. 12. The Effects of Father's Occupational Status on the Odds of Making School Transitions.

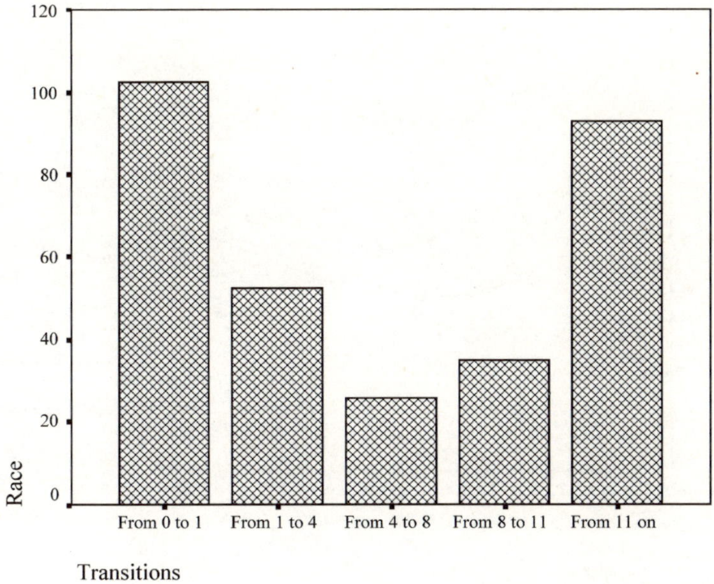

Fig. 13. The Effects of Race on the Odds of Making School Transitions.

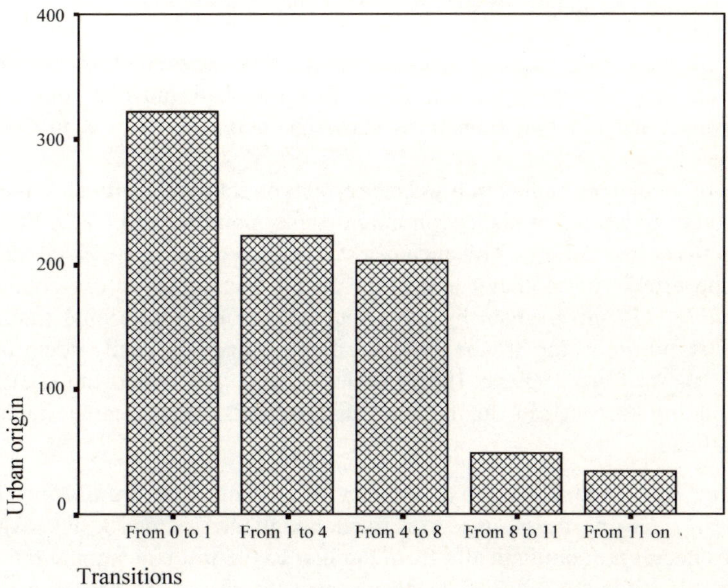

Fig. 14. The Effects of Urban Origin on the Odds of Making School Transitions.

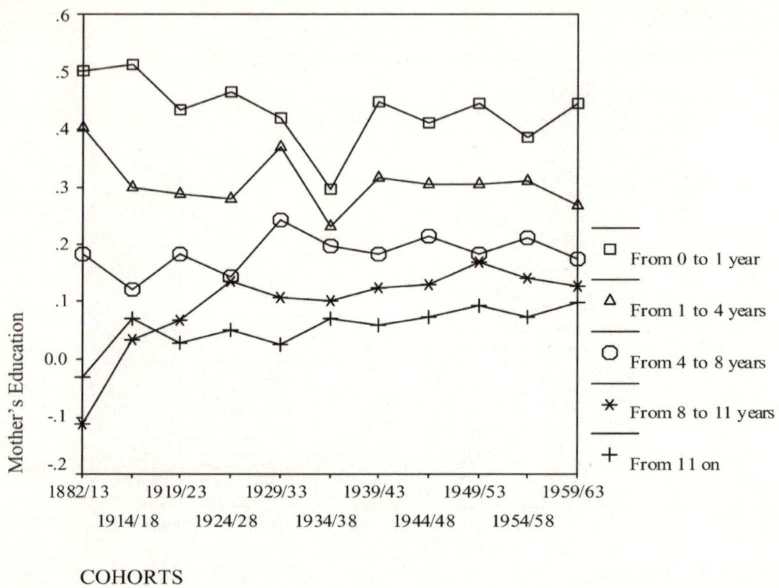

Fig. 15. Trends in Mother's Education Effects on School Transitions Controlling for Other Measures of Socioeconomic Origins.

By contrast, this increase goes down to 26% in predicting the chances of finishing elementary level, given that one had finished fourth grade. However, the last two transitions show an increasing pattern in the effect of race.

(c) Trends on mother's education and race effects on school transitions, controlling for other measures of socioeconomic origins, point toward a stability of the patterns across cohorts. This means a *stable* decreasing shape given by trends of the effects of mother's education on all transitions across cohort; and a *stable* "U" shape given by the effect of race on educational transitions. The stability of the trends in the effect of race is easily identified on the cohorts from 1939 to 1963 (Table 17, Fig. 17), when the coefficients estimating race effects in the two highest transitions became statistically significant.

Based on this evidence, I conclude that the Challenges From New Trends Hypothesis *is only* ***partially*** *supported*. Even though most of the social background measures decrease monotonically from the first to the last educational transition, the effect of race shows an initial decrease in the first three transitions, but it increases from the third to the forth transition and dramatically from the fourth to

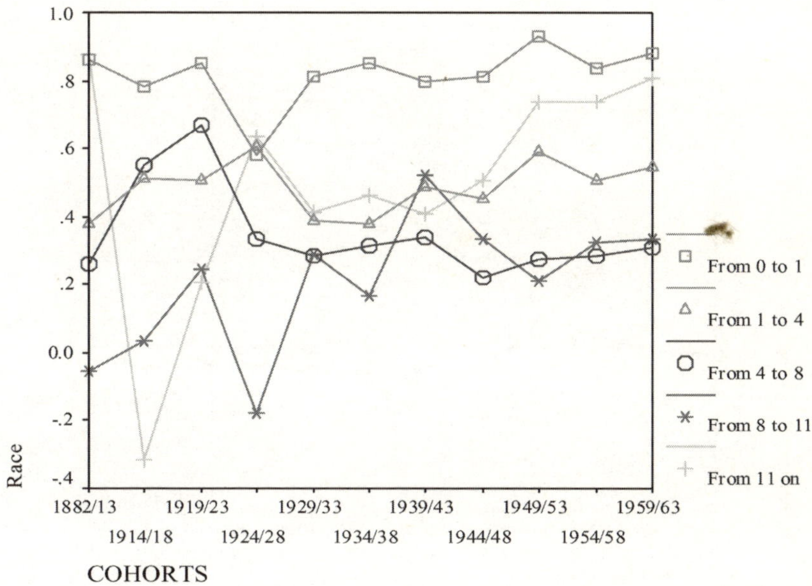

Fig. 16. Trends on Race Effects on School Transitions Controlling for Other Measures of Socioeconomic Origins – All Cohorts.

the fifth transition. According to Mare (1980, 1981), the decrease in the effects of social origin from one transition to the next can be explained by a decrease in the heterogeneity of socioeconomic origins that the population is exposed to as they move from lower to higher educational levels. Those who reach higher educational levels are *survivors* of a highly selective process.

This explanation, albeit while reasonable, does not explain the persistence of the effect of race on the probability of making the highest educational transitions: finishing secondary level and getting any kind of post secondary education. Non-Whites are also exposed to higher barriers than Whites at all school transitions. Although those who are at higher educational levels are more homogeneous in their socioeconomic background, they face a rise in the effects of ascripitive origins such as race. Being White or Asian increases one's chances of completing the first year of education by 102% relative to being Black or Mulatto. This figure is 52% for the chances of finishing fourth grade and 26% for completing elementary school. Completing secondary school is much harder than completing elementary school for Non-Whites, and having access to any kind of post secondary education is even harder. In fact, finishing at least one year of formal education after secondary school is almost as hard as completing the first year of education for Non-Whites in relation to whites and Asians. We should remember that the effect of race is

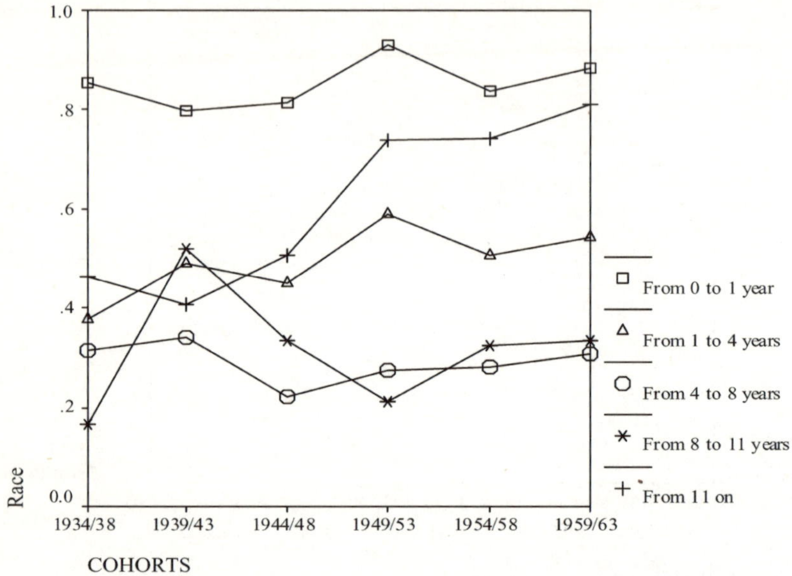

Fig. 17. Trends on Race Effects on School Transitions Controlling for Other Measures of Socioeconomic Origins – Selected Cohorts.

controlled by all other measures of socioeconomic origins as is demonstrated in Tables 5 through 9.

Figure 16 shows trends in mother's education effects in school transitions controlling other measures of socioeconomic origins. This pattern is stable over time, as predicted from this hypothesis. Shavit and Blossfeld (1993) show cross-nation comparison data of several countries. Among thirteen countries to which this model was applied, eleven showed the same pattern with little adjustment. I chose to look at the effect of mother's education to investigate the possible stability in this pattern because the effect of parental schooling is strongly associated with other family determinants of educational attainment (Mare, 1993). Other measures of socioeconomic origins show a similar pattern of stability across cohorts. Over 81 years, the effects of the determinants of educational stratification in Brazil did not decrease. Not even the process of industrialization, dramatically changing the socioeconomic structure of the country, altered patterns of educational stratification, as predicted by socioeconomic origin. The effect of race indicates even greater tendency toward educational inequality in a society in which the average level of education is around seven years.

Table 5. Trends in the Determinants of Educational Stratification Effects on School Transition 1 – from Zero to One Year.

Independent Variables	Cohort 1 (1882–1913)	Cohort 2 (1914–1918)	Cohort 3 (1919–1923)	Cohort 4 (1924–1928)	Cohort 5 (1929–1933)	Cohort 6 (1934–1938)
Part 1						
Mother's education	0.5014	0.5118	0.4339	0.4635	0.4202	0.2961
	(0.0648)	(0.0615)	(0.0507)	(0.0450)	(0.0387)	(0.0334)
Father's education	0.2749	0.2555	0.3581	0.3540	0.2907	0.3363
	(0.0494)	(0.0469)	(0.0404)	(0.0357)	(0.0312)	(0.0288)
Father's occupation	0.0491*	0.1302*	0.0850	0.1322	0.1365	0.0348*
	(0.0432)	(0.0511)	(0.0364)	(0.0297)	(0.0264)	(0.0154)
Father occupation (dummy)	0.1534*	0.3115*	0.3092	0.2555*	0.3356	0.2187*
	(0.1875)	(0.1947)	(0.1453)	(0.1192)	(0.1034)	(0.0726)
Race	0.8637	0.7841	0.8545	0.5797	0.8120	0.8530
	(0.0989)	(0.0936)	(0.0774)	(0.0649)	(0.0569)	(0.0525)
Gender	0.4742	0.3234	0.4305	0.4931	0.3917	0.3022
	(0.1004)	(0.0974)	(0.0844)	(0.0726)	(0.0643)	(0.0583)
Urban origin	1.4115	1.4339	1.6073	1.4030	1.2804	1.3410
	(0.1087)	(0.1067)	(0.0914)	(0.0786)	(0.0708)	(0.0649)
Intercept	−2.7201	−2.8817	−2.4918	−2.3308	−2.1391	−1.1453
	(0.3626)	(0.4267)	(0.3007)	(0.2471)	(0.2182)	(0.1319)
−2 Log likelihood	2621.238	2833.668	4114.344	5735.454	7380.023	8873.762
Goodness of fit	4549.847	37863.410	13279.779	9751.261	19441.306	1808502.91
χ^2	910.789	982.824	1523.597	1848.973	2162.560	2276.625
N	2638	2758	4074	5617	7303	8956

Table 5. *(Continued)*

Independent Variables	Cohort 7 (1939–1943)	Cohort 8 (1944–1948)	Cohort 9 (1949–1953)	Cohort 10 (1954–1958)	Cohort 11 (1959–1963)	Brazil
Part 2						
Mother's education	0.4470	0.4126	0.4437	0.3849	0.4460	0.4570
	(0.0347)	(0.0329)	(0.0334)	(0.0328)	(0.0379)	(0.0114)
Father's education	0.2944	0.3436	0.2999	0.2997	0.2873	0.3123
	(0.0281)	(0.0280)	(0.0276)	(0.0288)	(0.0330)	(0.0094)
Father's occupation	0.1324	0.1153	0.0778	0.0641	0.1747	0.1006
	(0.0217)	(0.0200)	(0.0178)	(0.0170)	(0.0272)	(0.0070)
Father's occupation (dummy)	0.6096	0.5611	0.5219	0.4043	0.5879	0.7405
	(0.0867)	(0.0821)	(0.0768)	(0.0766)	(0.1070)	(0.0281)
Race	0.7974	0.8151	0.9302	0.8363	0.8839	0.7063
	(0.0509)	(0.0499)	(0.0519)	(0.0548)	(0.0640)	(0.0172)
Gender	0.1995	0.1065*	−0.0522*	−0.1346*	−0.3012	−0.0439*
	(0.0560)	(0.0543)	(0.0557)	(0.0595)	(0.0704)	(0.0188)
Urban origin	1.3024	1.3238	1.4642	1.3475	1.1091	1.4387
	(0.0626)	(0.0631)	(0.0643)	(0.0654)	(0.0704)	(0.0212)
Intercept	−1.7902	−1.3808	−0.8027	−0.2993*	−0.8992	−1.4590
	(0.1797)	(0.1653)	(0.1482)	(0.1417)	(0.2220)	(0.0577)
−2 Log Likelihood	9471.392	102111.768	10181.686	9583.194	7477.596	83808.514
Goodness of fit	10576.8	20194.3	832970.1	28051.1	16688.0	5261951953
χ^2	2728.368	2989.785	3124.163	2608.633	2139.018	26536.881
N	10423	12457	14810	15722	14084	98842

Note: Standard error in parenthesis. Education: successfully completed years of education. Mother's education: successfully completed years of education. Father's education: successfully completed years of education. Father's Occupation: scale of occupational status. Lowest = 1, highest = 100. Father's occupation: dummy: answered = 1; did not answered = 0. Gender: Women = 0; Men = 1. Race: Whites and Asians = 1 Blacksand Mulattos = 0. Urban origin: Rural = 0; Urban = 1.

Source: PNAD-1988.

*$p > 0.01$.

Table 6. Trends in the Determinants of Educational Stratification Effects on School Transition 2 – from One to Four Years.

Independent Variables	Cohort 1 (1882–1913)	Cohort 2 (1914–1919)	Cohort 3 (1919–1923)	Cohort 4 (1924–1928)	Cohort 5 (1929–1933)	Cohort 6 (1934–1938)
Part 1						
Mother's education	0.4036	0.2985	0.2866	0.2779	0.3698	0.2299
	(0.0648)	(0.0575)	(0.0425)	(0.0370)	(0.0333)	(0.0274)
Father's education	0.0894*	0.2094	0.1321	0.1626	0.1016	0.1670
	(0.0390)	(0.0481)	(0.0305)	(0.0287)	(0.0242)	(0.0229)
Father's occupation	0.0099*	0.0216*	0.0458*	0.0706	0.0803	0.0602
	(0.0336)	(0.2757)	(0.0247)	(0.0182)	(0.0171)	(0.0143)
Father's occupation (dummy)	0.0552*	−0.1538*	−0.0968*	0.1868*	0.1249*	0.3292
	(0.2330)	(0.1958)	(0.1294)	(0.1007)	(0.0861)	(0.0733)
Race	0.3803*	0.5106	0.5064	0.6041	0.3883	0.3789
	(0.1579)	(0.1372)	(0.1046)	(0.0823)	(0.0709)	(0.0607)
Gender	−0.1954*	0.0541*	0.3336*	0.1375*	0.1582*	0.1150*
	(0.1491)	(0.1347)	(0.1056)	(0.0871)	(0.0763)	(0.0656)
Urban origin	0.9772	1.1307	0.7458	0.9230	1.2470	1.1875
	(0.1464)	(0.1335)	(0.1016)	(0.0841)	(0.0739)	(0.0640)
Intercept	−1.1717*	−1.6648	−1.5970	−1.7339	−1.6968	−1.4455
	(0.3146)	(0.3636)	(0.2222)	(0.1703)	(0.1538)	(0.1280)
−2 Log likelihood	1160.520	1423.546	2467.897	3736.210	4991.188	6670.596
Goodness of fit	995.405	5183.385	2466.774	3395.179	6189.027	8206.67
χ^2	249.955	371.408	448.013	753.771	1222.199	1303.126
N	1033	1310	2137	3340	4675	6142

Table 6. *(Continued)*

Independent Variables	Cohort 7 (1939–1943)	Cohort 8 (1944–1948)	Cohort 9 (1949–1953)	Cohort 10 (1954–1958)	Cohort 11 (1959–1963)	Brazil
Part 2						
Mother's education	0.3143	0.3054	0.3029	0.3085	0.2685	0.3112
	(0.0272)	(0.0241)	(0.0228)	(0.0232)	(0.0236)	(0.0088)
Father's education	0.1948	0.2129	0.2062	0.2325	0.2079	0.1760
	(0.0234)	(0.0215)	(0.0204)	(0.0215)	(0.0225)	(0.0075)
Father's occupation	0.0757	0.0748	0.0925	0.1045	0.0961	0.0822
	(0.0125)	(0.0119)	(0.0130)	(0.0145)	(0.0148)	(0.0049)
Father's occupation (dummy)	0.2235	0.4155	0.4301	0.4904	0.3330	0.4733
	(0.0655)	(0.0619)	(0.0605)	(0.0643)	(0.0678)	(0.0230)
Race	0.4887	0.4517	0.5884	0.5059	0.5441	0.4207
	(0.0553)	(0.0507)	(0.0476)	(0.0478)	(0.0521)	(0.0185)
Gender	0.1429*	−0.342*	01483	−0.0283*	0.0939*	−0.0270*
	(0.0595)	(0.0550)	(0.0516)	(0.0532)	(0.0597)	(0.0201)
Urban origin	0.9451	1.1662	1.3440	1.1922	1.1072	1.1721
	(0.0580)	(0.0548)	(0.0516)	(0.0522)	(0.0552)	(0.0198)
Intercept	−1.5336	−1.4317	−1.5868	−1.2691	−0.9574	−1.3968
	(0.1133)	(0.1054)	(0.1118)	(0.1216)	(0.1245)	(0.0421)
−2 Log likelihood	7969.9	9497.2	10971.5	11019.4	9522.1	71474.0
Goodness of fit	10258.110	1457.922	12011.843	37092.274	22881.530	119689.1
χ^2	1774.421	2371.175	3098.937	2895.281	2164.253	17151.072
N	7588	9687	12354	13667	12569	74501

Note: Standard error in parenthesis. Education: successfully completed years of education. Mother's education: successfully completed years of education. Father's education: successfully completed years of education. Father's Occupation: scale of occupational status. Lowest = 1, highest = 100. Father's occupation: dummy: answered = 1; did not answered = 0. Gender: Women = 0; Men = 1. Race: Whites and Asians = 1 Blacksand Mulattos = 0. Urban origin: Rural = 0; Urban = 1.

Source: PNAD-1988.

*$p > 0.01$.

Table 7. Trends in the Determinants of Educational Stratification Effects on School Transition 3 – from Four to Eight Years.

Independent Variables	Cohort 1 (1882–1913)	Cohort 2 (1914–1918)	Cohort 3 (1919–1923)	Cohort 4 (1924–1928)	Cohort 5 (1929–1933)	Cohort 6 (1934–1938)
Part 1						
Mother's education	0.1817	0.1225	0.1833	0.1445	0.2423	0.1979
	(0.0438)	(0.0384)	(0.0361)	(0.0276)	(0.0247)	(0.0219)
Father's education	0.0845	0.1479	0.1147	0.1542	0.1172	0.1148
	(0.0313)	(0.0283)	(0.0245)	(0.0212)	(0.0182)	(0.0168)
Father's occupation	0.0991*	0.0559*	0.0336*	0.0454	0.0090*	0.0352
	(0.0563)	(0.0457)	(0.0190)	(0.0132)	(0.0078)	(0.0086)
Father's occupation (dummy)	0.3105*	0.0497*	−0.0117*	−0.0369*	−0.0541*	0.3250
	(0.3321)	(0.2997)	(0.1927)	(0.1330)	(0.1022)	(0.8250)
Race	0.2608*	0.5492*	0.6720	0.3361*	0.2846	0.3138
	(0.2649)	(0.2360)	(0.1831)	(0.1272)	(0.0990)	(0.0815)
Gender	0.8681	0.6363	0.2794*	0.3822	0.3051	0.1402*
	(0.2112)	(0.1933)	(0.1500)	(0.1166)	(0.0954)	(0.0808)
Urban origin	0.9192	1.5191	1.3483	1.1647	1.1343	0.9706
	(0.2379)	(0.2596)	(0.1798)	(0.1315)	(0.1081)	(0.0880)
Intercept	−3.7491	−4.0800	−3.5453	−3.1493	−2.7133	−2.6918
	(0.5615)	(0.4761)	(0.2795)	(0.1935)	(0.1420)	(0.1211)
−2 Log likelihood	604.064	710.052	1218.928	2058.607	3048.324	4246.488
Goodness of fit	577.621	715.715	1287.773	2142.073	2957.806	4307.167
χ^2	135.197	211.922	335.225	522.625	744.075	867.422
N	591	738	1226	2011	2894	3887

Table 7. *(Continued)*

Independent Variables	Cohort 7 (1939–1943)	Cohort 8 (1944–1948)	Cohort 9 (1949–1953)	Cohort 10 (1954–1958)	Cohort 11 (1959–1963)	Brazil
Part 2						
Mother's education	0.1834	0.2135	0.1844	0.2118	0.1746	0.1976
	(0.0185)	(0.0159)	(0.0139)	(0.0131)	(0.0125)	(0.0056)
Father's education	0.1057	0.1229	0.1434	0.1470	0.1639	0.1232
	(0.0154)	(0.0136)	(0.0129)	(0.0124)	(0.0127)	(0.0048)
Father's occupation	0.0332	0.0574	0.0463	0.0280	0.0367	0.0360
	(0.0072)	(0.0072)	(0.0062)	(0.0057)	(0.0063)	(0.0025)
Father's occupation (dummy)	0.2837	0.3980	0.3841	0.3373	0.3247	0.4263
	(0.0708)	(0.0614)	(0.0520)	(0.0488)	(0.0500)	(0.0215)
Race	0.3407	0.2212	0.2739	0.2828	0.3093	0.2287
	(0.0702)	(0.0592)	(0.0498)	(0.0453)	(0.0458)	(0.0207)
Gender	0.1699*	−0.0798*	−0.0413*	−0.0479*	−0.1081*	−0.0439*
	(0.0697)	(0.602)	(0.0511)	(0.0481)	(0.0501)	(0.0212)
Urban origin	1.1527	0.9740	1.1227	1.1071	1.0735	1.1089
	(0.0759)	(0.0644)	(0.0556)	(0.0509)	(0.0517)	(0.0231)
Intercept	−2.5690	−2.4733	−2.3134	−2.0393	−2.0058	−2.3194
	(0.1049)	(0.0911)	(0.0788)	(0.0696)	(0.0731)	(0.0320)
−2 Log likelihood	5557.8	7662.4	10267.3	12056.6	11615.1	60095.7
Goodness of fit	5401.843	10740.905	9967.345	13679.931	1308.937	61957.939
χ^2	1252.662	1856.846	2462.795	2916.810	2645.563	1393.369
N	4996	6761	9184	10847	10360	53495

Note: Standard error in parenthesis. Education: successfully completed years of education. Mother's education: successfully completed years of education. Father's education: successfully completed years of education. Father's Occupation: scale of occupational status. Lowest = 1, highest = 100. Father's occupation: dummy: answered = 1; did not answered = 0. Gender: Women = 0; Men = 1. Race: Whites and Asians = 1 Blacksand Mulattos = 0. Urban origin: Rural = 0; Urban = 1.

Source: PNAD-1988.

*$p > 0.01$.

Table 8. Trends in the Determinants of Educational Stratification Effects on School Transition 4 – from Eight to Eleven Years.

Independent Variables	Cohort 1 (1882–1913)	Cohort 2 (1914–1918)	Cohort 3 (1919–1923)	Cohort 4 (1924–1928)	Cohort 5 (1929–1933)	Cohort 6 (1934–1938)
Part 1						
Mother's education	−0.1139*	0.0335*	0.0676*	0.1366	0.1063	0.1018
	(0.0663)	(0.0466)	(0.0449)	(0.0350)	(0.0291)	(0.0271)
Father's education	0.1380*	0.0364*	0.0678*	0.0497*	0.0403*	0.0638
	(0.0558)	(0.0326)	(0.0306)	(0.0237)	(0.0217)	(0.0210)
Father's occupation	0.6421*	−0.0050*	0.0536*	0.0225*	0.0285*	0.0197*
	(0.4510)	(0.0164)	(0.0368)	(0.0149)	(0.0134)	(0.0096)
Father's occupation (dummy)	2.7311*	1.3136*	0.2837*	0.5853	0.4919	0.2150*
	(1.1678)	(0.5679)	(0.3189)	(0.2163)	(0.1600)	(0.1256)
Race	−0.0565*	0.0359*	0.2456*	−0.1804*	0.2894*	0.1671*
	(0.4597)	(0.4072)	(0.3115)	(0.2090)	(0.1567)	(0.1299)
Gender	−0.1065*	−0.3147*	0.5234*	0.2742*	0.2174*	0.1218*
	(0.3317)	(0.2921)	(0.2309)	(0.1765)	(0.1427)	(0.1208)
Urban origin	−0.0609*	−0.6053*	0.4878*	0.2025*	0.3138*	0.3942
	(0.4185)	(0.5260)	(0.3217)	(0.2322)	(0.1852)	(0.1486)
Intercept	−5.0321*	0.7980*	−1.4965	−0.7813*	−1.0450	−0.7718
	(3.6919)	(0.6502)	(0.4958)	(0.3031)	(0.2342)	(0.1853)
−2 Log likelihood	220.180	291.858	486.949	832.975	1279.808	1731.826
Goodness of fit	181.618	234.056	396.110	680.548	1098.749	1430.471
χ^2	27.421	13.664	42.782	78.434	102.229	127.790
N	188	234	404	686	1051	1430

Table 8. *(Continued)*

Independent Variables	Cohort 7 (1939–1943)	Cohort 8 (1944–1948)	Cohort 9 (1949–1953)	Cohort 10 (1954–1958)	Cohort 11 (1959–1963)	Brazil
Part 2						
Mother's education	0.1227	0.1297	0.1678	0.1420	0.1261	0.1293
	(0.0228)	(0.0192)	(0.0170)	(0.0134)	(0.0124)	(0.0063)
Father's education	0.0785	0.0825	0.0787	0.0774	0.1121	0.0761
	(0.0191)	(0.0163)	(0.0144)	(0.0123)	(0.0121)	(0.0054)
Father's occupation	0.0129*	0.0127*	0.0211	0.0239	0.0127*	0.0197
	(0.0074)	(0.0067)	(0.0063)	(0.0058)	(0.0052)	(0.0025)
Father's occupation (dummy)	0.3146	0.05682	0.4242	0.4419	0.4373	0.4544
	(0.1068)	(0.0901)	(0.0730)	(0.0646)	(0.0647)	(0.0303)
Race	0.5199	0.3356	0.2135	0.3236	0.3351	0.3006
	(0.1072)	(0.0855)	(0.0716)	(0.0613)	(0.0612)	(0.0299)
Gender	0.1217*	−0.1407*	−0.1956	−0.1339*	−0.2543	−0.0894*
	(0.1029)	(0.0865)	(0.0707)	(0.0627)	(0.0635)	(0.0294)
Urban origin	0.2486*	0.5488	0.3637	0.4272	0.4689	0.3983
	(0.1281)	(0.1046)	(0.0889)	(0.0783)	(0.0803)	(0.0372)
Intercept	−0.9197	−0.9830	−0.7711	−1.0598	−1.3144	−1.0106
	(0.1602)	(0.1304)	(0.1109)	(0.0954)	(0.0968)	(0.0458)
−2 Log likelihood	2432.2	3407.8	595.1	6662.4	6704.1	29461.9
Goodness of fit	2118.839	3053.192	4570.220	5799.199	5740.164	25475.877
χ^2	252.185	404.584	631.511	827.196	975.880	3314.494
N	2119	3080	4653	5839	5692	25376

Note: Standard error in parenthesis. Education: successfully completed years of education. Mother's education: successfully completed years of education. Father's education: successfully completed years of education. Father's Occupation: scale of occupational status. Lowest = 1, highest = 100. Father's occupation: dummy: answered = 1; did not answered = 0. Gender: Women = 0; Men = 1. Race: Whites and Asians = 1 Blacksand Mulattos = 0. Urban origin: Rural = 0; Urban = 1.

Source: PNAD-1988.

$^*p > 0.01$.

Table 9. Trends in the Determinants of Educational Stratification Effects on School Transition 5 – from Eleven to Any Post Secondary Education.

Independent Variables	Cohort 1 (1882–1913)	Cohort 2 (1914–1918)	Cohort 3 (1919–1923)	Cohort 4 (1924–1928)	Cohort 5 (1929–1933)	Cohort 6 (1934–1938)
Part 1						
Mother's education	-0.0302^{*}	0.0699^{*}	0.0284^{*}	0.0505^{*}	0.0253^{*}	0.0696
	(0.0773)	(0.0614)	(0.0484)	(0.343)	(0.0297)	(0.0253)
Father's education	0.1797	0.0393^{*}	0.0453^{*}	0.0633^{*}	0.0153^{*}	0.0517^{*}
	(0.0648)	(0.0450)	(0.0334)	(0.0263)	(0.0233)	(0.0204)
Father's occupation	0.0551^{*}	0.0003^{*}	0.0107^{*}	-0.0034^{*}	0.0069^{*}	0.0021^{*}
	(0.0717)	(0.0158)	(0.0163)	(0.0094)	(0.0084)	(0.0075)
Father's occupation (dummy)	0.8577^{*}	0.2762^{*}	0.8433^{*}	0.4190^{*}	0.6202	0.7385
	(0.5975)	(0.5517)	(0.3735)	(0.2604)	(0.1855)	(0.1600)
Race	0.8778^{*}	-0.3153^{*}	0.2044^{*}	0.6354^{*}	0.4117^{*}	0.4650
	(0.6507)	(0.6010)	(0.4664)	(0.2811)	(0.2104)	(0.1762)
Gender	2.7977	2.2667^{*}	1.5137	1.0242	0.6025	0.2168^{*}
	(0.5982)	(0.4304)	(0.3214)	(0.2343)	(0.1729)	(0.1480)
Urban origin	0.0044^{*}	0.0025^{*}	0.4492^{*}	0.4057	0.3781^{*}	0.8187
	(0.5898)	(0.6055)	(0.5018)	(0.3184)	(0.2550)	(0.2291)
Intercept	−4.4534	-1.9273^{*}	−2.6303	−2.4519	−1.7821	−2.4585
	(1.1900)	(0.8034)	(0.6497)	(0.4227)	(0.3198)	(0.2803)
−2 Log likelihood	115.767	162.081	297.917	524.591	860.580	1147.130
Goodness of fit	110.815	153.056	253.835	422.703	665.627	921.328
χ^2	46.967	44.153	51.792	57.958	52.370	113.764
N	118	150	257	425	665	923

Table 9. *(Continued)*

Independent Variables	Cohort 7 (1939–1943)	Cohort 8 (1944–1948)	Cohort 9 (1949–1953)	Cohort 10 (1954–1958)	Cohort 11 (1959–1963)	Brazil
Part 2						
Mother's education	0.0596	0.0728	0.0936	0.0737	0.0980	0.0741
	(0.0200)	(0.0162)	(0.0131)	(0.0115)	(0.0120)	(0.0055)
Father's education	0.0434*	0.0362	0.0438	0.0760	0.0595	0.0490
	(0.0169)	(0.0141)	(0.0118)	(0.0110)	(0.0116)	(0.0049)
Father's occupation	0.0176	0.0212	0.0172	0.0185	0.0161	0.0167
	(0.0060)	(0.0053)	(0.0044)	(0.0041)	(0.0039)	(0.0018)
Father's occupation (dummy)	0.3403	0.3596	0.1656*	0.2615	0.2537	0.3156
	(0.1256)	(0.1032)	(0.0844)	(0.0798)	(0.0894)	(0.0372)
Race	0.4071	0.5055	0.7396	0.7407	0.8097	0.6569
	(0.1370)	(0.1059)	(0.1854)	(0.0800)	(−0.1449)	(0.0392)
Gender	0.4838	0.2721	0.1854*	0.0443*	−0.1449*	0.2414
	(0.1161)	(0.0943)	(0.0775)	(0.0733)	(0.0827)	(0.0344)
Urban origin	0.3792*	0.1137*	0.3271	0.4040	0.2340*	0.3005
	(0.1678)	(0.1375)	(0.1185)	(0.1162)	(0.1386)	(0.0532)
Intercept	−1.8194	−1.5405	−1.8798	−2.2586	−2.5855	−2.0817
	(0.2160)	(0.1683)	(0.1458)	(0.1394)	(0.1657)	(0.0653)
−2 Log likelihood	1817.631	2719.323	4042.916	4665.229	3819.702	20679.693
Goodness of fit	1423.030	2140.495	3258.290	3861.441	3400.367	16628.693
χ^2	153.578	226.012	436.814	622.778	572.328	2085.961
N	1422	2126	3232	3851	3392	16561

Note: Standard error in parenthesis. Education: successfully completed years of education. Mother's education: successfully completed years of education. Father's education: successfully completed years of education. Father's Occupation: scale of occupational status. Lowest = 1, highest = 100. Father's occupation: dummy: answered = 1; did not answered = 0. Gender: Women = 0; Men = 1. Race: Whites and Asians = 1 Blacksand Mulattos = 0. Urban origin: Rural = 0; Urban = 1.

Source: PNAD-1988.

$^{*}p > 0.01$.

7. CONCLUSIONS

The goal of this study was to analyze the process of educational stratification in Brazil in an attempt to answer two research questions: (1) what effect does economic development have on educational stratification? And (2) what is the role of race in this process? A strength of our study is that for the first time a cohort analysis is used to assess patterns that frame educational stratification trends for the adult population in Brazil.

The most important finding from this study is that industrialization and the socioeconomic transformations it brings about do not have an equalizing effect on educational stratification patterns. In fact, many ascriptive variables even strengthened their effect on educational attainment, unlike what had been predicted by Modernization Theory. This finding has important theoretical and policy implications given that Brazil had been seen within a context of a racial democracy, where all races have been assumed to live together with no conflict or social inequality directly associated to race relations, but rather to class relations. Most policies related to social inequality in Brazil rely on the idea that increasing rates of economic development alone will solve social inequality problems. Findings from this study explain the controversial relationship between the distribution of socioeconomic resources and economic development in Brazilian society.

More specifically, the overall conclusion from my analysis is that socioeconomic transformations brought about by the process of industrialization *lessen the direct effect of some of the determinants of educational stratification*, such as parental education, father's occupational status and gender. However, they *greatly increase the direct effect of some others determinants*, specifically race and rural-urban origin. At first one might conclude that barriers to educational attainment are fading in Brazil. Thus race and rural origin would be determinants that remained from old socioeconomic relations, and sooner or later they might decrease as the other determinants have done. Following this argument, the logical conclusion would be to conclude that the Brazilian educational expansion established paths toward meritocracy. However, focusing on the determinants of educational inequality as a separate process from educational expansion, it was possible to see a more precise picture of this process. There was in fact expansion of the educational system, but there is no increase in opportunity in access to education. In fact, in Brazil, as in other societies (Mare, 1980, 1981; Shavit & Blossfeld, 1993), access to lower educational levels is strongly associated with socioeconomic origins, but higher educational levels, on the contrary, are not associated with socioeconomic origins, but presumably with the ability and motivation one has. (Note that the presumed role of ability and motivation at higher transition points is an untested hypothesis.) This pattern shows a constant trend across cohorts. I conclude that socioeconomic transformations brought about by industrialization *have not decreased the effect of the determinants* of educational stratification in Brazil. In Brazil, these barriers show a pattern of stability over 81 years. This finding is of special importance to the understanding of educational inequality in Brazil, given the tendency for expansion in the Brazilian educational system not to reach universality, even at the lowest level. This means that socioeconomic origin is a strong determinant of educational inequality.

Finally, consider the effects of race on educational attainment. The socioeconomic status origin determinants of educational attainment show a single pattern of *strong effect on the lowest educational levels and a weaker effect on highest educational levels*. This pattern implies that, as individuals reach higher educational levels, their selectivity declines as they became more homogeneous in terms of socioeconomic background and presumably depend more on ability and motivation to go further. However, in the case of Brazil, race is the only determinant of educational stratification that does not show this pattern. Actually, race shows a "U" shape. The declining pattern in the effect of race is present from the first to the third transition points, but it then decreases, increasing at higher levels – dramatically at the fifth. This means that Non-Whites, face high barriers at the lowest levels of schooling, and then increasingly high barriers to access upper educational levels.

To conclude, socioeconomic transformations brought about by the process of industrialization *have lessened neither the effects of socioeconomic origins nor of race. Indeed there is compelling evidence that the negative effects of being Black or Mulatto have increased.*

NOTES

1. Our sample consists of heads of households and their spouses who were 25 years or older in 1988.
2. For a good review about educational inequality and its main research theoretical lines see Hurn (1993) and Karabel and Halsey (1977).
3. Featherman and Hauser (1978) show evidences that this pattern towards equality in educational attainments does not hold at post secondary levels in American society.
4. The other three social background characteristics that did not decline for the White group were: parental education, father's occupational status and intact family, a finding anticipated by Hauser and Featherman (1976).
5. It is important to recognize that Bowles and Gintis (1976) and Collins (1971, 1979) rely on different theoretical perspectives. Using a Neo Marxist view, Bowles and Gintis (1976) see the educational expansion as a consequence of the changing character of the social relations of production and the larger process of industrialization in modern societies. Educational expansion is a requirement of capitalist development. Employing Weber's ideas,

Collins (1971, 1979) sees educational expansion as a result of the competition among social groups for status and prestige, rather than the increasing needs of the society for more training. For this latter view, educational expansion can even be seen as an irrational process in which people are getting more educated to perform jobs that do not require such credentials.

6. See Note 5 above.

7. See Shavit and Blossefeld (1993, pp. 1–25) for a good summary of Mare's ideas.

8. Like the one used by Hauser and his associates described above.

9. During last century, higher education has got much more investment from the government than elementary and secondary education. Thus higher levels expanded at the expense of lower levels, despite the access to lower levels being far way from universal.

10. Oliveira et al. (1985) distinguishes two main chains in Brazilian thought about race. The first is directly connected to Gilberto Freyre's (1973) works and suggests the existence of a racial democracy as the basis of Brazilian society. The second is mostly influenced by the ideas of Florestan Fernandes (1964, 1972a, b) who gave theoretical support and empirical evidence about the deep inequality among racial groups based on the distribution of economic resources within class-antagonism-based society.

11. For the term Non-White I am referring basically to the Black and "Mulatto" population (mixed category that could be any mixture between Blacks, Whites and/or Brazilian Indians). About 45% of the Brazilian population perceives itself as Mulatto (pardo).

12. For a good review of the selected bibliography see Lovell (1991).

13. See Frank (1967) and Wallerstein (1974) as the main theoretical sources of this view. Also see F. Oliveira (1972) for an analysis of Brazilian development using this perspective.

14. Frank (1967, p. 9).

15. Wallerstein (1974) introduces the idea of a third part called *semi-periphery* when describing the logic of world the capitalist accumulation process, which he calls "the modern world-system."

16. Except for the first cohort that covers about 13 years due to the small number of survivors.

17. In the next section of this part we will present the variables and descriptive statistics from our data sample.

18. In Brazil, until the 1970s elementary school was divided into two levels. "Primário," ranging from 1st to 4th grades and "Ginásio" raging from 5th to 8th grades. We decided to maintain this division because it was in force during the lives of almost everyone in these cohorts. Here we call them lower and the upper elementary levels respectively.

19. There is another scale, the Socioeconomic Index of Brazilian occupations (SIBO) developed by Bills, Godfrey and Haller (1985). This is an occupational scale canonically weighted by each occupation's average income. We decide to use Valle Silva's scale, because it is based on the census of 1980, and then includes more up to date occupational categories than SIBO that is based on PNAD 1973. According to Kelley and Bills (1980), Valle Silva's scale and SIBO are highly correlated at 0.86.

20. See Duncan (1961).

21. For more detailed information about race classification history and procedures in Brazil see Araujo (1987).

22. The Portuguese terms used by IBGE in an exact English translation are: *White*, which means the color white, and should be applied for those who have a white skin color.

Black, which means the color black, and should be applied for who have a black skin color "*Mulatto*(*Pardo*)," which means an undefined color, and should be applied for those which have a mixed skin color. In a Portuguese-English and English-Portuguese dictionary, We found chestnut-brownish or dun-colored for the translation of the word "pardo." "*Amarelo*," which means the color yellow and is applied to Asian Brazilians.

23. We should remember that the variables, which represent the racial groups, are a set of dummy variables in which we set the value of 1 for those belonging to a specific racial group and zero for the rest, in each variable. Having a negative coefficient in the Black variable, for example, means that the fact of being black counts as a barrier to the access of educational attainment in relation to Whites.

REFERENCES

Andrews, G. (1992). Desigualdade racial no Brasil e nos estados unidos: Uma comparação estatística. *Estudos Afro-asiáticos*, *22*, 47–83.

Araujo, T. (1987). A classificação de 'cor' nas pesquisas do IBGE: Notas para uma discussão. *Cadernos de Pesquisa*, *63*, 14–16.

Azevedo, T. (1953). *Les elites de couleur dans une ville Brasiliene*. Paris: UNESCO.

Barcelos, L. C. (1992). Educação num quadro de desiqualdades raciais. *Estudos Afro-Asiáticos*, *23*, 37–69.

Bastide, R. (1952). Le problèm noir em Amerique Latine. *Bulletin Internacinal des Sciences Socilaes*, *4*(3), 459–467.

Becker, G. (1957). *The economics of racism*. Chicago: University of Chicago Press.

Beozzo, J. (1983). Situação do negro na sociedade Brasileira. *Revista de Cultura*, *77*(7), 485–497, setembro.

Bergmann, M. (1978). *Nasce um Povo*. Petrópolis: Vozes.

Bills, D., Godfrey, D., & Haller, A. (1985). A scale to measure the socioeconomic status of occupations in Brazil. *Rural Sociology*, *50*, 225–250.

Blau, P., & Duncan, O. (1967). *The American occupational structure*. New York: Wiley.

Boudon, R. (1974). *Education, opportunity and social inequality*. New York: Wiley.

Bourdieu, P. (1973). Cultural reproduction and social reproduction. In: R. Brown (Ed.), *Knowledge, Education and Cultural Change*. London: Tavistock.

Bourdieu, P., & Passeron, J.-C. (1977). *Reproduction in education, society and culture*. London: Sage.

Bowles, S., & Gintis, H. (1972). I.Q. in the U.S. class structure. *Social Policy*, *3*, 65–96.

Bowles, S., & Gintis, H. (1976). *Schooling in capitalist America*. New York: Basic Books.

Bowles, S., & Gintis, H. (2000). Does schooling raise earnings by making people smarter. In: *Meritocracy and Income Inequality*. Princeton: Princeton University Press.

Carvalho, J. M. (1980). *A construção da Ordem: A elite política imperial*. Rio de Janeiro: Editora Campos.

Castro, C. M. (1989). What is happening in Brazilian education? In: E. Bacha & H. Klein (Eds), *Social Change in Brazil, 1945–1985: The Incomplete Transition*. Albuquerque: University of New Mexico Press.

Collins, R. (1971). Functional and conflict theories of educational stratification. *American Sociological Review*, *36*, 1002–1019.

Collins, R. (1979). *The credential society*. New York: Academic Press.

Debreen, R. (1968). *On what is learned on school*. Massachusetts: Addison-Wesley.

Degler, C. (1971). *Neither black nor white: Slavery and race relations in Brazil and United States*. New York: Macmillian.

Dias, M. (1979). *Desigualdades sociais e oportunidades educacionais: A produção do Fracasso*. Rio de Janeiro. Thesis (M.Sc.) IUPERJ.

Duncan, D. (1961). A socioeconomic index of all occupations. In: A. Reiss (Ed.), *Occupations and Social Statusm*. New York: Free Press.

Edwards, R. (1979). *Contested terrain: The transformations of the workplace in the twentieth century*. New York: Basic Books.

Eriksson, R., & Goldthorp, J. (1992). *The constant flux: A study of class mobility in industrial societies*. Oxford: Clearendon Press.

Erikson, R., & Jonsson, J. (1996). The Swedish context: Educational reform and long-term changing in educational inequality. In: R. Erikson & J. Jonsson (Eds), *Can Education Be Equalized? The Swedish Case in Comparative Perspective*. Boulder: Westview Press.

Featherman, D., & Hauser, R. (1978). *Opportunity and change*. New York: Academic Press.

Featherman, D., Jones, F., & Hauser, R. (1975). Assumptions of mobility research in the United States. *Social Science Research*, *4*, 329–360.

Fernandes, F. (1964). A integração do negro à sociedade de classes. *Boletim da Faculdade de Filosofia, Ciências e Letras da Universidade de São Paulo: Sociologia I*, *12*.

Fernandes, F. (1969). The weight of the past. In: J. Franklin (Ed.), *Color and Race*. Boston: Beacon.

Fernandes, F. (1972a). *O negro no mundo dos brancos*. São Paulo: Difel.

Fernandes, F. (1972b). Representações coletivas sobre o negro: O negro na tradição oral. In: F. Fernandes (Ed.), *O Negro no Mundo dos Brancos*. São Paulo: Difel.

Figueira, V. (1988, janeiro/abril). Preconceito racial: Difusão e manutenção pela escola. *Intercâmbio*, *1*(1), 37–46.

Fotaine, P. (1985). *Race, class and power in Brazil*. Los Angeles: Center for Afro-American Studies.

Frank, A. (1967). *Capitalism and underdevelopment in Latin America: Historical studies of Chile and Brazil*. New York: Monthly Review Press.

Freyre, G. (1973). *Casa grande e senzala* (16ª ed.). Rio de janeiro: José Olimpio.

Fundação Carlos Chagas (1986). Raça negra e educação. *Cadernos Carlos Chagas*, *63*.

Gamoran, A. (1987). The stratification of high school learning opportunities. *Sociology of Education*, *60*, 135–155.

Gamoran, A., & Mare, R. (1989). Secondary school tracking and educational inequality: Compensation, reinforcement, or neutrality? *American Journal of Sociology*, *94*(5), 1146–1183.

Goldberg, A., & Cain, G. (1982). The causal analysis of cognitive outcomes in the Colleman, Hoffer and Kilgore report. *Sociology of Education*, *55*, 103–122.

Goldthorp, J. (1996). Problems of 'meritocracy'. In: R. Erikson & J. Jonsson (Eds), *Can Education Be Equalized? The Swedish Case in Comparative Perspective*. Boulder: Westview Press.

Gonsalves, L. (1985). *O silêncio: Um ritual pedagógico a favor da discriminação racial: Um estudo da discriminação racial como fator de seletividade na escola pública de 1 a 4 série*. Belo Horizonte, 1985. Thesis (M.Sc.), Faculdade de Educação. Universidade Federal de Minas Gerais.

Greenberg, S. (1980). *Race and state in capitalist development: Comparative perspectives*. New Haven: Yale University Press.

Haller, A., & Saraiva, H. (1991). Ascription and status transmission in Brazil. In: J. G. Scoville (Ed.), *Status Influences in Third World Labor Markets: Caste, Gender, and Custom*. Berlin and New York: Walter de Gruyter.

Haller, A., & Saraiva, H. (1992). The income effects of education in a developing country: Brazil – 1973 and 1982. In: R. Althauser & M. Wallace (Eds), *Research in Social Stratification and Mobility* (Vol. 11, pp. 295–336).

Halsey, A. H. (1977). Towards meritocracy? The case of Britain. In: J. Karabel & A. Halsey (Eds), *Power and Ideology in Education*. New York: Oxford University Press.

Hasenbalg, C. (1979). *Discriminação e desigualdades raciais no Brasil*. Rio de Janeiro: Edições Graal.

Hasenbalg, C. (1987). Desigualdades sociais e opportunidades occupacionais. *Cadernos de Pesquisa, 63*, 24–26.

Hasenbalg, C., & Valle Silva, N. (1988). *Estrutura social, mobilidade e raça*. Rio de Janeiro: IUPERJ.

Hasenbalg, C., & Valle Silva, N. (1990). Raça e oportunidades educacionais no Brasil. *Cadernos de Pesquisa, 73*, 5–12.

Hasenbalg, C., & Valle Silva, N. (1991). Raça e opportunidades educacionais no Brasil. In: P. Lovel (Ed.), *Desigualdade Racial no Brasil Contemporâneo*. Belo Horizonte: UFMG/CEDPLAR.

Hauser, R., & Featherman, D. (1976). Equality of schooling: Trends and perspectives. *Sociology of Education, 49*, 99–120.

Heyns, B. (1974). Social selection and stratification within schools. *American Journal of Sociology, 79*, 1434–1451.

Hoetink, H. (1973). *Slavery and race relations in the Americas: An inquiry into their nature and nexus*. New York: Harper & Row.

Holsinger, D. (1975). Education and the occupational attainment process in Brazil. *Comparative Education Review, 19*, 267–275.

Huntington, S. (1982/1983). Brazilan racial democracy: Reality or myth? *Humboldt Journal of Social Relations, 10*(1, Fall/Winter), 129–142.

Hurn, C. (1993). *The limits and possibilities of schooling: An introduction to the sociology of education* (3rd ed.). Boston: Allyn & Bacon.

Ianni, O. (1966). *Raças e classes sociais no Brasil*. Rio de Janeiro: Civilização Brasileira.

Instituto de Recursos Humanos João Pinheiro (1988). *Educação e discriminação dos negros*. Belo Horizonte: FAE/IRHJP.

Jencks, C. et al. (1972). *Inequality: A reassessment of the effect of family and schooling in America*. New York: Basic Books.

Karabel, J. (1972). Community colleges and social stratification. *Harvard Educational Review, 42*, 521–562.

Karabel, J., & Halsey, A. (1977). *Power and ideology in education*. New York: Oxford University Press.

Katsillis, J., & Rubinson, R. (1990). Cultural capital, student achievement, and educational reproduction: The case of Greece. *American Sociological Review, 55*, 270–279.

Kelley, J., & Bills, D. (1980). *The measurement of occupational status in Brazil: Comparison of various procedures*. UW-Madison, Brazil Projects – Department of Rural Sociology Technical Report 1980: 1. Mimeographed.

Kerckhoff, A. C. (1993). *Diverging pathways: Social structure and career deflections*. Cambridge: Cambridge University Press.

Kuo, H., & Hauser, R. (1995). Trends in family effects on the foundation of blacks and white brothers. *Sociology of Education, 68*(2), 136–160.

Langoni, C. (1973). *Distributição de renda e desenvolvimento econômico no Brasil*. Rio de Janeiro: Expressão e Cultura.

Lenski, G. (1966). *Power and privilege*. New York: McGraw-Hill.

Lipton, M. (1977). *Why poor people stay poor: Urban bias in world development*. Cambridge: Harvard University Press.

Lovell, P. (1991). *Desigualdade racial no Brazil contemporâneno*. Belo Horizonte, FMG/CEDPLAR.

Mare, R. (1980). Social background and the school continuation decision. *Journal of the American Statistical Association*, *75*, 295–305.

Mare, R. (1981). Changes and stability in educational stratification. *American Sociological Review*, *46*, 72–87.

Mare, R. (1993). Educational stratification on observed and unobserved components of family background. In: Y. Shavit & H. Blossfeld (Eds), *Persistent Inequality: Changing Educational Attainment in Thirteen Countries*. Boulder Co: Westview Press.

Mare, R., & Winship, C. (1988). Ethnic and racial patterns of educational attainment and school enrollment. In: G. Sandefur & M. Tienda (Eds), *Divided Opportunities: Minorities, Poverty and Social Policy*. New York: Plenum Press.

Mayer, J. et al. (1993). *School knowledge for the mass: World models and national primary circular categories in the twentieth century*. Washington, DC: Falmer.

Neves, J. A. (1997). Human capital, social classes, and the earnings determination process in Brazilian agriculture: 1973, 1982, and 1988. Thesis (Ph.D. Sociology), University of Wisconsin-Madison.

Oliveira, F. (1972). A economia Brasileira: Crítica à razão dualista. *Estudos CEBRAP*, *2*, 5–82.

Oliveira, L. E. et al. (1985). *O "lugar do negro" na força de trabalho*. Rio de Janeiro: Fundação Instituto Brasileiro de Geografia e Estatística – IBGE.

Olneck, M. (1979). The effects of education. In: C. Jencks et al. (Eds), *Who Gets Ahead? The Determinants of Economic Success in America*. New York: Basic Books.

Parsons, T. (1970). Equality and inequality in modern society, or social stratification revised. In: E. Laumann (Ed.), *Social Stratification: Research and Theory for the 1970s*. Indianapolis: Bobbs-Merril.

Pastore, J. (1982). *Inequality and social mobility in Brazil*. Madison: University of Wisconsin Press.

Pastore, J., & Haller, A. (1993). O que está acontecendo na mobilidade social no Brasil. In: J. Veloso & R. Albuquerque (Eds), *Pobreza e Mobilidade Social*. São Paulo: Nobel.

Pereira, J. B. (1967). *Cor, profissão e mobilidade: O negro e o rádio de São Paulo*. São Paulo: Pioneira.

Pinto, R. (1987). A educação do negro: Uma revisão da bilbiografia. *Cadernos de Pesquisa*, *62*, 3–34.

Ramos, ?. (1993). *Distribuição de rendimentos no Brasil*. Rio de Janeiro: IPEA.

Reich, M. (1994). The economics of racism. In: D. Grusky (Eds), *Social Stratification: Class Race and Gender in Sociological Perspective*. Boulder, CO: Westview Press.

Reeve, R. (1977). Race and socioeconomic mobility in a Brazilian town. *Luso Brazilian Review*, *14*, 236–253.

Ribeiro, M. (1984). *História da educação Brasileira: A organização escolar* (5th ed.). São Paulo: Editora Moraes.

Rosemberg, F. (1986). *Diagnóstico sobre a situação educacional de negros (pretos e pardos) no Estado de São Paulo* (v. 2) (mimeo). São Paulo, Fundação Carlos Chagas.

Rosemberg, F. (1987). Instrução, redimento, discriminação racial e de gênero. *Revista Brasileira de Estudos Pedagógicos*, *68*(159), 324–355.

Rosemberg, F. (1991). Raça e educação inicial. *Cadernos de Pesquisa*, *77*, 25–34.

Sewel, W., & Shah, V. (1967). Socioeconomic status. *Sociology of Education*, *40*, 1–23.

Shavit, Y. (1984). Tracking and ethnicity in Israeli secondary education. *American Sociological Education*, *49*, 210–220.

Shavit, Y., & Blossfeld, H. (1993). *Persistent inequality: A comparative study of educational attainment in thirteen countries*. Boulder, CO: Westview Press.

Shavit, Y., & Kraus, V. (1990). Educational stratification in Israel: A test of the industrialization and credentialism hypothesis. *Sociology of Education, 63*, 133–141.

Sorokin, P. (1927). *Social and cultural mobility*. Glenonce: Free Press.

Souza, A., & Valle Silva, N. (1994). Origem familiar, qualidade da educação e escolas públicas e particulares em São Paulo: Relações e effeitos nas transições escolares. *Pesquisa e Planejamento Econômico, 24*(1), 97–114.

Telles, E. (1992a). Who gets the formal sector jobs? Determinants of formal-informal sector participation in Brazilian metropolitan areas. *Work and Occupation, 19*, 108–127.

Telles, E. (1992b). Residential segregation by skin color in Brazil". *American Sociological Review, 57*, 186–197.

Telles, E. (1993). Urban labor market segmentation and income in Brazil. *Economic Development and Cultural Change, 41*, 231–249.

Thurow, L. (1975). *Generating inequality*. New York: Basic Books.

Treiman, D. (1968). *Occupational prestige and social structure*. Revised version of Ph.D. Dissertation at the University of Chicago: Markham.

Treiman, D. (1970). Industrialization and social stratification. In: E. Laumann (Ed.), *Social Stratification: Research and Theory for the 1970s*. Indianapolis: Bobbs Merrill.

Valle Silva, N. (1974). Posição social nas ocupações. Rio de Janeiro: Fundação IBGE. Mimeographed.

Valle Silva, N. (1980). O preco da cor: Diferenças raciais e distribuição da renda. *Pesquisa e Planejamento Econômico, 10*(1), 21–44.

Valle Silva, N. (1985). Atualização da escala socioeconômica de ocupações para 1980. Laboratório de Computação Científica. (Outubro). Mimeographed.

Valle Silva, N. (1992). Aspectos demográficos dos grupos raciais. *Estudos Afro-Asiáticos, 23*, 7–15.

Wallerstein, I. (1974). *The modern world-system I: Capitalist agriculture and the origins of the Europian world-economy in the sixteenth century*. New York: Academic Press.

Wilson, W. (1978). *The declining significance of race: Blacks and changing American institutions*. Chicago: The University of Chicago Press.

LABOR FORCE CLASSES AND THE EARNINGS DETERMINATION OF THE FARM POPULATION IN BRAZIL: 1973, 1982, AND 1988

Jorge Alexandre Neves

ABSTRACT

This paper assesses earnings differences among individuals of Brazilian farm social classes rooted in education and years of labor force experience. Very large subsamples of farm personnel in the National Household Sample Surveys of 1973, 1982, and 1988 provided the data. OLS regression showed large effects of each additional year of education and of labor force experience for the subsample as a whole. Five farm social classes were identified using a combination of class analysis and labor market segmentation criteria. They are Large Farmers (LF), Farm Managers (FM), Family Farmers (FF), Legally Protected (skilled) Farm Laborers (PFL), and Unprotected (unskilled) Farm Laborers (UFL). With or without taking into account the interactions of education and of years of job experience, the estimated earnings gains (EEG) (over those of the UFL class) to LF were found to be very large and increasing over the 15-year period. The EEGs of the FM class were also quite large and increasing over the period. Those of the FF fell. The EEG of the PFL class was moderately high; and it too rose. The class composition also changed. The

The Shape of Social Inequality: Stratification and Ethnicity in Comparative Perspective
Research in Social Stratification and Mobility, Volume 22, 423–475
© 2005 Published by Elsevier Ltd.
ISSN: 0276-5624/doi:10.1016/S0276-5624(04)22013-3

estimated size of the LF class fell from 6 to 3%. That of the FM class remained small at about 1%. The PFL gained from 3 to 7%, while that of the other two classes remained trendless – FF 32–30–32% and UFL 58–57–60%. It would appear the Brazilian farm social class structures moved toward a form of a Weberian ideal type of the capitalist system as large properties and their owners became more concentrated and the participation of more skilled laborers increased; and as the returns to education and experience increased, especially for large farmers, farm managers, and legally protected farm (skilled) laborers.

INTRODUCTION

This paper examines the comparative earnings levels of farm personnel in Brazil in three years – 1973, 1982, and 1988 – with special emphasis on the role of the labor force classes within the stratification system of the Brazilian countryside. Brazilian agriculture has always played an important role in the country's economy. Brazil is a big producer and exporter of agricultural staples. It is the largest producer of coffee, the second largest producer of soybeans, the second biggest producer of poultry, and the second largest producer of oranges, among others. Although manufacturing and services have become dominant in the Brazilian economy, agriculture is still a strategic sector with a promising growth. Analyses of the development potential of Brazilian agriculture – and socioeconomic factors associated with it – may provide important information for future planning. Similarly, they may contribute to a better understanding of the socioeconomic processes associated with agricultural production in other developing countries. Thus the present study of the process of earnings determination in Brazilian agriculture seeks to contribute to both social science knowledge and public policy.

The objective of my analyses is to propose a categorization of stratification of the labor force of farm population in Brazil and its role as a factor in the process of earnings determination in the agricultural sector of the Brazilian economy, as well as its interactions with human capital and certain other variables.

THE RESEARCH PROBLEM

The theoretical debate about earnings determination has been dominated by two general approaches: the individualistic and the structuralist. The former is represented by human capital theory and status attainment theory.[1,2] The structural approach includes two main theories, class analysis and labor market segmentation (or dual labor market, or dual economy) theory.[3,4]

The individualistic approach proposes that earnings vary with individual attributes, i.e. education, occupational status, training, age, experience, etc. In contrast, structuralists state that the relationship between individual attributes and earnings is mediated and modified by structural variables (social class and economic segmentation).

The Individualistic Approach

Mincer (1974) proposes that there are two main determinants of the distribution of earnings: (a) differences in accumulated human capital, i.e. length of schooling, quality of education, job training, experience, investment in health and nutrition, etc.; and (b) differences in rates of return to human capital. While the first factor consists unambiguously of individual attributes, the second is not necessarily an outcome of individual will or choice, or even inheritance. Thus, even in human capital theory we find room for a degree of structural or societal influence on earnings or income. Much of the research on earnings determination and distribution has been concentrated on differences in rates of return to human capital. Within the individualistic framework, Treiman (1970) proposes that the level of development (or industrialization) influences the respective effects of the occupational level of one's parents on one's own education. In the same way, Langoni (1973) proposed that Brazil's rapid economic development during the second half of the 1960s was the main factor responsible for the increase of income inequality between 1960 and 1970. His argument is that the process of development in Brazil comes along with more investment in capital-intensive technologies, and that capital and skilled labor are complementary. Thus, he concluded that Brazil was experiencing a rising rate of earnings returns to the investment in human capital, and that this was the main cause of the growing level of income inequality.

The human capital framework supplies us with other relevant predictions. For example, there is another possible variation in the rate of economic returns to education, which may play an important role in the agricultural sector in Brazil. Welch (1970) divided the effect of education on earnings into two: the *worker* or *direct* effect; and the *allocative* effect. The first suggests that schooling makes the individuals more productive, and so increases their earnings. The second is represented by the prediction that a portion of the earnings returns to schooling would be reflected in an efficient allocation of resources. This suggests that decision makers will have the highest earnings returns to the investment in human capital.

The Structuralist Approach

For structuralists, the central point is not only whether structural variables have significant and independent influences on earnings, but whether they constrain the way human capital influences earnings. As Wright and Perrone (1977, p. 37) state the problem:

> If class position is a critical mediating variable between social background and income, then it would be expected that class position would affect the ways in which background characteristics get transformed into income. That is, we hypothesize not only that class position has an independent impact on income from occupational position, but also it affects the extent to which background characteristics themselves can be 'cashed in' for income. In particular, the expectation is that class position will have a strong influence on the extent to which education influences income.

The same question can be stated for the labor market segmentation theory, by using labor market segment (or economic sector) as the intervening variable.

Hence, the goal of the present research is to analyze the process of earnings determination in the agricultural sector in Brazil, in particular the role played by the labor force stratification structure. The research is based on an analysis of the Brazilian farm labor force in 1973, 1982, and 1988. In terms of the analysis of changes over time, Brazilian agriculture has experienced relatively stable trends of growth, and a rising degree of land concentration.[5,6] These structural transformations, in connection with other factors, have forced important changes in the social stratification system of the agricultural labor force of Brazil.

ECONOMIC DEVELOPMENT, SOCIAL CHANGE, AND THE LABOR FORCE CLASS STRUCTURE IN BRAZILIAN AGRICULTURE

Brazil had one of the fastest growing economies in the 20th century. After World War II, its economy started to exhibit even higher rates of growth. In the period later called the *economic miracle* (1969–1973), the country achieved the highest rates of economic development in the world at that time. The 1980s, however, brought a deep and serious economic crisis to most countries of Latin America, and Brazil was not an exception. The 1980s' crisis – the so-called *debt crisis* – affected the performance of different sectors of Brazilian economy in varying degrees. While the industrial sector in general experienced a clear fall in its rates of growth (even showing negative growth rates during some years), the agricultural sector (although also negatively affected by the debt crisis) had a much better

Table 1. Average Yearly Growth Rates of Real Output, by Sector – Brazil: 1947–1992.

Year	Agriculture Crops	Livestock	Industry	Real GDP
1947–1950	4.4	6.2	11.0	6.8
1951–1954	3.0	9.4	7.2	6.8
1955–1958	5.6	1.5	9.9	6.5
1959–1962	5.7	4.9	10.0	7.7
1963–1966	3.0	4.7	3.1	3.1
1967–1970	5.1	2.3	10.1	8.2
1971–1976	5.5	6.3	14.0	12.2
1977–1981	4.8	5.1	5.5	5.4
1981–1986	3.9	−0.9	1.9	2.9
1987–1992	3.8	1.8	−2.2	0.4

Source: Baer (1995, p. 303).

performance, keeping sustained rates of growth during most years of the 1980s. This, however, was true only for crop production, the livestock complex having been more clearly constrained by the 1980s' economic crisis (see Table 1). The main reason for the relatively good performance of crop production during the 1980s was probably the expansion of soybean production in the Center-West region of the country, as well as the enormous increase of sugar-cane production due to the governmental program to substitute sugar-cane alcohol for gasoline as a response to the *petroleum crisis* of the 1970s.[7]

While these changes were going on the percentages of the labor force that was engaged in farming went from over 10% in 1973 to 7% in 1982 and to 6.6% in 1988, the years of the present analysis. This high post-World War II growth role of agriculture and its effect on the nation's overall economic growth rate occurred within an agricultural policy that has been called Conservative Modernization. This policy is the subject of the next paragraphs.

Conservative Modernization

Until the 1960s, the dominant interpretation by Latin American scholars held that Brazilian agriculture would never be able to modernize and become productive unless a deep land redistribution was imposed by the government. Many staff members of the United Nation's Economic Commission for Latin America (ECLA), which influenced the formation of the so-called *Latin American Structuralist School of Economics*, viewed the Brazilian economy as divided between two sectors. On the one side, they saw a *modern* and *efficient* segment, characterized by the growing industrial sector of the urban areas of the country. On the other side, they saw a *backward* and *traditional* economic segment in the rural areas. The agricultural sector, characterized by the *latifúndio-minifúndio* system, was seen as inefficient and non-responsive to demand increases. It was believed that there was no relationship between the two sectors, and that the *backward* agricultural economy represented an obstacle to a process of faster economic growth by the *modern* industrial sector. Based on this diagnosis, some ECLA's staff members proposed that a comprehensive program of land reform – breaking up large properties – was to be designed and applied in the Brazilian countryside.

Although many socioeconomic indicators and the recent resurgence of a strong political movement demanding a program of agrarian reform in Brazil indicate that it would probably be socially desirable to have a comprehensive project applied to change the land tenure structure (especially in the Northeastern region), reality has proved that the ECLA's prediction was wrong: Brazilian agriculture has been able to modernize and has not represented any obstacle to the capitalist development of the country's economy, even without the application of the proposed social reforms. The failure of ECLA's approach to Brazilian agriculture brought a reanalysis of the phenomenon by some Brazilian scholars. One of the most important revisions of the previous approach came with the *Neo-Marxist* analysis developed by Oliveira (1981). In his paper *Critic of the Dualistic Logic*, Oliveira criticized ECLA's approach by stating that in reality the *traditional* agricultural model has never represented an obstacle to the development of a *modern* and *integrated* capitalist economy in Brazil. In his work – later named the *articulation model* – Oliveira argues that instead of being isolated from the urban industrial sector, the agricultural production was completely *articulated* with the urban economy. They were just different sides of the same coin. The *plantation system on one side, and the household system of production on the other side characterized the latifúndio-minifúndio model of agricultural organization*. Even though this was supposed to be a *non-capitalist* or *semicapitalist* model of production, it was *functional* to the dominant model of development of *peripheral* capitalism. On the one hand, the *household system of production* was responsible for providing cheap food for the urban workforce – helping to keep the cost of labor low and to maintain a reserve of cheap labor that could be called by the industrial sector any time it was necessary. The *plantation system*, on the other hand, provided the necessary foreign currency to finance the process of Import Substitution Industrialization (ISI).

This *articulation* between the two sectors would be reproduced. Supporters of the *articulation model* now hold that the fast growth of the *urban informal economy* in Brazil is due to the incorporation of the rural *household production system* into the cities as a result of the increasing capacity of the Brazilian industrial sector to generate enough jobs.[8] At the same time, a considerable part of the agricultural sector has become *industrialized*, which has generated important *backward linkages*, by creating an enormous demand for industrialized inputs (machinery, fertilizers, pesticides), as well as important *forward linkages*, by providing the necessary raw materials for the establishment of new *agro-industrial complexes*.[9]

Indeed, over the long run, Brazilian agriculture has modernized, and has contributed – through *backward and forward linkages* – to the growth of the urban industrial sector. In the same way, as Graham, Gauthier and Barros (1987) show, Brazilian agriculture has been responsive to demand increases. Hence, Brazilian agriculture has participated in, and contributed to, the capitalist economic development of the country. This model of development initiated in the 1960s in the agricultural sector in Brazil – characterized by the combination of technological improvements with the absence of social reforms – has been named *conservative modernization*.[10] Whether this model of economic development and association between agriculture and industry will end up by guiding Brazil to levels of socioeconomic welfare similar to those of the western developed countries, or will only keep reproducing the association between a backward/low-wage sector and a more advanced/high-wage sector is not known. History has shown that it is possible for countries to rise within the system of economic stratification of the world nations. However, it also appears to have shown that social reforms, in particular land redistribution, are fundamental to better distribute the benefits of development. So far, Brazilian society has lacked these social reforms, especially the countryside.

Socioeconomic Trends in Farming

Opening New Lands

Despite the relatively stable trend of sustainable growth experienced by Brazilian agriculture (as we can see in Table 1), most of the increase of the agricultural output in Brazil has occurred on the *extension margin*. In other words, most of the growth has not been a consequence of increasing yields, but of the expansion of the cropped area, often while maintaining traditional technologies.[11] This has been possible due to the expansion of the agricultural frontier, first in the state of Paraná in the 1940s and 1950s, and later in the Center-West region in the 1960s and 1970s, and finally in the 1970s and 1980s in the Amazon region. These frontier expansion processes came as responses from peasants and farmers to explicit governmental policies (including tax incentives) and investments (such as the expansion of the highway system, as well as the construction of the new capital, Brasília, in the Center-West region) which created new incentives to frontier region in-migration.

Table 2. Annual Rate of Variation of the Intermediary Consumption Rate by Agriculture – Brazil: 1949–1980.

Year	Annual Rate of Variation of Intermediary Consumption Rate by Agriculture (%)
1949	1.0
1954	3.5
1959	1.6
1965	7.0
1968	5.3
1970	4.9
1975	4.5
1980	2.4

Source: Kageyama (1990, p. 121).

Many researchers have shown that it was only in the second half of the 1960s that the process of modernization of Brazilian agriculture started to accelerate.[12] Table 2 shows that in 1965 the Nation's rate of increase of the intermediary consumption rate[13] (intensity of the use of science-based products to raise productivity, an important indicator of technological modernization) experienced a significant *jump*, maintaining higher rates of increase into the 1970s.

Technological Modernization

Thus, since the mid-1960s modernization has become a second important trend in Brazilian agriculture. (The first is the expansion of the agricultural frontier.) However, this process of modernization has been restricted to the *export crops*. The *food crops* have remained in former modes of agricultural production. Some *food crops*, as it is the case of cassava, have even experienced declining levels of productivity. Using level of productivity as the main indicator, we see that, with the exception of soy, the yields of most *export crops* have increased.[14] Soy, which has become the most important crop in Brazil in recent years, and whose production system is extremely mechanized and also marked by other modern inputs (fertilizers, pesticides, etc.), has not had a significant rise in productivity, but this does not mean it has not become more modernized. The problem with soybeans is that its production has been extended to the poor and acidic soils of the *cerrado* ("savannah") region of Brazil's Center-West. The expansion of soy production to this region is in itself a consequence of the application of modern technologies to poor soils, resulting from important agricultural research developments from the Brazilian Corporation for Agricultural Research (EMBRAPA), as well as some Brazilian research universities. Hence, we can see that higher yields do not represent a perfect indicator of agricultural modernization, and that the expansion

Table 3. Rate of Intermediary Consumption by Agriculture, by State (Including the Federal District) – Brazil: 1970, 1980, and 1985.

State	1970	1980	1985
Rondônia	0.07	0.21	0.13
Acre	0.03	0.10	0.14
Amazonas	0.04	0.15	0.09
Roraima	0.08	0.36	0.18
Pará	0.10	0.15	0.12
Amapá	0.10	0.20	0.48
Maranhão	0.04	0.12	0.12
Piauí	0.07	0.20	0.16
Ceará	0.12	0.26	0.17
Rio Grande do Norte	0.16	0.31	0.20
Paraíba	0.11	0.26	0.17
Pernambuco	0.13	0.25	0.20
Alagoas	0.18	0.28	0.23
Sergipe	0.12	0.20	0.18
Bahia	0.06	0.16	0.15
Minas Gerais	0.15	0.29	0.24
Espírito Santo	0.11	0.26	0.23
Rio de Janeiro	0.19	0.27	0.22
São Paulo	0.27	0.38	0.29
Paraná	0.20	0.32	0.29
Santa Catarina	0.14	0.31	0.30
Rio Grande do Sul	0.24	0.35	0.32
Mato Grosso do Sul[a]		0.37	0.36
Mato Grosso	0.20	0.49	0.61
Goiás	0.14	0.32	0.34
Distrito Federal	0.38	0.72	0.57

Source: Brazilian Agricultural Censuses of 1970, 1980, and 1985.
[a] The state of Mato Grosso do Sul was still part of the state of Mato Grosso in 1973.

of the agricultural frontier in Brazil is itself due to interaction with technological modernization.[15]

The figures in Table 3, however, show that the process of modernization of Brazilian agriculture has not been a linear one. The process of modernization (measuring the rate of intermediary consumption by agriculture by the total value of production) was very clear for all states in the 1970s (comparing the 1980 figures with those from 1970). But this changed in the first half of the 1980s. The comparison of the figures from 1985 with those from 1980 (Table 3) shows that the intermediary consumption rate rose in only four states. In one state there was no change, and in twenty-one states it fell. Hence, the process of modernization of Brazilian agriculture – even though increasing in the long run – has not been linear.

In other words, the level of modernization grew up to the beginning of the 1980s, but fell in the middle of the decade. Our conclusion, therefore, is that the level of modernization is increasing in the long term, but falling in some years. The overall deep economic recession that Brazil was facing in the middle of the 1980s may have been responsible for this observed decline of the process of modernization of Brazilian agriculture. Firstly, economic difficulties experienced by the country may have created a negative environment for investments. In other words, given the *overall economic crisis*, most farmers reduced their expectations, and so did not invest in new equipment and training to deal with new technologies. Secondly, the *debt crisis* forced the government to make deep cuts in the agricultural credit programs. In short, the uncertainty about the economic future and the lack of credit were probably among the main reasons for the fall in expenditures with new technological inputs by Brazilian farmers.

Land Concentration

A third important trend experienced by Brazilian agriculture is the rising level of land concentration. Thiesenhusen and Melmed-Sanjak (1990) show that Brazil has experienced a continuous increase in the Gini coefficients of land concentration. It grew from 0.825 in 1940 to 0.838 in 1970, and to 0.853 in 1980. The data from Table 4 also support this conclusion.[16] From 1970 to 1980, levels of land concentration increased in twelve states, remained unchanged in three and decreased in ten. From 1980 to 1985, the trend was even more evident. Among the 26 states, 16 experienced an increase in the Gini coefficient, four remained unchanged and only 6 presented declining levels of land concentration. The enormous growth of production of export crops such as sugarcane and soybeans is probably responsible for this increase in the level of land concentration. Many farmers would have found it hard to compete in this new agribusiness environment, marked by *agro-industrial complexes* that work with low marginal costs of production. In the same way, increasing levels of concentration of credit accessibility accompanied the decreasing levels of credit availability through the 1980s. So without access to new credit lines, many farmers had to sell their lands or they lost them to the banks. Basically, only those farmers who were in some way *integrated* into these *agro-industrial complexes* were able to get credit to finance their production costs. In general, these are the farmers who hold larger plots of land.[17]

While in most *old* areas[18] of agricultural production in Brazil the level of land concentration has been rising, in the *new* areas[19] it has been falling. An important exception is the state of São Paulo, which – despite being a representative of the *old* agricultural areas – has experienced declining levels of land concentration. In the case of the *new* areas, our guess is that the rates of land concentration have

Table 4. Gini Coefficient of Land Concentration, by State (Including the Federal District) – Brazil: 1970, 1980, and 1985.

State	1970	1980	1985
Rondônia	0.69	0.65	0.63
Acre	0.63	0.69	0.62
Amazonas	0.73	0.87	0.81
Roraima	0.62	0.79	0.75
Pará	0.88	0.84	0.82
Amapá	0.87	0.85	0.88
Maranhão	0.93	0.93	0.93
Piauí	0.89	0.90	0.90
Ceará	0.79	0.78	0.83
Rio Grande do Norte	0.86	0.85	0.85
Paraíba	0.83	0.83	0.85
Pernambuco	0.84	0.83	0.84
Alagoas	0.84	0.85	0.89
Sergipe	0.86	0.85	0.87
Bahia	0.80	0.83	0.85
Minas Gerais	0.75	0.77	0.78
Espírito Santo	0.61	0.66	0.68
Rio de Janeiro	0.78	0.81	0.82
São Paulo	0.78	0.77	0.76
Paraná	0.71	0.74	0.75
Santa Catarina	0.66	0.68	0.69
Rio Grande do Sul	0.76	0.76	0.77
Mato Grosso do Sul[a]		0.87	0.86
Mato Grosso	0.93	0.92	0.90
Goiás	0.74	0.76	0.76
Distrito Federal	0.80	0.75	0.78

Source: Hoffmann and Graziano da Silva (1975, p. 251), Hoffmann (1990, p. 44), and Agricultural Census of 1985.

[a] The state of Mato Grosso do Sul was still part of the state of Mato Grosso in 1973.

decreased due to occupation by new colonists, many of whom are small-scale operators. Concerning São Paulo, we are not sure why the rate of land concentration has decreased. Sugar-cane expansion might be one of the main reasons for the rising levels of land concentration in the Northeastern states.

In order to observe some other important trends related to Brazilian agriculture, we are going to present some descriptive figures – see Table 5, for example – from the three Brazilian National Household Sample Surveys (PNADs) which will be assessed later in our statistical analysis. These data sets are from three different years: 1973, 1982, and 1988. (In a later section we will make detailed analyses of the characteristics – quality, advantages, limitations, etc. – of each of these data sets.)

Table 5. The Social Classes of Brazilian Farms – Number and Percentage of Agricultural Labor Force Participants 1973, 1982, and 1988.

Social Class	1973 Number	1973 %	1982 Number	1982 %	1988 Number	1988 %
Unprotected agricultural worker	29856	58.3	59230	60.4	14698	56.8
Protected agricultural worker	1698	3.3	4840	4.9	1800	7.0
Family farmer	16133	31.5	29639	30.2	8189	31.6
Farm manager	332	0.7	880	0.9	349	1.3
Large farmer	3163	6.2	3490	3.6	842	3.3
Total	51182	100.0	98079	100.0	25878	100.0

Notes: The "Family Farmer" Category Represents Those Farmers Who Are Self-Employed (Employing only Family Labor). The "Large Farmer" Category Represents Those Farmers Who Are Employers of Extra-Familiar Agricultural Workers.

Source: PNADs 1973, 1982, and 1988.

Table 5 shows another important aspect of the process of land concentration over the last decades. In our three household samples, the percentage of large farmers (employers) decreased from 6.2% in 1973 to 3.6% in 1982, and to 3.3% in 1988. Perhaps many medium-to-large size farmers who could not adapt to the new economic reality (the domination of the *agro-industrial complexes* and the lack of *easy* credit) went out of business.

Class Structure and Its Changes

Brazil's farm establishments can be seen as incorporating five basic social classes. These are:

(1) Owner-employers of larger enterprises; here called *Large Farmers.*
(2) *Managers* of such enterprises.
(3) Owner-operators or tenant-operators of family-sized farms; here called *Family Farmers.*
(4) Farm laborers holding job security under Brazilian law; here called *Protected Agricultural Workers.*
(5) Farm laborers lacking job security; here called *Unprotected Agricultural Workers.*

The activities and conditions of the two top classes are obvious and are not elaborated here. The other three classes require a bit of detail. We review them in connection with a discussion of what has been labeled "proletarianization."

Table 5 also presents some data on the fourth and last main socioeconomic trend experienced by Brazilian agriculture: changes of the structure of the social classes. Much research has been done in different areas of Brazil to describe this process ever since the end of World War II. Cabral (1987) describes the process of increasing proletarization of the labor force employed in the sugar-cane production of Northeastern Brazil. D'incao e Mello (1976) describes a similar process in the state of São Paulo. Most literature (like the ones just cited above) about "proletarianization" of the agricultural labor-force in Brazil has centered the analysis in the transition from a production system distinguished by the marked presence of *permanent resident laborers* (mostly sharecroppers) to a new one characterized by *seasonal laborers* or day laborers, the so-called *bóias frias* or *trabalhadores volantes*.[20] This transition is vital for the process of social formation and social change in Brazilian agriculture. However, there is another important point that has not been given the same attention. As Goodman, Sorj and Wilkinson (1985) suggest, one of the main manifestations of modernization in large agricultural properties is the emergence of semiskilled and skilled workers. Our data from Table 5 clearly support this statement. We can see that in our three household samples the proportions of protected agricultural workers and farm managers presented very clear patterns, having substantially increased over time. Protected agricultural workers represented 3.3% of Brazilian agricultural labor force in our 1973 sample, 4.9% in our 1982 sample, and 7.0% in our 1988 sample. The proportion of farm managers in the agricultural labor force grew from 0.7% in 1973 to 0.9% in 1982, and to 1.3% in 1988. Hence, we can see that the process of "proletarianization" of the agricultural labor force is marked not only by the *de-skilling* of the labor force (following the model proposed by Braverman, 1974, and applied by some researchers to explain the proletarization process in agriculture),[21] but also by the advent of a more skilled group of workers.

Unfortunately, the PNAD data do not permit us to differentiate between unskilled *permanent resident laborers* and *seasonal laborers*. In our analysis here both groups constitute the same social class category (unprotected agricultural workers). The main difference between these two groups of laborers is the way they are remunerated by their employers, as well as the time-space differences in their relationship with their employers. *Seasonal laborers* in general fit in the category of *wage labor* (i.e. all their on-farm remuneration comes from wages paid daily for each day worked). Unskilled *permanent resident laborers*, on the other hand, may receive wages, but at least a considerable part of their remuneration from the employer comes from the *rent* of the land cropped by the laborer for subsistence, or in sharecropping, or both. In the same way, while *permanent resident laborers* live on the employer's estate, and are employed by the same farmer the whole year long, *seasonal laborers* live in villages or towns outside the employer's estate, and most often are hired by many different farmers over the course of a year. A good example of the unskilled *permanent resident laborers* used to be the sugar-cane cutters of Northeastern Brazil up to the beginning of the 1960s. Those laborers used to live within their employer's property, had access to a small plot of land for subsistence purposes, and during the harvest time – four to six months – worked on the sugar-cane plantation, receiving as payments both the right to crop their plot, as well as wages in the form of currency or tickets to buy food, clothing, and other consumption goods in the stores owned by the employers.[22] Good examples of seasonal laborers are the *bóias frias* of São Paulo State. These workers, in general, live in an urban area (a city which is a bedroom community for the surrounding farms), and work for many different employers with a day-by-day informal employment relationship. They may harvest sugar-cane for part of the year, oranges during other part of the year, coffee a third period, and so forth, offering themselves to a *gato* (intermediary) in the morning; riding to a farm in back of the *gato's* truck; riding back to town at the end of the day; and being paid by the *gato*, rather than by the farm's management.[23]

Even though it would be useful to differentiate these two types of agricultural workers, this does not generate serious misunderstandings for an analysis of the process of earnings determination of the agricultural labor-force. Both categories of laborers stand at the bottom of the social stratification system of the agricultural labor-force, with by far the lowest levels of income.[24] Furthermore, even in 1973 most who were classified as Unprotected Farm Workers would have been seasonal laborers, rather than permanent resident laborers, a subclass which has been rapidly disappearing.[25]

It will be noted that the percentage of *unprotected agricultural workers* increased slightly from 1973 (58.3%) to 1982 (60.4%), and then fell a little in 1988 (56.8%). Indeed, in the long term the trend should be for the proportion of unprotected workers to decrease, given that mechanization should demand the replacement of unskilled laborers by others with the skills needed to be effective on modernized farms. However, the main explanation of the increase in the proportion of unprotected workers between 1970 and 1980 (in a period when Brazilian agriculture was modernizing rapidly) probably comes from the *boom* in the sugar-cane production caused by the *Programa Nacional do Álcool* (PROALCOOL). Pastore (1989), for example, points out that PROALCOOL directly and indirectly generated, in the first years of the 1980s, more than 1.5 million jobs. This was one of the main sources of job creation in Brazil at that time. Many of these were unskilled and legally unprotected sugar-cane cutters.

Finally, a last social class to be discussed is that of family farmers. As we said before, family farmers, by our definition, are those farmers (including owner-operators, tenant farmers, some sharecroppers, and contract farmers) who do not hire extra-familial labor. Small farmers are believed by many to be condemned to disappear. As Collins (1993, p. 54) presents it:

> Theories of agrarian transition have frequently been employed to address diversity in production relations; according to this approach different relations of production represent different 'stages' in a transition to fully capitalist agriculture. In such a view, agricultural enterprises are in the process of becoming fully integrated into the global economy, as they gradually shift from different kinds of share contracting and other 'pre-capitalist' production relations to the use of wage labor. Non-wage arrangements 'persist' where firms are not able to rationalize themselves for one reason or another, where profit margins are not high enough, or where the productive forces are insufficiently developed.

A considerable literature has shown, however, that under certain historical circumstances, sharecropping and/or subcontracting with small farmers may be an effective and rational option in some *agro-industrial complexes*.[26] In some regions of the world, sharecropping, for example, after being replaced by wage labor relations, is now being reintroduced in modern agribusiness enterprises. While Collins (1993) and Wells (1984) see this resurgence of sharecropping, as well as the strengthening of subcontracting, in some regions as a response from agribusiness to the political and economic constrains of the production systems of some crops, Graziano da Silva (1989) presents an analysis very much influenced by what we could call a Marxist *teleology*, i.e. he believes that these are forms of backward *production relations*, and so condemned to disappear with the development of the *forces of production*. Our data from Table 5 show that in Brazil the proportion of family farmers in the agricultural labor-force has been very stable, always just above 30%. Our data thus suggest that at least part of Brazilian family farmers have been able to adapt themselves to the new economic reality of *agro-industrial complexes*. Indeed, some important *agro-industrial complexes* in Brazil – like those that industrialize many different types of fruits and vegetables, as well as some livestock, particularly poultry and pigs – have been integrated into a central industrial plant system fed by the agricultural and livestock production of a large number of family farmers. This system has, in many cases, been very successful, allowing the agricultural sector to modernize, improving productivity and efficiency without the negative social impacts of other important *agro-industrial complexes*, in particular sugarcane and soybeans. Hence, we have good reasons to believe that family farmers are not necessarily condemned to extinction. They may still be efficient and competitive in some important modern subsectors of agricultural production.

HYPOTHESES

Based on the theoretical approaches described above, we specified our main research hypotheses as follows:

Hypothesis 1. From the individualistic approach, we will test the hypothesis that human capital factors have significant effects on earnings.

Hypothesis 2. From modernization theory (C. Langoni), we will test the hypothesis that the rate of earnings return to human capital is higher in areas with higher rates of capital inputs in the agricultural sector, and that the rate of return to human capital increases as a consequence of the process of modernization experienced by Brazilian agriculture.

Hypothesis 3. From the expected allocative effect of human capital on earnings, we will test the hypothesis that decision makers have the highest rates of earnings returns to human capital.

Hypothesis 4. From the structuralist approach, we will test the hypothesis that social class and labor market segmentation have significant independent effects on earnings.

Hypothesis 5. From the structuralist approach, we will test the hypothesis that the rate of earnings returns to human capital varies among social classes – and thus among the labor market segments they represent.

METHODOLOGY

Data

The data for this research come from three data sets of the Brazilian National Household Sample Survey (PNAD). The three PNADs used here – 1973, 1982, and 1988 – were designed especially for analyses of social stratification, mobility, education, and the labor market. PNAD–1982 has the largest sample (more than 1 million individuals for the country as a whole), followed by the 1973 sample (more than 300 thousand), and the 1988 sample (about 290 thousand). All the three data sets derive from stratified, multistage cluster samples of households. Given that our intention is to analyze the earnings determination of the labor force in the agricultural sector in Brazil, only those individuals who were economically active and were employed in the agricultural sector (including livestock production) appear in our analysis. Random subsamples of different sizes were taken from the three PNADs: 32,178 (1973), 69,561 (1982), and 19,089 (1988).

Given that PNADs do not include data about land concentration and capital inputs in agriculture, this information was obtained from the agricultural censuses of 1970, 1980, and 1985. The level of aggregation of the agricultural censuses data that we used is by state. Therefore, to each individual employed in agriculture we attribute the value of the level of agricultural land concentration of the state he or she works in, as well as the average rate of capital input (Intermediary Consumption Rate) in the same state. The data of the agricultural census of 1970 are used for the individuals of the PNAD data set of 1973, the 1980 agricultural census for the individuals in the PNAD of 1982, and the agricultural census of 1985 for the individuals of the PNAD of 1988.

These data sets are known to be of high quality. Much research has been done using the PNAD and census data sets produced by *Instituto Brasileiro de Geografia e Estatística* (the Brazilian Bureau of Census), and all social scientists who have used these data sets have pointed out their high quality. However, for the purposes of this research the PNAD data sets have a slight limitation: they do not include individuals from the more remote rural areas of the Brazilian Amazon Frontier, due to the enormous difficulties of access into the interior of this region. Nevertheless, the Brazilian Amazon, even though representing more than half of the country's territory, holds only about 11% of the Brazilian population, and its level of urbanization is quite high – over 70%. Besides that, our samples also include farmers of the Amazon inasmuch as some of the urban people sampled by the PNADs hold jobs in the agricultural sector. This is a bit of a problem of sample selectivity bias, but it is not great enough to affect the results.

Another problem is that some methodological analyses have shown that conclusions from stratified, multistage cluster samples cannot be interpreted as if they were simple random samples.[27] The best remedy for this has been said to be the one proposed by Goldberger and Cain (1982), and employed by many researchers. Goldberger and Cain (1982) argue that statistical estimations from stratified, multistage cluster samples in general understate the standard errors. Thus, they propose that we should use t ratio greater than 3.00 in statistical analyses based on data coming from this type of samples in order to achieve more reliable conclusions. This is the method employed in the present study.

Variables

Earnings Differences

This is our dependent variable. The original data of this variable are in units which are not comparable among the three years. But such comparisons are not essential to this analysis. What is essential is the increment in earnings that can be attributed statistically to a unit increase in each independent variable. Accordingly, the earnings data were calculated as the natural logarithm (ln) of monthly individual earnings divided by the number of hours worked per week. More specifically, in the PNADs of 1982 and 1988 individual earnings were presented in denominations of the official currency of Brazil in each year, but in the PNAD of 1973 this variable was presented in a group of 20 intervals. So, we followed the strategy of Haller and Saraiva (1992), i.e. we used the mean of each interval as an estimator for monthly earnings of each individual. This introduced a little error in the 1973 data, but it was inevitable given the circumstances. All regression coefficients were transformed by the formula $[(e^{b-1}) \times 100]$ to yield the percentage of increment of earnings (PIE) attributable to a unit increase in the independent variable. That is, our actual dependent variable is the PIE per unit increase in any given independent variable. This allows assessments of the differences in earnings among the various social classes and among other variables.

Education

The variable education is the number of years of education successfully completed.[28] For the PNAD of 1982, we have the actual number of years of education, varying from 0 to 16 years (for the agricultural labor force). For the PNAD of 1988, the variation goes from 0 to 17 years of schooling. For the PNAD of 1973, however, data on the *exact* number of years of education are not available. The original data on schooling for the 1973 sample were coded in the following way: I – no schooling at all; II – incomplete elementary school; III – complete elementary school; IV – incomplete middle school; V – complete middle school; VI – incomplete high-school; VII – complete high-school; VIII – incomplete college education; IX – complete college education. Here, we followed the strategy of Bills and Haller (1984), i.e. to use the following numbers to represent years of schooling: 0 (no schooling); 2 (incomplete elementary school); 4 (complete elementary school); 6 (incomplete middle school); 8 (complete middle school); 9.5 (incomplete high school); 11 (complete high school); 13 (incomplete college education); 15 (complete college education).[29] We are aware that this scheme incorporates a little unreliability of measurement, but less than the use of the original categorical coding would.

Experience

This variable, experience in the labor force, was constructed by subtracting the age of the individual in the year he or she started to work from his or her age when he or she was interviewed.[30] We had to make a choice between using experience or age, in order to avoid colinearity. Experience was chosen because it best represents the human capital approach in the statistical models.[31]

Experience Squared

This variable was introduced as a control variable, due to the fact that, on the average, experience yields positive earnings returns up to around age 50, diminishing after that.[32] In order to avoid colinearity, due to the high correlation between experience and experience squared, we transformed the variable experience, and than squared it to form the experience squared variable. In other words, our statistical models are *polynomial regression models*. Both variables, experience and experience squared, will be presented in this way in our statistical models. This strategy reduced the correlation between the two variables from more than 0.95 to around 0.55, but kept the same correlation between them and the other variables.[33]

Migration

The variable, migration, is dichotomous. It was constructed in the following way: (a) if the individual was living in the same state where he or she was born, the value for the variable is 0; (b) if the individual was living in a different state from the one where he or she was born, or if she or he was born outside Brazil, the value for the variable is 1. Unfortunately, the PNAD of 1982 does not provide data on birthplace. Because of this, in all relevant tables we present two types of models for 1973 and 1988: with and without migration as one of the regressors.

Gender

This is also a dichotomous variable. Its values are: 0 for women, and 1 for men. In our samples, the proportions of women were 24.2% in 1973, 20.0% in 1982, and 19.6% in 1988.[34]

Social Class and Labor Market Segmentation

In the analysis, we combined concepts from two seemingly different structural approaches (class analysis and labor market segmentation), to construct what we believe to be the best available representation of both the class structure and the segments of Brazilian farm labor force. As stated above, we divided the labor force into five social classes: *unprotected agricultural workers* (nonmanagerial-level employees who were not eligible for job security and other benefits); *protected agricultural workers* (nonmanagerial-level employees who were eligible for such benefits); *family farmers* (self-employed farmers); *farm managers* (managerial-level employees), and *large farmers* (owner-employers).[35] The distinction between the first and the second group is based on labor market segmentation theory, while the distinction between these two groups and the other classes, as well as between the other class groups themselves, is based on the "class analysis" approach. This class structure model appears in the multivariate regression analysis as four *dummy* variables. Unprotected agricultural workers form the reference group; each class is represented by a dummy variable.

These categories can be labeled social *classes of farmers*, or *classes*, for short. A few words about each may be useful. Let us take the two classes called *agricultural workers*. In Brazilian labor law, once an employee has been with a certain employer for more than three months, the employee is guaranteed an income at least equal to the legal minimum wage, plus fringe benefits – access to health services and freedom from arbitrary dismissal, among others. Those who have gained permanency are the workers who are employed at least three months with a given farm or with a company of farm enterprises, whether the individual has or does not have a written contract. Many do not. Those who are legally permanent are called *protected agricultural workers*, those who are not legally permanent are called *unprotected agricultural workers*. Many of the latter are day laborers. The class of *family farmers* consists of those who own or rent a farm and whose family members provide the labor. *Large* farmers are defined here as those farm owners who employ nonfamily labor. *Farm managers* are employees who supervise a farm's labor force, overseeing the routine work of the farm.

Land Concentration

This is a control variable and is represented by the *Gini Coefficients (times 100) of Land Concentration* for each state. We multiplied each figure by 100, in order to make it easier to interpret the regression coefficients.[36] Each individual is attributed to the score of his or her state. Gini coefficients presented in our study are calculated from data presented in the agricultural censuses of 1970, 1980, and 1985. They are all based on the *establishment* (or unit farm). (These figures understate the real degree of land concentration because any one owner may be the proprietor of more than one establishment – a common phenomenon in Brazil.)

Level of Modernization

This is also a control variable and indicates the level of technological modernization. Many different measures have been used as indicators of the level of modernization of agriculture in the Brazilian literature.[37] However, the most often used is also the most frequently supported as having the highest levels of validity. It is the *Intermediary Consumption Rate*.[38] This index is constructed in the following way: (a) the first step is to sum the total expenditures in intermediary industrialized goods (seeds, fertilizers, pesticides, livestock meals and medicines, machinery, manufactured wrappings, bags and boxes, etc.); and (b) the second step is to divide the result of the summation by the total value of production. As in the case of the Gini coefficients, we multiplied the *Intermediary Consumption Rate* figures by 100, in order to make easier to interpret them. So the resulting coefficients, like our Gini figures, vary from 0 to 100. These state-level scores – like those of the Gini's – are attributed to members of the sample in the corresponding state.

DESCRIPTIVE STATISTICS AND SPECIFICATION OF MODELS

Descriptive Statistics

In this section we provide descriptive statistics for all our variables, as well as the zero-order correlations among them. Tables 6–8 show the descriptive statistics.[39] Tables 9–11 present the zero-order correlation matrixes of all the variables to be included in our regression models. The correlation matrixes show that the transformation of the variable experience greatly reduces its correlation with experience squared to reasonable levels. They also show that among our independent variables education and some of the social class *dummy* variables

Table 6. Descriptive Statistics for Each Variable, Brazil – 1973.

Variable	Mean	Std. Deviation	Lowest	Highest
Education	1.57	1.66	0.00	15.00
Experience	20.90	16.64	0.00	60.00
Experience(b)	0.00	16.64	−20.90	39.10
Experience^2	713.70	912.96	0.00	3600.00
Experience(b)^2	276.77	316.92	0.01	1528.81
Migration	0.17	0.37	0.00	1.00
Gender	0.76	0.43	0.00	1.00
UnAgr. worker	0.58	0.54	0.00	1.00
Pr.Agr. worker	0.03	0.18	0.00	1.00
Family farmer	0.32	0.46	0.00	1.00
Manager	0.01	0.08	0.00	1.00
Large farmer	0.06	0.24	0.00	1.00
Gini × 100	78.13	6.83	61.00	93.00
Int. Con. × 100	15.15	7.22	3.00	38.00
Earnings	7.13	13.55	0.13	896.00
lnEarnings	1.58	0.77	−2.04	6.80

Notes: Education: Successfully Completed Years of Education; Experience: Years of Experience; Experience(b) represents the variable in the way it appears in the regression equations, given that they are polynomial regression equations [experience(b) = experience – experience/*n*]; Migration: Did not Migrate = 0; Migrated = 1; Gender: Women = 0; Men = 1; Protected Laborer: Unprotected Laborer = 0; Protected Laborer = 1; Family Farmer: Non-Family Farmer = 0; Family Farmer = 1; Manager: Non-Manager = 0; Manager = 1; Large Farmer: Non-Large Farmer = 0; Large Farmer = 1; Intermediary Consumption Rate: The Sum of all Industrial Inputs in the Agricultural Production (Index of Modernization) Divided by the Total Production Value (Multiplied by 100); Land Concentration: Gini Coefficient of Land Distribution (Multiplied by 100); Earnings: Individual Earnings; lnEarnings: The Natural Log of Individual Earnings.

Source: PNAD – 1973.

Table 7. Descriptive Statistics for Each Variable, Brazil – 1982.

Variable	Mean	Std. Deviation	Lowest	Highest
Education	1.76	2.28	0.00	16.00
Experience	23.27	16.33	0.00	60.00
Experience(b)	0.00	16.33	−23.27	36.73
Experience^2	808.11	962.49	0.00	3600.00
Experience(b)^2	266.77	300.52	0.07	1349.00
Gender	0.80	0.40	0.00	1.00
UnAgr. worker	0.60	0.66	0.00	1.00
Pr.Agr. worker	0.05	0.22	0.00	1.00
Family farmer	0.30	0.46	0.00	1.00
Manager	0.01	0.09	0.00	1.00
Large farmer	0.04	0.19	0.00	1.00
Gini × 100	80.01	6.47	65.00	93.00
Int. Con. × 100	27.57	8.84	10.00	72.00
Earnings	523.67	1459.38	1.67	66666.67
lnEarnings	5.78	0.83	0.51	11.11

Notes: Education: Successfully Completed Years of Education. Experience: Years of Experience. Experience(b) represents the variable in the way it appears in the regression equations, given that they are polynomial regression equations [experience(b) = experience – experience/*n*]. Migration: Did not Migrate = 0; Migrated = 1; Gender: Women = 0; Men = 1. Protected Laborer: Unprotected Laborer = 0; Protected Laborer = 1. Family Farmer: Non-Family Farmer = 0; Family Farmer = 1. Manager: Non-Manager = 0; Manager = 1. Large Farmer: Non-Large Farmer = 0; Large Farmer = 1. Intermediary Consumption Rate: The Sum of all Industrial Inputs in the Agricultural Production (Index of Modernization) Divided by the Total Production Value (Multiplied by 100). Land Concentration: Gini Coefficient of Land Distribution (Multiplied by 100). Earnings: Individual Earnings. lnEarnings: The Natural Log of Individual Earnings.

Source: PNAD – 1982.

have the highest correlation coefficients with our dependent variable (lnEarnings). Gender and intermediary consumption rate always show positive correlation coefficients with the natural log of earnings. Some variables, however, show some overtime variations in their correlation with the dependent variable. Migration has a positive correlation with the natural log of earnings in 1973, but a negative correlation in 1988. On the other hand, the Gini coefficient of land concentration exhibits a positive correlation with the natural log of earnings in 1973, but negative coefficients in 1982 and 1988.[40]

Models

Our models are based on OLS regression. As said before, the dependent variable is the natural logarithm of monthly earnings. The cross-sectional analysis will be

Table 8. Descriptive Statistics for Each Variable, Brazil – 1988.

Variable	Mean	Std. Deviation	Lowest	Highest
Education	1.92	2.50	0.00	17.00
Experience	23.72	15.75	0.00	60.00
Experience(b)	0.00	15.75	−23.72	36.28
Experience^2	810.71	935.39	0.00	3600.00
Experience(b)^2	248.17	284.88	0.078	1316.20
Migration	0.34	0.47	0.00	1.00
Gender	0.80	0.40	0.00	1.00
UnAgr. worker	0.57	0.65	0.00	1.00
Pr.Agr. worker	0.07	0.25	0.00	1.00
Family farmer	0.32	0.47	0.00	1.00
Manager	0.01	0.12	0.00	1.00
Large farmer	0.03	0.18	0.00	1.00
Gini × 100	81.56	6.57	62.00	93.00
Int. Con. × 100	24.08	10.39	9.00	61.00
Earnings	747.77	3240.83	4.15	200,000.00
lnEarnings	5.93	0.98	1.42	12.21

Notes: Education: Successfully Completed Years of Education; Experience: Years of Experience; Experience(b) represents the variable in the way it appears in the regression equations, given that they are polynomial regression equations [experience(b) = experience - experience/*n*]; Migration: Did not Migrate = 0; Migrated = 1; Gender: Women = 0; Men = 1; Protected Laborer: Unprotected Laborer = 0; Protected Laborer = 1; Family Farmer: Non-Family Farmer = 0; Family Farmer = 1; Manager: Non-Manager = 0; Manager = 1; Large Farmer: Non-Large Farmer = 0; Large Farmer = 1; Intermediary Consumption Rate: The Sum of all Industrial Inputs in the Agricultural Production (Index of Modernization) Divided by the Total Production Value (Multiplied by 100); Land Concentration: Gini Coefficient of Land Distribution (Multiplied by 100); Earnings: Individual Earnings; lnEarnings: The Natural Log of Individual Earnings.

Source: PNAD – 1988.

mainly based on the tests of the interaction terms between education and each of the contextual variables (class, land concentration, and level of agricultural modernization). For over time changes, *t-tests* are applied to test the equality between parameters of the earnings functions of the three different years. A *t* ratio greater than 3.00 will be used as the criterion.

Regression Models Estimated

Model 1

$$\text{lnEarnings} = \alpha + \beta_1 \text{Years of Education} + \varepsilon;$$

Model 2

$$\text{lnEarnings} = \alpha + \beta_1 \text{Years of Experience} + \beta_2 \text{Years of Experience Squared} + \varepsilon;$$

Table 9. Zero-Order Correlation Matrix of All Variables, Brazil – 1973.

Variables	Variables												
	Education	Exper.(b)	Exper.(b)2	Migration	Gender	PA. Worker	F. Farmer	Manager	L. Farmer	Gini × 100	ICon. × 100	Earnings	lnEarnings
Education	1.0000												
Exper. (b)	−0.2315	1.0000											
Exper. (b)2	−0.1132	0.5653	1.0000										
Migration	0.0379	0.1011	−0.0032	1.0000									
Gender	0.0662	0.0723	0.0394	0.0228	1.0000								
PA. Worker	0.0165	0.0027	−0.0399	0.0126	0.0662	1.0000							
F. Farmer	−0.1291	0.4116	0.1309	0.0515	0.1829	−0.1257	1.0000						
Manager	0.0623	0.0415	−0.0025	0.0275	0.0411	−0.0150	−0.0548	1.0000					
L. Farmer	0.1409	0.1831	0.0460	0.0368	0.1244	−0.0475	−0.1741	−0.0207	1.0000				
Gini × 100	−0.2552	0.0155	0.0264	−0.1111	−0.0517	0.0036	0.0849	0.0018	−0.0121	1.0000			
ICon .× 100	0.2997	−0.0114	−0.0338	0.2181	0.0579	0.1472	−0.1234	0.0336	0.0208	−0.3633	1.0000		
Earnings	0.2471	0.1239	0.0360	0.0983	0.0728	−0.0248	0.0100	0.0247	0.3325	−0.0737	0.1191	1.0000	
lnEarnings	0.2900	0.2555	0.0380	0.1518	0.1588	0.0046	0.2316	0.0491	0.3957	−0.1308	0.2024	0.6341	1.0000

Notes: Education: Successfully Completed Years of Education; Experience: Years of Experience; Experience(b) represents the variable in the way it appears in the regression equations, given that they are polynomial regression equations [experience(b) = experience - experience/*n*]; Migration: Did not Migrate = 0; Migrated = 1; Gender: Women = 0; Men = 1; Protected Laborer: Unprotected Laborer = 0; Protected Laborer = 1; Family Farmer: Non-Family Farmer = 0; Family Farmer = 1; Manager: Non-Manager = 0; Manager = 1; Large Farmer: Non-Large Farmer = 0; Large Farmer = 1; Intermediary Consumption Rate: The Sum of all Industrial Inputs in the Agricultural Production (Index of Modernization) Divided by the Total Production Value (Multiplied by 100); Land Concentration: Gini Coefficient of Land Distribution (Multiplied by 100); Earnings: Individual Earnings; lnEarnings: The Natural Log of Individual Earnings.

Source: PNAD – 1973.

Table 10. Zero-Order Correlation Matrix of All Variables, Brazil – 1982.

Variables	Variables												
	Education	Exper.(b)	Exper.(b)2	Migration	Gender	PA. Worker	F. Farmer	Manager	L. Farmer	Gini × 100	ICon. × 100	Earnings	lnEarnings
Education	1.0000												
Exper. (b)	−0.2745	1.0000											
Exper. (b)2	−0.1017	0.5653	1.0000										
Male	0.0075	0.0380	0.0317	1.0000									
PA. Worker	0.0305	0.0044	−0.0626	0.0525	1.0000								
F. Farmer	−0.0775	0.3625	0.1280	0.1297	−0.1499	1.0000							
Manager	0.0801	0.0302	−0.0080	0.0444	−0.0217	−0.0626	1.0000						
L. Farmer	0.1857	0.1415	0.0672	0.0823	−0.0438	−0.1264	−0.0183	1.0000					
Gini × 100	−0.2270	0.0390	−0.0098	−0.0019	−0.0208	0.1663	0.0102	−0.0369	1.0000				
ICon × 100	0.2228	−0.0434	−0.0040	0.0650	0.0912	−0.1704	0.0399	0.0515	−0.4029	1.0000			
Earnings	0.2419	0.0659	0.0205	0.0568	−0.0030	−0.0130	0.0623	0.3246	−0.0554	0.0728	1.0000		
lnEarnings	0.3077	0.1491	0.0022	0.1975	0.0933	0.0476	0.0979	0.4091	−0.1364	0.1873	0.5219	1.0000	

Notes: Education: Successfully Completed Years of Education; Experience: Years of Experience; Experience(b) represents the variable in the way it appears in the regression equations, given that they are polynomial regression equations [experience(b) = experience - experience/*n*]; Migration: Did not Migrate = 0; Migrated = 1; Gender: Women = 0; Men = 1; Protected Laborer: Unprotected Laborer = 0; Protected Laborer = 1; Family Farmer: Non-Family Farmer = 0; Family Farmer = 1; Manager: Non-Manager = 0; Manager = 1; Large Farmer: Non-Large Farmer = 0; Large Farmer = 1; Intermediary Consumption Rate: The Sum of all Industrial Inputs in the Agricultural Production (Index of Modernization) Divided by the Total Production Value (Multiplied by 100); Land Concentration: Gini Coefficient of Land Distribution (Multiplied by 100); Earnings: Individual Earnings; lnEarnings: The Natural Log of Individual Earnings.

Source: PNAD – 1982.

Table 11. Zero-Order Correlation Matrix of All Variables, Brazil – 1982.

Variables	Variables												
	Education	Exper.(b)	Exper.(b)2	Migration	Gender	PA. Worker	F. Farmer	Manager	L. Farmer	Gini × 100	ICon. × 100	Earnings	lnEarnings
Education	1.0000												
Exper. (b)	−0.1915	1.0000											
Exper. (b)2	−0.1111	0.55525	1.0000										
Migration	−0.0505	−0.3394	0.0808	1.0000									
Gender	0.0378	0.0377	0.0377	0.0138	1.0000								
PA. Worker	0.0505	−0.0228	−0.0669	−0.0283	0.0598	1.0000							
F. Farmer	−0.0068	0.3564	0.1124	−0.1977	0.1280	−0.1860	1.0000						
Manager	0.1436	0.0160	−0.0167	0.0036	0.0536	−0.0320	−0.0796	1.0000					
L. Farmer	0.2229	0.1347	0.0600	−0.0413	0.0660	−0.0501	−0.1248	−0.0214	1.0000				
Gini × 100	−0.2698	0.0086	0.0073	0.0162	−0.0109	−0.0817	0.1151	−0.0026	−0.0527	1.0000			
ICon .× 100	0.2263	−0.0215	−0.0241	0.2100	0.0819	0.0820	−0.0614	0.0598	0.0423	−0.2315	1.0000		
Earnings	0.2275	0.0483	0.0136	0.0114	0.0270	−0.0147	−0.0142	0.0298	0.3025	−0.0449	0.0546	1.0000	
lnEarnings	0.3848	0.1252	−0.0117	−0.0200	0.1636	0.1038	0.0618	0.1197	0.3772	−0.1908	0.2041	0.4218	1.0000

Notes: Education: Successfully Completed Years of Education; Experience: Years of Experience; Experience(b) represents the variable in the way it appears in the regression equations, given that they are polynomial regression equations [experience(b) = experience - experience/*n*]; Migration: Did not Migrate = 0; Migrated = 1; Gender: Women = 0; Men = 1; Protected Laborer: Unprotected Laborer = 0; Protected Laborer = 1; Family Farmer: Non-Family Farmer = 0; Family Farmer = 1; Manager: Non-Manager = 0; Manager = 1; Large Farmer: Non-Large Farmer = 0; Large Farmer = 1; Intermediary Consumption Rate: The Sum of all Industrial Inputs in the Agricultural Production (Index of Modernization) Divided by the Total Production Value (Multiplied by 100); Land Concentration: Gini Coefficient of Land Distribution (Multiplied by 100); Earnings: Individual Earnings; lnEarnings: The Natural Log of Individual Earnings.

Source: PNAD – 1988.

Model 3

$$\begin{aligned}\ln\text{Earnings} = \alpha &+ \beta_1\text{Years of Education} + \beta_2\text{Years of Experience}\\ &+ \beta_3\text{Years of Experience Squared} + \varepsilon;\end{aligned}$$

Model 4

$$\ln\text{Earnings} = \alpha + \beta_1\text{Migration} + \varepsilon;$$

Model 5

$$\begin{aligned}\ln\text{Earnings} = \alpha &+ \beta_1\text{Years of Education} + \beta_2\text{Years of Experience}\\ &+ \beta_3\text{Years of Experience Squared} + \beta_4\text{Migration} + \varepsilon;\end{aligned}$$

Model 6

$$\begin{aligned}\ln\text{Earnings} = \alpha &+ \beta_1\text{Years of Education} + \beta_2\text{Years of Experience}\\ &+ \beta_3\text{Years of Experience Squared} + \beta_4\text{Gender}\\ &+ \beta_5\text{Protected Agricultural Worker} + \beta_6\text{Family Farmer}\\ &+ \beta_7\text{Farm Manager} + \beta_8\text{Large Farmer}\\ &+ \beta_9\text{Gini Coefficient of Land Concentration}\\ &+ \beta_{10}\text{Intermediary Consumption Rate} + \varepsilon;\end{aligned}$$

Model 7

$$\begin{aligned}\ln\text{Earnings} = \alpha &+ \beta_1\text{Years of Education} + \beta_2\text{Years of Experience}\\ &+ \beta_3\text{Years of Experience Squared} + \beta_4\text{Migration}\\ &+ \beta_5\text{Gender} + \beta_6\text{Protected Agricultural Worker}\\ &+ \beta_7\text{Family Farmer} + \beta_8\text{Farm Manager} + \beta_9\text{Large Farmer}\\ &+ \beta_{10}\text{Gini Coefficient of Land Concentration}\\ &+ \beta_{11}\text{Intermediary Consumption Rate} + \varepsilon;\end{aligned}$$

Model 8

$$\begin{aligned}\ln\text{Earnings} = \alpha &+ \beta_1\text{Years of Education} + \beta_2\text{Years of Experience}\\ &+ \beta_3\text{Years of Experience Squared} + \beta_4\text{Gender}\\ &+ \beta_5\text{Protected Agricultural Worker} + \beta_6\text{Family Farmer}\\ &+ \beta_7\text{Farm Manager} + \beta_8\text{Large Farmer}\end{aligned}$$

$$+ \beta_9\text{Gini Coefficient of Land Concentration}$$
$$+ \beta_{10}\text{Intermediary Consumption Rate} + \beta_{11}(\text{Education} \times \text{Gender})$$
$$+ \beta_{12}(\text{Education} \times \text{Protected Agricultural Worker})$$
$$+ \beta_{13}(\text{Education} \times \text{Family Farmer})$$
$$+ \beta_{14}(\text{Education} \times \text{Farm Manager})$$
$$+ \beta_{15}(\text{Education} \times \text{Large Farmer})$$
$$+ \beta_{16}(\text{Education} \times \text{Gini Coefficient of Land Concentration})$$
$$+ \beta_{17}(\text{Education} \times \text{Intermediary Consumption Rate})$$
$$+ \beta_{18}(\text{Experience} \times \text{Gender})$$
$$+ \beta_{19}(\text{Experience} \times \text{Protected Agricultral Worker})$$
$$+ \beta_{20}(\text{Experience} \times \text{Family Farmer})$$
$$+ \beta_{21}(\text{Experience} \times \text{Farm Manager})$$
$$+ \beta_{22}(\text{Experience} \times \text{Large Farmer})$$
$$+ \beta_{23}(\text{Experience} \times \text{Gini Coefficient of Land Concentration})$$
$$+ \beta_{24}(\text{Experience} \times \text{Intermediary Consumption Rate}) + \varepsilon;$$

Model 9

$$\text{lnEarnings} = \alpha + \beta_1\text{Years of Education} + \beta_2\text{Years of Experience}$$
$$+ \beta_3\text{Years of Experience Squared} + \beta_4\text{Migration} + \beta_5\text{Gender}$$
$$+ \beta_6\text{Protected Agricultural Worker} + \beta_7\text{Family Farmer}$$
$$+ \beta_8\text{Farm Manager} + \beta_9\text{Large Farmer}$$
$$+ \beta_{10}\text{Gini Coefficient of Land Concentration}$$
$$+ \beta_{11}\text{Intermediary Consumption Rate} + \beta_{12}(\text{Education} \times \text{Gender})$$
$$+ \beta_{13}(\text{Education} \times \text{Protected Agricultural Worker})$$
$$+ \beta_{14}(\text{Education} \times \text{Family Farmer})$$
$$+ \beta_{15}(\text{Education} \times \text{Farm Manager})$$
$$+ \beta_{16}(\text{Education} \times \text{Large Farmer})$$
$$+ \beta_{17}(\text{Education} \times \text{Gini Coefficient of Land Concentration})$$

$$+ \beta_{18}(\text{Education} \times \text{Intermediary Consumption Rate})$$
$$+ \beta_{19}(\text{Experience} \times \text{Gender})$$
$$+ \beta_{20}(\text{Experience} \times \text{Protected Agricultral Worker})$$
$$+ \beta_{21}(\text{Experience} \times \text{Family Farmer})$$
$$+ \beta_{22}(\text{Experience} \times \text{Farm Manager})$$
$$+ \beta_{23}(\text{Experience} \times \text{Large Farmer})$$
$$+ \beta_{24}(\text{Experience} \times \text{Gini Coefficient of Land Concentration})$$
$$+ \beta_{25}(\text{Experience} \times \text{Intermediary Consumption Rate}) + \varepsilon;$$

RESULTS

Linear Combinations of Human Capital Variables

In the first part of the analysis of our empirical findings, we check the acceptability of three of the theoretical hypotheses presented above (Hypotheses 1, 2, and 4); the test of Hypothesis 3 is deferred and will follow that of Hypothesis 5, in the next section of the paper. Hypothesis 1 will be assessed by the observation of different regression models containing a varying number of human capital variables. These models will be basically the same for the three different years (1973, 1982, and 1988). Hypothesis 2, on the other hand, will be assessed by comparisons among the three different years. If this hypothesis is true, the rates of return to human capital variables will increase over time. We apply *t-tests* for the equality between parameters of the effect of human capital variables on earnings in the three different years. Hypothesis 4 tests whether the regression coefficients of the structural variables are significant in the three years.

Tables 12–14 show the regression coefficients and percentage increments to each additional year of education and of experience. These provide the tests of the hypotheses. *Every regression equation in these three tables is statistically significant*. They show that the earnings returns to education in all years, for each model, are positive, statistically significant, and high. The figures for education vary from a return of about a *9% increment* to income for each year of additional schooling (in Model 6 of Table 12) *to about 18.5%* (in Model 5 of Table 14). These findings are very similar to those from previous analyses based on the Brazilian labor force as a whole and for the urban labor force; and they are *markedly different* from the few previous analyses of earnings returns to schooling in Brazilian

Table 12. Unstandardized Regression Coefficients, Percentage Increments of Earnings Attributed to Unit Increments of Independent Variables (Human Capital Only), and Standardized Regression Coefficients, Brazil – 1973.

Independent Variables	Model 1	Model 2	Model 3	Model 4	Model 5	Model 6	Model 7
Education	0.13669*		0.16823*		0.16547*	0.08707*	0.08766*
	(14.647%)		(18.321%)		(17.995%)	(9.097%)	(9.162%)
	[0.28995]		[0.35654]		[0.35070]	[0.18454]	[0.18578]
Experience		0.02010*	0.02366*		0.02291*	0.00994*	0.00964*
		(2.030%)	(2.394%)		(2.317%)	(0.999%)	(0.969%)
		[0.42567]	[0.50101]		[0.48510]	[0.21047]	[0.20408]
Experience^2		−0.00054*	−0.00056*		−0.00054*	−0.00027*	−0.00026*
		(−0.054%)	(−0.056%)		(−0.054%)	(−0.027%)	(−0.026%)
		[−0.25083]	[−0.26102]		[−0.25106]	[−0.12750]	[−0.12310]
Migration				0.30393*	0.20292*		0.11330*
				(35.517%)	(22.497%)		(11.997%)
				[0.15184]	[0.10106]		[0.05642]
Gender (Male)						0.13138*	0.13147*
						(14.040%)	(14.050%)
						[0.31358]	[0.05606]
Protected agricultural worker						0.31358*	0.31727*
						(36.831%)	(37.337%)
						[0.08978]	[0.09084]
Family farmer						0.60398*	0.59904*
						(82.939%)	(82.037%)
						[0.39185]	[0.38865]
Farm manager						0.56066*	0.55135*
						(75.182%)	(73.559%)
						[0.07189]	[0.07069]
Large farmer						1.18710*	1.18200*
						(227.76%)	(226.09%)
						[0.44948]	[0.44755]

Gini coefficient of land concentration						−0.00310*	−0.00290*
						(−0.310%)	(−0.290%)
						[−0.02601]	[−0.02435]
Intermediary consumption rate						0.01881*	0.01735*
						(1.899%)	(1.750%)
						[0.17673]	[0.16307]
Intercept	1.38449*	1.63379*	1.38383*	1.52273*	1.34905*	0.90971*	0.89674*
R^2	0.0841	0.0992	0.2216	0.0231	0.2316	0.4245	0.4275
Adjusted R^2	0.0840	0.0992	0.2215	0.0230	0.2316	0.4243	0.4273
N	32178	31567	31567	32178	31567	31567	31567

Notes: Percentage Increment = $(e^b - 1) \times 100$; Numbers Between Brackets Are Standardized Regression Coefficients; Dependent Variable: lnEarnings; Education: Successfully Completed Years of Education; Experience: Number of Years Since Started to Work; Migration: Did not Migrated = 0; Migrated = 1; Educati × Experien: Interaction Term of Education and Experience.

Source: PNAD – 1973.

*$|t| > 3.00$.

Table 13. Unstandardized Regression Coefficients, Percentage Increments of Earnings Attributed to Unit Increments of Independent Variables (Human Capital Only), and Standardized Regression Coefficients, Brazil – 1982.

Independent Variables	Model 1	Model 2	Model 3	Model 6
Education	0.11245*		0.15328*	0.09379*
	(11.902%)		(16.565%)	(9.833%)
	[0.30774]		[0.41948]	[0.25667]
Experience		0.01383*	0.02485*	0.01650*
		(1.393%)	(2.516%)	(1.664%)
		[0.26846]	[0.48194]	[0.32006]
Experience^2		−0.00044*	−0.00068*	−0.00053*
		(−0.044%)	(−0.068%)	(−0.053%)
		[−0.17792]	[−0.27819]	[−0.21484]
Gender (Male)				0.32780*
				(38.791%)
				[0.12856]
Protected agricultural worker				0.41888*
				(52.026%)
				[0.12891]
Family farmer				0.22247*
				(24.916%)
				[0.13305]
Farm manager				0.62222*
				(86.306%)
				[0.08400]
Large farmer				1.31968*
				(274.222%)
				[0.34330]
Gini coefficient of land concentration				−0.00419*
				(−0.418%)
				[−0.03185]
Intermediary consumption rate				0.00923*
				(0.927%)
				[0.10014]
Intercept	5.58335*	5.84247*	5.59485*	5.28888*
R^2	0.0947	0.0396	0.2069	0.3456
Adjusted R^2	0.0947	0.0396	0.2069	0.3455
N	68607	69561	68607	68607

Notes: Percentage Increment = $(e^b - 1) \times 100$; Numbers Between Brackets Are Standardized Regression Coefficients; Dependent Variable: lnEarnings; Education: Successfully Completed Years of Education; Experience: Number of Years Since Started to Work; Educati × Experien: Interaction Term of Education and Experience.

Source: PNAD – 1982.

*$|t| > 3.00$.

Table 14. Unstandardized Regression Coefficients, Percentage Increments of Earnings Attributed to Unit Increments of Independent Variables (Human Capital Only), and Standardized Regression Coefficients, Brazil – 1988.

Independent Variables	Model 1	Model 2	Model 3	Model 4	Model 5	Model 6	Model 7
Education	0.14659*		0.16837*		0.16947*	0.10281*	0.10400*
	(15.788%)		(18.337%)		(18.468%)	(10.828%)	(10.960%)
	[0.38481]		[0.44199]		[0.44489]	[0.26990]	[0.27301]
Experience		0.01497*	0.02220*		0.02382*	0.01411*	0.01485*
		(1.508%)	(2.245%)		(2.411%)	(1.421%)	(1.496%)
		[0.23848]	[0.35371]		[0.37938]	[0.22475]	[0.23663]
Experience^2		−0.00053*	−0.00059*		−0.00065*	−0.00046*	−0.00048*
		(−0.053%)	(−0.059%)		(−0.065%)	(−0.046%)	(−0.048%)
		[−0.17034]	[−0.19013]		[−0.20908]	[−0.14711]	[−0.15592]
Migration				−0.04423*	0.13415*		0.06017*
				(−4.327%)	(14.356%)		(6.202%)
				[−0.02004]	[0.06078]		[0.02726]
Gender (Male)						0.28768*	0.28990*
						(33.333%)	(33.629%)
						[0.09597]	[0.09671]
Protected agricultural worker						0.49056*	0.49269*
						(63.323%)	(63.671%)
						[0.14681]	[0.14745]
Family farmer						0.28855*	0.28998*
						(33.449%)	(33.640%)
						[0.14613]	[0.14686]
Farm manager						0.70641*	0.70154*
						(102.67%)	(101.69%)
						[0.09665]	[0.09600]
Large farmer						1.52787*	1.52479*
						(360.84%)	(359.42%)
						[0.31864]	[0.31800]

Table 14. (*Continued*)

Independent Variables	Model 1	Model 2	Model 3	Model 4	Model 5	Model 6	Model 7
Gini coefficient of land concentration						−0.01120*	−0.01133*
						(−1.114%)	(−1.127%)
						[−0.07295]	[−0.7378]
Intermediary consumption rate						0.00852*	0.00773*
						(0.856%)	(0.776%)
						[0.09101]	[0.08263]
Intercept	5.64439*	6.00785*	5.67631*	5.93684*	5.64790*	6.00701*	6.01984*
R^2	0.1481	0.0318	0.2166	0.0004	0.2199	0.3378	0.3385
Adjusted R^2	0.1480	0.0317	0.2164	0.0003	0.2197	0.3375	0.3381
N	19087	19089	19087	19089	19087	19087	19087

Notes: Percentage Increment = $(e^b - 1) \times 100$; Numbers Between Brackets Are Standardized Regression Coefficients; Dependent Variable: lnEarnings; Education: Successfully Completed Years of Education; Experience: Number of Years Since Started to Work; Migration: Did not Migrated = 0; Migrated = 1.; Educati × Experien: Interaction Term of Education and Experience.

Source: PNAD – 1988.

*$|t| > 3.00$.

agriculture. These previous studies for Brazilian farm people *found no significant returns to additional years of education*. The enormous differences between our findings and those from the previous studies are probably due to three reasons. First, such studies cover only a few localities of the country. Second, they contain only family farmers. Their omission of large farmers, agricultural workers and managers is very serious, given that these last two groups of people presumably have their earnings completely determined by factors other than ownership of physical capital or land. Third, their samples are too small for detailed statistical analysis. Our own study is based on nationwide household probability samples taken at three points in time (1973, 1982 and 1988), and includes five different class/segment categories, as well as other key variables, each of which is statistically controlled.

However, the rates of earnings returns to schooling for the farmers' class groups – family farmers and large farmers – may have been slightly overestimated in the present study, given that we do not have any information about the amount of physical capital and land that each farmer utilizes. Hence, our guess is that the *real* rates of earnings return to education in Brazilian agriculture might be a trifle lower than estimated here, but considerably higher than had been previously estimated. In other words, though they might be a little lower than our findings indicate, *the impact of schooling on the earnings of the farm labor force is positive, statistically significant, and very large*.

Paralleling the above, our estimates of the earnings *returns to work experience* of the agricultural labor force in Brazil are also impressive. They *vary from about 1% per year* (Model 7 of Table 12) *to about 2.5%* (Model 3 of Table 13) *for each additional year of experience*. These figures are all statistically significant and possibly higher than those for the urban labor force. Considering that the number of years of experience ranges up to 60, 1% per year could amount to a great deal. This indicates that work experience plays a very important role in Brazilian agricultural economy.

Thus our first hypothesis is clearly supported by the empirical evidence. Human capital obviously has a substantial effect on the earnings levels of the farm labor force in Brazil, and this is net of all other variables measured.

About the temporal variations in the rates of return to human capital, we see that our findings provide little, if any, support for Hypothesis 2. The estimates of earnings returns to schooling tended to increase a little bit as time passed and the level of modernization/development rose. The regression coefficients of education in 1988 are, in general, significantly higher (at $|t| > 3.00$) than those from 1973 and 1982.[41] However, the figures from 1982 are not significantly higher than those of 1973. Concerning experience, we find that the earnings returns to each additional year of experience were highest in 1982. The support for the temporal aspect of the second hypothesis is too equivocal to permit its acceptance. It is rejected. (We shall check the *regional* aspect later.)

Thus even under conditions of extraordinarily tight statistical control, the percentage increments to earnings of each additional year of human capital accumulated (education and experience) was found to be high in each of the three years, and the increments noted for any one year are about the same as those noted for any other.

The evidence prescribed so far has assumed linear effects of the combination of human capital variables. But these are not the only ones predicted by current theory.

Linear Combinations of Farm Social Class

Hypothesis 4 is tested by data presented in Models 6 and 7 of Tables 12–14. The test of Hypothesis 3 is deferred to the next section. Each of the statistical models shows that all the *dummy* variables that together define the farm class structure show large and statistically significant direct effects on earnings. This is consistent support for the fourth hypothesis. Specifically, each of the four categories had higher average earnings than the reference group (Unprotected Agricultural Workers).

But their positions appear to have changed over the 1973–1988 period. While in 1973 the Family Farmer category was the second highest earnings group (just below the Large Farmer category), in 1982 and 1988 it was the second lowest (higher only than the reference group). In other words, in the 1980s the average earnings of those who belonged to the classes of Protected Agricultural Workers and Farm Managers gained higher average earnings (relative to the day laborers we have called "unprotected") than those who belonged to the Family Farmers class category. This is probably a result of the recent formation of a corps of skilled and semiskilled agricultural laborers, who now occupy positions as Protected Agricultural Workers and Farm Managers.[42] A possible explanation is that skilled workers have been becoming more and more necessary, and so the higher demand for them would explain the rising level of their earnings. On the other hand, the steep increase in the difference between the average earnings of the Family Farmer category and the reference group (from about 80% in 1973 to around 30% in the 1980s) may mean that the relative productivity of family-sized farms had fallen. Or it may mean that the prices for food for domestic consumption – produced by small operators – have been kept low while the prices for export products – produced by large operators – and their specialized labor – respond to the more profitable demands of the world market.[43]

Direct Effects: Other Variables

Migrants' earnings increment was around 11% in 1973 but half that in 1988. Other structural variables yield more or less predictable findings. Concerning gender, the average earnings of males were always higher than those of females. The difference, about 14% in 1973, rose to about 40% in 1982, and fell back a bit to around 30% in 1988.

The Gini Coefficient of Land Concentration's effect on earnings was negative in each year, and this negative effect increased from 1973 to 1982 (from approximately –0.3% to about –0.4%), and had more than doubled by 1988 (about −1.1%). Finally, concerning the level of technological modernization (Intermediary Consumption Rate), we see that its effect on earnings decreased from around 1.8% in 1973 to around 0.9% in 1982 and 0.8% in 1988. Thus while land concentration increasingly depressed worker's earnings over the period, on-farm technological modernization tended to raise them.

The impact of each of the latter two variables is quite large: recall that their ranges are from zero to 100, e.g. when an increase of one point on a 101-point scale yields an increase increment of a quarter of a percent, one is observing a large impact indeed.

Effects of Nonlinear Combinations of Human Capital with Farm Social Class and Other Structural Variables

In the previous section, we analyzed the direct effects (i.e. linear combinations) of human capital social class and other variables on earnings, using statistical models that included both sets of variables. In the present section, we will analyze the interactions between these factors. In other words, we will observe whether or not the social class and other variables modify the relationship between human capital and earnings.

The first hypothesis to be tested in this section is Hypothesis 5, which predicts that the rate of earnings returns to human capital varies between social classes/labor market segments, i.e. the social class/labor market positions modify the relationship between human capital variables and earnings. More specifically:

(a) Farm managers have a higher rate of earnings returns to human capital than agricultural workers.
(b) Farmers (both groups) have a lower rate of earnings returns to human capital than farm managers.

(c) Protected agricultural workers have a higher rate of earnings returns to human capital than unprotected agricultural workers.

Hypothesis 3 is to be tested next. From the conjecture on the allocative effect of education on earnings, we predict that decision makers (farmers and managers) will be found to have higher rates of earnings returns to human capital than will other classes.[44]

We will then reassess Hypothesis 2 (the *Modernization Theory Hypothesis*) using regional, rather than temporal, variations.

Hypothesis 5 generates three predictions. The first prediction states that *farm managers* should show higher rates of earnings returns to human capital than *agricultural workers*, in that human capital reinforces managers' authority over workers. The second prediction states that *farmers* (both *family* and *large*) should have lower rates of return to human capital than *managers*, given that their earnings, unlike managers, should be more a function of the amount of land and physical capital they own than their stock of human capital. Finally, the third prediction states that *protected agricultural workers* should show higher rates of return to human capital than *unprotected agricultural workers*, due to labor market segmentation, i.e. the former are employed in a more technologically advanced, skills-demanding, and unionized farming systems; the latter more often employed in low-technology, low-skills-demanding, nonunionized farming systems.

Hypothesis 3 predicts that *decision makers* should obtain higher rates of return to human capital than nondecision makers, given that they benefit twice from human capital: unlike the better educated and more experienced of the working class, they not only have this advantage but also a position that permits them to exercise their expertise. Thus, we should expect *family farmers*, *managers*, and *large farmers* to show the highest rates of return to human capital. This prediction is in agreement with the first expectation of Hypothesis 5, but not with the second.

We assess Hypotheses 5 then 3, by testing the statistical significance of the interaction terms between education and the social class variables, and between experience and the social class variables. Tables 15–17 present the necessary information.

Our findings are that:

(a) The first prediction from Hypothesis 5 is supported by the empirical data, wherein education is the human capital factor under consideration. *Managers* have net earnings returns to a year of additional schooling at least 5.6% higher than those both *unprotected* and *protected classes* of *agricultural workers*. However, when experience is the human capital factor under consideration, the first prediction from Hypothesis 5 does not find any empirical support. In none of the three years is the interaction term of experience and the *farm manager*

Table 15. Unstandardized Regression Coefficients, Percentage Increments of Earnings Attributed to Unit Increments of Independent Variables, and Standardized Regression Coefficients, Brazil – 1973.

Independent Variables	Model 8	Model 9
Education	0.09739*	0.08630
	(10.229%)	(9.013%)
	[0.20641]	[0.18290]
Experience	0.01826*	0.01761*
	(1.843%)	(1.777%)
	[0.38666]	[0.37293]
Experience^2	−0.00032*	−0.00031*
	(−0.032%)	(−0.310%)
	[−0.14863]	[−0.14508]
Migration		0.10574*
		(11.153%)
		[0.05266]
Gender (Male)	0.11883*	0.11957*
	(12.618%)	(12.701%)
	[0.05067]	[0.05098]
Protected agricultural worker	0.31617*	0.31796*
	(37.186%)	(37.432%)
	[0.09052]	[0.09104]
Family farmer	0.59520*	0.59092*
	(81.339%)	(80.565%)
	[0.38616]	[0.38338]
Farm manager	0.31246*	0.31233*
	(36.678%)	(36.661%)
	[0.04006]	[0.04005]
Large farmer	0.92355*	0.92518*
	(151.821%)	(152.232%)
	[0.34969]	[0.35031]
Gini coefficient of land concentration	−0.00217	−0.00223
	(−0.217%)	(−0.223%)
	[−0.01822]	[−0.01872]
Intermediary consumption rate	0.01779*	0.01654*
	(1.795%)	(1.668%)
	[0.16716]	[0.15537]
Education × Gender	0.01314	0.01253
	(1.323%)	(1.261%)
	[0.02780]	[0.02652]
Education × Protected agricultural worker	0.00451	0.00509
	(0.452%)	(0.510%)
	[0.00319]	[0.00361]
Education × Family farmer	0.00361	0.00315
	(0.362%)	(0.315%)
	[0.00549]	[0.00480]

Table 15. *(Continued)*

Independent Variables	Model 8	Model 9
Education × Farm manager	0.08755*	0.08454*
	(9.150%)	(8.821%)
	[0.04580]	[0.04423]
Education × Large farmer	0.06624*	0.06417*
	(6.848%)	(6.627%)
	[0.08939]	[0.08659]
Education × Gini coefficient of land concentration	−0.00073	−0.00055
	(−0.073%)	(−0.550%)
	[−0.11953]	[−0.09054]
Education × intermediary consumption rate	0.00080	0.00075
	(0.080%)	(0.075%)
	[0.03703]	[0.03463]
Experience × Gender	0.00262*	0.00254*
	(0.262%)	(0.254%)
	[0.05178]	[0.05010]
Experience × Protected agricultural worker	−0.00447*	−0.00453*
	(−0.446%)	(−0.452%)
	[−0.01894]	[−0.01918]
Experience × Family farmer	0.00210*	0.00216*
	(0.210%)	(0.216%)
	[0.03145]	[0.03235]
Experience × Farm manager	0.00546	0.00531
	(0.547%)	(0.532%)
	[0.01133]	[0.01103]
Experience × Large farmer	0.01208*	0.01191*
	(1.215%)	(1.198%)
	[0.08490]	[0.08373]
Experience × Gini coefficient of land concentration	−0.00015*	−0.00015*
	(−0.015%)	(−0.015%)
	[−0.25851]	[−0.24489]
Experience × Intermediary consumption rate	0.00002	0.00001
	(0.002%)	(0.001%)
	[0.00730]	[0.00286]
Intercept	0.89701*	0.90133*
R^2	0.4321	0.4347
Adjusted R^2	0.4317	0.4342
N	31567	31567

Notes: Percentage Increment = $(e^b - 1) \times 100$; Numbers Between Brackets Are Standardized Regression Coefficients; Dependent Variable: lnEarnings; Education: Successfully Completed Years of Education; Experience: Number of Years Since Started to Work; Educati × Experien: Interaction Term of Education and Experience.

Source: PNAD – 1973.

*$|t| > 3.00$.

Table 16. Unstandardized Regression Coefficients, Percentage Increments of Earnings Attributed to Unit Increments of Independent Variables, and Standardized Regression Coefficients, Brazil – 1982.

Independent Variables	Model 8
Education	0.09224*
	(9.663%)
	[0.25242]
Experience	0.02936*
	(2.980%)
	[0.56930]
Experience^2	−0.00058*
	(−0.058%)
	[−0.23907]
Gender (Male)	0.30442*
	(35.584%)
	[0.11939]
Protected agricultural worker	0.44189*
	(55.564%)
	[0.13599]
Family farmer	0.11213*
	(11.866%)
	[0.07304]
Farm manager	0.44407*
	(55.904%)
	[0.05995]
Large farmer	0.99858*
	(171.442%)
	[0.25977]
Gini coefficient of land concentration	−0.00186
	(−0.186%)
	[−0.01414]
Intermediary consumption rate	0.00815*
	(0.818%)
	[0.08843]
Education × Gender	0.00327**
	(0.328%)
	[0.00881]
Education × Protected agricultural worker	−0.00166
	(−0.166%)
	[−0.00161]
Education × Family farmer	0.03933*
	(4.011%)
	[0.07124]
Education × Farm manager	0.06463*
	(6.676%)
	[0.04821]

Table 16. (*Continued*)

Independent Variables	Model 8
Education × Large farmer	0.06133*
	(6.325%)
	[0.08967]
Education × Gini coefficient of land concentration	−0.00044
	(−0.044%)
	[−0.09476]
Education × Intermediary consumption rate	0.00031
	(0.031%)
	[0.02781]
Experience × Gender	0.00625*
	(0.627%)
	[0.11412]
Experience × Protected agricultural worker	−0.00550*
	(−0.548%)
	[−0.02363]
Experience × Family farmer	0.00760*
	(0.763%)
	[0.10084]
Experience × Farm manager	0.00156
	(0.156%)
	[0.00326]
Experience × Large farmer	0.01351*
	(1.360%)
	[0.00684]
Experience × Gini coefficient of land concentration	−0.00029*
	(−0.029%)
	[−0.45648]
Experience × Intermediary consumption rate	0.00009*
	(0.009%)
	[0.04762]
Intercept	5.20085*
R^2	0.3560
Adjusted R^2	0.3557
N	68607

Notes: Percentage Increment = $(e^b - 1) \times 100$; Numbers Between Brackets Are Standardized Regression Coefficients; Dependent Variable: lnEarnings; Education: Successfully Completed Years of Education; Experience: Number of Years Since Started to Work; Educati × Experien: Interaction Term of Education and Experience.

Source: PNAD – 1982.

*$|t| > 3.00$.

Table 17. Unstandardized Regression Coefficients, Percentage Increments of Earnings Attributed to Unit Increments of Independent Variables, and Standardized Regression Coefficients, Brazil – 1988.

Independent Variables	Model 8	Model 9
Education	0.04268	0.04702
	(4.360%)	(4.833%)
	[0.11204]	[0.12344]
Experience	0.02000*	0.02082*
	(2.020%)	(2.104%)
	[0.31866]	[0.33161]
Experience^2	–0.00052*	–0.00054*
	(–0.052%)	(–0.540%)
	[–0.16694]	[–0.17296]
Migration		0.04417*
		(4.516%)
		[0.02002]
Gender (Male)	0.29550*	0.29778*
	(34.380%)	(34.687%)
	[0.09858]	[0.09934]
Protected agricultural worker	0.55790*	0.55963*
	(74.700%)	(75.002%)
	[0.16696]	[0.16748]
Family farmer	0.21086*	0.21395*
	(23.474%)	(23.856%)
	[0.10679]	[0.10836]
Farm manager	0.52103*	0.52309*
	(68.376%)	(68.723%)
	[0.07129]	[0.07157]
Large farmer	1.27963*	1.28506*
	(259.531%)	(261.488%)
	[0.26687]	[0.26800]
Gini coefficient of land concentration	–0.01183*	–0.01187*
	(–1.176%)	(–1.180%)
	[–0.07705]	[–0.07733]
Intermediary consumption rate	0.00833*	0.00783*
	(0.836%)	(0.786%)
	[0.08904]	[0.08371]
Education × Gender	–0.01465	–0.01495
	(–1.454%)	(–1.484%)
	[–0.03797]	[–0.03875]
Education × Protected agricultural worker	–0.02389	–0.02445
	(–2.361%)	(–2.415%)
	[–0.02491]	[–0.02548]
Education × Family farmer	0.03073*	0.02992*
	(3.121%)	(3.037%)
	[0.05776]	[0.05622]

Table 17. (*Continued*)

Independent Variables	Model 8	Model 9
Education × Farm manager	0.05312*	0.05175*
	(5.456%)	(5.311%)
	[0.05071]	[0.04940]
Education × Large farmer	0.05203*	0.05040*
	(5.341%)	(5.169%)
	[0.07243]	[0.07017]
Education × Gini coefficient of land concentration	0.00071	0.00069
	(0.071%)	(0.069%)
	[0.14877]	[0.14340]
Education × intermediary consumption rate	−0.00015	−0.00017
	(−0.015%)	(−0.017%)
	[−0.01250]	[−0.01416]
Experience × Gender	0.00487*	0.00477*
	(0.488%)	(0.478%)
	[0.07299]	[0.07160]
Experience × Protected agricultural worker	−0.00764*	−0.00791*
	(−0.761%)	(−0.788%)
	[−0.03207]	[−0.03320]
Experience × Family farmer	0.00626*	0.00600*
	(0.628%)	(0.602%)
	[0.06800]	[0.06519]
Experience × Farm manager	0.00103	0.00049
	(0.103%)	(0.049%)
	[0.00205]	[0.00098]
Experience × Large farmer	0.00764*	0.00734*
	(0.767%)	(0.737%)
	[0.02984]	[0.02869]
Experience × Gini coefficient of land concentration	−0.00017	−0.00017
	(−0.017%)	(−0.017%)
	[−0.22019]	[−0.21776]
Experience × Intermediary consumption rate	0.00008**	0.00007**
	(0.008%)	(0.007%)
	[0.03265]	[0.02915]
Intercept	6.09578*	6.09769*
R^2	0.3445	0.3448
Adjusted R^2	0.3437	0.3440
N	19087	19087

Notes: Percentage Increment = $(e^b - 1) \times 100$; Numbers Between Brackets Are Standardized Regression Coefficients; Dependent Variable: lnEarnings; Education: Successfully Completed Years of Education; Experience: Number of Years Since Started to Work; Educati × Experien: Interaction Term of Education and Experience.

Source: PNAD – 1988.

*$|t| > 3.00$.

variable significant. (Logically, this should present a problem for class analysis theorists. But they may not yet have fixed it for they have only applied it to education. We ourselves decided to extend the hypothesis to experience. But it is perfectly consistent with class-analytic logic.) Therefore, we can conclude that one of the class analysis predictions is partially supported by our findings – strongly as concerning education, but not at all as concerning experience.

(b) The second prediction from Hypothesis 5 finds no empirical support from our figures. When we look at Tables 15–17, we see that, with the exception of 1973, the interaction terms of *education* and the *farm manager* variable, and *education* and the *large farmer* variable are not significantly different.[45] Concerning *experience*, we see that *family farmer* and *large farmer classes* both have higher rates of earnings returns to experience than *managers* in each of the three samples. These findings might indicate that the prediction from Hypothesis 3 is strongly supported by our data, i.e. the rates of earnings returns to human capital for decision makers are higher than for nondecision makers. (This would explain why we do not find much difference in the rates of returns to education for farmers and for farm managers.) However, the fact that the rate of returns to experience is so much higher for *farmers* may just be a consequence of the fact that as farmers become older they are able to buy more land and accumulate more physical capital. In the same way, similar rates of returns to education for farmers and managers may also be untrue. High rates of return to education for farmers might only be a consequence of the association between education and the amount of physical capital and land owned by the farmer, which could be causing spurious results. Indeed, there is no way for us to estimate which explanation is more reliable, given that we do not have information about amount of physical capital and land owned by each farmer. Therefore, *we conclude that our findings do not support the second prediction of Hypothesis 5 (managers have higher rates than farmers), but appear to support Hypothesis 3 (decision makers have higher return rates)*, although this conclusion is not certain, due to the lack of information about important control variables.

(c) The third prediction of Hypothesis 5 also fails to gain support from our statistical analysis. Our figures from Tables 15–17 show that the rates of return to *education* for *protected agricultural workers* are not significantly higher than the rates of return to *education* for *unprotected agricultural workers*, in any of our three samples. Concerning experience, we see that the regression coefficients for the interaction term between experience and the *dummy* variable of *protected agricultural worker* is always negative and statistically significant (at $|t| > 3.00$). This indicates that *the earnings returns to experience for protected workers are considerably lower than for unprotected workers.*

> This is the opposite of what we expected, when based on the prediction from the labor market segmentation theory. It was hypothesized that due to the fact that protected workers are unionized while unprotected workers are not, the former group would be able to impose recognition of seniority by employers while the latter would not. This prediction finds no support from our data. *Indeed this negative evidence is so striking that it may suggest an examination of the value of unions* as they operate in this context.

Finally, in order to reassess the *Modernization Theory Hypothesis* – which states that the rate of return to human capital should increase as the level of technological modernization rises – we test whether the interaction terms between human capital factors (education and experience) and the Intermediary Consumption Rate for agricultural production is positive in our three data samples. The figures in Tables 10–12 do not support this hypothesis. Most interaction terms between human capital factors (education and experience) and the Intermediary Consumption Rates are not significant (at $|t| > 3.00$).[46] Hence, we are now able to come to a more general judgment about the validity of the *Modernization Theory Hypothesis*. Like previous research, which assessed this hypothesis for the Brazilian labor force as a whole, our analysis finds very little empirical support for it.[47] Probably, the main problem with this prediction is that it does not take under consideration that the demand for skilled labor might be higher where the production system is dominated by more capital intensive technologies, but that the supply of skilled labor is also much higher in these regions. In other words, as the demand for skilled labor increases we also see a rise in the investments in human capital – more specifically, in education and vocational training.

CONCLUSIONS

The main goal of the present study was to analyze the earnings differences determination in the agricultural sector in Brazil and especially the role of the labor force class structure of the farm population in this process. Among the main causal factors analyzed here, we have: human capital, labor market segmentation, gender, class position, level of development, and land concentration. We not only observed the direct effects of each variable on earnings, but also estimated the interactions between variables; in particular the ways structural factors might mediate, and thus modify, the relation between human capital and earnings.

The first important finding from this study is that (as Goodman, Sorj & Wilkinson, 1985, predicted) *the process of modernization in Brazilian agriculture has caused an increase in the proportion of skilled and semiskilled labor.*[48] This finding is coherent with what was found by Rios and Oliveira (1991) in their study on the institutional aspects of class determination in Brazilian agriculture.

A second important finding from this study is that (contrary to previous reports) *the earnings returns to the human capital of the agricultural labor force in Brazil are positive and high*. This has important theoretical and policy implications, given that the impact of human capital investments on agricultural development appears to have been grossly understated in earlier, but less definitive, research.

Our second finding is related to the relationship between development/modernization and the rate of earnings returns to human capital. The overall conclusion about this is that *development/modernization does not seem to be an important modifier of the relation between human capital and earnings*. Both types of analysis conducted here – cross state and over time change analyses – have provided little if any empirical support for the *Modernization Theory Hypothesis*. The main reason may be that, even though capital and skilled labor may be complementary, the demand and the supply for skilled agricultural laborers vary together. In other words, in locations and years in which the demand for skilled labor is higher – due to more intensive use of capital – the educational and vocational training systems are also more able to provide a supply sufficient to fill the existing demand.

Besides human capital factors, we found that certain structural variables present very significant effects on earnings in Brazilian agriculture. We treated class division and labor market segmentation as a single variable to represent the social stratification system of the Brazilian farm population. We found that *social class has direct and independent effects, net of all other variables, on earnings*. More specifically, our findings show that: (a) *large farmers always have the highest earnings levels*, and that *this grew over the years*; (b) *farm managers*, with the exception of 1973, had the second highest levels of earnings; (c) *family farmers had the second highest level of earnings* in 1973, *but had fallen down to the fourth position* in 1982 and 1988; (d) *protected agricultural workers*, with the exception of 1973, *had the third highest level of earnings*; and (e) *unprotected agricultural workers were always found in the bottom of the earnings* stratification system.

Agricultural modernization of regions is another structural variable that was found to have a significant and independent effect on earnings, as was land concentration. However, while the former had a positive net effect (i.e. the higher the level of modernization the higher the average level of earnings) the latter had a negative net effect (i.e. the higher the level of land concentration the lower the average level of earnings). Even more important, given that we can compare the unstandardized regression coefficients for these two variables, we can say that up to 1982 the positive effect of agricultural modernization was greater than the negative effect of land concentration. However, by 1988 the negative effect of land concentration on earnings had become greater than the positive effect of agricultural modernization. This implies that in the process of socioeconomic change, Brazilian agriculture experienced forces acting in opposite directions on the earnings of the agricultural labor force. In turn this suggests that, on the whole, this process might no longer be improving the quality of life of that population.

Finally, in addition to the present strong evidence of direct effects of structural variables on earnings, net of human capital variables, we also found that most structural factors do not seem to work very well as modifiers of the relationship between human capital and earnings. Even though protected agricultural workers earn substantially more than unprotected workers, the average percentages of earnings returns to one additional year of education or experience are not higher for protected agricultural workers. In the same way, the earnings returns to human capital are not much different between farm managers and farmers. The only prediction of class analysis (about social class' modification of the relationship between human capital and earnings) which is well supported by our data analysis is that farm managers really have earnings returns to education that are higher than those of agricultural workers. However, given that farmers do too, we could say that this finding provides even more support for the analysis of the allocative effects of education on earnings, which predicts that decision makers should have higher earnings returns to schooling than nondecision makers. In the same way, land concentration and agricultural modernization levels do not seem to mediate the relationship between human capital factors and earnings.

NOTES

1. On human capital theory see: Becker (1964) and Mincer (1974). Concerning status attainment theory see: Blau and Duncan (1967), Sewell et al. (1969), Sewell et al. (1970), Haller and Portes (1973), Sewell and Hauser (1975), Featherman and Hauser (1978).

2. Although there are important differences between human capital and status attainment theories, both represent individualistic approaches to earnings determination. In the status attainment line the focus of empirical analysis has usually been on occupational status rather than earnings, although the theory clearly holds earnings as an important dependent variable (see Haller, 1981; Haller & Portes, 1973). Earnings or income appear explicitly as a dependent variable, among others, in: Sewell and Hauser (1975), Featherman and Hauser (1978), Otto and Haller (1979) and Haller and Saraiva (1992).

3. Concerning class analysis see: Wright and Perrone (1977), Wright (1979), Singer (1981) and Santos (2002). About labor market segmentation theory see: Doeringer and Piore (1971), Osterman (1975), Beck et al. (1978), Horan et al. (1980), Kalleberg et al. (1981), Tigges (1988) and Rios and Oliveira (1991).

4. As for the individualistic approach, there are important theoretical differences among the structuralist theories (including class analysis). However, their structural element permits to put them together in the same classification.

5. Schuh and Brandão (1991) show that, unlike the urban sector of Brazilian economy, agriculture kept growing by sustainable rates during the so-called *lost decade* of the 1980s. Indeed, the performance of Brazilian agriculture did not seem to be much affected by the 1980s crisis. They also show that while the average annual growth rate in the agricultural sector was 3.8% in the period 1965–1980, it was 3.5% in the period 1980–1988. Schuh and Brandão see three main reasons for the relatively good performance of Brazilian agriculture during the 1980s: the expansion of soy production in the *cerrado* area of the central region of the country; the *pró-álcool* program, which employed sugarcane-based alcohol as substitute for gasoline as a fuel for automobiles; and the improvement of agricultural research, especially by the Brazilian national organization of agricultural research (EMBRAPA). We would add another reason. With the debt crisis of the 1980s, Brazil needed to export as much as possible in order to achieve trade surplus. As a consequence, the Brazilian Federal Government made several currency devaluations, in order to improve exports. The most efficient sectors of Brazilian agriculture strongly benefited from this process.

6. Thiesenhusen and Melmed–Sanjak (1990) show a continuous trend for an increase in the Gini coefficients of land distribution in Brazil. It grew from 0.825 in 1940 to 0.838 in 1970, and to 0.853 in 1980. This was not only due to consolidation of smallholdings. The opening of huge tracts in formerly origin lands of the North and West accounts for part of it. Note that most analyses of land inequality are based on "establishments." Because many owners possess more than one establishment, even those high Gini coefficients may underestimate the true level of inequality.

7. See Schuh and Brandão (1991).

8. See Wood and Carvalho (1988).

9. See Kageyama (1990) and Santos (1994).

10. See Goodman et al. (1985) and Baer (1995).

11. See Schuh (1975) and Baer (1995).

12. See Schuh (1975), and Graham et al. (1987) and Baer (1995).

13. Intermediary consumption rate is given by the division of the total value of the sum of all industrialized agricultural inputs by the total value of production.

14. See Baer (1995).

15. Some other important examples of technological improvement of agricultural production also concern other *export crops*, such as coffee, sugarcane, cacao, and many different fruits such as oranges, grapes, etc.

16. Both the Gini coefficients calculated by Thiesenhusen and Melmed-Sanjak (1990) and those presented in Table 4 are based on the data from the agricultural censuses, which use the concept of *establishment*, instead of *property*, in their methodology. As a consequence, those figures probably understate the real degree of land concentration.

17. This is very much the case of soybeans, sugarcane, and citrus fruits (orange), for example. However, this model does not fit very well with other important and modern *agro-industrial complexes* in Brazil, such as the production of frozen poultry and related products, as well as the production of canned fruits and vegetables.

18. Areas farmed for a very long time. These are basically the states of the Northeast, Southeast, and South regions, within about 400–500 miles of the Atlantic coast.

19. Areas that had not been farmed until very recently. These are basically the states of the Center-West and the Amazon regions.

20. See Goodman et al. (1985).

21. The main argument behind this model is that the process of *industrialization* of agriculture – i.e. *bringing the factories to the fields* – is marked by a transition from a labor force characterized by *peasants* who have many more *skills* than the *workers* who will substitute them, given that *peasants* control all (or most) tasks related to the agricultural activity, while *workers* control only one task (or, some times, a few) related to their agricultural activity. This is the meaning of the process of *deskilling* of the agricultural labor force presented by this theoretical model. About this issue, see Collins (1993).

22. See Cabral (1987).

23. See D'Incao e Mello (1976).

24. See Singer (1981).

25. Apropos of this, Haller (1967 and 1970), in a study of changes in farm personnel in four *municípios* in the state of Rio de Janeiro between 1953 and 1962, reports that of his total samples of 582 (1953) and 576 (1962) the incidence of resident laborers dropped from 27 to 9% while that of the farm wage laborers grew from 26 to 32%. He also reports (personal communication) that during the *Military Repression*(1964–1965) in the *Agreste* region of rural Pernambuco, practically all resident laborers (*moradores*) were driven off the land – a report based on his personal interviews in 1968 with farm owners.

26. Wells (1984) and Collins (1993).

27. See Mare (1980) and Hasenbalg and Valle Silva (1991).

28. Successfully completed years of education in the case of Brazil is not the same as the number of years one has attended school. In most of the Brazilian educational system, if a student does not achieve a predetermined standard, he/she will fail. As a consequence, there are in Brazil, for example, children who have been attending school for five or six years but who have successfully completed only two years of education. In these data, they are recorded as having completed two years of schooling.

29. See Bills and Haller (1984).

30. The PNADs of 1973, 1982, and 1988 include information on individuals who are 10 years or older. Given that in Brazilian agriculture the use of young children in the labor force is frequent and is found in all regions of the country, we decided to select all individuals from 10 through 70 years old in our subsamples.

31. It is important to notice that we have avoided using a "*proxy*" for experience – like age, or age minus years of schooling minus 6 – as has been the case in many other studies.

32. In other words, the rate of earnings return to experience tends to decrease as the level of experience rises, becoming negative after some point (see Haller & Spenner, 1977).

33. For more details about *polynomial regression models*, see Neter et al. (1989).

34. It is important to notice the fall in the proportion of women in the agricultural labor force. However, we have not found any explanation for this in the literature.

35. The class category of family farmers includes tenant farmers. Regarding farm managers, a good way to construct this category might be by using information about supervision (as recommended in the class analysis approach; see Wright, 1979 and Wright & Perrone, 1977) i.e. to find out whether or not the employee has other laborers under her/his supervision. This information is not available in the PNADs; we identified managers from the occupational information, which is probably just as valid.

36. Gini coefficients vary from 0 to 1. As used herein, they will vary from 0 to 100.

37. See Graziano da Silva (1989).

38. See Ribeiro and Ghentever (1983).

39. The values for earnings and ln(earnings) vary substantially across years as a consequence of high rates of inflation, which forced constant changes in the official currency of the country.

40. We should note that land concentration and agricultural modernization always show a negative correlation. This happens in Brazil because some of the more backward regions of the country have many traditional large estates with agricultural activities of very low productivity.

41. To save space, we do not provide the tables with the *t-tests* for the equality of the differences between the regression coefficients of human capital factors in the three different years.

42. See Goodman et al. (1985).

43. This second possibility seems less likely, given that the difference between the earnings levels of all other class categories in the reference group increased from 1973 to the 1980s. However, it is possible that all class categories have experienced an increase in their earnings levels, but the inequality between the groups has increased even more markedly.

44. See Welch (1970).

45. The differences between them are not significant at $|t| > 3.00$.

46. The only exception is the interaction term between experience and intermediary consumption rate in Table 11 (1982), which is positive and statistically significant.

47. See Haller and Saraiva (1992).

48. See Goodman et al. (1985).

REFERENCES

Baer, W. (1995). *The Brazilian economy: Growth and development*. Westport: Prager.

Beck, E., Horan, P., & Tolbert, C., II (1978). Stratification in a dual economy: A sectoral model of earnings determination. *American Sociological Review*, *43*, 704–720.

Becker, G. (1964). *Human capital*. New York: Columbia University Press.

Bills, D., & Haller, A. (1984). Socioeconomic development and social stratification: Reassessing the Brazilian case. *Journal of Developing Areas*, *19*, 59–69.

Blau, P., & Duncan, O. (1967). *The American occupational structure*. New York: Wiley.

Braverman, H. (1974). *Labor and monopoly capital*. New York: Monthly Review Press.

Cabral, P. (1987). Tempo de Morada: A Constituição do Mercado de Trabalho Semi- Assalariado na Lavoura Canavieira Pernambucana. In: Y. Sampaio (Ed.), *Nordeste Rural: A Transição para o Capitalismo*. Recife: Editora Universitária/UFPE.

Collins, J. (1993). Gender, contracts and wage work: Agricultural restructuring in Brazil's São Francisco Valley. *Development and Change*, *24*, 53–82.

D'Incao e Mello, M. (1976). *O "Bóia Fria": Acumulação e Miséria*. Petrópolis: Vozes.

Doeringer, P., & Piore, M. (1971). *Internal labor markets and manpower analysis*. Lexington: Heath Lexington Books.

Featherman, D., & Hauser, R. (1978). *Opportunity and change*. New York: Academic Press.

Goldberger, A., & Cain, G. (1982). The causal analysis of cognitive outcomes in the Colleman, Hoffer, and Kilgore Report. *Sociology of Education*, *55*, 103–122.

Goodman, D., Sorj, B., & Wilkinson, J. (1985). Agroindústria, Políticas Públicas e Estruturas Sociais Rurais: Análises Recentes sobre a Agroindústria Brasileira. *Revista de Economia Política*, *5*(4), 31–56.

Graham, D., Gauthier, H., & Barros, J. (1987). Thirty years of agricultural growth in Brazil: Crop performance, regional profile, and recent policy review. *Economic Development and Cultural Change*, *36*, 1–34.

Graziano da Silva, J. (1989). *A Irrigação e a Problemática Fundiária do Nordeste*. Campinas: UNICAMP/PRONI.

Haller, A. (1967). Urban economic growth in rural stratification: Rio de Janeiro, 1953–1962. *America Latina*, *10*, 47–68.

Haller, A. (1970). Recent changes in rural stratification in Rio de Janeiro. In: Zimmerman & R. deWors (Eds), *Sociology of Underdevelopment*. Toronto: Copp-Clark.

Haller, A. (1981). Antecedents of income differences: Complementary hypotheses from conflicting theories?" Research Committee on Social Stratification, *International Sociological Association*. Paris.

Haller, A., & Portes, A. (1973). Status attainment processes. *Sociology of Education*, *46*, 51–91.

Haller, A., & Saraiva, H. (1992). The income effects of education in a developing country: Brazil – 1973 and 1982. In: R. Althauser & M. Wallace (Eds), *Research on Social Stratification and Mobility* (Vol. 11, pp. 295–336).

Haller, A., & Spenner, K. (1977). Occupational income differentiation in status attainment. *Rural Sociology*, *42*, 517–535.

Hasenbalg, C., & Valle Silva, N. (1991). Raça e Oportunidades Educacionais no Brasil. In: P. Lovell (Ed.), *Desigualdade Racial no Brasil Contemporâneo*. Belo Horizonte: UFMG/CEDEPLAR.

Hoffman, R. (1990). Distribuição da Renda e Pobreza na Agricultura Brasileira. In: C. Delagado, J. Gasques & C. Villa Verde (Eds), *Agricultura e Políticas Pública*. Rio de Janeiro: IPEA.

Hoffman, R., & Graziano, J. (1975). A Estrutura Agrária Brasileira. In: C. Contador (Ed.), *Tecnologia e Desenvolvimento Agrícola*. Rio de Janeiro: IPEA/IPES.

Horan, P., Beck, E., & Tolbert, C., II (1980). The market homogeneity assumption: On the theoretical foundations of empirical knowledge. *Social Science Quarterly*, *61*(2), 278–292.

Kageyama, A. (1990). O Novo Padrão Agrícola Brasileiro: do Complexo Rural aos Complexos Agroindustriais. In: Delgado et al. (Eds), *Agricultura e Políticas Públicas*. Brasília: IPEA.

Kalleberg, A., Wallace, M., & Althauser, R. (1981). Economic segmentation, worker power, and income inequality. *American Journal of Sociology*, *87*, 651–683.

Langoni, C. (1973). *Distribuição de Renda e Desenvolvimento Econômico no Brasil*. Rio de Janeiro: Expressão e Cultura.

Mare, R. (1980). Social background and school continuation decisions. *Journal of the American Statistical Association*, *75*, 295–305.

Mincer, J. (1974). *Schooling, experience and earnings*. New York: NBER/Columbia University Press.

Neter, J., Wasserman, W., & Kutner, M. (1989). *Applied linear regression models*. Boston: IRWIN.

Oliveira, F. (1981). *Elegia para uma Re(li)gião: SUDENE, Nordeste, Planejamento e Conflito de Classes*. Rio de Janeiro: Paz e Terra.

Osterman, P. (1975). An empirical study of labor market segmentation. *Industrial and Labor Relations Review*, *28*, 503–528.

Otto, L., & Haller, A. (1979). Evidence for a social psychological view of the status attainment process: Four studies compared. *Social Forces*, *57*, 887–914.

Pastore, J. (1989). Inequality and social mobility: Ten years later. In: E. Bacha & H. Klein (Eds), *Social Change in Brazil, 1945–1985: The Incomplete Transition*. Albuquerque: University of New Mexico Press.

Ribeiro, S., & Gheventer, B. (1983). Consumo Intermediário na Agricultural. *Revista Brasileira de Economia, 37*(1), 77–109.

Rios, E., & Oliveira, A. (1991). Posição na Ocupação do Setor Agrícola Brasileiro: Determinantes de Classe e Aspectos Institucionais. *Proceedings of the 19th National Meeting of Economics* (Brasília), pp. 245–258.

Santos, J. (2002). *Estrutura de Posições de Classe no Brasil: Mapeamento, Mudanças e Efeitos na Renda*. Belo Horizonte: Editora da UFMG.

Santos, R. (1994). Modernização da Agricultura: Um Projeto Industrial. *Revista de Sociologia Política, 2*, 122–145.

Schuh, E. (1975). A Modernização da Agricultura Brasileira: Uma Interpretação. In: C. Contador (Ed.), *Tecnologia e Desenvolvimento Agrícola*. Rio de Janeiro: IPEA/IPES.

Schuh, G., & Brandão, A. (1991). Latin American agriculture: The crises of the 1980s and the challenges of the 1990s. In: W. Baer, J. Petry & M. Simpson (Eds), *Latin America: The Crisis of the Eighties and the Opportunities of the Nineties*. Champaign: BEBR/University of Illinois.

Sewell, W., Haller, A., & Ohlendorf, G. (1970). The educational and early occupational status attainment process: A replication and revision. *American Sociological Review, 35*, 1014–1027.

Sewell, W., Haller, A., & Portes, A. (1969). The educational and early occupational attainment process. *American Sociological Review, 34*, 82–92.

Sewell, W., & Hauser, R. (1975). *Education, occupation, and earnings: Achievements in the early carrier*. New York: Academic Press.

Singer, P. (1981). *Dominação e Desigualdade: Estrutura de Classes e Repartição da Renda no Brasil*. Rio de Janeiro: Paz e Terra.

Thiesenhusen, W., & Melmed-Sanjak, J. (1990). Brazil's agrarian structure: Changes from 1970 through 1980a. *World Development, 18*(3), 393–415.

Tigges, L. (1988). Age, earnings, and change within the dual economy. *Social Forces, 66*(3), 676–698.

Treiman, D. (1970). Industrialization and social stratification. In: E. Laumann (Ed.), *Social Stratification: Research and Theory for the 1970s*. Indianapolis: Bobbs Merrill.

Welch, F. (1970). Education and production. *Journal of Political Economy, 78*(1), 35–59.

Wells, M. (1984). The resurgence of sharecropping: Historical anomaly or political strategy. *American Journal of Sociology, 90*(1), 1–29.

Wood, C., & Carvalho, J. (1988). *The demography of inequality in Brazil*. Cambridge: Cambridge University Press.

Wright, E. (1979). *Class structure and income determination*. London: Academic Press.

Wright, E., & Perrone, L. (1977). Marxist class categories and income inequality. *American Sociological Review, 42*, 32–55.

SET UP A CONTINUATION ORDER TODAY!

Did you know that you can set up a continuation order on all Elsevier-JAI series and have each new volume sent directly to you upon publication? For details on how to set up a **continuation order**, contact your nearest regional sales office listed below.

To view related series in Sociology, please visit:

www.elsevier.com/sociology

The Americas
Customer Service Department
11830 Westline Industrial Drive
St. Louis, MO 63146
USA
US customers:
Tel: +1 800 545 2522 (Toll-free number)
Fax: +1 800 535 9935
For Customers outside US:
Tel: +1 800 460 3110 (Toll-free number).
Fax: +1 314 453 7095
usbkinfo@elsevier.com

Europe, Middle East & Africa
Customer Service Department
Linacre House
Jordan Hill
Oxford OX2 8DP
UK
Tel: +44 (0) 1865 474140
Fax: +44 (0) 1865 474141
eurobkinfo@elsevier.com

Japan
Customer Service Department
2F Higashi Azabu, 1 Chome Bldg
1-9-15 Higashi Azabu, Minato-ku
Tokyo 106-0044
Japan
Tel: +81 3 3589 6370
Fax: +81 3 3589 6371
books@elsevierjapan.com

APAC
Customer Service Department
3 Killiney Road #08-01
Winsland House I
Singapore 239519
Tel: +65 6349 0222
Fax: +65 6733 1510
asiainfo@elsevier.com

Australia & New Zealand
Customer Service Department
30-52 Smidmore Street
Marrickville, New South Wales 2204
Australia
Tel: +61 (02) 9517 8999
Fax: +61 (02) 9517 2249
service@elsevier.com.au

30% Discount for Authors on All Books!

A 30% discount is available to Elsevier book and journal contributors on all books *(except multi-volume reference works).*

To claim your discount, full payment is required with your order, which must be sent directly to the publisher at the nearest regional sales office above.